D1596502

Foundations in Signal Processing,
Communications and Networking

Series Editors: W. Utschick, H. Boche, R. Mathar

Foundations in Signal Processing, Communications and Networking

Series Editors: W. Utschick, H. Boche, R. Mathar

Vol. 1. Dietl, G. K. E.
Linear Estimation and Detection in Krylov Subspaces, 2007
ISBN 978-3-540-68478-7

Vol. 2. Dietrich, F. A.
Robust Signal Processing for Wireless Communications, 2008
ISBN 978-3-540-74246-3

Frank A. Dietrich

Robust Signal Processing for Wireless Communications

With 88 Figures and 5 Tables

Series Editors:

Wolfgang Utschick
TU Munich
Associate Institute for Signal Processing
Arcisstrasse 21
80290 Munich, Germany

Holger Boche
TU Berlin
Dept. of Telecommunication Systems
Heinrich-Hertz-Chair for Mobile Communications
Einsteinufer 25
10587 Berlin, Germany

Rudolf Mathar
RWTH Aachen University
Institute of Theoretical Information Technology
52056 Aachen, Germany

Author:

Frank A. Dietrich
TU Munich
Associate Institute for Signal Processing
Arcisstrasse 21
80290 Munich, Germany
E-mail: Dietrich@mytum.de

ISBN 978-3-540-74246-3 e-ISBN 978-3-540-74249-4

DOI 10.1007/978-3-540-74249-4

Springer Series in
Foundations in Signal Processing, Communications and Networking ISSN 1863-8538

Library of Congress Control Number: 2007937523

Cover design: eStudioCalamar S.L., F. Steinen-Broo, Girona, Spain

Printed on acid-free paper

9 8 7 6 5 4 3 2 1

springer.com

To my family

Preface

Optimization of adaptive signal processing algorithms is based on a mathematical description of the signals and the underlying system. In practice, the structure and parameters of this model are not known perfectly. For example, this is due to a simplified model of reduced complexity (at the price of its accuracy) or significant estimation errors in the model parameters. The design of *robust signal processing algorithms* takes into account the parameter errors and model uncertainties. Therefore, its robust optimization has to be based on an explicit description of the uncertainties and the errors in the system's model and parameters.

It is shown in this book how *robust optimization* techniques and estimation theory can be applied to optimize robust signal processing algorithms for representative applications in wireless communications. It includes a review of the relevant *mathematical foundations* and literature. The presentation is based on the principles of estimation, optimization, and information theory: Bayesian estimation, Maximum Likelihood estimation, minimax optimization, and Shannon's information rate/capacity. Thus, where relevant, it includes the modern view of signal processing for communications, which relates algorithm design to performance criteria from information theory.

Applications in three key areas of signal processing for wireless communications at the *physical layer* are presented: Channel estimation and prediction, identification of correlations in a wireless fading channel, and linear and nonlinear precoding with incomplete channel state information for the broadcast channel with multiple antennas (multi-user downlink).

This book is written for research and development engineers in industry as well as PhD students and researchers in academia who are involved in the design of signal processing for wireless communications. Chapters 2 and 3, which are concerned with estimation and prediction of system parameters, are of general interest beyond communication. The reader is assumed to have knowledge in linear algebra, basic probability theory, and a familiarity with the fundamentals of wireless communications. The relevant notation is defined in Section 1.3. All chapters except for Chapter 5 can be read independently from each other since they include the necessary signal models. Chapter 5 additionally requires the signal model from Sections 4.1 and 4.2 and some of the ideas presented in Section 4.3.

Finally, the author would like to emphasize that the successful realization of this book project was enabled by many people and a very creative and excellent environment.

First of all, I deeply thank Prof. Dr.-Ing. Wolfgang Utschick for being my academic teacher and a good friend: His steady support, numerous intensive discussions, his encouragement, as well as his dedication to foster fundamental research on signal processing methodology have enabled and guided the research leading to this book.

I am indebted to Prof. Dr. Björn Ottersten from Royal Institute of Technology, Stockholm, for reviewing the manuscript and for his feedback. Moreover, I thank Prof. Dr. Ralf Kötter, Technische Universität München, for his support.

I would like to express my gratitude to Prof. Dr. techn. Josef A. Nossek for his guidance in the first phase of this work and for his continuous support. I thank my colleague Dr.-Ing. Michael Joham who has always been open to share his ideas and insights and has spent time listening; his fundamental contributions in the area of precoding have had a significant impact on the second part of this book.

The results presented in this book were also stimulated by the excellent working environment and good atmosphere at the Associate Institute for Signal Processing and the Institute for Signal Processing and Network Theory, Technische Universität München: I thank all colleagues for the open exchange of ideas, their support, and friendship. Thanks also to my students for their inspiring questions and their commitment.

Munich, August 2007 *Frank A. Dietrich*

Contents

Chapter 1
Introduction

Wireless communication systems are designed to provide high data rates reliably for a wide range of velocities of the mobile terminals. One important design approach to increase the spectral efficiency[1] envisions multiple transmit or receive antennas, i.e., Multiple-Input Multiple-Output (MIMO) systems, to increase the spectral efficiency.

This results in a larger number of channel parameters which have to be estimated accurately to achieve the envisioned performance. For increasing velocities, i.e., time-variance of the parameters, the estimation error increases and enhanced adaptive digital signal processing is required to realize the system. Improved concepts for *estimation* and *prediction* of the channel parameters together with *robust* design methods for *signal processing* at the physical layer can contribute to achieve these goals efficiently. They can already be crucial for small velocities.

Before giving an overview of the systematic approaches to this problem in three areas of physical layer signal processing which are presented in this book, we define the underlying notion of robustness.

1.1 Robust Signal Processing under Model Uncertainties

The design of adaptive signal processing relies on a model of the underlying physical or technical system. The choice of a suitable model follows the traditional principle: It should be as accurate as necessary and as simple as possible. But, typically, the complexity of signal processing algorithms increases with the model complexity. And on the other hand the performance degrades in case of model-inaccuracies.

---

[1] It is defined as the data rate normalized by the utilized frequency band.

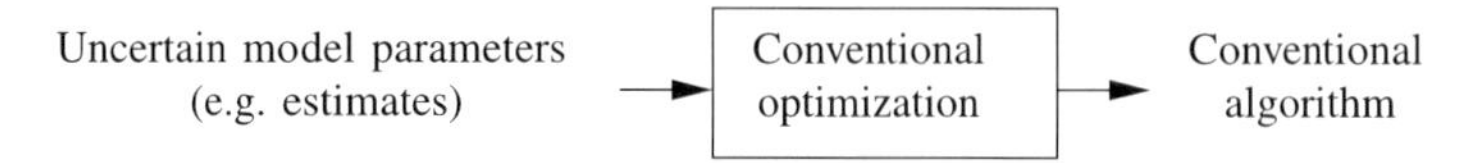

Fig. 1.1 Conventional approach to deal with model uncertainties: Treat the estimated parameters and the model as if they were true and perfectly known. The parameters for the signal processing algorithm are optimized under these idealized conditions.

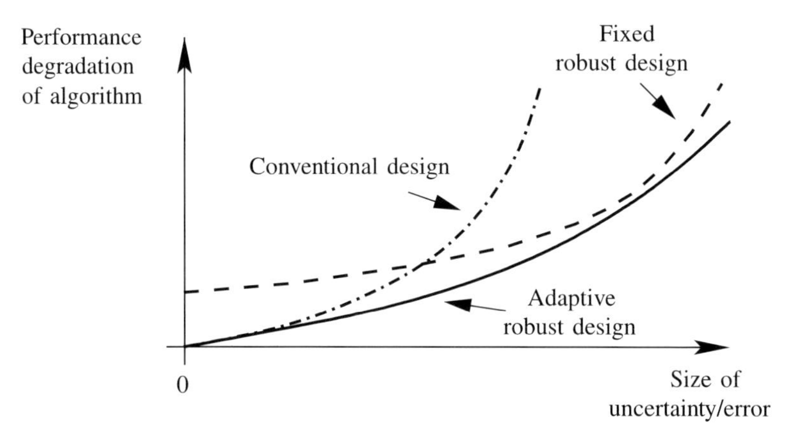

Fig. 1.2 Sensitivity of different design paradigms to the size of the model uncertainty or the parameter error.

For example, the following practical constraints lead to an imperfect characterization of the real system:

- To obtain an acceptable complexity of the model, some properties are not modeled explicitly.
- The model parameters, which may be time-variant, have to be estimated. Thus, they are not known perfectly.

Often a pragmatic design approach is pursued (Figure 1.1) which is characterized by two design steps:

- An algorithm is designed assuming the model is correct and its parameters are known perfectly.
- The model uncertainties are ignored and the estimated parameters are applied as if they were error-free.

It yields satisfactory results as long as the model errors are "small".

A *robust algorithm design* aims at minimizing the performance degradation due to model errors or uncertainties. Certainly, the first step towards a robust performance is an accurate parameter estimation which exploits all available information about the system. But in a second step, we would like to find algorithms which are robust, i.e., less sensitive, to the remaining model uncertainties.

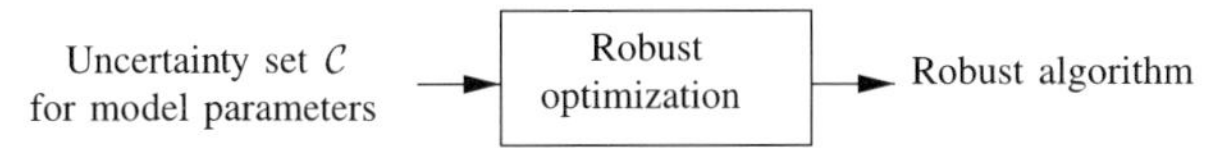

(a) General approach to robust optimization based on a set $\mathcal{C}$ describing the model/parameter uncertainties.

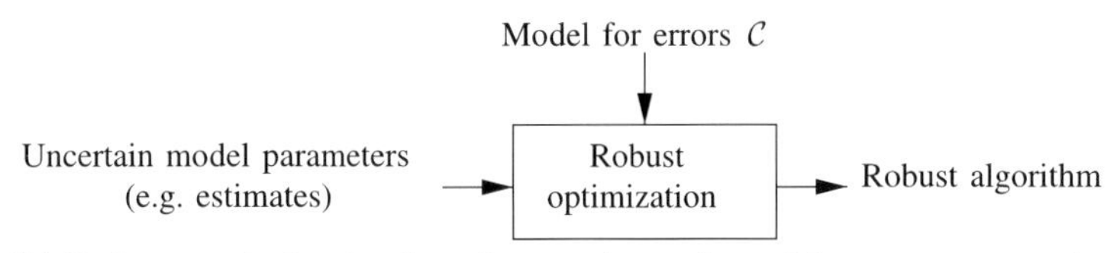

(b) Robust optimization based on estimated model parameters and a stochastic or deterministic description $\mathcal{C}$ of their errors.

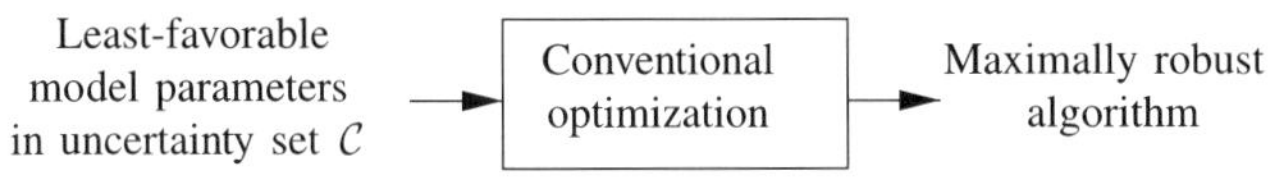

(c) Optimization for the least-favorable choice of parameters from an uncertainty set $\mathcal{C}$ for the considered optimality criterion of the algorithm: This parameterization of the standard algorithm (as in Figure 1.1) yields the maximally robust algorithm for important applications.

Fig. 1.3 General approach to robust optimization of signal processing algorithms under model uncertainties and two important special cases.

Sometimes suboptimum algorithms turn out to be less sensitive although they do not model the uncertainties explicitly: They give a *fixed robust design* which cannot adapt to the size of uncertainties (Figure 1.2).

An *adaptive robust design* of signal processing yields the optimum performance for a perfect model match (no model uncertainties) and an improved or in some sense optimum performance for increasing errors (Figure 1.2). Conceptually, this can be achieved by

- defining a mathematical model of the considered uncertainties and
- constructing an optimization problem which includes these uncertainties.

Practically, this corresponds to an enhanced interface between system identification and signal processing (Figure 1.3(a)). Now, both tasks are not optimized independently from each other but *jointly*.

In this book, we focus on three important types of uncertainties in the context of wireless communications:

1. Parameter errors with a stochastic error model,
2. parameter errors with a deterministic error model, and
3. unmodeled stochastic correlations of the model parameters.

The two underlying design paradigms are depicted in Figures 1.3(b) and 1.3(c), which are a special case of the general approach in Figure 1.3(a):

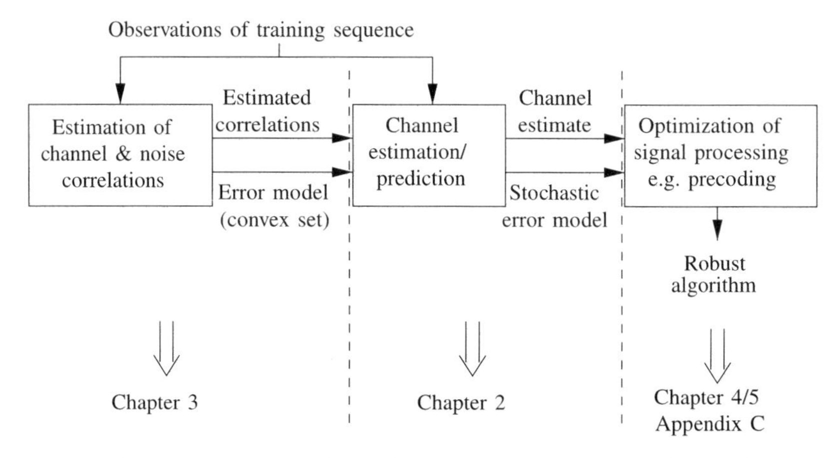

Fig. 1.4 Robust optimization of signal processing in the physical layer of a wireless communication system: Interfaces between signal processing tasks are extended by a suitable model for the parameter errors or uncertainties.

The first clearly shows the enhanced interface compared to Figure 1.1 and is suitable for treating parameter errors (Figure 1.3(b)). The second version guarantees a worst-case performance for the uncertainty set $\mathcal{C}$ employing a maxmin or minimax criterion: In a first step, it chooses the least-favorable model or parameters in $\mathcal{C}$ w.r.t. the conventional optimization criterion of the considered signal processing task. Thus, optimization of the algorithm is identical to Figure 1.1, but based on the worst-case model. In important practical cases, this is identical to the problem of designing a maximally robust algorithm (see Section 2.4).

Finally, we would like to emphasize that the systematic approach to robust design has a long history and many applications: Since the early 1960s, robust optimization has been treated systematically in mathematics, engineering, and other sciences. Important references are [97, 226, 122, 61, 16, 17, 171] which give a broader overview of the subject than this brief introduction. Other relevant classes of uncertainties and their applications are treated there.

1.2 Overview of Chapters and Selected Wireless Applications

The design of adaptive signal processing at the physical layer of a wireless communication system relies on *channel state information*. Employing the traditional stochastic model for the wireless fading channel, the channel state consists of the parameters of the channel parameters' probability distribution and their current or future realization. We treat three related fundamental signal processing tasks in the physical layer as shown in Figure 1.4:

- Estimation and prediction of the channel parameters (Chapter 2),
- estimation of the channel parameters' and noise correlations (Chapter 3), and
- linear and nonlinear precoding with partial channel state information for the broadcast channel (Chapters 4 and 5).[2]

We propose methods for a robust design of every task. Moreover, every chapter or main section starts with a survey of the underlying theory and literature with the intention to provide the necessary background.

- *Chapter 2: Channel Estimation and Prediction*
  The traditional approaches to *estimation* of the frequency-selective wireless channel using training sequences are introduced and compared regarding the achieved bias-variance trade-off (Section 2.2). This includes more recent techniques such as the reduced-rank Maximum Likelihood estimator and the matched filter.
  Prediction of the wireless channel gains in importance with the application of precoding techniques at the transmitter relying on channel state information. In Section 2.3, we focus on minimum mean square error (MMSE) prediction and discuss its performance for *band-limited* random sequences, which model the limited temporal dynamics of the wireless channel due to a maximum velocity in a communication system.
  Channel estimation as well as prediction relies on the knowledge of the channel parameters' probability distribution. It is either unknown, is only specified partially (e.g., by the mean channel attenuation or the maximum Doppler frequency), or estimates of its first and second order moments are available (Chapter 3). In Section 2.4, a general introduction to *minimax mean square error* (MSE) optimization is given: Its solution guarantees a worst-case performance for a given uncertainty set, proves that the Gaussian probability distribution is least-favorable, and provides the maximally robust estimator.
  We apply the minimax-results to robust MMSE channel estimation and robust MMSE prediction for band-limited random sequences. For example, for prediction of channels with a maximum Doppler frequency we provide the uncertainty set which yields the following predictor: MMSE prediction based on the rectangular and band-limited power spectral density; it is maximally robust for this uncertainty set.
- *Chapter 3: Estimation of Channel and Noise Covariance Matrices*
  Estimation of the covariance matrix for the channel parameters in space, delay, and time dimension together with the spatial noise correlations can be cast as the problem of estimating a structured covariance matrix. An

[2] This serves as an example for adaptive processing of the signal containing the information bearing data.

overview of the general Maximum Likelihood problem for structured covariance matrices is given in Section 3.1. The focus is on Toeplitz structure which yields an ill-posed Maximum Likelihood problem.
First, we present an iterative Maximum Likelihood solution based on a generalization of the expectation-maximization (EM) algorithm (Section 3.3). It includes the combined estimation of channel correlations in space, delay, and time as well as special cases.
If only the decimated *band-limited autocovariance sequence* can be estimated and the application requires the *interpolated* sequence, the decimated sequence has to be completed. As a constraint, it has to be ensured that the interpolated/completed sequence is still positive semidefinite. We propose a minimum-norm completion, which can be interpreted in the context of minimax MSE prediction (Section 3.4).
For estimation of correlations in space and delay dimensions, the Maximum Likelihood approach is rather complex. Least-squares approaches are computationally less complex and can be shown to be asymptotically equivalent to Maximum Likelihood. Including a positive semidefinite constraint we derive different suboptimum estimators based on this paradigm, which achieve a performance close to Maximum Likelihood (Section 3.5).

Chapters 2 and 3 deal with estimation of the channel state information (CSI). As signal processing application which deals with the transmission of data, we consider precoding of the data symbols for the wireless *broadcast channel.*[3] For simplicity, we treat the case of a single antenna at the receiver and multiple transmit antennas for a frequency flat channel[4]. Because precoding takes place at the transmitter, the availability of channel state information is a crucial question.

The last two chapters present *robust precoding* approaches based on MSE which can deal with partial channel state information at the transmitter. Its foundation is the availability of the channel state information from estimators of the channel realization and correlations which we present in Chapters 2 and 3.

- *Chapter 4: Linear Precoding with Partial Channel State Information*
  For linear precoding, we introduce different optimum and suboptimum performance measures, the corresponding approximate system models, and emphasize their relation to the information rate (Section 4.3). We elaborate on the relation between SINR-type measures to MSE-like expressions. It includes optimization criteria for systems with a common training channel instead of user-dedicated training sequences in the forward link.
  The central aspect of this new framework is the choice of an appropriate receiver model which controls the trade-off between performance and complexity.

[3] If restricted to be linear, precoding is also called preequalization or beamforming.

[4] This is a finite impulse response (FIR) channel model of order zero.

Linear precoding is optimized iteratively for the sum MSE alternating between the receiver models and the transmitter. This intuitive approach allows for an interesting interpretation. But optimization of sum performance measures does not give quality of service (QoS) guarantees to the receivers or provide fairness. We address these issues briefly in Section 4.6 which relies on the *MSE-dualities* between the broadcast and virtual multiple access channel developed in Section 4.5.

- *Chapter 5: Nonlinear Precoding with Partial Channel State Information*
  Nonlinear precoding for the broadcast channel, i.e., Tomlinson-Harashima precoding or the more general vector precoding, is a rather recent approach; an introduction is given in Section 5.1. We introduce novel robust performance measures which incorporate partial as well as statistical channel state information (Section 5.2). As for linear precoding, the choice of an appropriate receiver model is the central idea of this new framework. For statistical channel state information, the solution can be considered the first approach to *nonlinear beamforming* for communications.
  The potential of Tomlinson-Harashima precoding (THP) with only statistical channel state information is illustrated in examples. For channels with rank-one spatial correlation matrix, we show that THP based on statistical channel state information is not interference-limited: Every signal-to-interference-plus-noise (SINR) ratio is achievable as for the case of complete channel state information.
  Vector precoding and Tomlinson-Harashima precoding are optimized based on the sum MSE using the same paradigm as for linear precoding: Iteratively, we alternate between optimizing the receiver models and the transmitter (Section 5.3).
  As a practical issue, we address precoding for the training sequences to enable the implementation of the optimum receiver, which results from nonlinear precoding at the transmitter (Section 5.4). The numerical performance evaluation in Section 5.5 shows the potential of a robust design in many interesting scenarios.
- *Appendix*:
  In Appendix A, we briefly survey the necessary mathematical background. The completion of covariance matrices with unknown elements and the extension of band-limited sequences is reviewed in Appendix B. Moreover, these results are combined to prove a generalization to the band-limited trigonometric moment problem which is required to interpolate autocovariance sequences in Section 3.4.
  The robust optimization criteria which we apply for precoding with partial channel state information are discussed more generally in Appendix C from the perspective of estimation theory. The appendix concludes with detailed derivations, the definition of the channel scenarios for the performance evaluations, and a list of abbreviations.

1.3 Notation

Sets

Definitions for sets of numbers, vectors, and matrices are:

$\mathbb{Z}$	(positive and negative) integers
$\mathbb{R}$	real numbers
$\mathbb{R}_{+,0}$	real positive (or zero) numbers
$\mathbb{R}^M$	M-dimensional vectors in $\mathbb{R}$
$\mathbb{C}$	complex numbers
$\mathbb{C}^M$	M-dimensional vectors in $\mathbb{C}$
$\mathbb{C}^{M\times N}$	$M\times N$ matrices in $\mathbb{C}$
$\mathbb{S}^M$	$M\times M$ Hermitian matrices
$\mathbb{S}^M_{+,0}$	$M\times M$ positive semidefinite matrices (equivalent to $\boldsymbol{A}\succeq 0$)
$\mathbb{S}^M_{+}$	$M\times M$ positive definite matrices
$\mathbb{T}^M_{+,0}$	$M\times M$ positive semidefinite Toeplitz matrices
$\mathbb{T}^M_{+}$	$M\times M$ positive definite Toeplitz matrices
$\mathbb{T}^M_{\mathrm{c}}$	$M\times M$ circulant and Hermitian matrices
$\mathbb{T}^M_{\mathrm{c}+,0}$	$M\times M$ positive semidefinite circulant matrices
$\mathbb{T}^M_{\mathrm{c}+}$	$M\times M$ positive definite circulant matrices
$\mathbb{T}^M_{\mathrm{ext}+,0}$	$M\times M$ positive definite Toeplitz matrices with positive semidefinite circulant extension
$\mathbb{D}^M_{+,0}$	$M\times M$ positive semidefinite diagonal matrices
$\mathbb{T}^{M,N}_{+,0}$	$MN\times MN$ positive semidefinite block-Toeplitz matrices with blocks of size $M\times M$
$\mathbb{T}^{M,N}_{\mathrm{ext}+,0}$	$MN\times MN$ positive semidefinite block-Toeplitz matrices with blocks of size $M\times M$ and with positive semidefinite block-circulant extension
$\mathbb{L}^M$	M-dimensional rectangular Lattice $\tau\mathbb{Z}^M+\mathrm{j}\tau\mathbb{Z}^M$
$\mathbb{V}$	Voronoi region $\{x+\mathrm{j}y \mid x,y\in[-\tau/2,\tau/2)\}$

Matrix Operations

For a matrix $\boldsymbol{A}\in\mathbb{C}^{M\times N}$, we introduce the following notation for standard operations ($\boldsymbol{A}$ is assumed square or regular whenever required by the operation):

$\boldsymbol{A}^{-1}$	inverse	$\boldsymbol{A}^*$	complex conjugate
$\boldsymbol{A}^\dagger$	pseudo-inverse	$\boldsymbol{A}^\mathrm{T}$	transpose
$\mathrm{tr}[\boldsymbol{A}]$	trace	$\boldsymbol{A}^\mathrm{H}$	Hermitian (complex conjugate transpose)
$\det[\boldsymbol{A}]$	determinant		

The relation $\boldsymbol{A} \succeq \boldsymbol{B}$ is defined as $\boldsymbol{A} - \boldsymbol{B} \succeq 0$, i.e., the difference is positive semidefinite.

The operator $\boldsymbol{a} = \mathbf{vec}[\boldsymbol{A}]$ stacks the columns of $\boldsymbol{A}$ in a vector, and $\boldsymbol{A} = \mathbf{unvec}[\boldsymbol{a}]$ is the inverse operation. An $N \times N$ diagonal matrix with $d_i, i \in \{1, 2, \ldots, N\}$, on its diagonal is given by $\mathbf{diag}[d_1, d_2, \ldots, d_N]$. The Kronecker product $\otimes$, the Schur complement, and more detailed definitions are introduced in Appendix A.2.

Random Variables

Random vectors and matrices are denoted by lower and upper case sans-serif bold letters (e.g., a, A, h), whereas their realizations or deterministic variables are, e.g., $\boldsymbol{a}$, $\boldsymbol{A}$, $\boldsymbol{h}$. To describe their properties we define:

$\mathrm{p}_{\mathsf{a}}(\boldsymbol{a})$	probability density function (pdf) of random vector a evaluated for $\boldsymbol{a}$
$\mathrm{p}_{\mathsf{a}\mid\boldsymbol{y}}(\boldsymbol{a})$	conditional pdf of random vector a which is conditioned on the realization $\boldsymbol{y}$ of a random vector y
$\mathrm{E}_{\mathsf{a}}[\mathbf{f}(\mathsf{a})]$	expectation of $\mathbf{f}(\mathsf{a})$ w.r.t. random vector a
$\mathrm{E}_{\mathsf{a}\mid\boldsymbol{y}}[\mathbf{f}(\mathsf{a})]$	conditional expectation of $\mathbf{f}(\mathsf{a})$ w.r.t. random vector a
$\mathrm{E}_{\mathsf{a}}[\mathbf{f}(\mathsf{a})\mid\boldsymbol{y};\boldsymbol{x}]$	expectation w.r.t. random vector a for a conditional pdf with parameters $\boldsymbol{x}$
$\boldsymbol{\mu}_{\mathsf{a}} = \mathrm{E}_{\mathsf{a}}[\mathsf{a}]$	mean of a random vector a
$\boldsymbol{\mu}_{\mathsf{a}\mid\boldsymbol{y}} = \mathrm{E}_{\mathsf{a}\mid\boldsymbol{y}}[\mathsf{a}]$	conditional mean of random vector a
$\mathsf{a} \sim \mathcal{N}_{\mathrm{c}}(\boldsymbol{\mu}_{\mathsf{a}}, \boldsymbol{C}_{\mathsf{a}})$	random vector with complex Gaussian pdf (Appendix A.1)

Covariance matrices and conditional covariance matrices of a random vector a are $\boldsymbol{C}_{\mathsf{a}} = \mathrm{E}_{\mathsf{a}}[(\mathsf{a} - \boldsymbol{\mu}_{\mathsf{a}})(\mathsf{a} - \boldsymbol{\mu}_{\mathsf{a}})^{\mathrm{H}}]$ and $\boldsymbol{C}_{\mathsf{a}\mid\boldsymbol{y}} = \mathrm{E}_{\mathsf{a}\mid\boldsymbol{y}}[(\mathsf{a} - \boldsymbol{\mu}_{\mathsf{a}\mid\boldsymbol{y}})(\mathsf{a} - \boldsymbol{\mu}_{\mathsf{a}\mid\boldsymbol{y}})^{\mathrm{H}}]$. For a random matrix A, we define $\boldsymbol{C}_{\mathsf{A}} = \mathrm{E}_{\mathsf{A}}[(\mathsf{A} - \mathrm{E}_{\mathsf{A}}[\mathsf{A}])(\mathsf{A} - \mathrm{E}_{\mathsf{A}}[\mathsf{A}])^{\mathrm{H}}]$ and $\boldsymbol{C}_{\mathsf{A}^{\mathrm{H}}} = \mathrm{E}_{\mathsf{A}}[(\mathsf{A} - \mathrm{E}_{\mathsf{A}}[\mathsf{A}])^{\mathrm{H}}(\mathsf{A} - \mathrm{E}_{\mathsf{A}}[\mathsf{A}])]$, which are the sum of the covariance matrices for its columns and the complex conjugate of its rows, respectively. Correlation matrices are denoted by $\boldsymbol{R}_{\mathsf{a}} = \mathrm{E}_{\mathsf{a}}[\mathsf{a}\mathsf{a}^{\mathrm{H}}]$ and $\boldsymbol{R}_{\mathsf{A}} = \mathrm{E}_{\mathsf{A}}[\mathsf{A}\mathsf{A}^{\mathrm{H}}]$.

Other Definitions

A vector-valued function of a vector-valued argument $\boldsymbol{x}$ is written as $\mathbf{f}: \mathbb{C}^M \to \mathbb{C}^N, \boldsymbol{x} \mapsto \mathbf{f}(\boldsymbol{x})$ and a scalar function of vector-valued arguments as $\mathrm{f}: \mathbb{C}^M \to \mathbb{C}, \boldsymbol{x} \mapsto \mathrm{f}(\boldsymbol{x})$.

A sequence of vectors is $\boldsymbol{h}: \mathbb{Z} \to \mathbb{C}^M, q \mapsto \boldsymbol{h}[q]$. The (time-) index q is omitted if we consider a fixed time instance. For a stationary sequence of random vectors $\mathsf{h}[q]$, we define $\boldsymbol{\mu}_{\mathsf{h}} = \boldsymbol{\mu}_{\mathsf{h}[q]}$, $\boldsymbol{C}_{\mathsf{h}} = \boldsymbol{C}_{\mathsf{h}[q]}$, and $\mathrm{p}_{\mathsf{h}}(\boldsymbol{h}[q]) = \mathrm{p}_{\mathsf{h}[q]}(\boldsymbol{h}[q])$.

The standard convolution operator is $*$. The Kronecker sequence $\delta[i]$ is defined as $\delta[0] = 1$ and $\delta[i] = 0, i \neq 0$ and the imaginary unit is denoted j with $\mathrm{j}^2 = -1$.

Chapter 2
Channel Estimation and Prediction

Estimation and prediction of the wireless channel is a very broad topic. In this chapter, we treat only a selection of important aspects and methods in more detail. To clarify their relation to other approaches in the literature we first give a general overview. It also addresses topics which we do not cover in the sequel.

Introduction to Chapter

Approaches to estimation and prediction of the wireless channel can be classified according to the exploited information about the transmitted signal and the underlying channel model.

Blind (unsupervised) *methods* for channel estimation do not require a known training sequence to be transmitted but exploit the spatial or temporal structure of the channel or, alternatively, properties of the transmitted signal such as constant magnitude. The survey by Tong et al.[214] gives a good introduction and many references in this field. Application to wireless communications is discussed by Tugnait et al.[218] and Paulraj et al.[165].

In most communication systems (e.g., [215]) *training sequences* are defined and can be exploited for channel estimation. Two important issues are the optimum choice of training sequences as well as their placement and the number of training symbols (see [215] and references); the focus is on characterizing the trade-off between signaling overhead and estimation accuracy regarding system or link throughput and capacity. For systems which have already been standardized, these parameters are fixed and the focus shifts towards the development of training sequence based estimators. They can be distinguished by the assumed model for the communication channel (Figure 2.1):

- A *structured deterministic* model describes the physical channel properties explicitly. It assumes discrete paths parameterized by their delay, Doppler frequency, angle of arrival and of departure, and complex path attenuation.

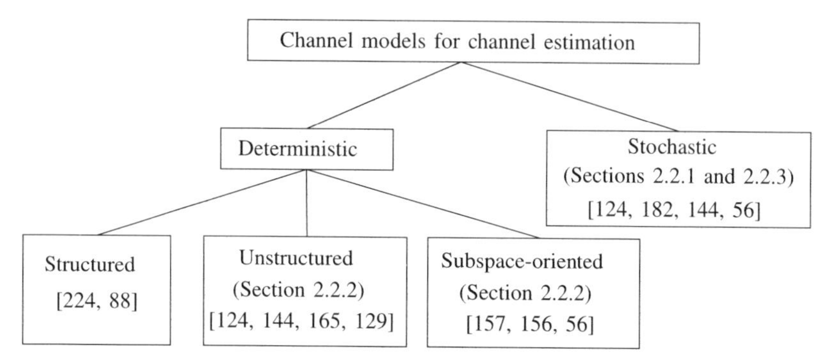

Fig. 2.1 Different categories of models for the communication channel and references to channel estimators based on the respective model.

This model is nonlinear in its parameters, which can be estimated, for example, using ESPRIT- and MUSIC-type algorithms [224, 88].

- An *unstructured deterministic* finite impulse response (FIR) model of the channel is linear in the channel parameters and leads to a linear (weighted) least-squares estimator which is a maximum likelihood (ML) estimator for Gaussian noise [124, 144, 165, 129]. The most popular low-complexity approximation in practice is the correlator [120, 56].
- The number of channel parameters is increased significantly in multiple-input multiple-output (MIMO) channels and with increasing symbol rate.[1] This results in a larger estimation variance of the ML estimator. At the same time these channels provide additional spatial and temporal structure, which can be exploited to improve the estimation accuracy. One point of view considers the (temporal) invariance of the angles and delays[2] and models it as an invariance of a low-dimensional subspace which contains the channel realization (*subspace-oriented deterministic model*). It considers the algebraic structure in contrast to the structured deterministic model from above, which approximates the physical structure.
  This approach has been pursued by many authors [205, 74], but the most advanced methods are the reduced-rank ML estimators introduced by Nicoli, Simeone, and Spagnolini [157, 156]. A detailed comparison with Bayesian methods is presented in [56]. Regarding channel estimation, the prominent rake-receiver can also be viewed as a subspace-oriented improvement of the correlator, which can also be generalized to space-time processing [56].
- The channel is modeled by an FIR filter whose parameters are a stationary (vector) random process with given mean and covariance matrix

[1] Assuming that the channel order increases with the symbol rate.

[2] The angles and delays of a wireless channel change slowly compared to the complex path attenuations (fading) [157, 229, 227] and are referred to as *long-term* or large-scale channel properties.

[14, 144]. This *stochastic model* allows for application of Bayesian estimation methods [124, 182] and leads to the conditional mean (CM) estimator in case of an mean square error (MSE) criterion. For jointly Gaussian observation and channel, it reduces to the linear minimum mean square error (LMMSE) estimator. Similar to the subspace approaches, the MMSE estimator exploits the *long-term* structure of the channel which is now described by the channel covariance matrix.

For *time-variant channels*, either interpolation of the estimated channel parameters between two blocks of training sequences (e.g. [9, 144]) or a decision directed approach [144] improves performance. Channel estimation can also be performed iteratively together with equalization and decoding to track the channel, e.g., [159].

Another important topic is *prediction* of the future channel state which we address in more detail in Sections 2.3 and 2.4.3. We distinguish three main categories of prediction algorithms for the channel parameters:

- Extrapolation based on a structured channel model with estimated Doppler frequencies,
- extrapolation based on the smoothness of the channel parameters over time, e.g., standard (nonlinear) extrapolation based on splines, and
- *MMSE prediction* based on a stochastic model with given autocovariance function [64, 63] (Section 2.3).

References to algorithms in all categories are given in Duel-Hallen et al.[63]. Here we focus on MMSE prediction as it has been introduced by Wiener and Kolmogorov [182, 234, 235] in a Bayesian framework [124].

If the Doppler frequency is bounded due to a maximum mobility in the system, i.e., the parameters' *power spectral densities* are *band-limited*, this property can be exploited for channel prediction as well as interpolation. Either the maximum Doppler frequency (maximum velocity) is known a priori by the system designer due to a specified environment (e.g., office or urban) or it can be estimated using algorithms surveyed in [211]. *Perfect prediction* (i.e, without error) of band-limited processes is possible given a non-distorted observation of an infinite number of past samples and a sampling rate above the Nyquist frequency [22, 162, 138, 27, 222]. But this problem is ill-posed, because the norm of the predictor is not bounded asymptotically for an infinite number of past observations [164, p. 380]. Thus, regularization is required. For about 50 years, the assumption of a rectangular power spectral density, which corresponds to prediction using sinc-kernels, has been known to be a good ad hoc assumption [35]. Optimality is only studied for the case of undistorted/noiseless past observations [34]. And a formal justification for its robustness is addressed for the "filtering" problem in [134], where it is shown to be the least-favorable spectrum.

For conditional mean or MMSE channel estimation and prediction, the probability distribution of the channel parameters and observation is required

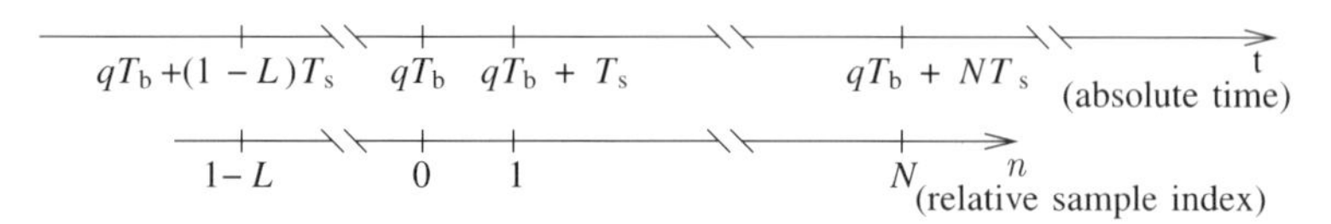

Fig. 2.2 On an absolute time scale the signals are sampled at $t = qT_b + nT_s$, where q is the block index and n the relative sample index. Within one block with index q we only consider the relative time scale in n.

to be known perfectly. Assuming a Gaussian distribution, only perfect knowledge of the first and second order moments is necessary. Estimators for the moments of the channel parameters are introduced in Chapter 3, but still the estimates will never be perfect. *Minimax MSE estimators*, which guarantee a worst case performance for a given uncertainty class of second order statistics, are introduced by Poor, Vastola, and Verdu [225, 226, 76, 75].[3] These results address the case of a linear minimax MSE estimator for an observation of infinite length. The finite dimensional case with uncertainty in the type of probability distribution and its second order moments is treated by Vandelinde [223] and Soloviov [197, 198, 199], where the Gaussian distribution is shown to be least-favorable and maximally robust.

Outline of Chapter

In this chapter we focus on training sequence based channel estimation, where we assume block-fading. An overview of different paradigms for channel estimation which are either based on an unstructured deterministic, subspace-based, or stochastic model, is given in Section 2.2 (see overview in Figure 2.1). In addition to the comparison, we present the matched filter for channel estimation. An introduction to multivariate and univariate MMSE prediction is given in Section 2.3, which emphasizes the perfect predictability of band-limited random sequences. After a brief overview of the general results on minimax MSE estimation in Section 2.4, we apply them to MMSE channel estimation and prediction. Therefore, we are able to consider errors in the estimated channel statistics. Given the channel's maximum Doppler frequency, different minimax MSE optimal predictors are proposed.

2.1 Model for Training Channel

A linear MIMO communication channel with K transmitters and M receivers is modeled by the time-variant impulse response $\boldsymbol{H}(t, \tau)$, where t denotes

[3] See also the survey [123] by Kassam and Poor with many references.

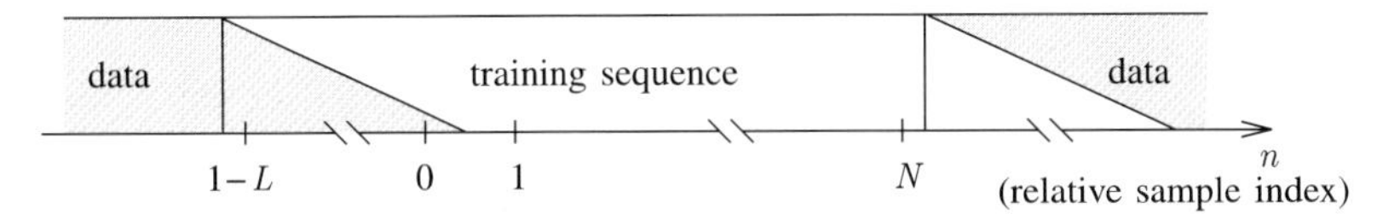

Fig. 2.3 The training sequence is time-multiplexed and transmitted periodically with period T_b. For a frequency selective channel with $L > 0$, the first L samples of the received training sequence are distorted by interblock interference.

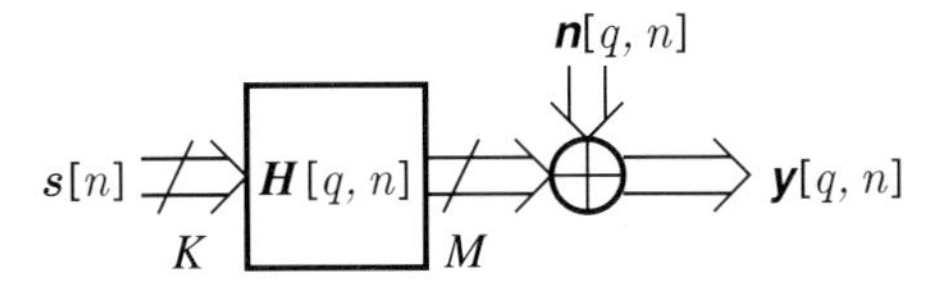

Fig. 2.4 Transmission of the nth symbol $\boldsymbol{s}[n] \in \mathbb{B}^K$ in block q over a frequency selective MIMO channel described by the finite impulse response (FIR) $\boldsymbol{H}[q,n] \in \mathbb{C}^{M\times K}$ (cf. (2.1) and (2.2)).

time and τ the delay. We describe the signals and channel by their complex baseband representations (e.g. [173, 129]).

Periodically, every transmitter sends a training sequence of N_t symbols which is known to the receivers. The period between blocks of training sequences is T_b. The receive signal is sampled at the symbol period T_s (Figure 2.2). The channel is assumed constant within a block of N_t symbols. Given an impulse response $\boldsymbol{H}(t,\tau)$ with finite support in τ, i.e., $\boldsymbol{H}(t,\tau) = \boldsymbol{0}_{M\times K}$ for $\tau < 0$ and $\tau > LT_\mathrm{s}$, the corresponding discrete time system during the qth block is the finite impulse response (FIR) of order L

$$\boldsymbol{H}[q,i] = \boldsymbol{H}(qT_\mathrm{b}, iT_\mathrm{s}) = \sum_{\ell=0}^{L} \boldsymbol{H}^{(\ell)}[q]\delta[i-\ell] \in \mathbb{C}^{M\times K}. \tag{2.1}$$

The training sequences are multiplexed in time with a second signal (e.g. a data channel), which interferes with the first L samples of the received training sequence as depicted in Figure 2.3. For all transmitters, the sequences are chosen from the (finite) set $\mathbb{B}$ and collected in $\boldsymbol{s}[n] \in \mathbb{B}^K$ for $n \in \{1-L, 2-L, \ldots, N\}$, with $N_\mathrm{t} = N + L$. In every block the same training sequence $\boldsymbol{s}[n]$ is transmitted. Including additive noise, the sampled receive signal $\boldsymbol{y}(t)$ during block q is (Figure 2.4)

$$\begin{aligned}\boldsymbol{y}[q,n] = \boldsymbol{y}(qT_\mathrm{b} + nT_\mathrm{s}) &= \boldsymbol{H}[q,n] * \boldsymbol{s}[n] + \boldsymbol{n}[q,n] \in \mathbb{C}^M,\ n = 1,2,\ldots,N \\ &= \sum_{\ell=0}^{L} \boldsymbol{H}^{(\ell)}[q]\boldsymbol{s}[n-\ell] + \boldsymbol{n}[q,n]. \end{aligned} \tag{2.2}$$

Note that the symbol index n is the relative sample index within one block. Only samples for $1 \leq n \leq N$ are not distorted due to intersymbol interference

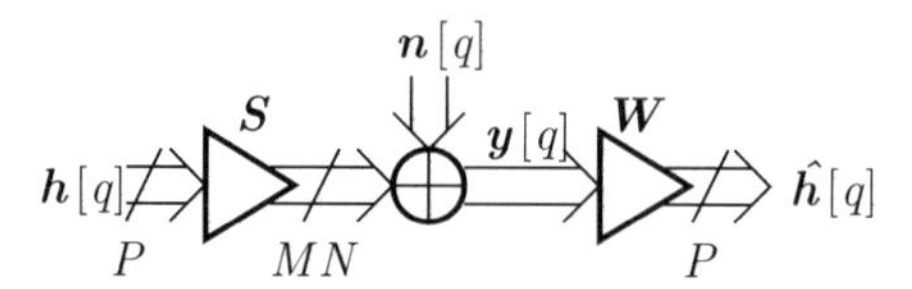

Fig. 2.5 Linear system model for optimization of channel estimators $\boldsymbol{W} \in \mathbb{C}^{P \times MN}$ for $\boldsymbol{h}[q]$ with $\hat{\boldsymbol{h}}[q] = \boldsymbol{W}\boldsymbol{y}[q]$ (Section 2.2).

from the other multiplexed signal (Figure 2.3). The relation between the absolute time $t \in \mathbb{R}$, the sampling of the signals at $t = qT_\mathrm{b} + nT_\mathrm{s}$, and the relative sample index within one block is illustrated in Figure 2.2.

The noise is modeled as a stationary, temporally uncorrelated, and zero-mean complex Gaussian random sequence $\boldsymbol{n}[q, n] \sim \mathcal{N}_\mathrm{c}(\mathbf{0}_M, \boldsymbol{C}_{\boldsymbol{n},\mathrm{s}})$. The spatial covariance matrix of the noise is $\boldsymbol{C}_{\boldsymbol{n},\mathrm{s}} = \mathrm{E}[\boldsymbol{n}[q,n]\boldsymbol{n}[q,n]^\mathrm{H}]$.

2.1.0.1 Model for one Block

With $n \in \{1, 2, \ldots, N\}$, we consider those N samples of the receive signal which are not influenced by the signal preceding the training block (Figure 2.3). The relevant samples of the receive signal are collected in $\boldsymbol{Y}[q] = [\boldsymbol{y}[q,1], \boldsymbol{y}[q,2], \ldots, \boldsymbol{y}[q,N]]$ (equivalently for $\boldsymbol{N}[q]$) and we obtain

$$\boldsymbol{Y}[q] = \boldsymbol{H}[q]\bar{\boldsymbol{S}} + \boldsymbol{N}[q] \in \mathbb{C}^{M \times N}, \tag{2.3}$$

where $\boldsymbol{H}[q] = [\boldsymbol{H}^{(0)}[q], \boldsymbol{H}^{(1)}[q], \ldots, \boldsymbol{H}^{(L)}[q]] \in \mathbb{C}^{M \times K(L+1)}$ and

$$\bar{\boldsymbol{S}} = \begin{bmatrix} \boldsymbol{s}[1] & \boldsymbol{s}[2] & \cdots & \boldsymbol{s}[1+L] & \cdots & \boldsymbol{s}[N] \\ \boldsymbol{s}[0] & \boldsymbol{s}[1] & & & & \boldsymbol{s}[N-1] \\ \vdots & & \ddots & & & \vdots \\ \boldsymbol{s}[1-L] & \cdots & & \boldsymbol{s}[1] & \cdots & \boldsymbol{s}[N-L] \end{bmatrix} \in \mathbb{B}^{K(L+1) \times N},$$

which has block Toeplitz structure. Defining $\boldsymbol{y}[q] = \mathbf{vec}[\boldsymbol{Y}[q]]$ we get the standard linear model in Figure 2.5 (cf. Eq. A.15)

$$\boxed{\boldsymbol{y}[q] = \boldsymbol{S}\boldsymbol{h}[q] + \boldsymbol{n}[q]} \tag{2.4}$$

with system matrix $\boldsymbol{S} = \bar{\boldsymbol{S}}^\mathrm{T} \otimes \boldsymbol{I}_M \in \mathbb{C}^{MN \times P}$. The channel is described by $P = MK(L+1)$ parameters $\boldsymbol{h}[q] = \mathbf{vec}[\boldsymbol{H}[q]] \in \mathbb{C}^P$, where $h_p[q]$ is the pth element of $\boldsymbol{h}[q]$.

We model the noise $\boldsymbol{n}[q] = \mathbf{vec}[\boldsymbol{N}[q]] \in \mathbb{C}^{MN}$ as $\boldsymbol{n}[q] \sim \mathcal{N}_\mathrm{c}(\mathbf{0}_{MN}, \boldsymbol{C}_{\boldsymbol{n}})$ with two different choices of covariance matrices:

1. $\boldsymbol{n}[q]$ is spatially and temporally uncorrelated with $\boldsymbol{C}_{\boldsymbol{n}} = c_n \boldsymbol{I}_{MN}$ and variance c_n, or

2. $\boldsymbol{n}[q]$ is temporally uncorrelated and spatially correlated with

$$\boldsymbol{C}_{\boldsymbol{n}} = \boldsymbol{I}_N \otimes \boldsymbol{C}_{\boldsymbol{n},\mathrm{s}}. \tag{2.5}$$

This can model an interference from spatially limited directions.

In wireless communications it is common [173] to choose a stochastic model for the channel parameters $\boldsymbol{h}[q]$. A stationary complex Gaussian model

$$\boldsymbol{h}[q] \sim \mathcal{N}_{\mathrm{c}}(\boldsymbol{\mu}_{\boldsymbol{h}}, \boldsymbol{C}_{\boldsymbol{h}}) \tag{2.6}$$

with $\boldsymbol{\mu}_{\boldsymbol{h}} = \mathrm{E}[\boldsymbol{h}[q]]$ and $\boldsymbol{C}_{\boldsymbol{h}} = \mathrm{E}[(\boldsymbol{h}[q] - \boldsymbol{\mu}_{\boldsymbol{h}})(\boldsymbol{h}[q] - \boldsymbol{\mu}_{\boldsymbol{h}})^{\mathrm{H}}]$ is standard. It is referred to as Rice fading because the magnitude distribution of the elements of $\boldsymbol{h}[q]$ is a Rice distribution [173]. In the sequel we assume $\boldsymbol{\mu}_{\boldsymbol{h}} = \boldsymbol{0}_P$, i.e., the magnitude of an element of $\boldsymbol{h}[q]$ is Rayleigh distributed. An extension of the results to non-zero mean channels is straightforward. Moreover, we assume that $\boldsymbol{h}[q]$ and $\boldsymbol{n}[q]$ are uncorrelated.

Depending on the underlying scenario, one can assume a channel covariance matrix $\boldsymbol{C}_{\boldsymbol{h}}$ with additional structure:

1. If the channel coefficients belonging to different delays are uncorrelated, e.g., due to uncorrelated scattering [14], $\boldsymbol{C}_{\boldsymbol{h}}$ is block-diagonal

$$\boldsymbol{C}_{\boldsymbol{h}} = \begin{bmatrix} \boldsymbol{C}_{\boldsymbol{h}^{(0)}} & & \\ & \ddots & \\ & & \boldsymbol{C}_{\boldsymbol{h}^{(L)}} \end{bmatrix} = \sum_{\ell=0}^{L} \boldsymbol{e}_{\ell+1}\boldsymbol{e}_{\ell+1}^{\mathrm{T}} \otimes \boldsymbol{C}_{\boldsymbol{h}^{(\ell)}}, \tag{2.7}$$

   where $\boldsymbol{C}_{\boldsymbol{h}^{(\ell)}} \in \mathbb{S}_{+,0}^{MK}$ is the covariance matrix of the ℓth channel delay $\boldsymbol{h}^{(\ell)}[q] = \mathbf{vec}[\boldsymbol{H}^{(\ell)}[q]] \in \mathbb{C}^{MK}$.
2. If the channel coefficients belonging to different transmit antennas are uncorrelated, e.g., they belong to different mobile terminals, the covariance matrix of the ℓth channel delay has the structure $\boldsymbol{C}_{\boldsymbol{h}^{(\ell)}} = \sum_{k=1}^{K} \boldsymbol{e}_k\boldsymbol{e}_k^{\mathrm{T}} \otimes \boldsymbol{C}_{\boldsymbol{h}_k^{(\ell)}}$, where $\boldsymbol{h}_k^{(\ell)}[q] \in \mathbb{C}^M$ with covariance matrix $\boldsymbol{C}_{\boldsymbol{h}_k^{(\ell)}}$ is the kth column of $\boldsymbol{H}^{(\ell)}[q]$. In case of a uniform linear array at the receiver and a narrowband signal (see Appendix E and [125]) $\boldsymbol{C}_{\boldsymbol{h}_k^{(\ell)}}$ has Toeplitz structure.

2.1.0.2 Model for Multiple Blocks

To derive predictors of the channel realization $\boldsymbol{h}[q]$ based on Q previously observed blocks of training sequences, we model observations and the corresponding channel vectors for an arbitrary block index q as realizations of random vectors

$$\boldsymbol{y}_{\mathrm{T}}[q] = \begin{bmatrix} \boldsymbol{y}[q-1] \\ \boldsymbol{y}[q-2] \\ \vdots \\ \boldsymbol{y}[q-Q] \end{bmatrix} \in \mathbb{C}^{QMN} \quad \text{and} \quad \boldsymbol{h}_{\mathrm{T}}[q] = \begin{bmatrix} \boldsymbol{h}[q-1] \\ \boldsymbol{h}[q-2] \\ \vdots \\ \boldsymbol{h}[q-Q] \end{bmatrix} \in \mathbb{C}^{QP}, \tag{2.8}$$

respectively. With (2.4) the corresponding signal model is

$$\boxed{\boldsymbol{y}_{\mathrm{T}}[q] = \boldsymbol{S}_{\mathrm{T}} \boldsymbol{h}_{\mathrm{T}}[q] + \boldsymbol{n}_{\mathrm{T}}[q] \in \mathbb{C}^{QMN}} \tag{2.9}$$

with $\boldsymbol{S}_{\mathrm{T}} = \boldsymbol{I}_Q \otimes \boldsymbol{S}$.

In addition to the covariance matrix $\boldsymbol{C}_{\boldsymbol{h}}$ in space and delay domain, we have to define the channel parameters' temporal correlations. In general, the temporal correlations (autocovariance sequences) are different for every element in $\boldsymbol{h}[q]$, and $\boldsymbol{C}_{\boldsymbol{h}}[i] = \mathrm{E}[\boldsymbol{h}[q]\boldsymbol{h}[q-i]^{\mathrm{H}}]$ does not have a special structure. Because $\boldsymbol{h}[q]$ is a stationary random sequence (equidistant samples of the continuous-time channel), the auto- and crosscovariance matrices $\boldsymbol{C}_{\boldsymbol{h}_{\mathrm{T},p}\boldsymbol{h}_{\mathrm{T},p'}} = \mathrm{E}[\boldsymbol{h}_{\mathrm{T},p}[q]\boldsymbol{h}_{\mathrm{T},p'}[q]^{\mathrm{H}}]$ of $\boldsymbol{h}_{\mathrm{T},p}[q] = [h_p[q-1], h_p[q-2], \ldots, h_p[q-Q]]^{\mathrm{T}} \in \mathbb{C}^Q$ are Toeplitz with first row $[c_{h_p h_{p'}}[0], c_{h_p h_{p'}}[1], \ldots, c_{h_p h_{p'}}[Q-1]]$ and $c_{h_p h_{p'}}[i] = \mathrm{E}[h_p[q]h_{p'}[q-i]^*]$ for $p, p' \in \{1, 2, \ldots, P\}$. Therefore, $\boldsymbol{C}_{\boldsymbol{h}_{\mathrm{T}}}$ is block-Toeplitz and can be written as

$$\boldsymbol{C}_{\boldsymbol{h}_{\mathrm{T}}} = \sum_{p=1}^{P} \sum_{p'=1}^{P} \boldsymbol{C}_{\boldsymbol{h}_{\mathrm{T},p}\boldsymbol{h}_{\mathrm{T},p'}} \otimes \boldsymbol{e}_p \boldsymbol{e}_{p'}^{\mathrm{T}}. \tag{2.10}$$

To obtain simplified models of the channel correlations we introduce the following definition:

Definition 2.1. The crosscovariance $c_{h_p h_{p'}}[i]$ of two parameters $h_p[q]$ and $h_{p'}[q-i]$, $p \neq p'$ is called *separable* if it can be factorized as $c_{h_p h_{p'}}[i] = a_p[i]b[p,p']$.[4] The sequence $a_p[i] = a_{p'}[i]$, which on p (or p') and the time difference i, describes the temporal correlations and $b[p,p']$ the crosscorrelation between different random sequences.

Thus, for a separable crosscovariance, we require the same normalized autocovariance sequence $c_{h_p}[i]/c_{h_p}[0] = c_{h_{p'}}[i]/c_{h_{p'}}[0]$ for $h_p[q]$ and $h_{p'}[q]$, $p \neq p'$. In other words, the crosscorrelations in space and delay are called separable from the temporal correlations if the corresponding physical processes are statistically independent and the temporal correlations are identical.[5]

---

[4] This definition is in analogy to the notion of separable transmitter and receiver correlations for MIMO channel models resulting in the "Kronecker model" for the spatial covariance matrix (see e.g. [231]) and is exploited in [192, 30].

[5] Separability according to this definition is a valid model for a physical propagation channel if the following conditions hold: The Doppler frequencies do not depend on the angles of arrival at the receiver or angles of departure at the transmitter, which determine the spatial correlations; and they can be modeled as statistically indepen-

Depending on the physical properties of the channel, the following special cases are of interest.

1. For $L = 0$, the channel from the kth transmitter to the M receive antennas is denoted $\boldsymbol{h}_k[q] \in \mathbb{C}^M$, which is the kth column of $\boldsymbol{H}[q]$. Its covariance matrix is $\boldsymbol{C}_{\boldsymbol{h}_k}$. For separability of the spatial and temporal correlation of $\boldsymbol{h}_k[q]$, the autocovariance sequence $c_k[i]$ is defined as $\mathrm{E}[\boldsymbol{h}_k[q]\boldsymbol{h}_k[q-i]^{\mathrm{H}}] = c_k[i]\boldsymbol{C}_{\boldsymbol{h}_k}$ with $c_k[0] = 1$. Additionally, we assume that channels of different transmitters are uncorrelated $\mathrm{E}[\boldsymbol{h}_k[q]\boldsymbol{h}_{k'}[q]^{\mathrm{H}}] = \boldsymbol{C}_{\boldsymbol{h}_k}\delta[k-k']$, i.e., $\boldsymbol{C}_{\boldsymbol{h}} = \sum_{k=1}^{K} \boldsymbol{e}_k\boldsymbol{e}_k^{\mathrm{T}} \otimes \boldsymbol{C}_{\boldsymbol{h}_k}$. This yields
$$\boldsymbol{C}_{\boldsymbol{h}_{\mathrm{T}}} = \sum_{k=1}^{K} \boldsymbol{C}_{\mathrm{T},k} \otimes \boldsymbol{e}_k\boldsymbol{e}_k^{\mathrm{T}} \otimes \boldsymbol{C}_{\boldsymbol{h}_k}, \tag{2.11}$$
where $\boldsymbol{C}_{\mathrm{T},k}$ is Toeplitz with first row $[c_k[0], c_k[1], \ldots, c_k[Q-1]]$. For example, this describes the situation of K mobile terminals with generally different velocity and geographical location and M receivers located in the same (time-invariant) environment and on the same object.
2. If all elements of $\boldsymbol{h}[q]$ have an identical autocovariance sequence $c[i]$ and the correlations are separable, we can write $\mathrm{E}[\boldsymbol{h}[q]\boldsymbol{h}[q-i]^{\mathrm{H}}] = c[i]\boldsymbol{C}_{\boldsymbol{h}}$. This defines $c[i]$ implicitly, which we normalize to $c[0] = 1$. These assumptions result in
$$\boldsymbol{C}_{\boldsymbol{h}_{\mathrm{T}}} = \boldsymbol{C}_{\mathrm{T}} \otimes \boldsymbol{C}_{\boldsymbol{h}}, \tag{2.12}$$
where $\boldsymbol{C}_{\mathrm{T}}$ is Toeplitz with first row equal to $[c[0], c[1], \ldots, c[Q-1]]$.

Separability of spatial and temporal correlations is not possible in general, e.g., when the mobile terminals have multiple antennas and are moving. In this case the Doppler spectrum as well as the spatial correlation are determined by the same angles of departure/arrival at the mobile terminal.

2.2 Channel Estimation

Based on the model of the receive signal[6]
$$\boldsymbol{y} = \boldsymbol{S}\boldsymbol{h} + \boldsymbol{n} \in \mathbb{C}^{MN} \tag{2.13}$$

---

dent random variables. As an example, consider a scenario in which the reflecting or scattering object next to the fixed receiver is not moving and the Doppler spectrum results from a ring of scatterers around the geographically well-separated mobile terminals with one antenna each [30].

[6] To develop estimators based on $\boldsymbol{y}[q]$ which estimate the current channel realization $\boldsymbol{h}[q]$, we omit the block index q of (2.4) in this section.

we discuss different approaches to optimize a linear estimator $\boldsymbol{W} \in \mathbb{C}^{P \times MN}$ for $\boldsymbol{h}$. Given a realization $\boldsymbol{y}$ of y, it yields the channel estimate

$$\hat{\boldsymbol{h}} = \boldsymbol{W}\boldsymbol{y} \in \mathbb{C}^P. \tag{2.14}$$

Depending on the model of $\boldsymbol{h}$, e.g., stochastic or deterministic, different estimation philosophies are presented in the literature (Figure 2.1). Here we give a survey of different paradigms. We emphasize their relation to classical regularization methods [154] regarding the inverse problem (2.13).

As the system matrix $\boldsymbol{S}$ in the inverse problem (2.13) can be chosen by the system designer via the design of training sequences (see references in [215]), it is rather well-conditioned in general. Thus, the purpose of regularization methods is to incorporate a priori information about $\boldsymbol{h}$ in order to improve the performance in case of many parameters, in case of low SNR, or for a small number of observations N of N_t training symbols.

2.2.1 Minimum Mean Square Error Estimator

In the Bayesian formulation of the channel estimation problem (2.13) we model the channel parameters $\boldsymbol{h}$ as a random vector h with given probability density function (pdf) $\mathrm{p}_{\mathsf{h}}(\boldsymbol{h})$. Thus, for example, correlations of h can be exploited to improve the estimator's performance. Given the joint pdf $\mathrm{p}_{\mathsf{h},\mathsf{y}}(\boldsymbol{h}, \boldsymbol{y})$ and the observation $\boldsymbol{y}$, we search a function $\mathbf{f} : \mathbb{C}^{MN} \to \mathbb{C}^P$ with $\hat{\boldsymbol{h}} = \mathbf{f}(\boldsymbol{y})$ such that $\hat{\boldsymbol{h}}$ is, on average, as close to the true channel realization $\boldsymbol{h}$ as possible.

Here we aim at minimizing the mean square error (MSE)

$$\mathrm{E}_{\mathsf{h},\mathsf{y}}\left[\|\mathsf{h} - \mathbf{f}(\mathsf{y})\|_2^2\right] = \mathrm{E}_{\mathsf{y}}\left[\mathrm{E}_{\mathsf{h}|\mathsf{y}}\left[\|\mathsf{h} - \mathbf{f}(\mathsf{y})\|_2^2\right]\right] \tag{2.15}$$

w.r.t. all functions $\mathbf{f}$ [124]. This is equivalent to minimizing the conditional MSE w.r.t. $\hat{\boldsymbol{h}} \in \mathbb{C}^P$ given the observation $\boldsymbol{y}$

$$\min_{\hat{\boldsymbol{h}} \in \mathbb{C}^P} \mathrm{E}_{\mathsf{h}|\boldsymbol{y}}\left[\|\mathsf{h} - \hat{\boldsymbol{h}}\|_2^2\right]. \tag{2.16}$$

Writing the conditional MSE explicitly, we obtain

$$\mathrm{E}_{\mathsf{h}|\boldsymbol{y}}\left[\|\mathsf{h} - \hat{\boldsymbol{h}}\|_2^2\right] = \mathrm{tr}\left[\boldsymbol{R}_{\mathsf{h}|\boldsymbol{y}}\right] + \|\hat{\boldsymbol{h}}\|_2^2 - \hat{\boldsymbol{h}}^{\mathrm{H}}\boldsymbol{\mu}_{\mathsf{h}|\boldsymbol{y}} - \boldsymbol{\mu}_{\mathsf{h}|\boldsymbol{y}}^{\mathrm{H}}\hat{\boldsymbol{h}} \tag{2.17}$$

with conditional correlation matrix $\boldsymbol{R}_{\mathsf{h}|\boldsymbol{y}} = \boldsymbol{C}_{\mathsf{h}|\boldsymbol{y}} + \boldsymbol{\mu}_{\mathsf{h}|\boldsymbol{y}}\boldsymbol{\mu}_{\mathsf{h}|\boldsymbol{y}}^{\mathrm{H}}$, conditional covariance matrix $\boldsymbol{C}_{\mathsf{h}|\boldsymbol{y}}$, and conditional mean $\boldsymbol{\mu}_{\mathsf{h}|\boldsymbol{y}}$ of the pdf $\mathrm{p}_{\mathsf{h}|\mathsf{y}}(\boldsymbol{h}|\boldsymbol{y})$.

The solution of (2.16)

$$\boxed{\hat{\boldsymbol{h}}_{\mathrm{CM}} = \boldsymbol{\mu}_{\boldsymbol{h}|\boldsymbol{y}} = \mathrm{E}_{\boldsymbol{h}|\boldsymbol{y}}[\boldsymbol{h}]} \tag{2.18}$$

is often called the minimum MSE (MMSE) estimator or the conditional mean (CM) estimator. Thus, in general $\mathbf{f}$ is a nonlinear function of $\boldsymbol{y}$. The MMSE

$$\mathrm{E}_{\boldsymbol{y}}\left[\mathrm{tr}\left[\boldsymbol{R}_{\boldsymbol{h}|\boldsymbol{y}}\right] - \|\boldsymbol{\mu}_{\boldsymbol{h}|\boldsymbol{y}}\|_2^2\right] = \mathrm{tr}\left[\mathrm{E}_{\boldsymbol{y}}\left[\boldsymbol{C}_{\boldsymbol{h}|\boldsymbol{y}}\right]\right] \tag{2.19}$$

is the trace of the mean error correlation matrix

$$\boldsymbol{C}_{\boldsymbol{h}-\hat{\boldsymbol{h}}_{\mathrm{CM}}} = \mathrm{E}_{\boldsymbol{y}}\left[\boldsymbol{C}_{\boldsymbol{h}|\boldsymbol{y}}\right] = \mathrm{E}_{\boldsymbol{h},\boldsymbol{y}}\left[(\boldsymbol{h} - \hat{\boldsymbol{h}}_{\mathrm{CM}})(\boldsymbol{h} - \hat{\boldsymbol{h}}_{\mathrm{CM}})^{\mathrm{H}}\right]. \tag{2.20}$$

So far in this section we have considered a general pdf $\mathrm{p}_{\boldsymbol{h},\boldsymbol{y}}(\boldsymbol{h}, \boldsymbol{y})$; now we assume $\boldsymbol{h}$ and $\boldsymbol{y}$ to be zero-mean and jointly complex Gaussian distributed. For this assumption, the conditional mean is a linear function of $\boldsymbol{y}$ and reads (cf. (A.2))

$$\hat{\boldsymbol{h}}_{\mathrm{MMSE}} = \boldsymbol{\mu}_{\boldsymbol{h}|\boldsymbol{y}} = \boldsymbol{W}_{\mathrm{MMSE}}\, \boldsymbol{y} \tag{2.21}$$

with linear estimator[7]

$$\boxed{\begin{aligned} \boldsymbol{W}_{\mathrm{MMSE}} &= \boldsymbol{C}_{\boldsymbol{h}\boldsymbol{y}} \boldsymbol{C}_{\boldsymbol{y}}^{-1} \\ &= \boldsymbol{C}_{\boldsymbol{h}} \boldsymbol{S}^{\mathrm{H}} (\boldsymbol{S} \boldsymbol{C}_{\boldsymbol{h}} \boldsymbol{S}^{\mathrm{H}} + \boldsymbol{C}_{\boldsymbol{n}})^{-1} \\ &= (\boldsymbol{C}_{\boldsymbol{h}} \boldsymbol{S}^{\mathrm{H}} \boldsymbol{C}_{\boldsymbol{n}}^{-1} \boldsymbol{S} + \boldsymbol{I}_P)^{-1} \boldsymbol{C}_{\boldsymbol{h}} \boldsymbol{S}^{\mathrm{H}} \boldsymbol{C}_{\boldsymbol{n}}^{-1}. \end{aligned}} \tag{2.22}$$

The first equality follows from the system model (2.13) and the second from the matrix inversion lemma (A.19) [124, p. 533]. The left side of this system of linear equations, i.e., the matrix to be "inverted", is smaller in the third line in (2.22). If $N > K(L+1)$ and $\boldsymbol{C}_{\boldsymbol{n}}^{-1}$ is given or $\boldsymbol{C}_{\boldsymbol{n}} = \boldsymbol{I}_N \otimes \boldsymbol{C}_{\boldsymbol{n},\mathrm{s}}$, the order of complexity to solve it is also reduced [216].

For the Gaussian assumption, the mean error correlation matrix in (2.20) is equivalent to the conditional error covariance matrix, as the latter is independent of $\boldsymbol{y}$. It is the Schur complement of $\boldsymbol{C}_{\boldsymbol{y}}$ in the covariance matrix of $[\boldsymbol{h}^{\mathrm{T}}, \boldsymbol{y}^{\mathrm{T}}]^{\mathrm{T}}$ (see A.17)

$$\boldsymbol{C}_{\boldsymbol{h}|\boldsymbol{y}} = \boldsymbol{C}_{\boldsymbol{h}} - \boldsymbol{W}_{\mathrm{MMSE}}\, \boldsymbol{C}_{\boldsymbol{h}\boldsymbol{y}}^{\mathrm{H}} = \mathrm{tr}\left[\mathbf{K}_{\boldsymbol{C}_{\boldsymbol{y}}}\left(\begin{bmatrix} \boldsymbol{C}_{\boldsymbol{h}} & \boldsymbol{C}_{\boldsymbol{h}\boldsymbol{y}} \\ \boldsymbol{C}_{\boldsymbol{h}\boldsymbol{y}}^{\mathrm{H}} & \boldsymbol{C}_{\boldsymbol{y}} \end{bmatrix}\right)\right]. \tag{2.23}$$

For every linear estimator $\boldsymbol{W}$, the MSE can be decomposed in the mean squared norm of the estimator's bias $\boldsymbol{h} - \boldsymbol{W}\boldsymbol{S}\boldsymbol{h}$ and the variance of the estimator:

---

[7] The channel covariance matrix $\boldsymbol{C}_{\boldsymbol{h}}$ is determined by the angles of arrival and departure, delays, and mean path loss. These are long-term or large-scale properties of the channel and change slowly compared to the realization $\boldsymbol{h}$. Thus, in principle it is not necessary to estimate these parameters directly as in [224] to exploit the channel's structure for an improved channel estimate $\hat{\boldsymbol{h}}$.

$$\mathrm{E}_{\boldsymbol{h},\boldsymbol{y}}\left[\|\boldsymbol{h}-\boldsymbol{W}\boldsymbol{y}\|_2^2\right] = \mathrm{E}_{\boldsymbol{h},\boldsymbol{n}}\left[\|\boldsymbol{h}-\boldsymbol{W}(\boldsymbol{S}\boldsymbol{h}+\boldsymbol{n})\|_2^2\right] \tag{2.24}$$

$$= \underbrace{\mathrm{E}_{\boldsymbol{h}}\left[\|\boldsymbol{h}-\boldsymbol{W}\boldsymbol{S}\boldsymbol{h}\|_2^2\right]}_{\text{mean squared norm of Bias}} + \underbrace{\mathrm{E}_{\boldsymbol{n}}\left[\|\boldsymbol{W}\boldsymbol{n}\|_2^2\right]}_{\text{Variance}}. \tag{2.25}$$

The MMSE estimator finds the best trade-off among both contributions because it exploits the knowledge about the channel and noise covariance matrix $\boldsymbol{C}_{\boldsymbol{h}}$ and $\boldsymbol{C}_{\boldsymbol{n}}$, respectively.

Due to its linearity, $\boldsymbol{W}_{\mathrm{MMSE}}$ is called linear MMSE (LMMSE) estimator and could also be derived from (2.16) restricting $\mathbf{f}$ to the class of linear functions as in (2.24). In Section 2.4 we discuss the robust properties of the LMMSE estimator for general (non-Gaussian) pdf $\mathrm{p}_{\boldsymbol{h},\boldsymbol{y}}(\boldsymbol{h},\boldsymbol{y})$.

The estimate $\hat{\boldsymbol{h}}_{\mathrm{MMSE}}$ is also the solution of the regularized weighted least-squares problem

$$\hat{\boldsymbol{h}}_{\mathrm{MMSE}} = \operatorname*{argmin}_{\boldsymbol{h}} \|\boldsymbol{y}-\boldsymbol{S}\boldsymbol{h}\|^2_{\boldsymbol{C}_{\boldsymbol{n}}^{-1}} + \|\boldsymbol{h}\|^2_{\boldsymbol{C}_{\boldsymbol{h}}^{-1}}, \tag{2.26}$$

which aims at simultaneously minimizing the weighted norm of the approximation error and of $\boldsymbol{h}$ [24]. In the terminology of [154], $\boldsymbol{C}_{\boldsymbol{h}}$ describes the smoothness of $\boldsymbol{h}$. Thus, smoothness and average size of $\boldsymbol{h}$ — relative to $\boldsymbol{C}_{\boldsymbol{n}}$ — is the a priori information exploited here.

2.2.2 Maximum Likelihood Estimator

Contrary to the MMSE estimator, the Maximum Likelihood (ML) approach treats $\boldsymbol{h}$ as a deterministic parameter. At first we assume to have no additional a priori information about it.

Together with the noise covariance matrix $\boldsymbol{C}_{\boldsymbol{n}}$, the channel parameters $\boldsymbol{h}$ define the stochastic model of the observation $\boldsymbol{y}$,

$$\boldsymbol{y} \sim \mathcal{N}_{\mathrm{c}}(\boldsymbol{S}\boldsymbol{h}, \boldsymbol{C}_{\boldsymbol{n}}). \tag{2.27}$$

The ML approach aims at finding the parameters of this distribution which present the best fit to the given observation $\boldsymbol{y}$ and maximize the likelihood

$$\max_{\boldsymbol{h}\in\mathbb{C}^P, \boldsymbol{C}_{\boldsymbol{n}}\in\mathcal{C}_{\mathrm{n}}} \mathrm{p}_{\boldsymbol{y}}(\boldsymbol{y};\boldsymbol{h},\boldsymbol{C}_{\boldsymbol{n}}). \tag{2.28}$$

The complex Gaussian pdf is (see Appendix A.1)

$$\mathrm{p}_{\boldsymbol{y}}(\boldsymbol{y};\boldsymbol{h},\boldsymbol{C}_{\boldsymbol{n}}) = \left(\pi^{MN}\det\boldsymbol{C}_{\boldsymbol{n}}\right)^{-1}\exp\left(-\left(\boldsymbol{y}-\boldsymbol{S}\boldsymbol{h}\right)^{\mathrm{H}}\boldsymbol{C}_{\boldsymbol{n}}^{-1}\left(\boldsymbol{y}-\boldsymbol{S}\boldsymbol{h}\right)\right). \tag{2.29}$$

Restricting the noise to be temporally uncorrelated and spatially correlated, the corresponding class of covariance matrices is

$$\mathcal{C}_{\mathrm{n}} = \left\{\boldsymbol{C}_{\boldsymbol{n}} | \boldsymbol{C}_{\boldsymbol{n}} = \boldsymbol{I}_N \otimes \boldsymbol{C}_{\boldsymbol{n},\mathrm{s}}, \boldsymbol{C}_{\boldsymbol{n},\mathrm{s}} \in \mathbb{S}_{+,0}^{M}\right\}, \tag{2.30}$$

for which (2.28) yields [129]

$$\boxed{\begin{aligned} &\hat{\boldsymbol{h}}_{\mathrm{ML}} = \boldsymbol{W}_{\mathrm{ML}}\, \boldsymbol{y}, \quad \boldsymbol{W}_{\mathrm{ML}} = \boldsymbol{S}^{\dagger} \\ &\hat{\boldsymbol{C}}_{\boldsymbol{n},\mathrm{s}} = \frac{1}{N}\left(\boldsymbol{Y} - \hat{\boldsymbol{H}}_{\mathrm{ML}}\bar{\boldsymbol{S}}\right)\left(\boldsymbol{Y} - \hat{\boldsymbol{H}}_{\mathrm{ML}}\bar{\boldsymbol{S}}\right)^{\mathrm{H}} \end{aligned}} \tag{2.31}$$

with $\hat{\boldsymbol{H}}_{\mathrm{ML}} = \mathbf{unvec}\left[\hat{\boldsymbol{h}}_{\mathrm{ML}}\right] \in \mathbb{C}^{M \times K(L+1)}$. The estimates exist only if $\boldsymbol{S}$ has full column rank, i.e., necessarily $MN \geq P$. If $\boldsymbol{C}_{\boldsymbol{n}} = c_n \boldsymbol{I}_{MN}$, we have $\hat{c}_n = \frac{1}{MN}\|\boldsymbol{y} - \boldsymbol{S}\hat{\boldsymbol{h}}_{\mathrm{ML}}\|_2^2$. Introducing the orthogonal projector $\boldsymbol{P} = \bar{\boldsymbol{S}}^{\mathrm{H}}(\bar{\boldsymbol{S}}\bar{\boldsymbol{S}}^{\mathrm{H}})^{-1}\bar{\boldsymbol{S}}$ on the row space of $\bar{\boldsymbol{S}}$ and the projector $\boldsymbol{P}^{\perp} = \boldsymbol{I}_N - \boldsymbol{P}$ on the space orthogonal to the row space of $\bar{\boldsymbol{S}}$, we can write [129]

$$\hat{\boldsymbol{C}}_{\boldsymbol{n},\mathrm{s}} = \frac{1}{N}\boldsymbol{Y}\boldsymbol{P}^{\perp}\boldsymbol{Y}^{\mathrm{H}} \tag{2.32}$$

$$\hat{c}_n = \frac{1}{MN}\|\boldsymbol{Y}\boldsymbol{P}^{\perp}\|_{\mathrm{F}}^2. \tag{2.33}$$

The space orthogonal to the row space of $\bar{\boldsymbol{S}}$ contains only noise ("noise subspace"), because $\boldsymbol{Y}\boldsymbol{P}^{\perp} = \boldsymbol{N}\boldsymbol{P}^{\perp}$ (cf. Eq. 2.3).

In contrast to (2.21), $\hat{\boldsymbol{h}}_{\mathrm{ML}}$ in (2.31) is an unbiased estimate of $\boldsymbol{h}$, whereas the bias of $\hat{\boldsymbol{C}}_{\boldsymbol{n},\mathrm{s}}$ and $\hat{c}_n$ is

$$\mathrm{E}_{\boldsymbol{n}}\left[\hat{\boldsymbol{C}}_{\boldsymbol{n},\mathrm{s}}\right] - \boldsymbol{C}_{\boldsymbol{n},\mathrm{s}} = \left(\frac{N - K(L+1)}{N} - 1\right)\boldsymbol{C}_{\boldsymbol{n},\mathrm{s}} \tag{2.34}$$

$$\mathrm{E}_{\boldsymbol{n}}[\hat{c}_n] - c_n = \left(\frac{N - K(L+1)}{N} - 1\right)c_n. \tag{2.35}$$

It results from the scaling of the estimates in (2.32) and (2.33) by N and MN, respectively. On average, $\hat{\boldsymbol{C}}_{\boldsymbol{n},\mathrm{s}}$ and $\hat{c}_n$ underestimate the true noise covariance matrix, because $\mathrm{E}[\hat{\boldsymbol{C}}_{\boldsymbol{n},\mathrm{s}}] \preceq \boldsymbol{C}_{\boldsymbol{n},\mathrm{s}}$. Note that the rank of $\boldsymbol{P}^{\perp}$ is $N - K(L+1)$. Thus, for unbiased estimates, a scaling with $N - K(L+1)$ and $M[N - K(L+1)]$ instead of N and MN, respectively, would be sufficient. These unbiased estimators are derived in Chapter 3.

For the general class $\mathcal{C}_{\mathrm{n}} = \{\boldsymbol{C}_{\boldsymbol{n}} | \boldsymbol{C}_{\boldsymbol{n}} \in \mathbb{S}_{+,0}^{MN}\}$ and if we assume $\boldsymbol{C}_{\boldsymbol{n}}$ to be known, the ML problem (2.28) is equivalent to the weighted least-squares problem [124]

$$\min_{\boldsymbol{h}} \|\boldsymbol{y} - \boldsymbol{S}\boldsymbol{h}\|_{\boldsymbol{C}_{\boldsymbol{n}}^{-1}}, \tag{2.36}$$

which is solved by

$$\hat{\boldsymbol{h}}_{\mathrm{ML}} = \boldsymbol{W}_{\mathrm{ML}}\boldsymbol{y}, \quad \boldsymbol{W}_{\mathrm{ML}} = \left(\boldsymbol{S}^{\mathrm{H}}\boldsymbol{C}_{\boldsymbol{n}}^{-1}\boldsymbol{S}\right)^{-1}\boldsymbol{S}^{\mathrm{H}}\boldsymbol{C}_{\boldsymbol{n}}^{-1}. \tag{2.37}$$

Note that $\boldsymbol{S}^{\mathrm{H}}\boldsymbol{C}_{\boldsymbol{n}}^{-1}\boldsymbol{y}$ is a sufficient statistic for $\boldsymbol{h}$. Applying Tikhonov regularization [154, 24, 212] to (2.36), we obtain (2.26) and the LMMSE estimator (2.21).

For (2.31) and (2.37), the covariance matrix of the zero-mean estimation error $\boldsymbol{\varepsilon} = \hat{\boldsymbol{h}} - \boldsymbol{h}$ is

$$\boldsymbol{C}_{\boldsymbol{\varepsilon}} = \left(\boldsymbol{S}^{\mathrm{H}}\boldsymbol{C}_{\boldsymbol{n}}^{-1}\boldsymbol{S}\right)^{-1} \tag{2.38}$$

and the MSE is $\mathrm{E}[\|\boldsymbol{\varepsilon}\|_2^2] = \mathrm{tr}[\boldsymbol{C}_{\boldsymbol{\varepsilon}}]$, since $\hat{\boldsymbol{h}}_{\mathrm{ML}}$ is unbiased.

The relationship between $\boldsymbol{W}_{\mathrm{ML}}$ from (2.37) and the MMSE estimate (2.21) is given by

$$\begin{aligned}\boldsymbol{W}_{\mathrm{MMSE}} &= \left(\boldsymbol{C}_{\boldsymbol{\varepsilon}}\boldsymbol{C}_{\boldsymbol{h}}^{-1} + \boldsymbol{I}_P\right)^{-1}\boldsymbol{W}_{\mathrm{ML}} &(2.39)\\ &= \boldsymbol{C}_{\boldsymbol{\varepsilon}}^{1/2}\left(\boldsymbol{C}_{\boldsymbol{\varepsilon}}^{1/2}\boldsymbol{C}_{\boldsymbol{h}}^{-1}\boldsymbol{C}_{\boldsymbol{\varepsilon}}^{1/2} + \boldsymbol{I}_P\right)^{-1}\boldsymbol{C}_{\boldsymbol{\varepsilon}}^{-1/2}\boldsymbol{W}_{\mathrm{ML}}\\ &= \boldsymbol{C}_{\boldsymbol{\varepsilon}}^{1/2}\boldsymbol{V}\underbrace{\left(\boldsymbol{\Sigma} + \boldsymbol{I}_P\right)^{-1}\boldsymbol{\Sigma}}_{\text{matrix of filter factors } \boldsymbol{F}}\boldsymbol{V}^{\mathrm{H}}\underbrace{\boldsymbol{C}_{\boldsymbol{\varepsilon}}^{-1/2}}_{\text{whitening}}\boldsymbol{W}_{\mathrm{ML}} &(2.40)\end{aligned}$$

with the eigenvalue decomposition of $\boldsymbol{C}_{\boldsymbol{\varepsilon}}^{-1/2}\boldsymbol{C}_{\boldsymbol{h}}\boldsymbol{C}_{\boldsymbol{\varepsilon}}^{-1/2} = \boldsymbol{V}\boldsymbol{\Sigma}\boldsymbol{V}^{\mathrm{H}}$ and $[\boldsymbol{\Sigma}]_{i,i} \geq [\boldsymbol{\Sigma}]_{i+1,i+1}$. Based on this decomposition of $\boldsymbol{W}_{\mathrm{MMSE}}$, the MMSE estimator can be interpreted as follows: The first stage is an ML channel estimator followed by a whitening of the estimation error. In the next stage, the basis in $\mathbb{C}^P$ is changed and the components are weighted by the diagonal elements of $\left(\boldsymbol{\Sigma} + \boldsymbol{I}_P\right)^{-1}\boldsymbol{\Sigma}$, which describe the reliability of the estimate in every dimension. The ith diagonal element of $\boldsymbol{\Sigma}$ may be interpreted as the signal to noise ratio in the one-dimensional subspace spanned by the ith column of $\boldsymbol{C}_{\boldsymbol{\varepsilon}}^{1/2}\boldsymbol{V}$.

2.2.2.1 Reduced-Rank Maximum Likelihood Estimator

In contrast to this "soft" weighting of the subspaces, the *reduced-rank ML* (RML) solution proposed in [191, 157] (for $\boldsymbol{C}_{\boldsymbol{n}} = \boldsymbol{I}_N \otimes \boldsymbol{C}_{\boldsymbol{n},\mathrm{s}}$) results in the new matrix of filter factors $\boldsymbol{F}$ in (2.40), chosen as

$$\boldsymbol{F}_{\mathrm{RML}} = \begin{bmatrix}\boldsymbol{I}_R & \\ & \boldsymbol{0}_{P-R\times P-R}\end{bmatrix}, \tag{2.41}$$

which selects the strongest R dimensions and truncates the weakest $P-R$ dimensions in the basis $\boldsymbol{C}_{\boldsymbol{\varepsilon}}^{1/2}\boldsymbol{V}$.[8] The resulting RML estimator

$$\boxed{\boldsymbol{W}_{\mathrm{RML}} = \boldsymbol{C}_{\boldsymbol{\varepsilon}}^{1/2}\boldsymbol{V}\boldsymbol{F}_{\mathrm{RML}}\boldsymbol{V}^{\mathrm{H}}\boldsymbol{C}_{\boldsymbol{\varepsilon}}^{-1/2}\boldsymbol{W}_{\mathrm{ML}}} \tag{2.42}$$

is of reduced rank $R \leq P$.

---

[8] The eigenvalues are assumed to be sorted in decreasing magnitude.

Truncating the singular value decomposition of the system matrix $\boldsymbol{S}$ is a standard technique for regularizing the solution of inverse problems [154, 205]. Here, a non-orthogonal basis is chosen which describes additional a priori information available for this problem via $\boldsymbol{C}_{\boldsymbol{h}}$.

The relation of the RML to the MMSE estimator was established in [56], which allows for the heuristic derivation of the RML estimator given here. In the original work [191, 157], a rigorous derivation of this estimator is presented. It assumes a reduced-rank model of dimension R for $\boldsymbol{h} = \boldsymbol{U}_R \boldsymbol{\xi}$ with $\boldsymbol{\xi} \in \mathbb{C}^R$. The optimum choice for $\boldsymbol{U}_R \in \mathbb{C}^{P \times R}$ is the non-orthogonal basis $\boldsymbol{U}_R = \boldsymbol{C}_{\boldsymbol{\varepsilon}}^{1/2} \boldsymbol{V} \boldsymbol{F}$. This basis does not change over B blocks of training sequences, whereas $\boldsymbol{\xi}$ is assumed to change from block to block. Moreover, based on the derivation in [191, 157] for finite B, the estimate $\hat{\boldsymbol{C}}_{\boldsymbol{h}} = \frac{1}{B} \sum_{q=1}^{B} \hat{\boldsymbol{h}}_{\mathrm{ML}}[q] \hat{\boldsymbol{h}}_{\mathrm{ML}}[q]^{\mathrm{H}}$ of $\boldsymbol{C}_{\boldsymbol{h}}$ is optimum, where $\hat{\boldsymbol{h}}_{\mathrm{ML}}[q]$ is the ML estimate (2.31) based on $\boldsymbol{y}[q]$ (2.4) in block q.

In practice, the choice of R is a problem, as $\boldsymbol{C}_{\boldsymbol{h}}$ is generally of full rank. Choosing $R < P$ results in a bias. Thus, the rank R has to be optimized to achieve a good bias-variance trade-off (see Section 2.2.4, also [56, 157]). The MMSE estimator's model of $\boldsymbol{h}$ is more accurate and results in a superior performance.

2.2.3 Correlator and Matched Filter

For orthogonal columns in $\boldsymbol{S}$ and white noise, i.e., $\boldsymbol{S}^{\mathrm{H}} \boldsymbol{S} = N c_{\mathrm{s}} \boldsymbol{I}_P$ and $\boldsymbol{C}_{\boldsymbol{n}} = c_n \boldsymbol{I}_{MN}$, the ML estimator in (2.31) and (2.37) simplifies to

$$\boxed{\boldsymbol{W}_{\mathrm{C}} = \frac{1}{N c_{\mathrm{s}}} \boldsymbol{S}^{\mathrm{H}},} \tag{2.43}$$

which is sometimes called *"correlator"* in the literature. It is simple to implement, but if $\boldsymbol{S}^{\mathrm{H}} \boldsymbol{S} = N c_{\mathrm{s}} \boldsymbol{I}_P$ does not hold, the channel estimate becomes biased and the MSE saturates for high SNR. Orthogonality of $\boldsymbol{S}$ can be achieved by appropriate design of the training sequences, e.g., choosing orthogonal sequences for $L = 0$ or with the choice of [201] for $L > 0$. The correlator does not exploit any additional a priori information of the channel parameters such as correlations.

The *matched filter* (MF) aims at maximizing the cross-correlation between the output of the estimator $\hat{\boldsymbol{h}} = \boldsymbol{W} \boldsymbol{y}$ and the signal $\boldsymbol{h}$ to be estimated. The correlation is measured by $|\mathrm{E}_{\boldsymbol{h},\boldsymbol{n}}[\hat{\boldsymbol{h}}^{\mathrm{H}} \boldsymbol{h}]|^2 = |\mathrm{tr}[\boldsymbol{R}_{\boldsymbol{h}\hat{\boldsymbol{h}}}]|^2$, which is additionally normalized by the noise variance in $\hat{\boldsymbol{h}}$ to avoid $\|\boldsymbol{W}\|_{\mathrm{F}} \to \infty$. This can be interpreted as a generalization of the SNR which is defined in the literature for estimating a scalar signal [108, p.131]. Thus, the MF assumes a stochastic model for $\boldsymbol{h}$ with correlation matrix[9] $\boldsymbol{R}_{\boldsymbol{h}}$.

The optimization problem reads

$$\max_{\boldsymbol{W}} \frac{\left|\mathrm{E}_{\boldsymbol{h},\boldsymbol{n}}\left[\hat{\boldsymbol{h}}^{\mathrm{H}}\boldsymbol{h}\right]\right|^2}{\mathrm{E}_{\boldsymbol{n}}\left[\|\boldsymbol{W}\boldsymbol{n}\|_2^2\right]} \tag{2.44}$$

and is solved by

$$\boxed{\boldsymbol{W}_{\mathrm{MF}} = \alpha \boldsymbol{R}_{\boldsymbol{h}} \boldsymbol{S}^{\mathrm{H}} \boldsymbol{C}_{\boldsymbol{n}}^{-1},\ \alpha \in \mathbb{C}.} \tag{2.45}$$

The solution is unique up to a complex scaling α because the cost function is invariant to α. A possible choice is $\alpha = \frac{c_n}{N c_{\mathrm{s}}}$ to achieve convergence to the correlator (2.43) in case $\boldsymbol{R}_{\boldsymbol{h}} = \boldsymbol{I}_P$ and $\boldsymbol{C}_{\boldsymbol{n}} = c_n \boldsymbol{I}_{MN}$, where $c_n = \mathrm{tr}[\boldsymbol{C}_{\boldsymbol{n}}]/(MN)$ and c_{s} is defined by $c_{\mathrm{s}} = \|\boldsymbol{s}[n]\|_2^2/K$ for training sequences with constant $\|\boldsymbol{s}[n]\|_2$.

The MF produces a sufficient statistic $\boldsymbol{S}^{\mathrm{H}}\boldsymbol{C}_{\boldsymbol{n}}^{-1}\boldsymbol{y}$ for $\boldsymbol{h}$ in the first stage before weighting the estimate by the correlation matrix $\boldsymbol{R}_{\boldsymbol{h}}$. This weighting $\boldsymbol{R}_{\boldsymbol{h}} = \boldsymbol{U}\boldsymbol{\Lambda}\boldsymbol{U}^{\mathrm{H}}$ corresponds to changing the basis with $\boldsymbol{U}$, which is followed by a weighting by the eigenvalues (signal variances in the transformed domain) and a transformation back to the original basis. Thus, the estimate is amplified more in those dimensions for which the channel parameters are stronger.

For high noise variance, the MMSE estimator (2.22) scaled by c'_n which is defined via $\boldsymbol{C}_{\boldsymbol{n}} = c'_n \boldsymbol{C}'_{\boldsymbol{n}}$ and $\boldsymbol{C}'_{\boldsymbol{n}} = \lim_{c'_n \to \infty} \boldsymbol{C}_{\boldsymbol{n}}/c'_n \succ 0$, converges to the MF in (2.45) with $\alpha = 1$ and $\boldsymbol{C}_{\boldsymbol{n}} = \boldsymbol{C}'_{\boldsymbol{n}}$:

$$\boldsymbol{W}_{\mathrm{MF}}\Big|_{\alpha=1, \boldsymbol{C}_{\boldsymbol{n}}=\boldsymbol{C}'_{\boldsymbol{n}}} = \lim_{c'_n \to \infty} c'_n \boldsymbol{W}_{\mathrm{MMSE}}. \tag{2.46}$$

If the diagonal of $\boldsymbol{C}_{\boldsymbol{n}}$ is constant and equal to c'_n, then we have $\boldsymbol{C}'_{\boldsymbol{n}} = \boldsymbol{I}_{MN}$. In systems operating at low SNR and with short training sequences, the MF is a low-complexity implementation of the MMSE estimator with similar performance.

Although the correlator for channel estimation is often called a MF, the relation of the MF based on (2.44) to the MMSE estimator (2.21) in (2.46) suggests using this terminology for (2.45).

2.2.4 Bias-Variance Trade-Off

The MSE can be decomposed as

---

[9] Although it is equal to the covariance matrix for our assumption of zero-mean $\boldsymbol{h}$, in general the correlation matrix has to be used here.

$$
\begin{aligned}
\mathrm{E}_{\boldsymbol{h},\boldsymbol{n}}\left[\|\boldsymbol{\varepsilon}\|_2^2\right] &= \mathrm{E}_{\boldsymbol{h},\boldsymbol{n}}\left[\|\boldsymbol{h}-\boldsymbol{W}(\boldsymbol{S}\boldsymbol{h}+\boldsymbol{n})\|_2^2\right] \\
&= \mathrm{E}_{\boldsymbol{h}}\left[\|(\boldsymbol{I}_P-\boldsymbol{W}\boldsymbol{S})\boldsymbol{h}\|_2^2\right] + \mathrm{E}_{\boldsymbol{n}}\left[\|\boldsymbol{W}\boldsymbol{n}\|_2^2\right] \\
&= \underbrace{\mathrm{tr}\left[(\boldsymbol{I}_P-\boldsymbol{W}\boldsymbol{S})\boldsymbol{C}_{\boldsymbol{h}}(\boldsymbol{I}_P-\boldsymbol{W}\boldsymbol{S})^{\mathrm{H}}\right]}_{\text{mean squared norm of bias}} + \underbrace{c_n\|\boldsymbol{W}\|_{\mathrm{F}}^2}_{\text{variance}},
\end{aligned}
\tag{2.47}
$$

where the first term is the mean of the squared norm of the bias and the second the variance of the estimate $\hat{\boldsymbol{h}} = \boldsymbol{W}\boldsymbol{y}$ with $\boldsymbol{C}_{\boldsymbol{n}} = c_n\boldsymbol{I}_{MN}$. The estimators presented in the previous sections aim at a reduced variance compared to the ML estimator, at the price of introducing and increasing the bias.

Assuming orthogonal training sequences with $\boldsymbol{S}^{\mathrm{H}}\boldsymbol{S} = Nc_{\mathrm{s}}\boldsymbol{I}_P$, the basic behavior of the estimators w.r.t. to this bias-variance trade-off can be explained. With this assumption and the EVD $\boldsymbol{C}_{\boldsymbol{h}} = \boldsymbol{U}\boldsymbol{\Lambda}\boldsymbol{U}^{\mathrm{H}}$, $\boldsymbol{\Lambda} = \mathbf{diag}[\lambda_1,\ldots,\lambda_P]$, the estimators read

$$
\begin{aligned}
\boldsymbol{W}_{\mathrm{ML}} &= \boldsymbol{W}_{\mathrm{C}} = \frac{1}{Nc_{\mathrm{s}}}\boldsymbol{S}^{\mathrm{H}} \\
\boldsymbol{W}_{\mathrm{MMSE}} &= \boldsymbol{U}\underbrace{\left(\boldsymbol{\Lambda}+\frac{c_n}{Nc_{\mathrm{s}}}\boldsymbol{I}_P\right)^{-1}\boldsymbol{\Lambda}}_{\text{Optimal weighting of subspaces}}\boldsymbol{U}^{\mathrm{H}}\boldsymbol{W}_{\mathrm{ML}} \\
\boldsymbol{W}_{\mathrm{MF}} &= \frac{1}{\lambda_1}\boldsymbol{U}\boldsymbol{\Lambda}\boldsymbol{U}^{\mathrm{H}}\boldsymbol{W}_{\mathrm{ML}} \\
\boldsymbol{W}_{\mathrm{RML}} &= \boldsymbol{U}\boldsymbol{F}_{\mathrm{RML}}\boldsymbol{U}^{\mathrm{H}}\boldsymbol{W}_{\mathrm{ML}}.
\end{aligned}
$$

The scaling for the MF is chosen as $\alpha = c_n/(\lambda_1 Nc_{\mathrm{s}})$ to obtain a reduced variance compared to the ML estimator. The correlator and ML estimator are identical in this case.

For all estimators, the mean bias and variance as defined in (2.47) are given in Table 2.1. The MMSE, MF, and RML estimators reduce the bias compared to the ML estimator. By definition, the MMSE estimator achieves the optimum trade-off, whereas the performance of the RML approach critically depends on the choice of the rank R. The bias-variance trade-off for the MF estimator cannot be adapted to the channel correlations, number of training symbols N, and SNR. Thus, the MF approach only yields a performance improvement for large correlation, small N, and low SNR where it approaches the MMSE estimator.

To assess the estimators' MSE numerically,[10] all estimators $\boldsymbol{W}$ are scaled by a scalar $\beta = \mathrm{tr}[\boldsymbol{W}\boldsymbol{S}\boldsymbol{C}_{\boldsymbol{h}}]^*/\mathrm{tr}[\boldsymbol{W}\boldsymbol{S}\boldsymbol{C}_{\boldsymbol{h}}\boldsymbol{S}^{\mathrm{H}}\boldsymbol{W}^{\mathrm{H}} + \boldsymbol{W}\boldsymbol{C}_{\boldsymbol{n}}\boldsymbol{W}^{\mathrm{H}}]$ minimizing the

[10] In the figures the MSE $\mathrm{E}_{\boldsymbol{h},\boldsymbol{n}}[\|\boldsymbol{\varepsilon}\|_2^2]$ is normalized by $\mathrm{tr}[\boldsymbol{C}_{\boldsymbol{h}}]$ to achieve the same maximum normalized MSE equal to one, for all scenarios.

Channel Estimator	Mean Bias	Variance
ML/Correlator	0	$P \frac{c_n}{N c_{\mathrm{s}}}$
Matched Filter	$\sum_{p=1}^{P} \lambda_p \left(\frac{\lambda_p}{\lambda_1} - 1\right)^{-1}$	$\sum_{p=1}^{P} \left(\frac{\lambda_p}{\lambda_1}\right)^2 \frac{c_n}{N c_{\mathrm{s}}}$
MMSE	$\sum_{p=1}^{P} \lambda_p \left(\frac{1}{1+\frac{c_n}{\lambda_p N c_{\mathrm{s}}}} - 1\right)^2$	$\sum_{p=1}^{P} \frac{1}{\left(1+\frac{c_n}{\lambda_p N c_{\mathrm{s}}}\right)^2} \frac{c_n}{N c_{\mathrm{s}}}$
Red.-Rank ML (RML)	$\sum_{p=R+1}^{P} \lambda_p$	$R \frac{c_n}{N c_{\mathrm{s}}}$

Table 2.1 Mean bias and variance (2.47) of channel estimators for $\boldsymbol{S}^{\mathrm{H}}\boldsymbol{S} = N c_{\mathrm{s}} \boldsymbol{I}_P$ and $\boldsymbol{C_n} = c_n \boldsymbol{I}_{MN}$.

MSE, i.e., $\min_\beta \mathrm{E}_{\boldsymbol{h},\boldsymbol{n}}\left[\|\boldsymbol{h} - \beta \boldsymbol{W}\boldsymbol{y}\|_2^2\right]$. Thus, we focus on the estimators' capability to estimate the channel vector $\boldsymbol{h}$ up to its norm.[11]

First, we consider a scenario with $M = K = 8$, $L = 0$ (frequency-flat), and $N = 16$ in Figure 2.6. The binary training sequences satisfy $\boldsymbol{S}^{\mathrm{H}}\boldsymbol{S} = N c_{\mathrm{s}} \boldsymbol{I}_P$. The covariance matrix $\boldsymbol{C_h}$ is given by the model in Appendix E with mean angles of arrival $\bar{\varphi} = [-45^\circ, -32.1^\circ, -19.3^\circ, -6.4^\circ, 6.4^\circ, 19.3^\circ, 32.1^\circ, 45^\circ]$ and Laplace angular power spectrum with spread $\sigma = 5^\circ$.

Due to a small N, the ML estimator requires an SNR $= c_{\mathrm{s}}/c_n$ at a relative MSE of 10^{-1} which is about 5 dB larger than for the MMSE estimator. With the RML estimator, this gap can be closed over a wide range of SNR reducing the rank from $R = P = 64$ for the ML estimator to $R = 40$. For reference, we also give the RML performance with the optimum rank w.r.t. MSE obtained from a brute-force optimization. The MF estimator is only superior to the ML estimator at very low SNR, as its bias is substantial.

In Figures 2.7 and 2.8, performance in a frequency selective channel with $L = 4$, $M = 8$, $K = 1$, and $N = 16$ is investigated. The covariance matrix $\boldsymbol{C_h}$ is given by the model in Appendix E with mean angles of arrival $\bar{\varphi} = [-45^\circ, -22.5^\circ, 0^\circ, 22.5^\circ, 45^\circ]$ per delay and Laplace angular power spectrum with spread $\sigma = 5^\circ$.

In this channel scenario the MF estimator reduces the MSE compared to the correlator (with comparable numerical complexity) for an interesting SNR region (Figures 2.7). Asymptotically, for high SNR it is outperformed by the correlator. The RML estimator achieves a performance close to the MMSE approach if the rank is chosen optimally for every SNR (Figure 2.8). For full rank ($R = P = 40$), the RML is equivalent to the ML estimator. For a fixed rank of $R = 5$ and $R = 10$, the MSE saturates at high SNR due to the

[11] If the application using the channel estimates is sensitive to a scaling of $\hat{\boldsymbol{h}}$, estimators are proposed in the literature (e.g. [15]) which also introduce a proper scaling for ML estimators and do not rely on the knowledge of $\boldsymbol{C_h}$.

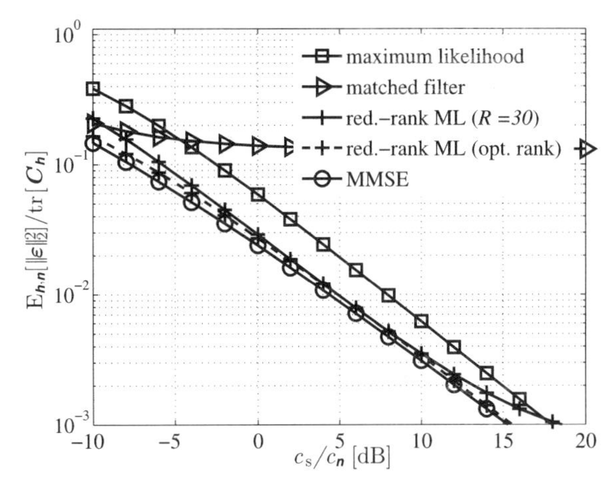

Fig. 2.6 Normalized MSE of channel estimators for $M = K = 8$, $L = 0$ (frequency-flat), and $N = 16$.

Channel Estimator	Order of complexity
ML	$K^3(L+1)^3$
Matched Filter	$M^3K^2(L+1)^2N$
MMSE	$M^3K^3(L+1)^3$
Red.-Rank ML (RML)	$M^3K^3(L+1)^3$

Table 2.2 Order of computational complexity of channel estimators assuming $\boldsymbol{C_n} = c_n\boldsymbol{I}_{MN}$.

bias. At low SNR a smaller rank should be chosen than at high SNR. Thus, an adaptation of the rank R to the SNR is required. The MMSE estimator performs the necessary bias-variance trade-off based on $\boldsymbol{C_h}$ and $\boldsymbol{C_n}$.

The correlator and MF estimator are simple estimators with a low and comparable numerical complexity. As discussed in [56], only a suboptimum estimate of $\boldsymbol{C_h}$ is required for the MF. Regarding the possibilities to reduce the complexity of the RML and MMSE estimators with efficient implementations and approximations, it is difficult to draw a final conclusion for their complexity. Roughly, the complexity of the MMSE and RML estimators is also similar [56], but the RML approach requires an additional EVD of $\boldsymbol{C_h}$. The ML estimator is considerably less complex. The order of required floating point operations is given in Table 2.2 for $\boldsymbol{C_n} = c_n\boldsymbol{I}_{MN}$.[12] For the MF, RML, and MMSE estimators, the covariance matrix $\boldsymbol{C_h}$ has to be estimated (Section 3). This increases their complexity compared to the ML estimator. A low-complexity tracking algorithm for the MMSE estimator is introduced

[12] This assumption influences only the order of the matched filter.

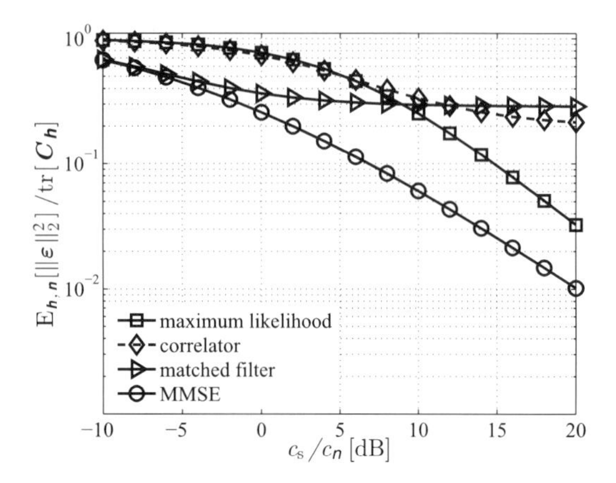

Fig. 2.7 Normalized MSE of channel estimators for $M = 8$, $K = 1$, $L = 4$ (frequency-selective), and $N = 16$.

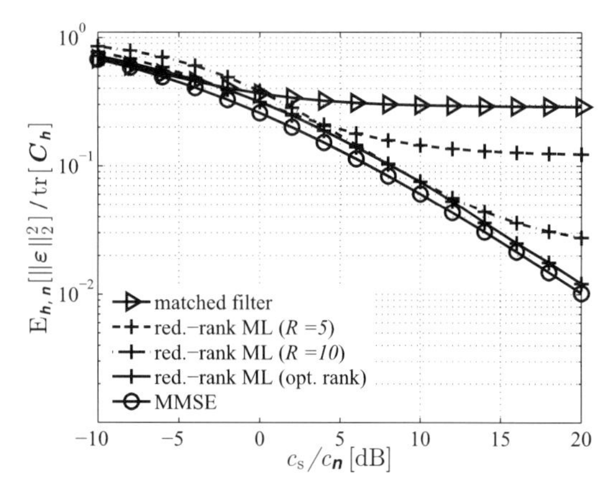

Fig. 2.8 Normalized MSE of channel estimators for $M = 8$, $K = 1$, $L = 4$ (frequency-selective), and $N = 16$. Comparison with reduced-rank ML estimators.

in [141], whereas tracking of the projector on the R-dimensional subspace is possible for the RML (e.g. [219]).

2.3 Channel Prediction

The observation of Q previous realizations for the channel parameters $\boldsymbol{h}_\mathrm{T}[q] = [\boldsymbol{h}[q-1]^\mathrm{T}, \boldsymbol{h}[q-2]^\mathrm{T}, \ldots, \boldsymbol{h}[q-Q]^\mathrm{T}]^\mathrm{T} \in \mathbb{C}^{QP}$ is described by the stochastic signal model (2.9)

$$\boldsymbol{y}_\mathrm{T}[q] = \boldsymbol{S}_\mathrm{T}\boldsymbol{h}_\mathrm{T}[q] + \boldsymbol{n}_\mathrm{T}[q] \in \mathbb{C}^{QMN}. \tag{2.48}$$

They are observed indirectly via known training sequences which determine $\boldsymbol{S}_{\mathrm{T}}$. This observation is disturbed by additive noise $\boldsymbol{n}_{\mathrm{T}}[q]$. To estimate $\boldsymbol{h}[q]$ based on $\boldsymbol{y}_{\mathrm{T}}[q]$, we first assume that the joint pdf $\mathrm{p}_{\boldsymbol{h},\boldsymbol{y}_{\mathrm{T}}}(\boldsymbol{h}[q], \boldsymbol{y}_{\mathrm{T}}[q])$ is given.

2.3.1 Minimum Mean Square Error Prediction

Given the stochastic model $\mathrm{p}_{\boldsymbol{h},\boldsymbol{y}_{\mathrm{T}}}(\boldsymbol{h}[q], \boldsymbol{y}_{\mathrm{T}}[q])$, Bayesian parameter estimation gives the framework for designing optimum predictors. As in Section 2.2.1, we choose the MSE as average measure of the estimation quality. The main ideas are reviewed in Section 2.2.1 and can be applied directly to the problem of channel prediction. First we treat the general case of predicting the channel vector $\boldsymbol{h}[q] \in \mathbb{C}^P$ before focusing on the prediction of a single channel parameter in $\boldsymbol{h}[q]$. The restriction to the prediction of a scalar allows us to include standard results about prediction of parameters or time series from Wiener and Kolmogorov (see references in [182, chapter 10]) such as asymptotic performance results for $Q \to \infty$.[13]

2.3.1.1 Multivariate Prediction

As discussed in Section 2.2.1, the MMSE estimate $\hat{\boldsymbol{h}}[q]$ is the (conditional) mean of the conditional pdf $\mathrm{p}_{\boldsymbol{h}|\boldsymbol{y}_{\mathrm{T}}}(\boldsymbol{h}[q]|\boldsymbol{y}_{\mathrm{T}}[q])$. If $\mathrm{p}_{\boldsymbol{h},\boldsymbol{y}_{\mathrm{T}}}(\boldsymbol{h}[q], \boldsymbol{y}_{\mathrm{T}}[q])$ is the zero-mean complex Gaussian pdf, the conditional mean is linear in $\boldsymbol{y}_{\mathrm{T}}[q]$

$$\hat{\boldsymbol{h}}[q] = \mathrm{E}_{\boldsymbol{h}|\boldsymbol{y}_{\mathrm{T}}}[\boldsymbol{h}[q]] = \boldsymbol{W}\boldsymbol{y}_{\mathrm{T}}[q]. \tag{2.49}$$

Thus, the assumption of a linear estimator is not a restriction and $\boldsymbol{W}$ can also be obtained from

$$\min_{\boldsymbol{W} \in \mathbb{C}^{P \times QMN}} \mathrm{E}_{\boldsymbol{h},\boldsymbol{y}_{\mathrm{T}}}\left[\|\boldsymbol{h}[q] - \boldsymbol{W}\boldsymbol{y}_{\mathrm{T}}[q]\|_2^2\right]. \tag{2.50}$$

For the model (2.48), the estimator $\boldsymbol{W} \in \mathbb{C}^{P \times QMN}$, which can also be interpreted as a MIMO FIR filter of length Q, reads [124]

$$\begin{aligned} \boldsymbol{W} &= \boldsymbol{C}_{\boldsymbol{h}\boldsymbol{y}_{\mathrm{T}}}\boldsymbol{C}_{\boldsymbol{y}_{\mathrm{T}}}^{-1} \\ &= \boldsymbol{C}_{\boldsymbol{h}\boldsymbol{h}_{\mathrm{T}}}\boldsymbol{S}_{\mathrm{T}}^{\mathrm{H}}\left(\boldsymbol{S}_{\mathrm{T}}\boldsymbol{C}_{\boldsymbol{h}_{\mathrm{T}}}\boldsymbol{S}_{\mathrm{T}}^{\mathrm{H}} + \boldsymbol{C}_{\boldsymbol{n}_{\mathrm{T}}}\right)^{-1} \end{aligned} \tag{2.51}$$

where $\boldsymbol{C}_{\boldsymbol{h}\boldsymbol{h}_{\mathrm{T}}} = \mathrm{E}[\boldsymbol{h}[q]\boldsymbol{h}_{\mathrm{T}}[q]^{\mathrm{H}}]$. It can be decomposed into a ML channel estimation stage (2.37), which also provides a sufficient statistic for $\{\boldsymbol{h}[i]\}_{i=q-Q}^{q-1}$,

[13] The general case of vector or multivariate prediction is treated in [234, 235]. The insights which can be obtained for the significantly simpler scalar or univariate case are sufficient in the context considered here.

and a prediction stage with $\boldsymbol{W}_{\mathrm{P}} \in \mathbb{C}^{P \times QP}$

$$\boxed{\boldsymbol{W} = \underbrace{\boldsymbol{C}_{\boldsymbol{h}\boldsymbol{h}_{\mathrm{T}}} \left(\boldsymbol{C}_{\boldsymbol{h}_{\mathrm{T}}} + (\boldsymbol{S}_{\mathrm{T}}^{\mathrm{H}} \boldsymbol{C}_{\boldsymbol{n}_{\mathrm{T}}}^{-1} \boldsymbol{S}_{\mathrm{T}})^{-1} \right)^{-1}}_{\boldsymbol{W}_{\mathrm{P}}} (\boldsymbol{I}_Q \otimes \boldsymbol{W}_{\mathrm{ML}}) .} \tag{2.52}$$

The advantage of this decomposition is twofold: On the one hand, it allows for a less complex implementation of $\boldsymbol{W}$ exploiting the structure of the related system of linear equations (2.51). On the other hand, we can obtain further insights separating the tasks of channel estimation and prediction; we focus on the prediction stage $\boldsymbol{W}_{\mathrm{P}}$ of the estimator $\boldsymbol{W}$ in the sequel. Denoting the ML channel estimates by $\tilde{\boldsymbol{h}}[q] = \boldsymbol{W}_{\mathrm{ML}} \boldsymbol{y}[q]$ and collecting them in $\tilde{\boldsymbol{h}}_{\mathrm{T}}[q] = [\tilde{\boldsymbol{h}}[q-1]^{\mathrm{T}}, \tilde{\boldsymbol{h}}[q-2]^{\mathrm{T}}, \ldots, \tilde{\boldsymbol{h}}[q-Q]^{\mathrm{T}}]^{\mathrm{T}} \in \mathbb{C}^{QP}$ we can rewrite (2.48)

$$\hat{\boldsymbol{h}}[q] = \boldsymbol{W}_{\mathrm{P}} \tilde{\boldsymbol{h}}_{\mathrm{T}}[q] = \boldsymbol{W}_{\mathrm{P}} (\boldsymbol{I}_Q \otimes \boldsymbol{W}_{\mathrm{ML}}) \boldsymbol{y}_{\mathrm{T}}[q]. \tag{2.53}$$

Example 2.1. Consider the case of white noise $\boldsymbol{C}_{\boldsymbol{n}_{\mathrm{T}}} = c_n \boldsymbol{I}_{QMN}$ and orthogonal training sequences $\boldsymbol{S}_{\mathrm{T}}^{\mathrm{H}} \boldsymbol{S}_{\mathrm{T}} = N c_{\mathrm{s}} \boldsymbol{I}_{QP}$, where c_{s} denotes the power of a single scalar symbol of the training sequence. We assume $\boldsymbol{h}[q] = [h_1[q], h_2[q], \ldots, h_P[q]]^{\mathrm{T}}$ to be uncorrelated, i.e., $\boldsymbol{C}_{\boldsymbol{h}} = \mathbf{diag}[c_{h_1}[0], c_{h_2}[0], \ldots, c_{h_P}[0]]$, with autocovariance sequence $c_{h_p}[i] = \mathrm{E}[h_p[q] h_p[q-i]^*]$ for its pth element. Thus, we obtain $\boldsymbol{C}_{\boldsymbol{h}_{\mathrm{T}}} = \sum_{p=1}^{P} \boldsymbol{C}_{\boldsymbol{h}_{\mathrm{T},p}} \otimes \boldsymbol{e}_p \boldsymbol{e}_p^{\mathrm{T}}$ from (2.10), where $\boldsymbol{C}_{\boldsymbol{h}_{\mathrm{T},p}} \in \mathbb{T}_{+,0}^{Q}$ is Toeplitz with first row $[c_{h_p}[0], c_{h_p}[1], \ldots, c_{h_p}[Q-1]]$. With $\boldsymbol{c}_{h_p \boldsymbol{h}_{\mathrm{T},p}} = [c_{h_p}[1], c_{h_p}[2], \ldots, c_{h_p}[Q]]$, the crosscovariance matrix in (2.52) reads

$$\boldsymbol{C}_{\boldsymbol{h}\boldsymbol{h}_{\mathrm{T}}} = \sum_{p=1}^{P} \boldsymbol{c}_{h_p \boldsymbol{h}_{\mathrm{T},p}} \otimes \boldsymbol{e}_p \boldsymbol{e}_p^{\mathrm{T}}. \tag{2.54}$$

Applying these assumptions to the predictor $\boldsymbol{W}_{\mathrm{P}}$ from (2.52), we get

$$\boldsymbol{W}_{\mathrm{P}} = \sum_{p=1}^{P} \underbrace{\boldsymbol{c}_{h_p \boldsymbol{h}_{\mathrm{T},p}} \left(\boldsymbol{C}_{\boldsymbol{h}_{\mathrm{T},p}} + \frac{c_n}{N c_{\mathrm{s}}} \right)^{-1}}_{\boldsymbol{w}_p^{\mathrm{T}}} \otimes \boldsymbol{e}_p \boldsymbol{e}_p^{\mathrm{T}}, \tag{2.55}$$

i.e., the prediction stage is decoupled for every parameter $h_p[q]$, and we can write

$$\hat{h}_p[q] = \boldsymbol{w}_p^{\mathrm{T}} \tilde{\boldsymbol{h}}_{\mathrm{T},p}[q] \tag{2.56}$$

with $\tilde{\boldsymbol{h}}_{\mathrm{T},p}[q] = [\tilde{h}_p[q-1], \tilde{h}_p[q-2], \ldots, \tilde{h}_p[q-Q]]^{\mathrm{T}}$. □

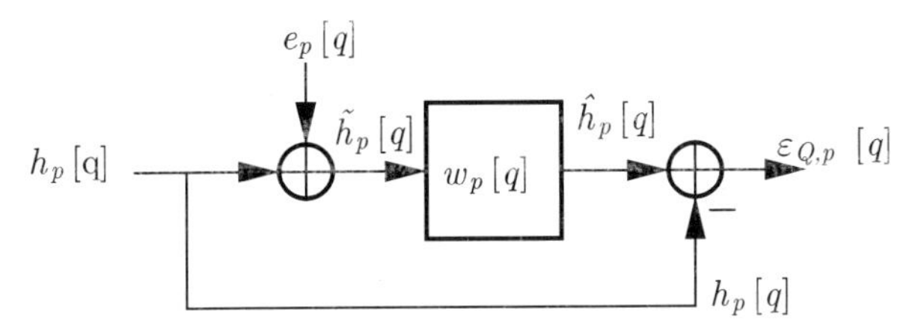

Fig. 2.9 Linear prediction of $h_p[q]$ one step ahead based on Q past noisy observations $\tilde{h}_p[q]$ and the FIR filter $w_p[q]$ with non-zero coefficients $\boldsymbol{w}_p = [w_p[1], w_p[2], \ldots, w_p[Q]]^{\mathrm{T}}$: $\hat{h}_p[q] = \sum_{\ell=1}^{Q} w_p[\ell]\tilde{h}_p[q-\ell]$.

2.3.1.2 Univariate Prediction

The prediction problem can also be treated separately for every parameter $h_p[q]$, i.e., it is understood as a univariate prediction problem. This is equivalent to restricting the structure of $\boldsymbol{W}$ in optimization problem (2.50) to $\boldsymbol{W} = \boldsymbol{W}_{\mathrm{P}}(\boldsymbol{I}_Q \otimes \boldsymbol{W}_{\mathrm{ML}})$ with

$$\boldsymbol{W}_{\mathrm{P}} = \sum_{p=1}^{P} \boldsymbol{w}_p^{\mathrm{T}} \otimes \boldsymbol{e}_p \boldsymbol{e}_p^{\mathrm{T}}. \tag{2.57}$$

With this restriction on the domain of $\boldsymbol{W}$, problem (2.50) is equivalent to

$$\min_{\{\boldsymbol{w}_p\}_{p=1}^{P}} \sum_{p=1}^{P} \mathrm{E}_{h_p, \tilde{\boldsymbol{h}}_{\mathrm{T},p}}\left[\left|h_p[q] - \boldsymbol{w}_p^{\mathrm{T}} \tilde{\boldsymbol{h}}_{\mathrm{T},p}[q]\right|^2\right]. \tag{2.58}$$

Of course, the MMSE achieved by this solution is larger than for the general multivariate prediction (2.50) because the correlations $\boldsymbol{C}_{\boldsymbol{h}}$ and $\boldsymbol{C}_{\boldsymbol{\varepsilon}}$ (2.38) are not exploited. The MMSEs are equal for the assumptions made in Example 2.1 at the end of the previous subsection. The advantage of this solution is its simplicity and, consequently, reduced complexity.

The estimate of $h_p[q]$ is $\hat{h}_p[q] = \boldsymbol{w}_p^{\mathrm{T}} \tilde{\boldsymbol{h}}_{\mathrm{T},p}[q]$ (2.56). It is based on the estimated channel parameters in Q previous blocks $\tilde{\boldsymbol{h}}_{\mathrm{T},p}[q] = [\tilde{h}_p[q-1], \tilde{h}_p[q-2], \ldots, \tilde{h}_p[q-Q]]^{\mathrm{T}}$, which we model as

$$\tilde{h}_p[q] = h_p[q] + e_p[q] \tag{2.59}$$

with $e_p[q] = \boldsymbol{e}_p^{\mathrm{T}} \boldsymbol{W}_{\mathrm{ML}} \boldsymbol{n}[q]$, $e_p[q] \sim \mathcal{N}_{\mathrm{c}}(0, c_{e_p})$, and

$$c_{e_p} = \left[(\boldsymbol{S}^{\mathrm{H}} \boldsymbol{C}_{\boldsymbol{n}}^{-1} \boldsymbol{S})^{-1}\right]_{p,p} \geq \frac{c_n}{N c_{\mathrm{s}}}, \tag{2.60}$$

where the inequality holds for $\boldsymbol{C}_{\boldsymbol{n}} = c_n \boldsymbol{I}_{MN}$. The last inequality gives a lower bound, which is achieved for $\boldsymbol{S}^{\mathrm{H}} \boldsymbol{S} = N c_{\mathrm{s}} \boldsymbol{I}_P$.

The terms in the cost function (2.58) are decoupled for every parameter and the solution is

$$\boxed{\boldsymbol{w}_p^{\mathrm{T}} = \boldsymbol{c}_{h_p \boldsymbol{h}_{\mathrm{T},p}} \left(\boldsymbol{C}_{\boldsymbol{h}_{\mathrm{T},p}} + c_{e_p} \boldsymbol{I}_Q \right)^{-1}.} \tag{2.61}$$

The MMSE for the univariate predictor of $h_p[q]$ is

$$\mathrm{E}\left[|\varepsilon_{Q,p}[q]|^2\right] = c_{h_p}[0] - \boldsymbol{w}_p^{\mathrm{T}} \boldsymbol{c}_{\boldsymbol{h}_{\mathrm{T},p} h_p} \tag{2.62}$$

with estimation error $\varepsilon_{Q,p}[q] = \hat{h}_p[q] - h_p[q]$ (Figure 2.9).

Considering prediction based on observations of all past samples of the sequence $\tilde{h}_p[q]$, i.e., for a predictor length $Q \to \infty$, we arrive at the classical univariate prediction theory of Wiener and Kolmogorov [182]. For this asymptotic case, the MMSE of the predictor can be expressed in terms of the power spectral density

$$\mathrm{S}_{h_p}(\omega) = \sum_{\ell=-\infty}^{\infty} c_{h_p}[\ell] \exp(-\mathrm{j}\omega\ell) \tag{2.63}$$

of $h_p[q]$ with autocovariance sequence $c_{h_p}[\ell] = \mathrm{E}[h_p[q] h_p[q-\ell]^*]$ and reads [196]

$$\boxed{\begin{aligned} \mathrm{E}\left[|\varepsilon_{\infty,p}[q]|^2\right] &= c_{e_p} \left[\exp\left(\frac{1}{2\pi} \int_{-\pi}^{\pi} \ln\left(1 + \frac{\mathrm{S}_{h_p}(\omega)}{c_{e_p}}\right) \mathrm{d}\omega \right) - 1 \right] \\ &\leq \mathrm{E}\left[|\varepsilon_{Q,p}[q]|^2\right]. \end{aligned}} \tag{2.64}$$

It is a lower bound for $\mathrm{E}[|\varepsilon_{Q,p}[q]|^2]$ in (2.62).

In case the past observations of the channel $\tilde{h}_p[q]$ are undistorted, i.e., $c_{e_p} \to 0$ and $\tilde{h}_p[q] = h_p[q]$, the MMSE (2.64) converges to [196]

$$\mathrm{E}\left[|\varepsilon_{\infty,p}[q]|^2\right] = \exp\left(\frac{1}{2\pi} \int_{-\pi}^{\pi} \ln\left(\mathrm{S}_{h_p}(\omega)\right) \mathrm{d}\omega \right). \tag{2.65}$$

An interpretation for this relation is given by Scharf [182, p.431 and 434]: $\mathrm{E}[|\varepsilon_{\infty,p}[q]|^2]/c_{h_p}[0]$ is a measure for the flatness of the spectrum because in the case of $\mathrm{S}_{h_p}(\omega) = c_{h_p}[0], -\pi \leq \omega \leq \pi$, we have $\mathrm{E}[|\varepsilon_{\infty,p}[q]|^2]/c_{h_p}[0] = 1$, i.e., it is a white and not predictable random sequence. Moreover, the exponent in (2.65) determines the entropy rate of a stationary Gaussian random sequence [40, p. 274].

2.3.2 Properties of Band-Limited Random Sequences

In practice, the time-variance of the channel parameters is limited, since the mobility or velocity of the transmitter, receiver, or their environment is limited. For a sufficiently high sampling rate, autocovariance sequences with a band-limited power spectral density are an accurate model. The set of band-limited power spectral densities of $h_p[q]$ with maximum frequency $\omega_{\max,p} = 2\pi f_{\max,p}$ and variance $c_{h_p}[0]$ is denoted

$$\mathcal{B}_{h_p,f_{\max,p}} = \Bigg\{ \mathrm{S}_{h_p}(\omega) \Bigg| \mathrm{S}_{h_p}(\omega) = 0 \text{ for } \omega_{\max,p} < |\omega| \leq \pi, \\ \omega_{\max,p} = 2\pi f_{\max,p} < \pi, \mathrm{S}_{h_p}(\omega) \geq 0, c_{h_p}[0] = \frac{1}{2\pi} \int_{-\omega_{\max,p}}^{\omega_{\max,p}} \mathrm{S}_{h_p}(\omega) \mathrm{d}\omega \Bigg\}. \tag{2.66}$$

In wireless communications, $f_{\max,p}$ is the maximum Doppler frequency of the channel normalized by the sampling period, i.e., the block period T_b in our case.

Random processes with band-limited power spectral density ($\omega_{\max,p} < \pi$) belong to the class of deterministic processes (in contrast to regular processes) because [169]

$$\frac{1}{2\pi} \int_{-\pi}^{\pi} \ln\left(\mathrm{S}_{h_p}(\omega)\right) \mathrm{d}\omega = -\infty. \tag{2.67}$$

Applying this property to (2.65) we obtain $\mathrm{E}[|\varepsilon_{\infty,p}[q]|^2] = 0$ if $c_{e_p} \to 0$, i.e., deterministic processes are perfectly predictable. For $c_{e_p} > 0$, the power spectral density of $\tilde{h}_p[q]$ is not band-limited anymore and perfect prediction is impossible.

At first glance, this property is surprising: Intuitively, we do not expect a "regular" random process to have enough structure such that perfect prediction could be possible. In contrast to random processes, the values of deterministic functions can be predicted perfectly as soon as we know the class of functions and their parameters. This is one reason why the formal classification into regular and deterministic random processes was introduced.

In the literature about prediction (or extrapolation) of band-limited random sequences this classical property of perfect predictability, which goes back to Szegö-Kolmogorov-Krein [233], seems to be less known and alternative proofs are given: For example, Papoulis [162],[164, p. 380], [163] shows that a band-limited random process can be expressed only by its past samples if sampled at a rate greater than the Nyquist rate. But Marvasti [138] states that these results go back to Beutler [22]. In [27], an alternative proof to Papoulis based on the completeness of a set of complex exponentials on a finite interval is given. These references do not refer to the classical property of perfect predictability for deterministic random processes, however, the new proofs do not require knowledge about the shape of the power spectral density.

Consequently, perfect prediction based on the MSE criterion requires perfect knowledge of the power spectral density, but perfect prediction can be achieved without any knowledge about the power spectral density except that it belongs to the set $\mathcal{B}_{h_p,f_{\max,p}}$ with given maximum frequency. Algorithms and examples are given in [27, 222]. The practical problem with perfect prediction is that the norm of the prediction filter is not bounded as $Q \to \infty$ [164, p. 198]. Thus, perfect prediction is an ill-posed problem in the sense of Hadamard [212].

The necessary a priori information for prediction can be even reduced further as in [28, 153, 44]: Prediction algorithms for band-limited sequences are developed which do not require knowledge about the bandwidth, but the price is a larger required sampling rate.

The following two observations provide more understanding of the perfect predictability of band-limited random processes.

1. A realization of an ergodic random process with band-limited power spectral density is also band-limited with probability one, i.e., every realization itself has a rich structure.
   The argumentation is as follows: The realization of a random process has finite power (second order moment) but is generally not of finite energy, i.e., it is not square summable. Thus, its Fourier transform does not exist [178, 169] and the common definition of band-limitation via the limited support of the Fourier transform of a sequence is not applicable. An alternative definition of band-limitation is reported in [178]: A random sequence with finite second order moment is called band-limited if it passes a filter $g[q]$, whose Fourier transform is $\mathrm{G}(\omega) = 1, |\omega| \leq \omega_{\max}$, zero elsewhere, without distortion for almost all realizations. Employing this definition we can show that the realization of a random process with band-limited power spectral density is band-limited in the generalized sense with probability one.[14]
   This result is important because all realizations can now be treated equivalently to deterministic band-limited sequences — except that they are power-limited and not energy-limited. Thus, the results in the literature, e.g., [27, 222, 28, 153], for deterministic signals are also valid for random processes with probability one.

[14] *Proof:* Consider an ergodic random sequence $h[q]$ with power spectral density $\mathrm{S}_h(\omega)$ band-limited to $\omega_{\max}$, i.e., $\mathrm{S}_h(\omega)\mathrm{G}(\omega) = \mathrm{S}_h(\omega)$. The filter $g[q]$ is defined above. Because $\mathrm{E}[|h[q] - \sum_{\ell=-\infty}^{\infty} g[\ell]h[q-\ell]|^2] = 0$, i.e., the output of $g[q] * h[q]$ converges to $h[q]$ in the mean-square sense, the filter output is also equal to the random sequence $h[q]$ with probability one [200, p. 432]. Thus, almost all realization $h[q]$ of $h[q]$ pass the filter undistorted, i.e., are band-limited in this generalized sense.

2. For a sequence $h[q]$ band-limited to $\omega_{\max} < \pi$, we construct a high pass filter $f[q]$ with cut-off frequency $\omega_{\max}$ and $f[0] = 1$. The sequence at the filter output $e[q] = \sum_{\ell=0}^{Q} f[\ell]h[q-\ell]$ has "small" energy, i.e., $e[q] \approx 0$. For $Q \to \infty$, the stop-band of the filter $f[q]$ can be designed with arbitrary accuracy, which results in $e[q] = 0$. We can rewrite this equation in $h[q] = -\sum_{\ell=1}^{\infty} f[\ell]h[q-\ell]$, i.e., the low-pass prediction filter $w[q] = -f[q], q = 1, 2, \ldots, \infty$, achieves perfect prediction [222]. Thus, it can be shown that perfect prediction for band-limited (equivalently for bandpass) processes can be achieved knowing only the maximum frequency $\omega_{\max}$ if the sampling rate is strictly larger than the Nyquist rate [222]. The design of a high-pass filter is one possible approach.

Example 2.2. Let us consider the rectangular power spectral density

$$\mathrm{S_R}(\omega) = \begin{cases} \frac{\pi}{\omega_{\max,p}} c_{h_p}[0], & |\omega| \leq \omega_{\max,p} \\ 0, & \omega_{\max,p} < |\omega| \leq \pi \end{cases}, \tag{2.68}$$

which yields the prediction error (2.64)

$$\mathrm{E}\left[|\varepsilon_{\infty,p}[q]|^2\right] = c_{e_p}\left[\left(1 + \frac{\pi c_{h_p}[0]}{c_{e_p}\omega_{\max,p}}\right)^{\omega_{\max,p}/\pi} - 1\right]. \tag{2.69}$$

For $\omega_{\max,p} = \pi$, we get $\mathrm{E}[|\varepsilon_{\infty,p}[q]|^2] = c_{h_p}[0]$, whereas for $\omega_{\max,p} < \pi$ and $c_{e_p} \to 0$ perfect prediction is possible, i.e., $\mathrm{E}[|\varepsilon_{\infty,p}[q]|^2] = 0$. □

2.4 Minimax Mean Square Error Estimation

When deriving the MMSE channel estimator and predictor in Sections 2.2.1 and 2.3.1 we assume perfect knowledge of the joint probability distribution for the observation and the parameters to be estimated. For the Gaussian distribution, the estimators are linear in the observation and depend only on the first and second order moments which are assumed to be known perfectly.

The performance of the Bayesian approach depends on the quality of the a priori information in terms of this probability distribution. The probability distribution is often assumed to be Gaussian for simplicity; the covariance matrices are either estimated or a fixed "typical" or "most likely" covariance matrix is selected for the specified application scenario. It is the *hope* of the engineer that the errors are small enough such that the performance degradation from the optimum design for the true parameters is small.

The minimax approach to MSE estimation aims at a *guaranteed MSE* performance for a given class of probability distributions optimizing the estimator for the worst case scenario. Thus, for a well chosen class of covariance matrices, e.g., based on a model of the estimation errors or restriction to possible scenarios, a robust estimator is obtained. If a large class is chosen, the resulting robust estimator is very conservative, i.e., the guaranteed MSE may be too large.

An important question is: For which class of probability distributions is the linear MMSE solution assuming a Gaussian distribution minimax robust?[15] This will provide more justification for this standard approach and is discussed below.

2.4.1 General Results

Two cases are addressed in the literature:

- A minimax formulation for the classical Wiener-Kolmogorov theory is given by Vastola and Poor [225], which treats the case of univariate linear estimation based on an observation of infinite dimension. See also the review of Kassam and Poor [123] or the more general theory by Verdu and Poor [226].
- Multivariate minimax MSE estimation based on an observation of finite dimension is discussed by Vandelinde [223], where robustness w.r.t. the type of probability distribution is also considered. Soloviov [199, 198, 197] gives a rigorous derivation of these results.

We briefly review both results in this section.

2.4.1.1 Observation of Infinite Dimension

Given the sequence of scalar noisy observations of the random sequence $h[q]$ (2.59)

$$\tilde{h}[q'] = h[q'] + e[q'], \; q' = q-1, q-2, \ldots, \tag{2.70}$$

we estimate $h[q+d]$ with the causal IIR filter $w[\ell]$ $(w[\ell] = 0, \ell \leq 0)$[16]

$$\hat{h}[q+d] = \sum_{\ell=1}^{\infty} w[\ell] \tilde{h}[q-\ell]. \tag{2.71}$$

The estimation task is often called prediction if $d = 0$ (Figure 2.9), filtering if $d = -1$, and smoothing in case $d < -1$. Formulating this problem in the frequency domain [200, 225] the MSE reads

[15] This is called an inverse minimax problem [226].

[16] In [225], the more general case of estimating the output of a general linear filter applied to the sequence $h[q]$ is addressed.

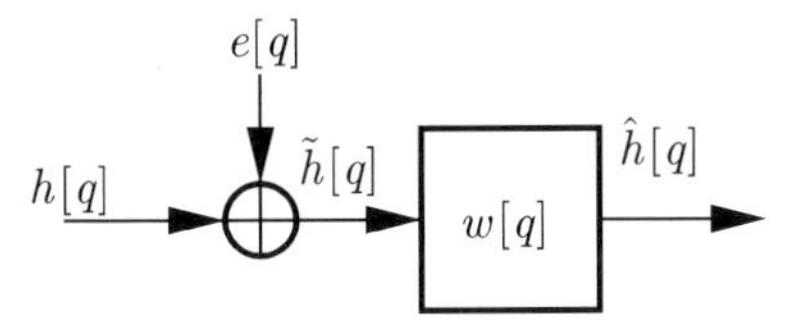

Fig. 2.10 Linear prediction of $h[q]$ based on Q past noisy observations $\tilde{h}[q]$.

$$\begin{aligned}
\mathrm{F}(\mathrm{W}, \mathrm{S}_h, \mathrm{S}_e) &= \mathrm{E}\left[|\varepsilon_\infty[q]|^2\right] = \mathrm{E}\left[\left|h[q+d] - \hat{h}[q+d]\right|^2\right] \\
&= \frac{1}{2\pi}\int_{-\pi}^{\pi}\left(|\exp(-\mathrm{j}\omega d) - \mathrm{W}(\omega)|^2\,\mathrm{S}_h(\omega) + |\mathrm{W}(\omega)|^2\mathrm{S}_e(\omega)\right)\mathrm{d}\omega.
\end{aligned} \tag{2.72}$$

The transfer function W of $w[\ell]$ is

$$\mathrm{W}(\omega) = \sum_{\ell=1}^{\infty} w[\ell]\exp(-\mathrm{j}\omega\ell) \in \mathcal{H}_+^2. \tag{2.73}$$

It is chosen from the space of causal and square-integrable transfer functions with $\int_{-\infty}^{\infty}|\mathrm{W}(\omega)|^2\mathrm{d}\omega < \infty$, which is called Hardy space [225] and denoted by $\mathcal{H}_+^2$. The power spectral densities of the uncorrelated signal $h[q]$ and noise $e[q]$ are

$$\mathrm{S}_h(\omega) = \sum_{\ell=-\infty}^{\infty} c_h[\ell]\exp(-\mathrm{j}\omega\ell) \in \mathcal{C}_h \tag{2.74}$$

$$\mathrm{S}_e(\omega) = \sum_{\ell=-\infty}^{\infty} c_e[\ell]\exp(-\mathrm{j}\omega\ell) \in \mathcal{C}_e. \tag{2.75}$$

Both power spectral densities are not known perfectly, but are assumed to belong to the *uncertainty classes* $\mathcal{C}_h$ and $\mathcal{C}_e$, respectively.

An estimator W is called *robust* if there exists a finite upper bound on the MSE $\mathrm{F}(\mathrm{W}, \mathrm{S}_h, \mathrm{S}_e)$ over all $\mathrm{S}_h \in \mathcal{C}_h$ and $\mathrm{S}_e \in \mathcal{C}_e$. Thus, the performance

$$\max_{\mathrm{S}_h\in\mathcal{C}_h, \mathrm{S}_e\in\mathcal{C}_e} \mathrm{F}(\mathrm{W}, \mathrm{S}_h, \mathrm{S}_e)$$

can be guaranteed.

The *most robust* transfer function W^{R} minimizes this upper bound and is the solution of the minimax problem

$$\min_{\mathrm{W}\in\mathcal{H}_+^2}\ \max_{\mathrm{S}_h\in\mathcal{C}_h, \mathrm{S}_e\in\mathcal{C}_e} \mathrm{F}(\mathrm{W}, \mathrm{S}_h, \mathrm{S}_e). \tag{2.76}$$

The solutions $\mathrm{S}_h^{\mathrm{L}}$ and $\mathrm{S}_e^{\mathrm{L}}$ to the maxmin problem, which is the dual problem to (2.76),

$$\mathrm{F}(\mathrm{W}^{\mathrm{L}}, \mathrm{S}_h^{\mathrm{L}}, \mathrm{S}_e^{\mathrm{L}}) = \max_{\mathrm{S}_h \in \mathcal{C}_h, \mathrm{S}_e \in \mathcal{C}_e} \min_{\mathrm{W} \in \mathcal{H}_+^2} \mathrm{F}(\mathrm{W}, \mathrm{S}_h, \mathrm{S}_e) \tag{2.77}$$

are termed *least-favorable* for causal estimation for the given uncertainty classes. If the values of (2.76) and (2.77) are equal, then $\mathrm{S}_h^{\mathrm{L}}$ and $\mathrm{S}_e^{\mathrm{L}}$ are least favorable if and only if $\mathrm{S}_h^{\mathrm{L}}$, $\mathrm{S}_e^{\mathrm{L}}$, and W^{L} form a saddle-point solution. This means they satisfy

$$\mathrm{F}(\mathrm{W}^{\mathrm{L}}, \mathrm{S}_h, \mathrm{S}_e) \leq \mathrm{F}(\mathrm{W}^{\mathrm{L}}, \mathrm{S}_h^{\mathrm{L}}, \mathrm{S}_e^{\mathrm{L}}) \leq \mathrm{F}(\mathrm{W}, \mathrm{S}_h^{\mathrm{L}}, \mathrm{S}_e^{\mathrm{L}}), \tag{2.78}$$

and W^{L} is a most robust causal transfer function.

The following theorem by Vastola and Poor [225] states the conditions on $\mathcal{C}_h$ and $\mathcal{C}_e$, under which the optimal transfer function for a least favorable pair of power spectral densities is most robust.

Theorem 2.1 (Vastola, Poor [225]). *If the spectral uncertainty classes $\mathcal{C}_h$ and $\mathcal{C}_e$ are such that the following three conditions hold:*

1. $\sup_{\mathrm{S}_h \in \mathcal{C}_h} \frac{1}{2\pi} \int_{-\pi}^{\pi} \mathrm{S}_h(\omega) \mathrm{d}\omega < \infty$,
2. *$\mathcal{C}_h$ and $\mathcal{C}_e$ are convex,*
3. *At least one of the following properties holds for some $\varepsilon > 0$:*

   a. *Every $\mathrm{S}_h \in \mathcal{C}_h$ satisfies $\mathrm{S}_h(\omega) \geq \varepsilon > 0$,*
   b. *Every $\mathrm{S}_e \in \mathcal{C}_e$ satisfies $\mathrm{S}_e(\omega) \geq \varepsilon > 0$,*

then the power spectral densities $\mathrm{S}_h^{\mathrm{L}} \in \mathcal{C}_h$ and $\mathrm{S}_e^{\mathrm{L}} \in \mathcal{C}_e$ and their optimal transfer function W^{L} form a saddle-point solution to the minimax problem (2.76), if and only if $\mathrm{S}_h^{\mathrm{L}}, \mathrm{S}_e^{\mathrm{L}}$ are least favorable for causal estimation, i.e., they solve (2.77). In this case W^{L} is a most robust transfer function.

The maximization in (2.77) is easier to solve than the original minimax problem (2.76) if closed-form expressions for the minimum in (2.77) are available, e.g., (2.64) and (2.65). Among others (see [225]), Snyders [196] gives a general method for finding these expressions. With this theorem and closed-form expressions, often a closed-form robust estimator W^{R} can be found (see examples in [196]).

2.4.1.2 Observation of Finite Dimension

Next we treat the case of estimating a parameter vector $\boldsymbol{h} \in \mathbb{C}^P$ based on an N_{y}-dimensional observation $\boldsymbol{y} \in \mathbb{C}^{N_{\mathrm{y}}}$. The joint probability distribution is not known exactly, but belongs to an uncertainty class $\mathcal{P}$ of possible probability distributions.

$$\mathcal{P} = \left\{ \mathrm{p}_{\boldsymbol{h},\boldsymbol{y}}(\boldsymbol{h},\boldsymbol{y}) \middle| \boldsymbol{\mu}_{\boldsymbol{z}} = \mathbf{0}_{P+N_\mathrm{y}}, \boldsymbol{C}_{\boldsymbol{z}} = \begin{bmatrix} \boldsymbol{C}_{\boldsymbol{h}} & \boldsymbol{C}_{\boldsymbol{hy}} \\ \boldsymbol{C}_{\boldsymbol{yh}} & \boldsymbol{C}_{\boldsymbol{y}} \end{bmatrix} \in \mathcal{C} \subseteq \mathbb{S}_{+,0}^{P+N_\mathrm{y}} \right\} \tag{2.79}$$

is the set of all zero mean distributions of $\boldsymbol{z} = [\boldsymbol{h}^\mathrm{T}, \boldsymbol{y}^\mathrm{T}]^\mathrm{T} \in \mathbb{C}^{P+N_\mathrm{y}}$ having a covariance matrix $\boldsymbol{C}_{\boldsymbol{z}}$ from an uncertainty class $\mathcal{C}$ of positive semidefinite matrices.[17] Although we present the case of zero mean $\boldsymbol{z}$ and non-singular $\boldsymbol{C}_{\boldsymbol{y}}$, the derivations in [199, 198, 197] include non-zero mean and singular $\boldsymbol{C}_{\boldsymbol{y}}$ and in [197] uncertainties in the mean are discussed. The estimator $\hat{\boldsymbol{h}} = \mathbf{f}(\boldsymbol{y})$ is chosen from the class of measurable functions $\mathcal{F} = \{\mathbf{f} : \mathbb{C}^{N_\mathrm{y}} \to \mathbb{C}^P, \hat{\boldsymbol{h}} = \mathbf{f}(\boldsymbol{y})\}$.

The most robust estimator $\mathbf{f}^\mathrm{R}$ is the solution of the minimax problem

$$G_{\min} = \min_{\mathbf{f}\in\mathcal{F}} \max_{\mathrm{p}_{\boldsymbol{h},\boldsymbol{y}}(\boldsymbol{h},\boldsymbol{y})\in\mathcal{P}} \mathrm{E}_{\boldsymbol{h},\boldsymbol{y}}[\|\boldsymbol{h} - \hat{\boldsymbol{h}}\|_2^2]. \tag{2.80}$$

The least-favorable probability distribution $\mathrm{p}^\mathrm{L}_{\boldsymbol{h},\boldsymbol{y}}(\boldsymbol{h},\boldsymbol{y})$ is defined by

$$G_{\min} = \mathrm{G}\left(\mathbf{f}^\mathrm{L}, \mathrm{p}^\mathrm{L}_{\boldsymbol{h},\boldsymbol{y}}(\boldsymbol{h},\boldsymbol{y})\right) = \max_{\mathrm{p}_{\boldsymbol{h},\boldsymbol{y}}(\boldsymbol{h},\boldsymbol{y})\in\mathcal{P}} \min_{\mathbf{f}\in\mathcal{F}} \mathrm{E}_{\boldsymbol{h},\boldsymbol{y}}[\|\boldsymbol{h} - \hat{\boldsymbol{h}}\|_2^2]. \tag{2.81}$$

As for the infinite dimensional case the optimal values of both problems are identical if and only if $\mathbf{f}^\mathrm{L}$ and $\mathrm{p}^\mathrm{L}_{\boldsymbol{h},\boldsymbol{y}}(\boldsymbol{h},\boldsymbol{y})$ form a saddle point[18], i.e.,

$$\mathrm{E}_{\mathrm{p}_{\boldsymbol{h},\boldsymbol{y}}(\boldsymbol{h},\boldsymbol{y})}[\|\boldsymbol{h} - \mathbf{f}^\mathrm{L}(\boldsymbol{y})\|_2^2] \le \mathrm{E}_{\mathrm{p}^\mathrm{L}_{\boldsymbol{h},\boldsymbol{y}}(\boldsymbol{h},\boldsymbol{y})}[\|\boldsymbol{h} - \mathbf{f}^\mathrm{L}(\boldsymbol{y})\|_2^2] \le \mathrm{E}_{\mathrm{p}^\mathrm{L}_{\boldsymbol{h},\boldsymbol{y}}(\boldsymbol{h},\boldsymbol{y})}[\|\boldsymbol{h} - \mathbf{f}(\boldsymbol{y})\|_2^2]. \tag{2.82}$$

The equality of (2.80) and (2.81) (minimax equality) holds, because of the following arguments:

1. The MSE in (2.80) is a linear function in the second order moments and convex in the parameters of $\mathbf{f}^\mathrm{L}$ if $\mathbf{f}^\mathrm{L}$ is a linear function. Due to the minimax theorem [193], problems (2.80) and (2.81) are equal.
2. For fixed $\boldsymbol{C}_{\boldsymbol{z}} = \boldsymbol{C}^\mathrm{L}_{\boldsymbol{z}}$ ($\mathcal{C} = \{\boldsymbol{C}^\mathrm{L}_{\boldsymbol{z}}\}$), the left inequality in (2.82) is an equality because the MSE only depends on the second order moments of $\boldsymbol{h}$ and $\boldsymbol{y}$ for linear $\mathbf{f}^\mathrm{L}$. For general $\mathcal{C}$, the left inequality is true if we choose $\boldsymbol{C}^\mathrm{L}_{\boldsymbol{z}}$ according to (2.83).
3. For all linear estimators and all distributions $\mathrm{p}_{\boldsymbol{h},\boldsymbol{y}}(\boldsymbol{h},\boldsymbol{y})$, $\mathrm{p}^\mathrm{L}_{\boldsymbol{h},\boldsymbol{y}}(\boldsymbol{h},\boldsymbol{y})$ and $\mathbf{f}^\mathrm{L}$ form a saddle point.
4. For arbitrary nonlinear estimators $\mathbf{f}$ (measurable functions in $\mathcal{F}$), the linear estimate is optimum for a Gaussian distribution $\mathrm{p}^\mathrm{L}_{\boldsymbol{h},\boldsymbol{y}}(\boldsymbol{h},\boldsymbol{y})$ and the right inequality in (2.82) holds. Therefore, the Gaussian distribution is least-favorable.

---

[17] Without loss of generality, we implicitly restrict $\boldsymbol{h}$ and $\boldsymbol{y}$ to have uncorrelated and identically distributed real and imaginary parts.

[18] $\mathrm{E}_{\mathrm{p}_{\boldsymbol{h},\boldsymbol{y}}(\boldsymbol{h},\boldsymbol{y})}[\bullet]$ is the expectation operator given by the probability distribution $\mathrm{p}_{\boldsymbol{h},\boldsymbol{y}}(\boldsymbol{h},\boldsymbol{y})$.

This yields the following theorem due to Soloviov.

Theorem 2.2 (Soloviov [199]). *If $\mathcal{C}$ is a compact and convex set of positive semidefinite matrices, then*

$$G_{\min} = \mathrm{G}\left(\mathbf{f}^{\mathrm{L}}, \mathrm{p}^{\mathrm{L}}_{\boldsymbol{h},\boldsymbol{y}}(\boldsymbol{h}, \boldsymbol{y})\right) = \max_{\boldsymbol{C}_{\boldsymbol{z}} \in \mathcal{C}} \mathrm{tr}\left[\mathbf{K}_{\boldsymbol{C}_{\mathrm{y}}}(\boldsymbol{C}_{\boldsymbol{z}})\right] = \max_{\boldsymbol{C}_{\boldsymbol{z}} \in \mathcal{C}} \mathrm{tr}\left[\boldsymbol{C}_{\boldsymbol{h}} - \boldsymbol{C}_{\boldsymbol{hy}}\boldsymbol{C}_{\boldsymbol{y}}^{-1}\boldsymbol{C}_{\boldsymbol{yh}}\right]. \tag{2.83}$$

The least favorable solution $\mathrm{p}^{\mathrm{L}}_{\boldsymbol{h},\boldsymbol{y}}(\boldsymbol{h}, \boldsymbol{y})$ in the class $\mathcal{P}$ is the normal distribution

$$\boldsymbol{z} = [\boldsymbol{h}^{\mathrm{T}}, \boldsymbol{y}^{\mathrm{T}}]^{\mathrm{T}} \sim \mathcal{N}_{\mathrm{c}}(\mathbf{0}_{P+N_{\mathrm{y}}}, \boldsymbol{C}^{\mathrm{L}}_{\boldsymbol{z}}), \tag{2.84}$$

where $\boldsymbol{C}^{\mathrm{L}}_{\boldsymbol{z}}$ is any solution of (2.83). The maximally robust solution of the original problem (2.80) is given by the linear estimator

$$\hat{\boldsymbol{h}} = \mathbf{f}^{\mathrm{R}}(\boldsymbol{y}) = \boldsymbol{C}^{\mathrm{L}}_{\boldsymbol{hy}}\boldsymbol{C}^{\mathrm{L},-1}_{\boldsymbol{y}}\boldsymbol{y} \tag{2.85}$$

in case $\boldsymbol{C}^{\mathrm{L}}_{\boldsymbol{y}}$ is non-singular.

The Schur complement in (2.83) is concave in $\boldsymbol{C}_{\boldsymbol{z}}$. For example, given an estimate $\hat{\boldsymbol{C}}_{\boldsymbol{z}}$ and the constraints $\|\boldsymbol{C}_{\boldsymbol{z}} - \hat{\boldsymbol{C}}_{\boldsymbol{z}}\|_{\mathrm{F}} \leq \varepsilon$ and $\boldsymbol{C}_{\boldsymbol{z}} \succeq 0$, problem (2.83) can be formulated as a semidefinite program [24, p. 168] introducing P slack variables and applying Theorem A.2 to the diagonal elements of the Schur complement. Numerical optimization tools are available for this class of problems (e.g., [207]) that may be too complex for online adaptation in wireless communications at the moment. But if we can choose $\mathcal{C}$ such that it has a maximum element $\boldsymbol{C}^{\mathrm{L}}_{\boldsymbol{z}} \in \mathcal{C}$, in the sense that $\boldsymbol{C}^{\mathrm{L}}_{\boldsymbol{z}} \succeq \boldsymbol{C}_{\boldsymbol{z}}, \forall \boldsymbol{C}_{\boldsymbol{z}} \in \mathcal{C}$, the solution is very simple: If the set $\mathcal{C}$ has a maximum element $\boldsymbol{C}^{\mathrm{L}}_{\boldsymbol{z}} \in \mathcal{C}$, then it follows from Theorem A.1 that $\boldsymbol{C}^{\mathrm{L}}_{\boldsymbol{z}}$ is a solution to (2.83).

We conclude with a few remarks on minimax estimators and the solution given by the theorem are in order:

1. Minimax robust estimators guarantee a certain performance $G_{\min}$, but may be very conservative if the set $\mathcal{C}$ is "large". For example, this is the case when the nominal or most likely parameter from $\mathcal{C}$ in the considered application deviates significantly from the least-favorable parameter. But a certain degree of conservatism is always the price for robustness and can be reduced by choosing the set $\mathcal{C}$ as small as possible. On the other hand, if the least-favorable parameter happens to be the most likely or the nominal parameter, then there is no performance penalty for robustness.
2. The theorem gives the justification for using a linear MMSE estimator, although the nominal joint pdf is not Gaussian: The Gauss pdf is the worst case distribution.
3. The theorem also shows the importance of the constraint on the second order moments: The efficiency of the linear estimator is determined by this

constraint and not the Gaussian assumption [223]. This is due to the fact that any deviation from the Gaussian pdf with identical constraint set $\mathcal{C}$ on the second order moments $\boldsymbol{C}_{\boldsymbol{z}}$ does *not increase* the cost in (2.82).

4. The difficulty lies in defining the constraint set $\mathcal{C}$. Obtaining an accurate characterization of this set based on a finite number of observations is critical for a robust performance, whereas the assumption of a Gaussian pdf is already the worst-case model. [19]

2.4.2 Minimax Mean Square Error Channel Estimation

The general results on minimax MSE estimation based on an observation of finite dimension are now applied to obtain robust MMSE channel estimators (Section 2.2). In channel estimation we have $\boldsymbol{C}_{\boldsymbol{y}} = \boldsymbol{S}\boldsymbol{C}_{\boldsymbol{h}}\boldsymbol{S}^{\mathrm{H}} + (\boldsymbol{I}_N \otimes \boldsymbol{C}_{\boldsymbol{n},\mathrm{s}})$ and $\boldsymbol{C}_{\boldsymbol{h}\boldsymbol{y}} = \boldsymbol{C}_{\boldsymbol{h}}\boldsymbol{S}^{\mathrm{H}}$ (2.13). Thus, the covariance matrix $\boldsymbol{C}_{\boldsymbol{z}}$ of $\boldsymbol{z} = [\boldsymbol{h}^{\mathrm{T}}, \boldsymbol{y}^{\mathrm{T}}]^{\mathrm{T}} \in \mathbb{C}^{P+MN}$ (2.79) is determined by $\boldsymbol{C}_{\boldsymbol{h}}$ and $\boldsymbol{C}_{\boldsymbol{n},\mathrm{s}}$.

We consider two types of uncertainty sets for $\boldsymbol{C}_{\boldsymbol{h}}$ and $\boldsymbol{C}_{\boldsymbol{n}}$. The first assumes that only upper bounds on the 2-norm[20] of the covariance matrices are available:

$$\mathcal{C}_{\boldsymbol{h}}^{(1)} = \left\{\boldsymbol{C}_{\boldsymbol{h}} \middle| \, \|\boldsymbol{C}_{\boldsymbol{h}}\|_2 \leq \alpha_{\boldsymbol{h}}^{(1)}, \boldsymbol{C}_{\boldsymbol{h}} \succeq 0\right\}, \tag{2.86}$$

$$\mathcal{C}_{\boldsymbol{n}}^{(1)} = \left\{\boldsymbol{C}_{\boldsymbol{n}} \middle| \boldsymbol{C}_{\boldsymbol{n}} = \boldsymbol{I}_N \otimes \boldsymbol{C}_{\boldsymbol{n},\mathrm{s}}, \|\boldsymbol{C}_{\boldsymbol{n},\mathrm{s}}\|_2 \leq \alpha_{\boldsymbol{n}}^{(1)}, \boldsymbol{C}_{\boldsymbol{n},\mathrm{s}} \succeq 0\right\}. \tag{2.87}$$

The effort to parameterize these sets via the upper bounds $\alpha_{\boldsymbol{h}}^{(1)}$ and $\alpha_{\boldsymbol{n}}^{(1)}$ is very small.[21] Because we employed the 2-norm, both sets have a maximum element[22] given by $\boldsymbol{C}_{\boldsymbol{h}}^{\mathrm{L}} = \alpha_{\boldsymbol{h}}^{(1)}\boldsymbol{I}_P$ and $\boldsymbol{C}_{\boldsymbol{n}}^{\mathrm{L}} = \alpha_{\boldsymbol{n}}^{(1)}\boldsymbol{I}_M$, respectively. Defining $\mathcal{C}^{(1)} = \{\boldsymbol{C}_{\boldsymbol{z}}|\boldsymbol{C}_{\boldsymbol{h}} \in \mathcal{C}_{\boldsymbol{h}}^{(1)}, \boldsymbol{C}_{\boldsymbol{n}} \in \mathcal{C}_{\boldsymbol{n}}^{(1)}\}$ and due to the linear structure of $\boldsymbol{C}_{\boldsymbol{z}}$ in $\boldsymbol{C}_{\boldsymbol{h}}$ and $\boldsymbol{C}_{\boldsymbol{n}}$, the maximum element $\boldsymbol{C}_{\boldsymbol{z}}^{\mathrm{L}}$ of $\mathcal{C}^{(1)}$ is determined by $\boldsymbol{C}_{\boldsymbol{h}}^{\mathrm{L}}$ and $\boldsymbol{C}_{\boldsymbol{n}}^{\mathrm{L}}$.

The minimax MSE channel estimator is obtained from

$$\boxed{\min_{\boldsymbol{W}} \max_{\boldsymbol{C}_{\boldsymbol{z}} \in \mathcal{C}^{(1)}} \mathrm{E}_{\boldsymbol{h},\boldsymbol{y}}\left[\|\boldsymbol{h} - \boldsymbol{W}\boldsymbol{y}\|_2^2\right].} \tag{2.88}$$

---

[19] The example from [223] illustrates this: Consider a distribution of a Gauss mixed with a Cauchy distribution of infinite variance, where the latter occurs only with a small probability. To obtain a reliable estimate of the large variance would require many observations.

[20] The 2-norm $\|\boldsymbol{A}\|_2$ of a matrix $\boldsymbol{A}$ is defined as $\max_{\boldsymbol{x}} \|\boldsymbol{A}\boldsymbol{x}\|_2/\|\boldsymbol{x}\|_2$ which is its largest singular value.

[21] They may be found exploiting $\mathrm{tr}[\boldsymbol{C}_{\boldsymbol{h}}]/P \leq \|\boldsymbol{C}_{\boldsymbol{h}}\|_2 \leq \|\boldsymbol{C}_{\boldsymbol{h}}\|_{\mathrm{F}} \leq \mathrm{tr}[\boldsymbol{C}_{\boldsymbol{h}}]$. Thus, an estimate of the variances of all channel parameters is sufficient and $\alpha_{\boldsymbol{h}}^{(1)}$, in this example, can be chosen proportional to $\mathrm{tr}[\boldsymbol{C}_{\boldsymbol{h}}]$.

[22] W.r.t. the Loewner partial order [96], i.e., $\boldsymbol{A} \succeq \boldsymbol{B}$ if and only if $\boldsymbol{A} - \boldsymbol{B} \succeq 0$.

According to Theorem 2.2, it is the optimum estimator for the least-favorable parameters. With Theorem A.1 the least-favorable parameters w.r.t. $\mathcal{C}^{(1)}$ are $\boldsymbol{C}_{\boldsymbol{z}}^{\mathrm{L}}$ and the minimax robust estimator is given by (cf. Eq. 2.22)

$$\boxed{\boldsymbol{W}_{\mathrm{MMSE}}^{(1)} = \left(\boldsymbol{S}^{\mathrm{H}}\boldsymbol{S} + \frac{\alpha_{\boldsymbol{n}}^{(1)}}{\alpha_{\boldsymbol{h}}^{(1)}}\boldsymbol{I}_P\right)^{-1}\boldsymbol{S}^{\mathrm{H}}.} \tag{2.89}$$

If estimates $\hat{\boldsymbol{C}}_{\boldsymbol{h}}$ and $\hat{\boldsymbol{C}}_{\boldsymbol{n},\mathrm{s}}$ of the covariance matrices and a model for the estimation errors are available, the uncertainty sets can be restricted to

$$\mathcal{C}_{\boldsymbol{h}}^{(2)} = \left\{\boldsymbol{C}_{\boldsymbol{h}} \middle| \boldsymbol{C}_{\boldsymbol{h}} = \hat{\boldsymbol{C}}_{\boldsymbol{h}} + \boldsymbol{E}, \|\boldsymbol{E}\|_2 \leq \alpha_{\boldsymbol{h}}^{(2)}, \boldsymbol{C}_{\boldsymbol{h}} \succeq 0\right\} \tag{2.90}$$

$$\mathcal{C}_{\boldsymbol{n}}^{(2)} = \left\{\boldsymbol{C}_{\boldsymbol{n}} \middle| \boldsymbol{C}_{\boldsymbol{n}} = \boldsymbol{I}_N \otimes \boldsymbol{C}_{\boldsymbol{n},\mathrm{s}}, \boldsymbol{C}_{\boldsymbol{n},\mathrm{s}} = \hat{\boldsymbol{C}}_{\boldsymbol{n},\mathrm{s}} + \boldsymbol{E}_{\boldsymbol{n}}, \|\boldsymbol{E}_{\boldsymbol{n}}\|_2 \leq \alpha_{\boldsymbol{n}}^{(2)}, \boldsymbol{C}_{\boldsymbol{n}} \succeq 0\right\}. \tag{2.91}$$

Because Theorem 2.2 requires compact and convex sets, the errors have to be modeled as being bounded.[23] Again we choose the 2-norm to obtain the uncertainty classes $\mathcal{C}_{\boldsymbol{h}}^{(2)}$ and $\mathcal{C}_{\boldsymbol{n}}^{(2)}$ with maximum elements which are $\boldsymbol{C}_{\boldsymbol{h}}^{\mathrm{L}} = \hat{\boldsymbol{C}}_{\boldsymbol{h}} + \alpha_{\boldsymbol{h}}^{(2)}\boldsymbol{I}_P$ and $\boldsymbol{C}_{\boldsymbol{n}}^{\mathrm{L}} = \boldsymbol{I}_N \otimes \boldsymbol{C}_{\boldsymbol{n},\mathrm{s}}^{\mathrm{L}}$ with $\boldsymbol{C}_{\boldsymbol{n},\mathrm{s}}^{\mathrm{L}} = \hat{\boldsymbol{C}}_{\boldsymbol{n},\mathrm{s}} + \alpha_{\boldsymbol{n}}^{(2)}\boldsymbol{I}_M$. With (2.22) this yields the robust estimator

$$\begin{aligned}\boldsymbol{W}_{\mathrm{MMSE}}^{(2)} = &\left[\left(\hat{\boldsymbol{C}}_{\boldsymbol{h}} + \alpha_{\boldsymbol{h}}^{(2)}\boldsymbol{I}_P\right)\boldsymbol{S}^{\mathrm{H}}\left(\boldsymbol{I}_N \otimes (\hat{\boldsymbol{C}}_{\boldsymbol{n},\mathrm{s}} + \alpha_{\boldsymbol{n}}^{(2)}\boldsymbol{I}_M)^{-1}\right)\boldsymbol{S} + \boldsymbol{I}_P\right]^{-1} \\ &\times \left(\hat{\boldsymbol{C}}_{\boldsymbol{h}} + \alpha_{\boldsymbol{h}}^{(2)}\boldsymbol{I}_P\right)\boldsymbol{S}^{\mathrm{H}}\left(\boldsymbol{I}_N \otimes (\hat{\boldsymbol{C}}_{\boldsymbol{n},\mathrm{s}} + \alpha_{\boldsymbol{n}}^{(2)}\boldsymbol{I}_M)^{-1}\right).\end{aligned} \tag{2.92}$$

Both robust estimators $\boldsymbol{W}_{\mathrm{MMSE}}^{(1)}$ and $\boldsymbol{W}_{\mathrm{MMSE}}^{(2)}$ converge to the ML channel estimator (2.31) as the 2-norm of $\boldsymbol{C}_{\boldsymbol{h}}$ and of the estimation error of the estimate $\hat{\boldsymbol{C}}_{\boldsymbol{h}}$, respectively, increase: $\alpha_{\boldsymbol{h}}^{(1)} \to \infty$ and $\alpha_{\boldsymbol{h}}^{(2)} \to \infty$. This provides a degree of freedom to ensure a performance of the MMSE approach for estimated second order moments which is always better or, in the worst case, equal to the ML channel estimator. The upper bounds in (2.90) and (2.91) can be chosen proportional to or as a monotonically increasing function of a rough estimate for the first order moment of $\|\boldsymbol{E}\|_2^2 \leq \|\boldsymbol{E}\|_{\mathrm{F}}^2$.[24]

---

[23] For stochastic estimation errors, the true covariance matrices are only included in the set with a certain probability. This probability is another design parameter and gives an upper bound on the probability that the worst case MSE is exceeded (outage probability).

[24] A stochastic model for the estimation error of sample mean estimators for covariance matrices is derived in [238].

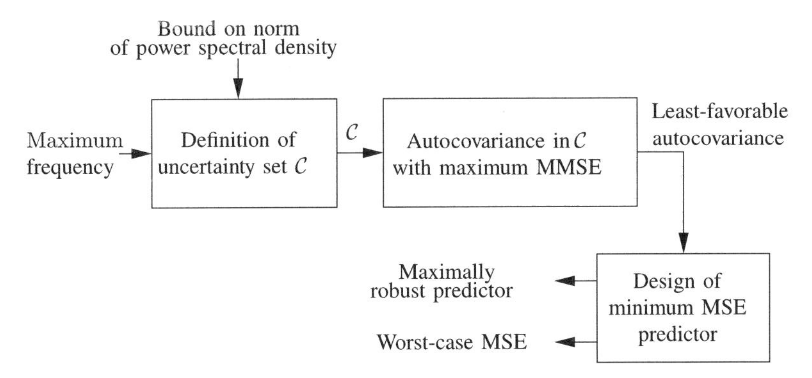

Fig. 2.11 Minimax MSE prediction of band-limited random sequences with a norm constraint on their power spectral density.

2.4.3 Minimax Mean Square Error Channel Prediction

The MMSE predictor in Section 2.3.1 requires the knowledge of the autocovariance sequence of the channel parameters. In Section 2.3.2 we introduce the set $\mathcal{B}_{h,f_{\max}}$ (2.66) of power spectral densities corresponding to band-limited autocovariance sequences with given maximum (Doppler-) frequency $f_{\max}$ and variance $c_h[0]$. We would like to obtain minimax MSE predictors for similar classes $\mathcal{C}$ of autocovariance sequences, which require *only* knowledge of $f_{\max}$ and a bound on the norm of their power spectral densities. This robust solution does *not* require knowledge of the autocovariance sequence and is maximally robust, i.e., minimizes the worst-case MSE (Figure 2.11).

The case of an observation of infinite dimension $Q \to \infty$ and of finite dimension Q are treated separately. The derivations are based on the model (2.56) and (2.59), but in the sequel we omit the index p denoting the pth channel parameter.

2.4.3.1 Minimax MSE Prediction for $Q \to \infty$ and L_1-Norm

The models for $Q \to \infty$ are given by (2.70) and (2.71) for $d = 0$. With (2.72) and (2.76), the minimax robust predictor is given by

$$\boxed{\min_{\mathrm{W}\in\mathcal{H}_+^2} \max_{\mathrm{S}_{\boldsymbol{h}}(\omega)\in\mathcal{B}_{\boldsymbol{h},f_{\max}}} \mathrm{F}(\mathrm{W},\mathrm{S}_h,\mathrm{S}_e) = \min_{\mathrm{W}\in\mathcal{H}_+^2} \max_{\mathrm{S}_{\boldsymbol{h}}(\omega)\in\mathcal{B}_{\boldsymbol{h},f_{\max}}} \mathrm{E}\left[|\varepsilon_\infty[q]|^2\right].} \quad (2.93)$$

For simplicity, we assume white noise $\mathsf{e}[q]$ with power spectral density $\mathrm{S}_e(\omega) = c_e$, i.e, known variance $c_e > 0$. Because $c_e > 0$, we have $\mathrm{S}_e(\omega) > 0$. Moreover, the set $\mathcal{B}_{h,f_{\max}}$ (2.66) is convex and $c_h[0]$ is bounded. Therefore, the conditions of Theorem 2.1 are satisfied.[25] Thus, the least-favorable power spectral density $\mathrm{S}_{h}^{\mathrm{L}}(\omega)$ and the corresponding optimum predictor W^{L} form a saddle point for (2.93). Applying the closed-form expression from [196] (see (2.64)) to (2.93), we obtain

$$\begin{aligned}\max_{\mathrm{S}_{\boldsymbol{h}}(\omega)\in\mathcal{B}_{\boldsymbol{h},f_{\max}}} \min_{\mathrm{W}\in\mathcal{H}_{+}^{2}} \mathrm{F}(\mathrm{W},\mathrm{S}_{h},\mathrm{S}_{e}) \\ = \max_{\mathrm{S}_{\boldsymbol{h}}(\omega)\in\mathcal{B}_{\boldsymbol{h},f_{\max}}} c_e\left[\exp\left(\frac{1}{2\pi}\int_{-\pi}^{\pi}\ln\left(1+\frac{\mathrm{S}_{h}(\omega)}{c_e}\right)\mathrm{d}\omega\right)-1\right].\end{aligned} \tag{2.94}$$

This is equivalent to maximizing the spectral entropy [123]

$$\max_{\mathrm{S}_{\boldsymbol{h}}(\omega)\in\mathcal{B}_{\boldsymbol{h},f_{\max}}} \frac{1}{2\pi}\int_{-\pi}^{\pi}\ln\left(1+\frac{\mathrm{S}_{h}(\omega)}{c_e}\right)\mathrm{d}\omega, \tag{2.95}$$

which is a variational problem. To solve it we follow the steps in [134], where the least-favorable power spectral density for the filtering problem $d=-1$ is derived. We compare this cost function with the spectral entropy of the rectangular power spectral density $\mathrm{S}_{\mathrm{R}}(\omega)$ (2.68) with cut-off frequency $f_{\max}$

$$\frac{1}{2\pi}\int_{-\pi}^{\pi}\ln\left(1+\frac{\mathrm{S}_{h}(\omega)}{c_e}\right)\mathrm{d}\omega-\frac{1}{2\pi}\int_{-\pi}^{\pi}\ln\left(1+\frac{\mathrm{S}_{\mathrm{R}}(\omega)}{c_e}\right)\mathrm{d}\omega \tag{2.96}$$

$$=\frac{1}{2\pi}\int_{-\omega_{\max}}^{\omega_{\max}}\ln\left(1+\frac{\mathrm{S}_{h}(\omega)-\frac{\pi}{\omega_{\max}}c_h[0]}{c_e\left(1+\frac{\pi c_h[0]}{\omega_{\max}c_e}\right)}\right)\mathrm{d}\omega \tag{2.97}$$

$$\leq\frac{1}{2\pi}\int_{-\omega_{\max}}^{\omega_{\max}}\frac{\mathrm{S}_{h}(\omega)-\frac{\pi}{\omega_{\max}}c_h[0]}{c_e\left(1+\frac{\pi c_h[0]}{\omega_{\max}c_e}\right)}\mathrm{d}\omega=0, \tag{2.98}$$

where the inequality is due to $\ln(x+1)\leq x$ for $x>-1$. Equality holds if and only if $\mathrm{S}_{h}(\omega)=\mathrm{S}_{\mathrm{R}}(\omega)$. The optimum value of (2.93) and (2.94) is (cf. Eq. 2.69)

$$\max_{\mathrm{S}_{\boldsymbol{h}}(\omega)\in\mathcal{B}_{\boldsymbol{h},f_{\max}}} \min_{\mathrm{W}} \mathrm{F}(\mathrm{W},\mathrm{S}_{h},\mathrm{S}_{e})=c_e\left[\left(1+\frac{\pi c_h[0]}{c_e\omega_{\max}}\right)^{\omega_{\max}/\pi}-1\right]. \tag{2.99}$$

This is the upper bound for the prediction error for all autocovariance sequences with power spectral density in $\mathcal{B}_{h,f_{\max}}$. This argumentation yields the following theorem.

---

[25] The class $\mathcal{B}_{h,f_{\max}}$ in (2.66) could also be extended to contain all power spectral densities with L_1 norm (variance $c_h[0]$) upper-bounded by β. The following argumentation shows that the least-favorable power spectral density w.r.t. this extended set has an L_1 norm equal to $c_h^{\mathrm{L}}[0]=\beta$.

Theorem 2.3. *The worst-case or least-favorable autocovariance sequence for problem* (2.93) *is the inverse Fourier transform of* $\mathrm{S}_{h}^{\mathrm{L}}(\omega)=\mathrm{S}_{\mathrm{R}}(\omega)$

$$\boxed{c_h^{\mathrm{L}}[\ell]=c_h[0]\mathrm{sinc}(2f_{\max}\ell)=c_h[0]\frac{\sin(2\pi f_{\max}\ell)}{2\pi f_{\max}\ell}.} \tag{2.100}$$

The corresponding random sequence is the least predictable with maximum entropy rate [40] in this class. $\mathrm{S}_{\mathrm{R}}(\omega)$ is the maximally flat power spectral density in $\mathcal{B}_{h,f_{\max}}$, since $\mathrm{E}[|\varepsilon_{\infty}[q]|^2]/c_h[0]$ is a measure for the flatness of the spectrum [182].

It is often mentioned in the literature that this choice of power spectral density yields results close to optimum performance in various applications (see [144, p. 658], [9]), but no proof is given for its robustness or optimality for prediction: Franke, Vastola, Verdu, and Poor [123, 75, 76, 225, 226] either discuss the general minimax MSE problem or other cases of minimax MSE prediction (for $c_e=0$ and no band-limitation); in [134] it is shown that the rectangular power spectral density is also least-favorable for the causal filtering problem (2.76) with $d=-1$ in (2.71).

2.4.3.2 Minimax MSE Prediction for Finite Q and L_1-Norm

For prediction based on Q past observations $\{\tilde{h}[q-\ell]\}_{\ell=1}^{Q}$, the MSE $\mathrm{E}[|\varepsilon_Q[q]|^2]$ depends only on $Q+1$ samples $\{c_h[\ell]\}_{\ell=0}^{Q}$ of the autocovariance sequence. We define the set $\mathcal{C}_{h,f_{\max}}^{1}$ of partial sequences $\{c_h[\ell]\}_{\ell=0}^{Q}$ for which a power spectral density $\mathrm{S}_h(\omega)$ with support limited to $[-2\pi f_{\max}, 2\pi f_{\max}]$ and L_1-norm $c_h[0]=\|\mathrm{S}_h(\omega)\|_1$ bounded by β exists:

$$\begin{aligned}\mathcal{C}_{h,f_{\max}}^{1}=\Big\{\{c_h[\ell]\}_{\ell=0}^{Q}\Big|\exists \mathrm{S}_h(\omega)\geq 0: c_h[\ell]=\frac{1}{2\pi}\int_{-\pi}^{\pi}\mathrm{S}_h(\omega)\exp(\mathrm{j}\omega\ell)\mathrm{d}\omega,\\ c_h[0]=\frac{1}{2\pi}\int_{-\pi}^{\pi}\mathrm{S}_h(\omega)\mathrm{d}\omega\leq\beta, \mathrm{S}_h(\omega)=0 \text{ for } 2\pi f_{\max}<|\omega|\leq\pi\Big\}.\end{aligned} \tag{2.101}$$

Thus, the partial sequence can be extended to a sequence with a band-limited power spectral density (Appendix B.2). With Theorem 2.2, the maximally robust predictor is the solution of

$$\min_{\boldsymbol{w}}\max_{\{c_{\boldsymbol{h}}[\ell]\}_{\ell=0}^{Q}\in\mathcal{C}_{\boldsymbol{h},f_{\max}}^{1}}\mathrm{E}\big[|\varepsilon_Q[q]|^2\big]=\max_{\{c_{\boldsymbol{h}}[\ell]\}_{\ell=0}^{Q}\in\mathcal{C}_{\boldsymbol{h},f_{\max}}^{1}}\min_{\boldsymbol{w}}\mathrm{E}\big[|\varepsilon_Q[q]|^2\big]. \tag{2.102}$$

The equivalent maxmin-optimization problem can be written as (cf. (2.62))

$$\max_{\{c_{\boldsymbol{h}}[\ell]\}_{\ell=0}^{Q}\in\mathcal{C}_{\boldsymbol{h},f_{\max}}^{1}}\beta-\boldsymbol{c}_{\boldsymbol{h}_{\mathrm{T}}h}^{\mathrm{H}}\boldsymbol{C}_{\tilde{\boldsymbol{h}}_{\mathrm{T}}}^{-1}\boldsymbol{c}_{\boldsymbol{h}_{\mathrm{T}}h} \tag{2.103}$$

with $\boldsymbol{C}_{\tilde{\boldsymbol{h}}_\mathrm{T}} = \boldsymbol{C}_{\boldsymbol{h}_\mathrm{T}} + c_e \boldsymbol{I}_Q$, $\boldsymbol{C}_{\boldsymbol{h}_\mathrm{T}} \in \mathbb{T}^Q_{+,0}$ with first row $[c_h[0], \ldots, c_h[Q-1]]$, and $\boldsymbol{c}_{\boldsymbol{h}_\mathrm{T}h} = [c_h[1]^*, \ldots, c_h[Q]^*]^\mathrm{T}$.

The set $\mathcal{C}^1_{h,f_\mathrm{max}}$ can be characterized equivalently by Theorem B.3 from Arun and Potter [6], which yields positive semidefinite constraints on the Toeplitz matrices $\boldsymbol{C}_\mathrm{T} \in \mathbb{T}^{Q+1}_{+,0}$ with first row $[c_h[0], \ldots, c_h[Q]]$ and $\boldsymbol{C}'_\mathrm{T} \in \mathbb{T}^Q_{+,0}$ with first row $[c'_h[0], \ldots, c'_h[Q-1]]$, where $c'_h[\ell] = c_h[\ell-1] - 2\cos(2\pi f_\mathrm{max}) c_h[\ell] + c_h[\ell+1]$ and $c_h[0] = c^\mathrm{L}_h[0] = \beta$. The least-favorable variance is $c^\mathrm{L}_h[0] = \beta$.

Therefore, problem (2.103) can be reformulated as

$$\boxed{\max_{\{c_h[\ell]\}_{\ell=1}^Q} \beta - \boldsymbol{c}^\mathrm{H}_{\boldsymbol{h}_\mathrm{T}h} \boldsymbol{C}^{-1}_{\tilde{\boldsymbol{h}}_\mathrm{T}} \boldsymbol{c}_{\boldsymbol{h}_\mathrm{T}h} \quad \text{s.t.} \quad \boldsymbol{C}_\mathrm{T} \succeq 0, \boldsymbol{C}'_\mathrm{T} \succeq 0.} \tag{2.104}$$

It can be transformed into a semidefinite program[26] which has a unique solution and can be solved numerically by SeDuMi [207], for example.

If the partial sequence $c_h[0] = \beta, c_h[\ell] = 0$, for $1 \leq \ell \leq Q$, is an element of $\mathcal{C}^1_{h,f_\mathrm{max}}$, the maximally robust predictor is trivial $\boldsymbol{w} = \boldsymbol{0}_Q$ and the maximum MSE (2.103) equals $c^\mathrm{L}_h[0] = \beta$. Whether this partial sequence has a band-limited positive semidefinite extension, i.e., belongs to $\mathcal{C}^1_{h,f_\mathrm{max}}$, depends on the values of Q and f_max. Its existence can be verified by Theorem B.3, which results in the necessary condition $f_\mathrm{max} \geq 1/3$ (in the case of $Q \geq 2$). For example, for $Q = 3$ it exists for all $f_\mathrm{max} \geq 0.375$ (Example B.1). The minimum f_max for its existence increases as Q increases. For $Q \to \infty$, the autocovariance sequence which is zero for $\ell \neq 0$ is not band-limited anymore and its power spectral density is not an element of $\mathcal{B}_{h,f_\mathrm{max}}$ for $f_\mathrm{max} < 0.5$ and $c_h[0] = \beta$.

2.4.3.3 Minimax MSE Prediction for Finite Q and L_∞-Norm

Obviously, for finite Q, uncertainty set $\mathcal{C}^1_{h,f_\mathrm{max}}$, and $f_\mathrm{max} < 0.5$, the least-favorable autocovariance sequence differs from (2.100) and the predictor corresponding to a rectangular power spectral density is not maximally robust.[27]

---

[26] Introducing the slack variable t we have

$$\min_{t,\{c_h[\ell]\}_{\ell=1}^Q} t \quad \text{s.t.} \quad \boldsymbol{c}^\mathrm{H}_{\boldsymbol{h}_\mathrm{T}h} \boldsymbol{C}^{-1}_{\tilde{\boldsymbol{h}}_\mathrm{T}} \boldsymbol{c}_{\boldsymbol{h}_\mathrm{T}h} \leq t, \boldsymbol{C}_\mathrm{T} \succeq 0, \boldsymbol{C}'_\mathrm{T} \succeq 0, \tag{2.105}$$

which is equivalent to the semidefinite program

$$\min_{t,\{c_h[\ell]\}_{\ell=1}^Q} t \quad \text{s.t.} \quad \begin{bmatrix} t & \boldsymbol{c}^\mathrm{H}_{\boldsymbol{h}_\mathrm{T}h} \\ \boldsymbol{c}_{\boldsymbol{h}_\mathrm{T}h} & \boldsymbol{C}_{\tilde{\boldsymbol{h}}_\mathrm{T}} \end{bmatrix} \succeq 0, \boldsymbol{C}_\mathrm{T} \succeq 0, \boldsymbol{C}'_\mathrm{T} \succeq 0 \tag{2.106}$$

by means of Theorem A.2.

[27] The robustness of a given predictor $\boldsymbol{w}$, e.g., a good heuristic, can be evaluated by solving

It has often been stated in the literature that a rectangular power spectral density is a "good" choice for parameterizing an MMSE predictor for band-limited signals. But for which uncertainty class is this choice maximally robust?

We change the definition of the uncertainty set and derive a maximally robust predictor which depends on the (least-favorable) autocovariance sequence $\{c_h^{\mathrm{L}}[\ell]\}_{\ell=0}^{Q}$ related to a band-limited rectangular power spectral density.

In the proofs of [153, 27] for perfect predictability of band-limited random sequences, where both references consider the case of no noise $c_e = 0$ in the observation, the class of autocovariance sequences with band-limited and *bounded* power spectral density is considered. This motivates us to define the uncertainty set

$$\mathcal{C}_{h,f_{\max}}^{\infty} = \Bigg\{ \{c_h[\ell]\}_{\ell=0}^{Q} \Bigg| \exists \mathrm{S}_h(\omega) \geq 0 : c_h[\ell] = \frac{1}{2\pi} \int_{-\pi}^{\pi} \mathrm{S}_h(\omega) \exp(\mathrm{j}\omega\ell) \mathrm{d}\omega,$$
$$\mathrm{S}_h(\omega) \leq \gamma, \mathrm{S}_h(\omega) = 0 \text{ for } \omega_{\max} < |\omega| \leq \pi, \omega_{\max} = 2\pi f_{\max} \Bigg\}, \quad (2.108)$$

which is based on a bounded infinity norm $\mathrm{L}_\infty : \|\mathrm{S}_h(\omega)\|_\infty = \max_\omega \mathrm{S}_h(\omega)$ of the band-limited power spectral density $\mathrm{S}_h(\omega)$. The upper bound on the variance of $h[q]$ is $c_h^{\mathrm{L}}[0] = 2 f_{\max}\gamma$, which results from the upper bound γ on the L_∞ norm.

Instead of applying a minimax theorem from above, we solve the minimax problem

$$\boxed{\min_{\boldsymbol{w}} \max_{\{c_h[\ell]\}_{\ell=0}^{Q} \in \mathcal{C}_{h,f_{\max}}^{\infty}} \mathrm{E}\left[|\varepsilon_Q[q]|^2\right]} \quad (2.109)$$

directly, where $\boldsymbol{w} = [w[1], w[2], \ldots, w[Q]]^{\mathrm{T}} \in \mathbb{C}^Q$.[28] With (2.56) and (2.59) (omitting the index p), the estimate of $h[q]$ reads

$$\hat{h}[q] = \sum_{\ell=1}^{Q} w[\ell] \tilde{h}[q-\ell] = \sum_{\ell=1}^{Q} w[\ell] h[q-\ell] + \sum_{\ell=1}^{Q} w[\ell] e[q-\ell], \quad (2.110)$$

where we assume white noise $e[q]$ with variance c_e. Similar to (2.72), the MSE $\mathrm{E}[|\varepsilon_Q[q]|^2]$ of the estimation error $\varepsilon_Q[q] = \hat{h}[q] - h[q]$ can be written as

---

$$\max_{\{c_h[\ell]\}_{\ell=0}^{Q} \in \mathcal{C}_{h,f_{\max}}^{1}} \mathrm{E}\left[|\varepsilon_Q[q]|^2\right] \quad (2.107)$$

numerically.

[28] Problem (2.109) can also be solved using the minimax Theorem 2.2 together with Theorem A.1.

$$\mathrm{E}\left[|\varepsilon_Q[q]|^2\right] = \frac{1}{2\pi}\int_{-\omega_{\max}}^{\omega_{\max}} \mathrm{S}_h(\omega)\left|1 - \sum_{\ell=1}^{Q} w[\ell]\exp(-\mathrm{j}\omega\ell)\right|^2 \mathrm{d}\omega + c_e \|\boldsymbol{w}\|_2^2 . \tag{2.111}$$

Because $\mathrm{S}_h(\omega) \leq \gamma$, we can bound the MSE by

$$\begin{aligned}\mathrm{E}\left[|\varepsilon_Q[q]|^2\right] &\leq \gamma\frac{1}{2\pi}\int_{-\omega_{\max}}^{\omega_{\max}} \left|1 - \sum_{\ell=1}^{Q} w[\ell]\exp(-\mathrm{j}\omega\ell)\right|^2 \mathrm{d}\omega + c_e \|\boldsymbol{w}\|_2^2 \qquad (2.112)\\ &= \gamma\, 2f_{\max} + \gamma \sum_{\ell_1=1}^{Q}\sum_{\ell_2=1}^{Q} w[\ell_1]w[\ell_2]^*\, 2f_{\max}\mathrm{sinc}(2f_{\max}(\ell_1-\ell_2))\\ &\quad - \gamma\sum_{\ell=1}^{Q} w[\ell]\, 2f_{\max}\mathrm{sinc}(2f_{\max}\ell) - \gamma\sum_{\ell=1}^{Q} w[\ell]^*\, 2f_{\max}\mathrm{sinc}(2f_{\max}\ell)\\ &\quad + c_e\|\boldsymbol{w}\|_2^2\\ &= \bar{c}_{\varepsilon_Q}\left(\{w[\ell]\}_{\ell=1}^{Q}\right), \qquad (2.113)\end{aligned}$$

which is convex in $\boldsymbol{w}$. The bound on the MSE is achieved with equality for the rectangular power spectral density $\mathrm{S}_\mathrm{R}(\omega)$ with variance $c_h^\mathrm{L}[0] = 2f_{\max}\gamma$. Due to the convexity of $\mathcal{C}_{h,f_{\max}}^{\infty}$, Theorem 2.2 holds, and $\mathrm{S}_\mathrm{R}(\omega)$ is the least-favorable power spectral density. This is similar to $Q \to \infty$ and $\mathcal{B}_{h,f_{\max}}$ (2.93), but now the variance of the uncertainty set (2.108) is not fixed.

The necessary and sufficient conditions for the minimum of the upper bound $\bar{c}_{\varepsilon_Q[q]}(\{w[\ell]\}_{\ell=1}^{Q})$ form a system of Q linear equations

$$\begin{aligned}\frac{\partial \bar{c}_{\varepsilon_Q[q]}\left(\{w[\ell]\}_{\ell=1}^{Q}\right)}{\partial w[\ell]^*} &= 2f_{\max}\gamma\left(\sum_{\ell_1=1}^{Q} w[\ell_1]\mathrm{sinc}(2f_{\max}(\ell_1-\ell)) - \mathrm{sinc}(2f_{\max}\ell)\right)\\ &\quad + c_e w[\ell] = 0,\ \ell = 1,2,\ldots,Q. \qquad (2.114)\end{aligned}$$

In matrix-vector notation it reads[29]

$$\left(\boldsymbol{C}_{\boldsymbol{h}_\mathrm{T}}^\mathrm{L} + c_e\boldsymbol{I}_Q\right)\boldsymbol{w} = \boldsymbol{c}_{\boldsymbol{h}_\mathrm{T}h}^\mathrm{L}, \tag{2.115}$$

where $\boldsymbol{c}_{\boldsymbol{h}_\mathrm{T}h}^\mathrm{L} = c_h^\mathrm{L}[0][\mathrm{sinc}(2f_{\max}), \mathrm{sinc}(2f_{\max}2), \ldots, \mathrm{sinc}(2f_{\max}Q)]^\mathrm{T}$ and $\boldsymbol{C}_{\boldsymbol{h}_\mathrm{T}}^\mathrm{L} \in \mathbb{T}_{+,0}^{Q}$ is Toeplitz with first row $c_h^\mathrm{L}[0]\cdot[1, \mathrm{sinc}(2f_{\max}), \ldots, \mathrm{sinc}(2f_{\max}(Q-1))]$.

We collect these results in a theorem:

Theorem 2.4. *The solution of* (2.109) *is identical to* (2.61) *with the least-favorable autocovariance sequence*

---

[29] Applying the Levinson algorithm [170] it can be solved with a computational complexity of order Q^2.

$$\boxed{c_h^{\mathrm{L}}[\ell] = 2f_{\max}\gamma\,\mathrm{sinc}(2f_{\max}\ell)} \tag{2.116}$$

corresponding to the least-favorable rectangular power spectral density $\mathrm{S_R}(\omega)$.
Over the class $\mathcal{C}_{h,f_{\max}}^{\infty}$ *the achievable MSE is upper bounded by*

$$\begin{aligned}
\min_{\boldsymbol{w}} \mathrm{E}\left[|\varepsilon_Q[q]|^2\right] &= c_h[0] - \boldsymbol{c}_{\boldsymbol{h}_{\mathrm{T}}h}^{\mathrm{H}} \left(\boldsymbol{C}_{\boldsymbol{h}_{\mathrm{T}}} + c_e\boldsymbol{I}_Q\right)^{-1} \boldsymbol{c}_{\boldsymbol{h}_{\mathrm{T}}h} \\
&\leq \min_{\boldsymbol{w}} \max_{\{c_h[\ell]\}_{\ell=0}^{Q} \in \mathcal{C}_{h,f_{\max}}^{\infty}} \mathrm{E}\left[|\varepsilon_Q[q]|^2\right] \\
&= \left(2f_{\max}\gamma - \boldsymbol{c}_{\boldsymbol{h}_{\mathrm{T}}h}^{\mathrm{L}\,\mathrm{T}} \left(\boldsymbol{C}_{\boldsymbol{h}_{\mathrm{T}}}^{\mathrm{L}} + c_e\boldsymbol{I}_Q\right)^{-1} \boldsymbol{c}_{\boldsymbol{h}_{\mathrm{T}}h}^{\mathrm{L}}\right) \\
&\forall\ \{c_h[\ell]\}_{\ell=0}^{Q} \in \mathcal{C}_{h,f_{\max}}^{\infty}.
\end{aligned} \tag{2.117}$$

Obviously, it can also be shown in a similar fashion that the rectangular power spectral density is least-favorable for the finite dimensional filtering and smoothing/interpolation problem and the same uncertainty class. Knowing the literature, this result is not very surprising, but confirms the empirical observations of many authors: Interpolation (extrapolation) with the sinc-kernel as in (2.115) or the parameterization of an MMSE estimator with (2.116) is widely considered a "good choice" [35, 9]. To our knowledge, the optimality and minimax robustness of this choice has not been shown explicitly before. Many publications dealing with extrapolation or interpolation of band-limited signals assume that a noiseless observation ($c_e = 0$) of the signal is available [34]. For the noiseless case, it is well known that perfect prediction (interpolation) can be achieved with (2.115) for $Q \to \infty$. But it is also clear that the system of equations (2.115) becomes ill-conditioned (for $c_e = 0$) as Q grows [195, 194]. Thus, standard regularization methods can and must be applied [154].[30]

Moreover, the minimax robustness of the rectangular power spectral density for $\mathcal{C}_{h,f_{\max}}^{\infty}$ shows that the guaranteed MSE (2.117) for this robust predictor is very conservative if the nominal power spectral density deviates significantly from the rectangular power spectral density. An example is the power spectral density with band-pass shape in Figure 2.12, which is included in the uncertainty set $\mathcal{C}_{h,f_{\max}}^{\infty}$ with parameters $\gamma = 50$ for $f_{\max} = 0.1$. If the uncertainty class $\mathcal{C}_{h,f_{\max}}^{\infty}$ should also contain these power spectral densities, the upper bound γ must be large. Thus, the guaranteed MSE (2.117) will be large in general. A smaller MSE can only be guaranteed by choosing the more appropriate uncertainty class $\mathcal{C}_{h,f_{\max}}^{1}$.

It is straightforward to *generalize* our results for the uncertainty class $\mathcal{C}_{h,f_{\max}}^{\infty}$ to

---

[30] For example, when predicting a random sequence, a Tikhonov regularization [212] of the approximation problem in equation (1.2) of [153] would also lead to (2.113). But in the cost function (2.113), the regularization parameter is determined by the considered uncertainty class and the noise model.

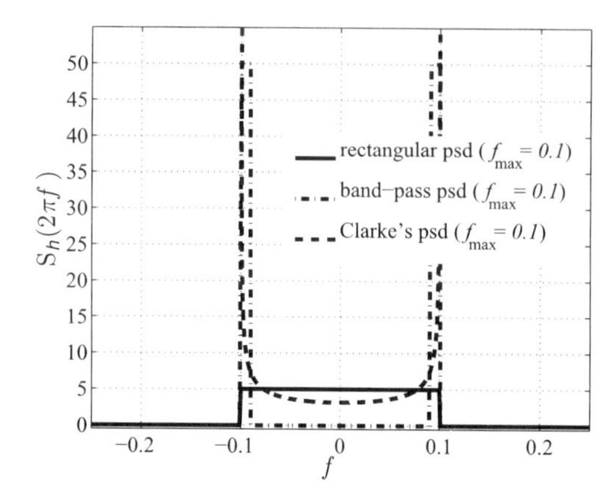

Fig. 2.12 Rectangular, band-pass, and Clarke's power spectral density (psd) $S_h(2\pi f)$ [206, p. 40] with maximum frequency $f_{\max} = 0.1$ ($\omega_{\max} = 2\pi f_{\max}$) and $c_h[0] = 1$.

$$\mathcal{C}_h^{\text{gen}} = \left\{ \{c_h[\ell]\}_{\ell=0}^{Q} \,\middle|\, \exists \mathrm{S}_h(\omega) \geq 0 : c_h[\ell] = \frac{1}{2\pi}\int_{-\pi}^{\pi} \mathrm{S}_h(\omega)\exp(\mathrm{j}\omega\ell)\mathrm{d}\omega, \right.$$
$$\left. \mathrm{L}(\omega) \leq \mathrm{S}_h(\omega) \leq \mathrm{U}(\omega) \right\}, \quad (2.118)$$

with arbitrary upper and lower bounds $\mathrm{U}(\omega) \geq 0$ and $\mathrm{L}(\omega) \geq 0$. The least-favorable power spectral density is given by $\mathrm{S}_h^{\mathrm{L}}(\omega) = \mathrm{U}(\omega)$, since the MSE can be bounded from above, similar to (2.113). Thus, in case of a nominal power spectral density deviating significantly from the rectangular power spectral density, the uncertainty class can be easily modified to obtain a less conservative predictor. For example, piece-wise constant power spectral densities can be chosen whose additional parameters can be adapted measuring the signal power in different predefined frequency bands.

2.4.3.4 Performance Evaluation

The performance of the minimax robust predictors (2.104) and (2.115) is evaluated numerically for the power spectral densities shown in Figures 2.12 and 2.13 and for different maximum frequencies $f_{\max}$. In the figures, "Minimax MSE L_1" denotes the robust predictor based on (2.104) (L_1-norm) and "Minimax MSE L_∞" based on (2.115). They are compared with the MMSE predictor assuming perfect knowledge of the autocovariance sequence and finite Q ("MMSE (perfect knowledge)"), for $Q = 5$. The MMSE for prediction with perfect knowledge and $Q \to \infty$ is denoted "MMSE ($Q \to \infty$)". As a reference, we give the MSE if the outdated observation is used as an estimate $\hat{h}[q] = \tilde{h}[q-1]$ ("Outdating"). This simple "predictor" serves as an upper limit for the MSE, which should not be exceeded by a more sophisticated approach. The upper bounds for the MSEs of the minimax predictors given by (2.104)

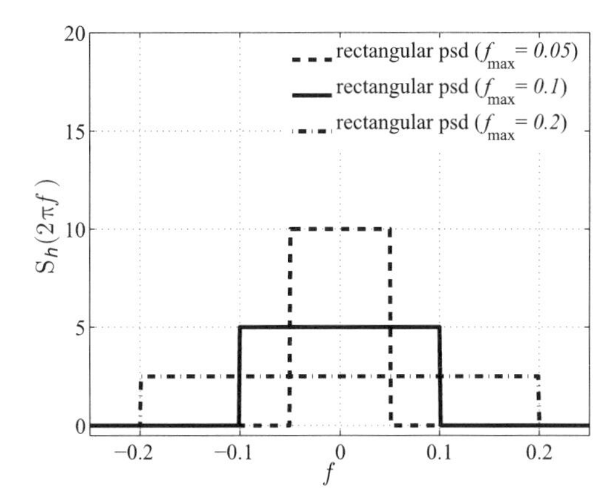

Fig. 2.13 Rectangular power spectral density (psd) $S_h(2\pi f)$ with maximum frequencies $f_{\max} \in \{0.05, 0.1, 0.2\}$ and $c_h[0] = 1$.

and (2.117) ("Minimax MSE L$_1$ (upper bound)" and "Minimax MSE L$_\infty$ (upper bound)") are guaranteed for the corresponding minimax predictor if the considered power spectral density belongs to $\mathcal{C}^1_{h,f_{\max}}$ or $\mathcal{C}^\infty_{h,f_{\max}}$, respectively. For $\mathcal{C}^1_{h,f_{\max}}$, we choose $\beta = 1$ which is the true variance $c_h[0] = 1$. When we use Clarke's power spectral density, which is not bounded and does not belong to $\mathcal{C}^\infty_{h,f_{\max}}$ for finite γ, the choice of $\gamma = 2/f_{\max}$ yields only an approximate characterization of the considered nominal power spectral density (see Figure 2.12). At first we assume perfect knowledge of $f_{\max}$.

Figure 2.14 illustrates that prediction based on $Q = 5$ past observations $\{\tilde{h}[q-\ell]\}_{\ell=1}^{Q}$ is already close to the asymptotic MSE for $Q \to \infty$.[31] The result is given for Clarke's power spectral density ($f_{\max} = 0.1$, $c_h[0] = 1$) and $c_e = 10^{-3}$, but $Q = 5$ is also a good choice w.r.t. performance and numerical complexity for other parameters and power spectral densities. We choose $Q = 5$ in the sequel.

Because Clarke's power spectral density is rather close to a rectangular power spectral density (Figure 2.12), except for $f \approx \pm f_{\max}$, the MSEs of the minimax predictors ($\gamma = 2/f_{\max}$) are close to the MMSE predictor with perfect knowledge (Figure 2.15). The graph for perfect knowledge is not shown, because it cannot be distinguished from the minimax MSE performance based on $\mathcal{C}^1_{h,f_{\max}}$. Furthermore, the minimax MSE based on $\mathcal{C}^\infty_{h,f_{\max}}$ is almost identical to the upper bound for $\mathcal{C}^1_{h,f_{\max}}$. The upper bounds on the MSEs (2.104) and (2.117) are rather tight for low $f_{\max}$.

A different approach to the prediction or interpolation of band-limited signals is based on *discrete prolate spheroidal sequences* (DPSS) [195, 48]. Prolate spheroidal sequences can be shown to be the basis which achieves the most compact representation of band-limited sequences concentrated in

---

[31] The MSE for $Q \to \infty$ is given by (2.64) with Equations (27)-(30) in [134].

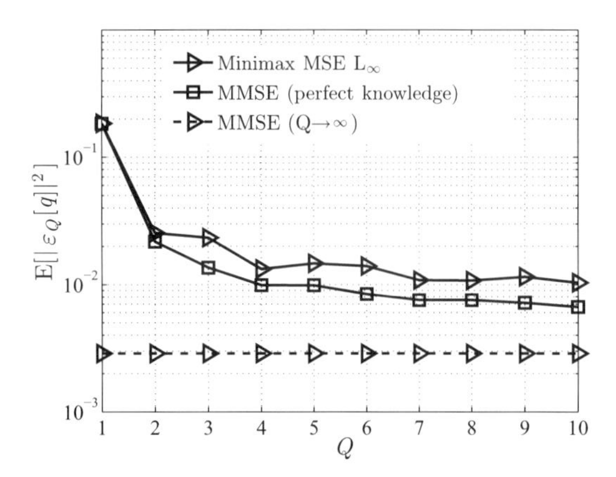

Fig. 2.14 MSE $\mathrm{E}[|\varepsilon_Q[q]|^2]$ for prediction of $h[q]$ for different number Q of observations $\{\tilde{h}[q-\ell]\}_{\ell=1}^{Q}$. (Model: Clarke's power spectral density (Figure 2.12) with $c_h[0]=1$, $c_e=10^{-3}$, $f_{\max}=0.1$)

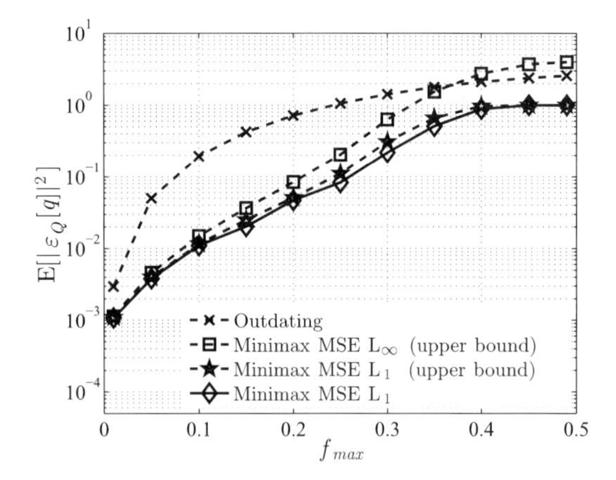

Fig. 2.15 MSE $\mathrm{E}[|\varepsilon_Q[q]|^2]$ vs. maximum frequency $f_{\max}$ for prediction of $h[q]$ based on $Q=5$ observations $\{\tilde{h}[q-\ell]\}_{\ell=1}^{Q}$. (Model: Clarke's power spectral density (Figure 2.12) with $c_h[0]=1$, $c_e=10^{-3}$, $Q=5$)

the time domain with a fixed number of basis vectors, i.e., a reduced dimension. But the norm of the resulting linear filters becomes unbounded as the observation length increases. This requires regularization methods [154] to decrease the norm and amplification of the observation noise. A truncated singular valued decomposition [48] or a reduction of the dimension [242] are most common.

We use two approaches as references in Figure 2.16: "DPSS (fixed dimension)" chooses the approximate dimension $\lceil 2f_{\max}Q\rceil+1$ of the signal subspace of band-limited and approximately time-limited signal, which is the time-bandwidth dimension [48, 195], and uses the algorithm in Section V of

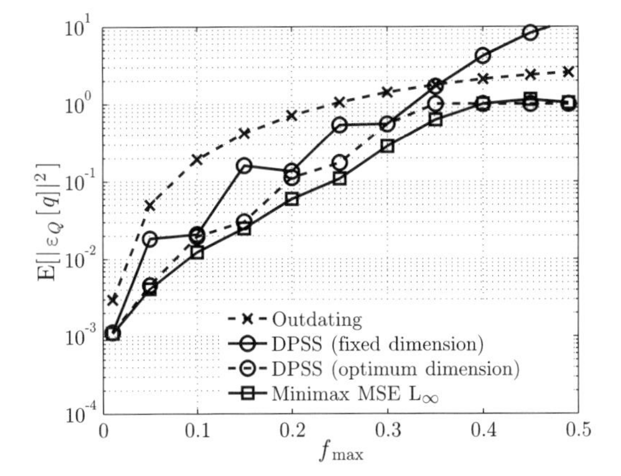

Fig. 2.16 MSE $\mathrm{E}[|\varepsilon_Q[q]|^2]$ vs. maximum frequency $f_{\max}$ for prediction of $h[q]$ based on $Q = 5$ observations $\{\tilde{h}[q-\ell]\}_{\ell=1}^{Q}$: Comparison of minimax prediction (2.115) with prediction based on discrete prolate spheroidal sequences (DPSS) [48, Sec. V]. (Model: Clarke's power spectral density (Figure 2.12) with $c_h[0] = 1$, $c_e = 10^{-3}$, $Q = 5$)

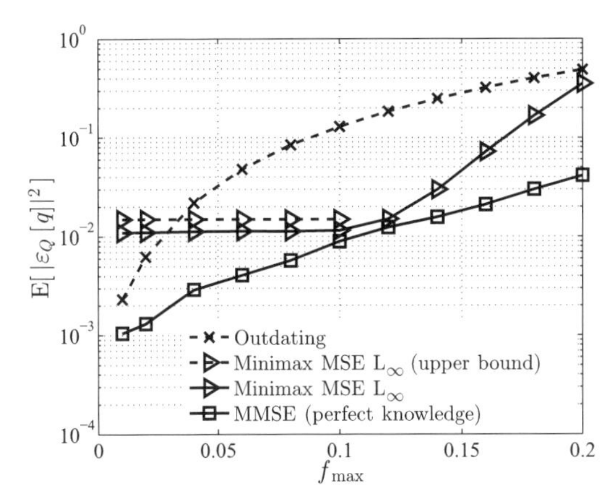

Fig. 2.17 MSE $\mathrm{E}[|\varepsilon_Q[q]|^2]$ vs. maximum frequency $f_{\max}$ for prediction of $h[q]$ based on $Q = 5$ observations $\{\tilde{h}[q-\ell]\}_{\ell=1}^{Q}$: Sensitivity of minimax prediction (2.115) designed for a fixed maximum frequency $f_{\max}^{\mathrm{fix}} = 0.1$ and $\gamma = 20$ to a different nominal maximum frequency $f_{\max}$. (Model: Rectangular power spectral density (Figure 2.13) with $c_h[0] = 1$, $c_e = 10^{-3}$, $Q = 5$)

[48] without any additional regularization. This choice of dimension for the subspace is also proposed in [242], which solves an interpolation problem. The performance of the DPSS based predictor depends on the selected dimension of the signal subspace, which is given from a bias-variance trade-off. For "DPSS (optimum dimension)", the MSE is minimized over all all possible dimensions in the set $\{1, 2, \ldots, Q\}$, i.e., we assume the genie-aided optimum choice of dimension w.r.t. the bias-variance trade-off. The minimax robust predictor for $\mathcal{C}_{h,f_{\max}}^{\infty}$ determines its regularization parameter from the noise

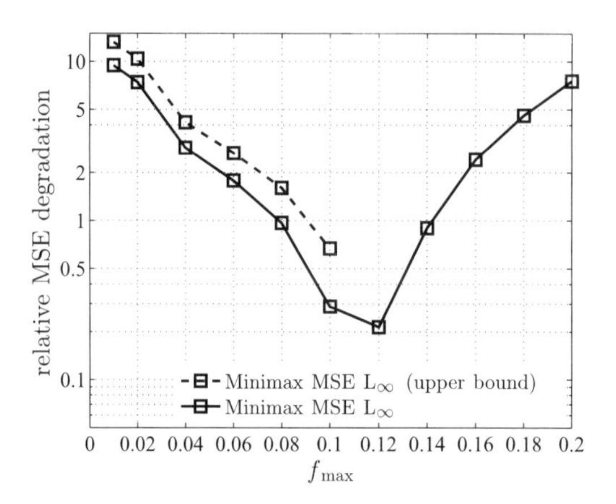

Fig. 2.18 Relative degradation of the MSE $\mathrm{E}[|\varepsilon_Q[q]|^2]$ for minimax prediction (2.115) designed as for Figure 2.17 w.r.t. the MMSE predictor with perfect knowledge of the power spectral density and the nominal maximum frequency $f_{\max}$. (Model: Rectangular power spectral density (Figure 2.13) with $c_h[0] = 1$, $c_e = 10^{-3}$, $Q = 5$)

and signal uncertainty class. Besides its simplicity, its MSE is smaller than for the DPSS based approach (Figure 2.16).

Next we investigate the sensitivity of the minimax robust predictor (2.115) for $\mathcal{C}^{\infty}_{h,f_{\max}}$ to the knowledge of $f_{\max}$. As shown in Figure 2.13, we fix $f_{\max}$ to $f^{\text{fix}}_{\max} = 0.1$ for designing the robust predictor, but $f_{\max}$ of the nominal rectangular power spectral density changes from 0.02 to 0.2. Choosing $\gamma = 20$, the uncertainty set contains the nominal power spectral density for $f_{\max} \in [0.025, 0.1]$ because $c_h[0] = 1 = 2f_{\max}\gamma$ (2.116) has to be satisfied. Thus, the upper bound (2.117) is the guaranteed performance only in the interval $[0.025, 0.1]$ and in Figure 2.17 we give the upper bound only for $f_{\max} \leq 0.1$. The shape of the graphs is also typical for other choices of $f^{\text{fix}}_{\max}$ and types of power spectral densities. For a deviation in MSE (of the mismatched minimax predictor relative to the MMSE with perfect knowledge of the power spectral density) which is smaller than two, the nominal maximum frequency is allowed to deviate from the design point of $f_{\max}$ by -40% or $+60\%$ (Figure 2.18). The relative degradation is approximately symmetric w.r.t. $f_{\max} = 0.1$. Overestimating $f_{\max}$ is less critical because the worst-case MSE is given by (2.117).

Finally, we choose the band-pass power spectral density from Figure 2.12 with support in the intervals $[0.9f_{\max}, f_{\max}]$ and $[-f_{\max}, -0.9f_{\max}]$ and variance $c_h[0] = 1$. This power spectral density deviates significantly from the rectangular power spectral density (Figure 2.12). For example, for $f_{\max} = 0.1$ we have to set $\gamma = 50$ to include the power spectral density into $\mathcal{C}^{\infty}_{h,f_{\max}}$. The MSEs of the minimax predictors as well as their upper bounds increase with $f_{\max}$, as for Clarke's power spectral density (Figure 2.19). But the MSE of the MMSE predictor with perfect knowledge of the power spectral density is

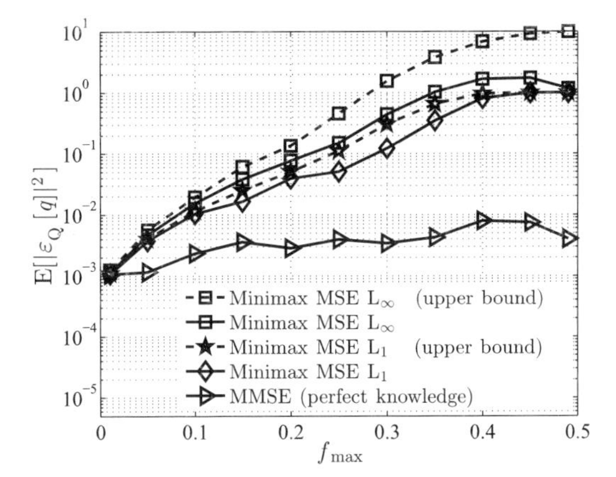

Fig. 2.19 MSE $\mathrm{E}[|\varepsilon_Q[q]|^2]$ vs. maximum frequency $f_{\max}$ for prediction of $h[q]$ based on $Q = 5$ observations $\{\tilde{h}[q-\ell]\}_{\ell=1}^{Q}$. (Model: Band-pass power spectral density (Figure 2.12) with $c_h[0] = 1$, $c_e = 10^{-3}$, $Q = 5$, $\gamma = 1/(f_{\max} - 0.9 f_{\max})/2$)

almost independent of $f_{\max}$. Both minimax predictors are very conservative. For example, the mismatch of the band-pass to the worst-case power spectral density of the uncertainty class $\mathcal{C}_{h,f_{\max}}^{\infty}$ which is rectangular is significant. The choice of a more detailed model for the power spectral density $\mathrm{S}_h(\omega)$ or direct estimation of the autocovariance sequence $c_h[\ell]$, which is presented in Chapter 3, reduces the degradation significantly.

2.4.3.5 Summary of Minimax MSE Prediction

For wireless communications, the L_1-norm is the more appropriate norm for the channel parameters' power spectral density $\mathrm{S}_h(\omega)$ because the variance $c_h[0]$ can be estimated directly. The related robust predictor from (2.104) yields a smaller MSE than its counterpart for the L_∞-norm. Its major disadvantage is the computational complexity for solving (2.104). In general, it is considerably larger (by one order in Q) than for the linear system of equations in (2.115). But, since the results depend only on β and $f_{\max}$, they could be computed offline with sufficient accuracy. On the other hand, the approach based on the L_∞-norm can be modified easily to decrease the size of the corresponding uncertainty class such as in (2.118). This is very important in practice to improve performance in scenarios deviating significantly from the rectangular power spectral density.

Only for $Q \to \infty$, the rectangular power spectral density is least-favorable for $\mathcal{C}_{h,f_{\max}}^{1}$ (Theorem 2.3). Whereas the corresponding partial autocovariance sequence is always least-favorable for $\mathcal{C}_{h,f_{\max}}^{\infty}$ (Theorem 2.4) with generally larger worst-case MSE. Thus, this prominent "robust" choice of the MMSE predictor is not maximally robust for $\mathcal{C}_{h,f_{\max}}^{1}$ but for $\mathcal{C}_{h,f_{\max}}^{\infty}$.

To conclude, we illustrate the robust solutions for the uncertainty sets $\mathcal{C}^1_{h,f_{\max}}$ and $\mathcal{C}^\infty_{h,f_{\max}}$ by an example.

Example 2.3. Consider the case of $Q = 1$, i.e., $\hat{h}[q] = w\tilde{h}[q-1]$. For $\mathcal{C}^1_{h,f_{\max}}$, the robust predictor reads $w = c^{\mathrm{L}}_h[1]/(\beta + c_e)$ with the least-favorable correlation coefficients $c^{\mathrm{L}}_h[0] = \beta$ and $c^{\mathrm{L}}_h[1] = \max(\beta\cos(2\pi f_{\max}), 0)$.[32] The worst-case MSE

$$\max_{\{c_h[\ell]\}_{\ell=0}^1 \in \mathcal{C}^1_{h,f_{\max}}} c_h[0] - |c_h[1]|^2 (c_h[0] + c_e)^{-1} = \\ = \begin{cases} \beta\left(1 - \frac{\beta}{\beta + c_e}\cos^2(2\pi f_{\max})\right), & f_{\max} < 1/4 \\ \beta, & f_{\max} \geq 1/4 \end{cases} \tag{2.119}$$

is bounded by β, which is reached for $f_{\max} \geq 1/4$.

For $\mathcal{C}^\infty_{h,f_{\max}}$, we get $w = 2f_{\max}\gamma \mathrm{sinc}(2f_{\max}) / (2f_{\max}\gamma + c_e)$ and the maximum guaranteed MSE

$$\max_{\{c_h[\ell]\}_{\ell=0}^1 \in \mathcal{C}^\infty_{h,f_{\max}}} c_h[0] - |c_h[1]|^2 (c_h[0] + c_e)^{-1} = \\ = 2f_{\max}\gamma\left(1 - \frac{2f_{\max}\gamma}{2f_{\max}\gamma + c_e}\mathrm{sinc}^2(2f_{\max})\right) \leq 2f_{\max}\gamma \tag{2.120}$$

is bounded by $2f_{\max}\gamma > c_h[0]$.

The maximally robust predictors w differ for $f_{\max} < 0.5$ and the maximum MSE for $\mathcal{C}^\infty_{h,f_{\max}}$ is larger than for $\mathcal{C}^1_{h,f_{\max}}$ for $2f_{\max}\gamma > \beta$. □

---

[32] $c^{\mathrm{L}}_h[1]$ results from $\boldsymbol{C}'_{\mathrm{T}} \succeq 0$ and maximizes the MMSE in (2.119).

Chapter 3
Estimation of Channel and Noise Covariance Matrices

Introduction to Chapter

Estimation of channel and noise covariance matrices is equivalent to estimation of the structured covariance matrix of the receive signal [103, 54, 41, 220]. We start with an overview of the theory behind this problem, previous approaches, and selected aspects.

Estimation of a general *structured covariance matrix* has been studied intensively in the literature:

- The properties of the *Maximum Likelihood* (ML) problem for estimation of a structured covariance matrix are presented by Burg et al.[29] and, in less detail for a linear structure, by Anderson [4, 5]. Both references also propose algorithmic solutions for the generally nonlinear likelihood equation.[1]
- Uniqueness of a ML estimate for doubly symmetric structure is established in [176]. Other structures are treated in [46, 23].
- Robust ML estimation in case of an uncertain probability distribution is discussed by Williams et al.[118].

Covariance matrices with *Toeplitz structure* are an important class in many applications. Several properties of the corresponding ML estimation problem have been described:

- Conditions for existence of an ML estimate are presented by Fuhrmann and Miller [81, 78, 77, 79]. The existence of local maxima of the likelihood function is established by [45]. The EM algorithm is applied to this problem by [81, 149, 47, 150]. This algorithm is generalized to block-Toeplitz matrices in [80, 12, 11]. Finally, estimation of a Toeplitz covariance matrix together with the (non-zero) mean of the random sequence is addressed in [151].

[1] The likelihood equation is the gradient of the likelihood function set to zero.

- Least-squares approximation of sample covariance matrices in the class of Toeplitz matrices is another approach to include information about their structure (see [105] and references). This is based on the extended invariance principle (EXIP) introduced by Stoica and Söderström [204], which is applied to estimation of Toeplitz matrices in [133]. The issue of indefinite estimates in case of least-squares (LS) approximation is mentioned in [13] and, briefly, in [133].

Estimation of *channel and noise covariance matrices* has attracted little attention in wireless communications. In context of CDMA systems an estimator for the channel covariance matrix is presented in [33]. Also for a CDMA system, estimators for the channel and the noise covariance matrix are obtained solving a system of linear equations in [228] which generally yields indefinite estimates. A first solution to the ML problem for estimating the channel covariance matrix and noise variance is given in [60]. It is based on the "expectation-conditional maximization either algorithm". A different approach based on the *space-alternating expectation maximization* (SAGE) algorithm, which also covers the case of spatially correlated noise, is introduced in [220]. The estimate of the noise variance in this approach is superior to [60]. The details of this iterative ML solution and a comparison with [60] are given in Section 3.3 and 3.6, respectively. A computationally less complex estimator based on LS approximation of the sample covariance matrix and the EXIP of [204] yields indefinite estimates of the channel covariance matrix for a small number of observations [54]; this can be avoided including a positive semidefinite constraint, which results in an efficient algorithm [103, 54] (the least-squares criterion was proposed independently by [41] for uncorrelated noise). A computationally very efficient and recursive implementation together with the MMSE channel estimator is possible if the positive semidefinite constraint is not included in the optimization [141].

The previous references address estimation of channel correlations in the space and delay dimensions. But except identification of a classical autoregressive model for the temporal channel correlations [67] or estimation of the maximum Doppler frequency for a given class of autocovariance sequences [211], no results are known to the author which estimate the temporal correlations in wireless communications. We also address this problem in this chapter.

Another interesting problem in estimation of covariance matrices is a scenario where only some elements of the covariance matrix can be inferred from the observations. The remaining unknown elements are to be completed. Consider a Toeplitz covariance matrix with some unspecified diagonals: Obviously, this is an interpolation problem in the class of autocovariance sequences. But it can also be understood as a (covariance) *matrix completion problem*. In this field, similar problems have been addressed already: General conditions for existence and uniqueness of a positive semidefinite completion are provided in [131, 130]; algorithms for maximum entropy completion are derived by Lev-Ari et al.[131].

When estimating channel and noise correlations we assume *stationarity* of the channel and noise process within the observation time. In a wireless channel the channel parameters are only approximately stationary. This can be quantified defining a *stationarity time* [140]. The larger the stationarity time of the channel the more observations of a periodically transmitted training sequence are available to estimate the covariance matrices. On the other hand, the coherence time of the channel is a measure for the time span between two approximately uncorrelated channel realizations. The ratio of stationarity time and coherence time gives the approximate number of uncorrelated channel realizations available for estimation of the covariance matrices. A profound theory for non-stationary channels and the associated stationarity time is developed by Matz [140] for single-input single-output systems. First steps towards a generalization to multiple-input multiple-output (MIMO) channels are presented in [91]. Moreover, only few channel measurement campaigns address this topic [94, 227]. Based on an ad-hoc and application-oriented definition of stationarity, Viering [227] estimates that approximately 100 uncorrelated channel realization are available in urban/suburban scenarios and only 10 realization in an indoor scenario (within a stationarity time period according to his definition). On the other hand, in an 8×8 MIMO system with 64 channel parameters a channel covariance matrix with $64^2 = 4096$ independent and real-valued parameters has to be estimated. Therefore, it is necessary to derive computationally efficient estimators which perform well for a small number of observations.

Outline of Chapter

In Section 3.1 we give a review of previous results on ML estimation of structured covariance matrices, which provide the foundation for our estimation problem. The necessary assumptions and the system model for estimation of the channel and noise covariance matrices are introduced in Section 3.2. Based on the SAGE algorithm for ML estimation we propose estimators for the channel correlations in space, delay, and time domain as well as the noise correlations (Section 3.3). A new approach to the completion of a partially known or observable band-limited autocovariance sequence is presented in Section 3.4 with application to MMSE prediction. Efficient estimators based on LS approximation and the EXIP, which we derive in Section 3.5, provide almost optimum estimates for the channel correlation in space and delay domain as well as the spatial noise correlations. Finally, the performance of all estimators is compared in Section 3.6.

3.1 Maximum Likelihood Estimation of Structured Covariance Matrices: A Review

3.1.1 General Problem Statement and Properties

Given B statistically independent observations $\boldsymbol{x}[q], q = 1, 2, \ldots, B$, of the M-dimensional random vector $\boldsymbol{x}$, we would like to estimate its covariance matrix $\boldsymbol{C}_{\boldsymbol{x}} = \mathrm{E}[\boldsymbol{x}\boldsymbol{x}^{\mathrm{H}}] \in \mathbb{M} \subseteq \mathbb{S}_{+}^{M}$. If the set of admissible covariance matrices $\mathbb{M}$ is a proper subset of the set of all positive semidefinite matrices $\mathbb{S}_{+,0}^{M}$ ($\mathbb{M} \subset \mathbb{S}_{+,0}^{M}$), we call it a *structured covariance matrix.*

Assuming $\boldsymbol{x} \sim \mathcal{N}_{\mathrm{c}}(\boldsymbol{0}_M, \boldsymbol{C}_{\boldsymbol{x}})$ the *Maximum Likelihood* (ML) method maximizes the likelihood for given observations $\{\boldsymbol{x}[q]\}_{q=1}^{B}$

$$\hat{\boldsymbol{C}}_{\boldsymbol{x}} = \underset{\boldsymbol{C}_{\boldsymbol{x}} \in \mathbb{M}}{\operatorname{argmax}} \prod_{q=1}^{B} \mathrm{p}_{\boldsymbol{x}}(\boldsymbol{x}[q]; \boldsymbol{C}_{\boldsymbol{x}}). \tag{3.1}$$

It is often chosen due to its asymptotic properties:

- The estimate $\hat{\boldsymbol{C}}_{\boldsymbol{x}}$ is asymptotically (for large B) unbiased and consistent.
- The estimate $\hat{\boldsymbol{C}}_{\boldsymbol{x}}$ is asymptotically efficient, i.e., the variances for all matrix elements achieve the Cramér-Rao lower bound [168, 124].

For the zero-mean complex Gaussian assumption, the log-likelihood function reads (A.1)

$$\ln\left[\prod_{q=1}^{B} \mathrm{p}_{\boldsymbol{x}}(\boldsymbol{x}[q]; \boldsymbol{C}_{\boldsymbol{x}})\right] = -MB\ln[\pi] - B\ln[\det \boldsymbol{C}_{\boldsymbol{x}}] - B\,\mathrm{tr}\left[\tilde{\boldsymbol{C}}_{\boldsymbol{x}}\boldsymbol{C}_{\boldsymbol{x}}^{-1}\right]. \tag{3.2}$$

The sample covariance matrix $\tilde{\boldsymbol{C}}_{\boldsymbol{x}} = \frac{1}{B}\sum_{q=1}^{B} \boldsymbol{x}[q]\boldsymbol{x}[q]^{\mathrm{H}}$ is a sufficient statistic for estimating $\boldsymbol{C}_{\boldsymbol{x}}$.

The *properties* of optimization problem (3.1) for different choices of $\mathbb{M}$ and assumptions on $\tilde{\boldsymbol{C}}_{\boldsymbol{x}}$, which can be found in the literature, are summarized in the sequel:

1. If $\tilde{\boldsymbol{C}}_{\boldsymbol{x}} \in \mathbb{S}_{+}^{M}$, i.e., $B \geq M$ necessarily, and $\mathbb{M} \subseteq \mathbb{S}_{+,0}^{M}$ is closed, then (3.1) has positive definite solutions $\hat{\boldsymbol{C}}_{\boldsymbol{x}}$ [29]. But if $\tilde{\boldsymbol{C}}_{\boldsymbol{x}}$ is singular, the *existence* of a positive definite solution is not guaranteed, although very often (3.1) will yield a positive definite estimate (see Section 3.1.2).
   In [81, 78, 149] conditions are given under which the likelihood function (3.2) is unbounded above, i.e., the solution of (3.1) is possibly singular. A necessary condition is that $\tilde{\boldsymbol{C}}_{\boldsymbol{x}}$ must be singular. Moreover, whether the likelihood function for realizations $\{\boldsymbol{x}[q]\}_{q=1}^{B}$ of the random vector $\boldsymbol{x}$ is unbounded depends on the particular realizations observed, i.e., the *probability* of the likelihood to be unbounded may be non-zero.

2. If $\boldsymbol{C}_{\boldsymbol{x}}$ does not have structure, i.e., $\mathbb{M} = \mathbb{S}_{+,0}^{M}$, the solution is unique and given by the sample covariance matrix [29]

$$\hat{\boldsymbol{C}}_{\boldsymbol{x}} = \tilde{\boldsymbol{C}}_{\boldsymbol{x}}. \tag{3.3}$$

The *likelihood equation* in $\boldsymbol{C}_{\boldsymbol{x}}$ follows from the necessary condition that the gradient of the log-likelihood function w.r.t. $\boldsymbol{C}_{\boldsymbol{x}}$ must be zero and reads

$$\boxed{\tilde{\boldsymbol{C}}_{\boldsymbol{x}} = \boldsymbol{C}_{\boldsymbol{x}}.} \tag{3.4}$$

If $\mathbb{M} = \mathbb{S}_{+,0}^{M}$, we obtain (3.3). For $\boldsymbol{C}_{\boldsymbol{x}} \in \mathbb{M} \subset \mathbb{S}_{+,0}^{M}$, it does not have a solution in general.[2] Very often (3.4) is only satisfied with probability zero, but the probability increases with B (see Section 3.5). If $\tilde{\boldsymbol{C}}_{\boldsymbol{x}} \in \mathbb{M}$ and we have a unique parameterization of $\mathbb{M}$, then the unstructured ML estimate $\hat{\boldsymbol{C}}_{\boldsymbol{x}} = \tilde{\boldsymbol{C}}_{\boldsymbol{x}}$ is also the ML estimate for the restricted class of covariance matrices $\boldsymbol{C}_{\boldsymbol{x}} \in \mathbb{M}$ due to the invariance principle [124, 46].
3. In general, the necessary conditions, i.e., the likelihood equation, for a solution of (3.1) based on the gradient form a nonlinear system of equations. For example, the necessary conditions for c_i in case of a linear structure $\boldsymbol{C}_{\boldsymbol{x}} = \sum_{i=1}^{V} c_i \boldsymbol{S}_i$ are [4, 5]

$$\operatorname{tr}\left[\boldsymbol{C}_{\boldsymbol{x}}^{-1}(\boldsymbol{C}_{\boldsymbol{x}} - \tilde{\boldsymbol{C}}_{\boldsymbol{x}})\boldsymbol{C}_{\boldsymbol{x}}^{-1}\boldsymbol{S}_i\right] = 0, \ i = 1, 2, \ldots, V. \tag{3.5}$$

Iterative solutions were proposed based on the Newton-Raphson algorithm [4] or inverse iterations [29].
4. According to the *extended invariance principle* (EXIP) for parameter estimation [204], an asymptotic (for large B) ML estimate of a structured covariance matrix in $\mathbb{M}$ can be obtained from a weighted least-squares approximation of the sample covariance matrix $\tilde{\boldsymbol{C}}_{\boldsymbol{x}}$. For a complex Gaussian distribution of $\boldsymbol{x}$ (3.2) and ML estimation, the asymptotically equivalent problem reads [133]

$$\boxed{\min_{\boldsymbol{C}_{\boldsymbol{x}} \in \mathbb{M}} \|\tilde{\boldsymbol{C}}_{\boldsymbol{x}} - \boldsymbol{C}_{\boldsymbol{x}}\|_{\tilde{\boldsymbol{C}}_{\boldsymbol{x}}^{-1}}^{2} = \min_{\boldsymbol{C}_{\boldsymbol{x}} \in \mathbb{M}} \operatorname{tr}\left[(\tilde{\boldsymbol{C}}_{\boldsymbol{x}} - \boldsymbol{C}_{\boldsymbol{x}})\tilde{\boldsymbol{C}}_{\boldsymbol{x}}^{-1}(\tilde{\boldsymbol{C}}_{\boldsymbol{x}} - \boldsymbol{C}_{\boldsymbol{x}})\tilde{\boldsymbol{C}}_{\boldsymbol{x}}^{-1}\right].} \tag{3.6}$$

Here, the weighting matrices $\tilde{\boldsymbol{C}}_{\boldsymbol{x}}^{-1}$ correspond to the inverses of consistent estimates for the error covariance matrix of $\tilde{\boldsymbol{c}}_{\boldsymbol{x}} = \mathbf{vec}[\tilde{\boldsymbol{C}}_{\boldsymbol{x}}]$ which is $\boldsymbol{C}_{\tilde{\boldsymbol{c}}_{\boldsymbol{x}}} = \frac{1}{B}(\boldsymbol{C}_{\boldsymbol{x}}^{\mathrm{T}} \otimes \boldsymbol{C}_{\boldsymbol{x}})$.[3]
In general, the EXIP aims at a simple solution of the ML problem (3.1) by first extending the set $\mathbb{M}$ (here to $\mathbb{S}_{+,0}^{M}$) such that a simple solution (here:

[2] It is not the likelihood equation for this constrained set $\mathbb{M}$.

[3] Precisely, (3.6) results from (A.15) and the consistent estimate $\tilde{\boldsymbol{C}}_{\boldsymbol{x}}$ of $\boldsymbol{C}_{\boldsymbol{x}}$ which yield $\|\tilde{\boldsymbol{c}}_{\boldsymbol{x}} - \boldsymbol{c}_{\boldsymbol{x}}\|_{\boldsymbol{C}_{\tilde{\boldsymbol{c}}_{\boldsymbol{x}}}^{-1}}^{2} = B\|\tilde{\boldsymbol{C}}_{\boldsymbol{x}} - \boldsymbol{C}_{\boldsymbol{x}}\|_{\tilde{\boldsymbol{C}}_{\boldsymbol{x}}^{-1}}^{2}$ for $\boldsymbol{c}_{\boldsymbol{x}} = \mathbf{vec}[\boldsymbol{C}_{\boldsymbol{x}}]$.

$\tilde{\boldsymbol{C}}_{\boldsymbol{x}}$) is obtained. In the second step, a weighted least squares approximation is performed; the weighting matrix describes the curvature of the original (ML) cost function.

Let us consider a few more examples for classes of covariance matrices $\mathbb{M}$ for which the solution of (3.1) has been studied.

Example 3.1. If the eigenvectors of $\boldsymbol{C}_{\boldsymbol{x}}$ are known, i.e., $\mathbb{M} = \{\boldsymbol{C}_{\boldsymbol{x}} | \boldsymbol{C}_{\boldsymbol{x}} = \boldsymbol{U}_{\boldsymbol{x}} \boldsymbol{\Lambda}_{\boldsymbol{x}} \boldsymbol{U}_{\boldsymbol{x}}^{\mathrm{H}}, \boldsymbol{\Lambda}_{\boldsymbol{x}} \in \mathbb{D}_{+,0}^{M}\}$, and only the eigenvalues have to be estimated, the ML solution is $\hat{\boldsymbol{C}}_{\boldsymbol{x}} = \boldsymbol{U}_{\boldsymbol{x}} \hat{\boldsymbol{\Lambda}}_{\boldsymbol{x}} \boldsymbol{U}_{\boldsymbol{x}}$ with $[\hat{\boldsymbol{\Lambda}}_{\boldsymbol{x}}]_{i,i} = [\boldsymbol{U}_{\boldsymbol{x}}^{\mathrm{H}} \tilde{\boldsymbol{C}}_{\boldsymbol{x}} \boldsymbol{U}_{\boldsymbol{x}}]_{i,i}$.

For circulant covariance matrices $\boldsymbol{C}_{\boldsymbol{x}}$, which can be defined as

$$\mathbb{M} = \{\boldsymbol{C}_{\boldsymbol{x}} | \boldsymbol{C}_{\boldsymbol{x}} \in \mathbb{T}_{+}^{M}, \boldsymbol{C}_{\boldsymbol{x}}^{-1} \in \mathbb{T}_{+}^{M}\} = \mathbb{T}_{\mathrm{c}+}^{M}, \tag{3.7}$$

this estimate is equivalent to the periodogram estimate of the power spectral density [46] because the eigenvectors of a circulant matrix form the FFT matrix $[\boldsymbol{U}_{\boldsymbol{x}}]_{m,n} = \exp(-\mathrm{j}2\pi(m-1)(n-1)/M)/\sqrt{M}, m,n \in \{1,2,\ldots,M\}$. Note that a (wide-sense) periodic random sequence with period M, i.e., its first and second order moments are periodic [200, p. 408], is described by a circulant covariance matrix $\boldsymbol{C}_{\boldsymbol{x}}$. □

Example 3.2. If $\tilde{\boldsymbol{C}}_{\boldsymbol{x}} \in \mathbb{S}_{+}^{M}$, $\mathbb{M} \subseteq \mathbb{S}_{+,0}^{M}$ is closed, and $\{\boldsymbol{C}_{\boldsymbol{x}}^{-1} | \boldsymbol{C}_{\boldsymbol{x}} \in \mathbb{M} \cap \mathbb{S}_{+}^{M}\}$ is convex, then (3.1) has a unique positive definite solution [176].

An example for $\mathbb{M}$ with these properties is the set of positive semidefinite doubly symmetric matrices, i.e., matrices symmetric w.r.t. their main and anti-diagonal. A subset for this choice of $\mathbb{M}$ is the set of positive semidefinite Toeplitz matrices. But for Toeplitz matrices the set of inverses of Toeplitz matrices is not a convex set [176]. □

Example 3.3. For the important class of Toeplitz covariance matrices $\mathbb{M} = \mathbb{T}_{+,0}^{M}$, no analytical solution to (3.1) is known. In [45] it is proved that (3.2) can have local maxima for $M = 3$ and examples are given for $M \in \{3,4,5\}$. The local maxima may be close to the global optimum. Thus, a good initialization is required for iterative algorithms. We discuss the question of existence of a positive definite estimate $\hat{\boldsymbol{C}}_{\boldsymbol{x}}$ in the next section. □

3.1.2 Toeplitz Structure: An Ill-Posed Problem and its Regularization

For a better understanding of the issues associated with estimation of covariance matrices with Toeplitz structure, we first summarize some important properties of Toeplitz matrices.

- Consider a Hermitian Toeplitz matrix with first row $[c[0], c[1], \ldots, c[M-1]]$. The *eigenvalues* of a Hermitian Toeplitz matrix can be bounded from below and above by the (essential) infimum and supremum, respectively, of a function whose inverse discrete Fourier transform yields the matrix elements $\{c[i]\}_{i=0}^{M-1}$. If a power spectral density which is zero only on an interval of measure zero [85], i.e., only at a finite number of points, the corresponding positive semidefinite Toeplitz matrix is regular. Conversely, if a positive semidefinite Toeplitz matrix is singular, its elements can be associated with a power spectral density which is zero on an interval of measure greater than zero [85]. Random sequences with such a power spectral density are referred to as "deterministic" random sequences (see also Section 2.3.2). But the following result from Caratheodory is even stronger.

- According to the *representation theorem of Caratheodory* (e.g. [81, 86]) every positive semidefinite Toeplitz matrix has a unique decomposition

$$\boldsymbol{C}_{\mathsf{x}} = \sum_{i=1}^{N} p_i \boldsymbol{\gamma}_i \boldsymbol{\gamma}_i^{\mathrm{H}} + \sigma^2 \boldsymbol{I}_M, \tag{3.8}$$

where $p_i > 0$, $N < M$, $\sigma^2 \geq 0$, and

$$\boldsymbol{\gamma}_i = [1, \exp(\mathrm{j}\omega_i), \exp(\mathrm{j}2\omega_i), \ldots, \exp(\mathrm{j}(M-1)\omega_i)]^{\mathrm{T}} \tag{3.9}$$

with frequencies $\omega_i \in [-\pi, \pi]$. For singular $\boldsymbol{C}_{\mathsf{x}}$, we have $\sigma^2 = 0$.
For $\mathrm{rank}[\boldsymbol{C}_{\mathsf{x}}] = R < M$, there exists a unique extension of $\{c[i]\}_{i=0}^{M-1}$ on $\mathbb{Z}$ composed of R complex exponentials [6, 163] which is a valid covariance sequence. The associated power spectral density consists only of N discrete frequency components $\{\omega_i\}_{i=1}^{N}$.[4]

From the both properties we conclude: Assuming that $\boldsymbol{C}_{\mathsf{x}}$ describes a stochastic process which is not composed by fewer than M complex exponentials or is a regular (non-deterministic) stochastic process, we would like to ensure a positive definite estimate $\hat{\boldsymbol{C}}_{\mathsf{x}}$.

For estimating positive definite covariance matrix $\boldsymbol{C}_{\mathsf{x}}$ with Toeplitz structure, several theorems were presented by Fuhrmann et al.[81, 78, 79, 149, 77]. The relevant results are reviewed in the sequel.

Theorem 3.1 (Fuhrmann [77]). *For statistically independent $\{\mathbf{x}[q]\}_{q=1}^{B}$ with $\mathbf{x} \sim \mathcal{N}(\mathbf{0}_M, \boldsymbol{C}_{\mathsf{x}})$ and $\boldsymbol{C}_{\mathsf{x}} \in \mathbb{T}_{+}^{M}$, the ML estimate (3.1) of $\boldsymbol{C}_{\mathsf{x}}$ yields a positive definite solution if and only if $B \geq \lceil M/2 \rceil$.*

If $M \geq 3$, we already need $B \geq 2$ statistically independent observations. But if $\boldsymbol{C}_{\mathsf{x}}$ contains the samples of the temporal autocovariance function of a random process, the adequate choice to describe the observations is $B = 1$ because successive samples of the random process are correlated and statistically independent observations are often not available. The next theo-

[4] This theorem has been, for example, in array processing and spectral estimation [203, 105].

rem treats this case for the example of estimating the very simple choice of $\boldsymbol{C}_{\mathsf{x}} = c\boldsymbol{I}_M$, which is a Toeplitz covariance matrix.

Theorem 3.2 (Fuhrmann and Barton [79]). *For $B = 1$, $\mathsf{x}[1] \in \mathbb{C}^M$, and $\boldsymbol{C}_{\mathsf{x}} = c\boldsymbol{I}_M$ the ML estimate is given by (3.1) with $\mathbb{M} = \mathbb{T}^M_{+,0}$.[5] The probability $\mathcal{P}$ that the ML estimate is positive definite is upper bounded as*

$$\mathcal{P} \leq \frac{2^{-(M-2)}}{(M-1)!}. \tag{3.10}$$

As the length M of the data vector increases we get

$$\lim_{M\to\infty} \mathcal{P} = 0. \tag{3.11}$$

The bound on the probability is one for $M = 2$ as required by the previous theorem. For large data length M, it is very likely that we do not obtain a positive definite estimate, i.e., even in the asymptotic case $M \to \infty$ the estimate of the covariance matrix does not converge to the true positive definite covariance matrix.[6]

For a real-valued random sequence $\mathsf{x}[q] \in \mathbb{R}^M$, more general results are available:

Theorem 3.3 (Fuhrmann and Miller [81, 78]). *For $B = 1$ and $\mathsf{x}[q] \sim \mathcal{N}(\mathbf{0}_M, \boldsymbol{C}_{\mathsf{x}})$ with circulant $\boldsymbol{C}_{\mathsf{x}} \in \mathbb{T}^M_{c+}$, the probability $\mathcal{P}$ that the ML estimate (3.1) is positive definite for $\mathbb{M} = \mathbb{T}^M_{+,0}$ is bounded by*

$$\mathcal{P} \leq \frac{1}{2^{M-2}}. \tag{3.12}$$

For $\boldsymbol{C}_{\mathsf{x}}$ with general Toeplitz structure, a similar result is given in [81], but the bound decreases more slowly with M. Due to the closeness of Toeplitz to circulant matrices [81, 85] it can be expected that the probability of having a positive definite estimate for $\boldsymbol{C}_{\mathsf{x}} \in \mathbb{T}^M_{+,0}$ and $\mathbb{M} = \mathbb{T}^M_{+,0}$ does not deviate significantly from the bound given for circulant covariance matrices.

The ML problem (3.1) is typically ill-posed because the probability of obtaining the required positive definite solution becomes arbitrarily small for increasing M. Thus, a regularization of the problem becomes necessary. Fuhrmann and Miller [81] propose to restrict the set $\mathbb{M}$ from Toeplitz matrices $\mathbb{T}^M_{+,0}$ to contain only $M \times M$ Toeplitz matrices with positive semidefinite circulant extension of period $\bar{M} > M$. We denote this set as $\mathbb{T}^M_{\text{ext}+,0} \subset \mathbb{T}^M_{+,0}$. By this definition we mean that for all $\boldsymbol{C}_{\mathsf{x}} \in \mathbb{T}^M_{\text{ext}+,0}$ a positive semidefinite circulant matrix in $\mathbb{T}^{\bar{M}}_{c+,0}$ exists whose upper-left $M \times M$ block is $\boldsymbol{C}_{\mathsf{x}}$. With this constrained set $\mathbb{T}^M_{\text{ext}+,0}$ the ML estimate (3.1) has the desired positive definite property as stated by the next theorem.

[5] Although $c\boldsymbol{I}_M \subset \mathbb{T}^M_{+,0}, c \geq 0$, optimization is performed over $\mathbb{M} = \mathbb{T}^M_{+,0}$.

[6] Note that the bound decreases exponentially with M.

Theorem 3.4 (Fuhrmann and Miller [81]). *For $B \geq 1$ and $\mathbf{x}[q] \sim \mathcal{N}(\mathbf{0}_M, \boldsymbol{C}_{\mathbf{x}})$ with $\boldsymbol{C}_{\mathbf{x}} \in \mathbb{S}_{+}^{M}$, the probability $\mathcal{P}$ that the ML estimate* (3.1) *is positive definite for $\mathbb{M} = \mathbb{T}_{\text{ext}+,0}^{M}$ is one.*

For this reason, $\mathbb{M} = \mathbb{T}_{\text{ext}+,0}^{M}$ is a good choice in (3.1) if $\boldsymbol{C}_{\mathbf{x}} \in \mathbb{T}_{+}^{M}$.

For example, the minimal (Hermitian) circulant extension with period $\bar{M} = 2M - 1 = 5$ of the Toeplitz matrix

$$\boldsymbol{C}_{\mathbf{x}} = \begin{bmatrix} c[0] & c[1] & c[2] \\ c[1]^* & c[0] & c[1] \\ c[2]^* & c[1]^* & c[0] \end{bmatrix} \tag{3.13}$$

is given by

$$\boldsymbol{C}_{\bar{\mathbf{x}}} = \begin{bmatrix} c[0] & c[1] & c[2] & c[2]^* & c[1]^* \\ c[1]^* & c[0] & c[1] & c[2] & c[2]^* \\ c[2]^* & c[1]^* & c[0] & c[1] & c[2] \\ c[2] & c[2]^* & c[1]^* & c[0] & c[1] \\ c[1] & c[2] & c[2]^* & c[1]^* & c[0] \end{bmatrix} \in \mathbb{T}_{\text{c}}^{5}, \tag{3.14}$$

which is indefinite in general. Choosing $\bar{M} > 2M - 1$ the circulant extension is not unique, e.g., generated padding with zeros. The question arises, for which $\bar{M}$ a positive semidefinite circulant extension exists and how it can be constructed. This will also give an answer to the question, whether we introduce a bias when restricting $\mathbb{M}$ in this way.

The eigenvalue decomposition of the $\bar{M} \times \bar{M}$ circulant matrix $\boldsymbol{C}_{\bar{\mathbf{x}}}$ is

$$\boldsymbol{C}_{\bar{\mathbf{x}}} = \boldsymbol{V}\boldsymbol{D}\boldsymbol{V}^{\text{H}}, \tag{3.15}$$

where $\boldsymbol{D} = \mathbf{diag}[d_1, d_2, \ldots, d_{\bar{M}}]$ is a diagonal matrix with nonnegative elements and the columns of $\boldsymbol{V} = [\boldsymbol{V}_M^{\text{T}}, \bar{\boldsymbol{V}}_M^{\text{T}}]^{\text{T}}$, with $[\boldsymbol{V}]_{k,\ell} = \exp(-\text{j}2\pi(k-1)(\ell-1)/\bar{M})$ and $\boldsymbol{V}_M \in \mathbb{C}^{M \times \bar{M}}$, are the eigenvectors. Therefore, we can represent all $\boldsymbol{C}_{\mathbf{x}} \in \mathbb{T}_{\text{ext}+,0}^{M}$ as

$$\boldsymbol{C}_{\mathbf{x}} = \boldsymbol{V}_M \boldsymbol{D} \boldsymbol{V}_M^{\text{H}} = \sum_{\ell=1}^{\bar{M}} d_\ell \boldsymbol{v}_{M,\ell} \boldsymbol{v}_{M,\ell}^{\text{H}}, \tag{3.16}$$

where $\boldsymbol{v}_{M,\ell} = [1, \exp(-\text{j}2\pi(\ell-1)/\bar{M}), \ldots, \exp(-\text{j}2\pi(M-1)(\ell-1)/\bar{M})]^{\text{T}}$ is the ℓth column of $\boldsymbol{V}_M$. This parameterization shows a close relation to (3.8) from the theorem of Caratheodory: Here, we have $\omega_\ell = -2\pi(\ell-1)/\bar{M}$ and $\bar{M} > M$, i.e., a restriction to a fixed set of equidistant discrete frequencies.

The extension of $\mathbf{x}$ to dimension $\bar{M}$ is denoted $\bar{\mathbf{x}} \in \mathbb{C}^{\bar{M}}$, where the elements of $\bar{\mathbf{x}}$ form a period of a periodic random sequence with autocovariance sequence $c[\ell]$. $\{c[\ell]\}_{\ell=0}^{\bar{M}-1}$ is the extension of $\{c[\ell]\}_{\ell=0}^{M-1}$ to one full period of the periodic autocovariance sequence of the elements in $\boldsymbol{x}$.[7]

Dembo et al.[47] and Newsam et al.[155] give sufficient conditions on $\boldsymbol{C}_{\boldsymbol{x}}$ to be element of $\mathbb{T}_{\text{ext}+,0}^{M}$: They derive lower bounds on $\bar{M}$ which depend on the observation length M and the minimum eigenvalue of $\boldsymbol{C}_{\boldsymbol{x}}$ for positive definite $\boldsymbol{C}_{\boldsymbol{x}}$. The bound increases with decreasing smallest eigenvalue and for a singular $\boldsymbol{C}_{\boldsymbol{x}}$ a periodic extension does not exist. Their constructive proofs also provide algorithms which construct an extension. But the extension of $\{c[\ell]\}_{\ell=0}^{M-1}$ is not unique. And if the bounds are loose, the computational complexity of the corresponding ML estimator increases. It has not been addressed in the literature, how a positive semidefinite circulant extension can be constructed or chosen that would be a good initialization for the algorithms in Section 3.3. We address this issue at the end of Sections 3.3.3 and 3.3.5.

3.2 Signal Models for Estimation of Channel and Noise Covariance Matrices

For the purpose of estimating channel and noise covariance matrices, we define the underlying signal model and its probability distribution in this section. It follows from the detailed derivations presented in Section 2.1. Therefore, we only summarize the necessary definitions in the sequel.

The observation in block q is (cf. (2.4))

$$\boldsymbol{y}[q] = \boldsymbol{S}\boldsymbol{h}[q] + \boldsymbol{n}[q] \in \mathbb{C}^{MN}, \tag{3.17}$$

where $\boldsymbol{S} = \bar{\boldsymbol{S}}^{\mathrm{T}} \otimes \boldsymbol{I}_M$ contains the training sequence in a wireless communication system.[8] As in (2.2) and (2.3) we partition it into

$$\boldsymbol{n}[q] = [\boldsymbol{n}[q,1]^{\mathrm{T}}, \boldsymbol{n}[q,2]^{\mathrm{T}}, \ldots, \boldsymbol{n}[q,N]^{\mathrm{T}}]^{\mathrm{T}} \tag{3.18}$$

and equivalently for $\boldsymbol{y}[q]$.

The zero-mean additive noise $\boldsymbol{n}[q]$ is complex Gaussian distributed with covariance matrix

$$\boldsymbol{C}_{\boldsymbol{n}} = \boldsymbol{I}_N \otimes \boldsymbol{C}_{\boldsymbol{n},\mathrm{s}}, \tag{3.19}$$

i.e., uncorrelated in time and correlated over the M receiver dimensions with $\boldsymbol{C}_{\boldsymbol{n},\mathrm{s}} = \mathrm{E}[\boldsymbol{n}[q,n]\boldsymbol{n}[q,n]^{\mathrm{H}}]$. As as special case, we also consider spatially and temporally uncorrelated noise with $\boldsymbol{C}_{\boldsymbol{n}} = c_n \boldsymbol{I}_M$.

---

[7] $[c[0], c[1], \ldots, c[M-1]]$ is the first row of $\boldsymbol{C}_{\boldsymbol{x}}$.

[8] In general the rows of $\bar{\boldsymbol{S}} \in \mathbb{C}^{K(L+1)\times N}$ form a (known) basis of the signal subspace in $\mathbb{C}^N$ (compare with (2.3)).

Summarizing the observations of B blocks, we obtain

$$\boldsymbol{y}_{\mathrm{T}}[q] = \boldsymbol{S}_{\mathrm{T}}\boldsymbol{h}_{\mathrm{T}}[q] + \boldsymbol{n}_{\mathrm{T}}[q] \in \mathbb{C}^{BMN} \tag{3.20}$$

with definitions $\boldsymbol{y}_{\mathrm{T}}[q] = [\boldsymbol{y}[q-\ell_1]^{\mathrm{T}}, \boldsymbol{y}[q-\ell_2]^{\mathrm{T}}, \ldots, \boldsymbol{y}[q-\ell_B]^{\mathrm{T}}]^{\mathrm{T}}$, $\boldsymbol{n}_{\mathrm{T}}[q] = [\boldsymbol{n}[q-\ell_1]^{\mathrm{T}}, \boldsymbol{n}[q-\ell_2]^{\mathrm{T}}, \ldots, \boldsymbol{n}[q-\ell_B]^{\mathrm{T}}]^{\mathrm{T}}$, $\boldsymbol{h}_{\mathrm{T}}[q] = [\boldsymbol{h}[q-\ell_1]^{\mathrm{T}}, \boldsymbol{h}[q-\ell_2]^{\mathrm{T}}, \ldots, \boldsymbol{h}[q-\ell_B]^{\mathrm{T}}]^{\mathrm{T}} \in \mathbb{C}^{BP}$, and $\boldsymbol{S}_{\mathrm{T}} = \boldsymbol{I}_B \otimes \boldsymbol{S}$ as in (2.8) and (2.9). We assume that the observations are sorted according to $\ell_i < \ell_j$ for $i < j$, e.g., $\ell_i = i$. Furthermore, it follows that the noise covariance matrix is $\boldsymbol{C}_{\boldsymbol{n}_{\mathrm{T}}} = \boldsymbol{I}_{BN} \otimes \boldsymbol{C}_{\boldsymbol{n},\mathrm{s}}$. The channel parameters $\boldsymbol{h}_{\mathrm{T}}[q]$ are assumed to be zero mean for simplicity.

The channel covariance matrix $\boldsymbol{C}_{\boldsymbol{h}_{\mathrm{T}}}$ is *block Toeplitz*, i.e, $\boldsymbol{C}_{\boldsymbol{h}_{\mathrm{T}}} \in \mathbb{T}_{+,0}^{P,B}$, if the observations $\boldsymbol{h}[q-\ell_i]$ are equidistant.[9] More detailed models are discussed in Section 2.1.

For estimation of the channel and noise covariance matrix, we focus on two models for the observations and the corresponding channel covariance matrix:

1. **Model 1:** I statistically independent observations of $\boldsymbol{y}_{\mathrm{T}}[q]$ are available. As an example, we choose $\ell_i = i$ and $q \in \mathbb{O}_{\mathrm{T}} = \{1, B+1, \ldots, (I-1)B+1\}$ in the sequel if not stated otherwise. Thus, we assume that adjacent observations of frames containing B blocks are statistically independent. If a frame $\boldsymbol{y}_{\mathrm{T}}[q]$ containing B blocks is not well separated from the next frame $\boldsymbol{y}_{\mathrm{T}}[q+1]$, this assumption serves as an approximation to reduce the complexity of estimation algorithms in the next section. The associated channel covariance matrix $\boldsymbol{C}_{\boldsymbol{h}_{\mathrm{T}}}$ to be estimated is block Toeplitz as in (2.10).
2. **Model 2:** This is a special case of model 1. Although some aspects and derivations will be redundant, we introduce it with its own notation due to its importance for estimating the channel correlations in space and delay domain. Estimation of the covariance matrices is based on B statistically independent observations of $\boldsymbol{y}[q]$ (3.17) for $q \in \mathbb{O} = \{1, 2, \ldots, B\}$ $(I = 1)$. This corresponds to the assumption of a block diagonal covariance matrix $\boldsymbol{C}_{\boldsymbol{h}_{\mathrm{T}}} = \boldsymbol{I}_B \otimes \boldsymbol{C}_{\boldsymbol{h}}$ with identical blocks due to stationarity of the channel. We would like to estimate the covariance matrix $\boldsymbol{C}_{\boldsymbol{h}}$ which has no additional structure besides being positive semidefinite.

For ML estimation, the observations $\{\boldsymbol{y}_{\mathrm{T}}[q]\}_{q=1}^{I}$ and $\{\boldsymbol{y}[q]\}_{q=1}^{B}$ are zero-mean and complex Gaussian. The log-likelihood function for *model 1* reads (cf. (A.1))

[9] $\ell_i - \ell_{i-1}$ is constant for all $2 \leq i \leq B$.

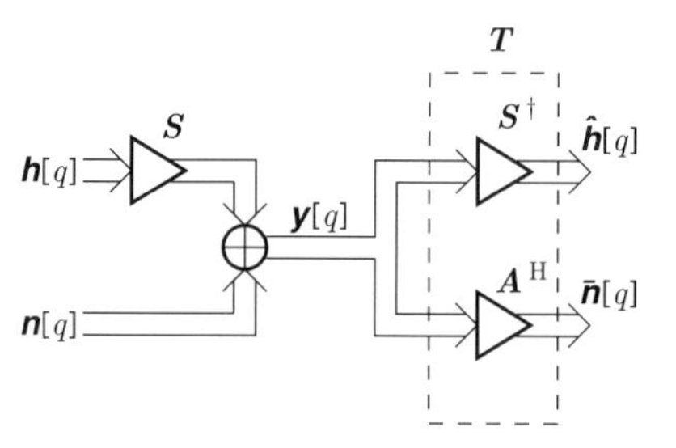

Fig. 3.1 System model for estimation of $\boldsymbol{C_h}$ and $\boldsymbol{C}_{\boldsymbol{n},\mathrm{s}}$ with hidden data spaces $\boldsymbol{h}[q]$ and $\boldsymbol{n}[q]$ (Section 3.3.1); transformation of the observation $\boldsymbol{y}[q]$ with $\boldsymbol{T}$ yields the sufficient statistic $\hat{\boldsymbol{h}}[q]$ and $\bar{\boldsymbol{n}}[q]$.

$$\begin{aligned}
\mathrm{L}_{\boldsymbol{y}_\mathrm{T}}(\{\boldsymbol{y}_\mathrm{T}[q]\}_{q\in\mathbb{O}_\mathrm{T}};\boldsymbol{C}_{\boldsymbol{h}_\mathrm{T}},\boldsymbol{C}_{\boldsymbol{n},\mathrm{s}}) &= \ln\left[\prod_{q\in\mathbb{O}_\mathrm{T}} \mathrm{p}_{\boldsymbol{y}_\mathrm{T}}(\boldsymbol{y}_\mathrm{T}[q];\boldsymbol{C}_{\boldsymbol{h}_\mathrm{T}},\boldsymbol{C}_{\boldsymbol{n}_\mathrm{T}})\right] \\
&= -IBMN\ln\pi - I\ln\det\left[\boldsymbol{S}_\mathrm{T}\boldsymbol{C}_{\boldsymbol{h}_\mathrm{T}}\boldsymbol{S}_\mathrm{T}^\mathrm{H}+\boldsymbol{C}_{\boldsymbol{n}_\mathrm{T}}\right] \\
&- I\mathrm{tr}\left[\tilde{\boldsymbol{C}}_{\boldsymbol{y}_\mathrm{T}}\left(\boldsymbol{S}_\mathrm{T}\boldsymbol{C}_{\boldsymbol{h}_\mathrm{T}}\boldsymbol{S}_\mathrm{T}^\mathrm{H}+\boldsymbol{C}_{\boldsymbol{n}_\mathrm{T}}\right)^{-1}\right] \qquad (3.21)
\end{aligned}$$

with $\boldsymbol{C}_{\boldsymbol{n}_\mathrm{T}} = \boldsymbol{I}_{BN}\otimes\boldsymbol{C}_{\boldsymbol{n},\mathrm{s}}$ and a sufficient statistic given by the sample covariance matrix

$$\tilde{\boldsymbol{C}}_{\boldsymbol{y}_\mathrm{T}} = \frac{1}{I}\sum_{q\in\mathbb{O}_\mathrm{T}} \boldsymbol{y}_\mathrm{T}[q]\boldsymbol{y}_\mathrm{T}[q]^\mathrm{H}. \qquad (3.22)$$

The covariance matrix $\boldsymbol{C}_{\boldsymbol{y}_\mathrm{T}}$ of the observation has the structure $\boldsymbol{C}_{\boldsymbol{y}_\mathrm{T}} = \boldsymbol{S}_\mathrm{T}\boldsymbol{C}_{\boldsymbol{h}_\mathrm{T}}\boldsymbol{S}_\mathrm{T}^\mathrm{H}+\boldsymbol{I}_N\otimes\boldsymbol{C}_{\boldsymbol{n},\mathrm{s}}$.

For *model 2*, we have

$$\begin{aligned}
\mathrm{L}_{\boldsymbol{y}}(\{\boldsymbol{y}[q]\}_{q\in\mathbb{O}};\boldsymbol{C_h},\boldsymbol{C}_{\boldsymbol{n},\mathrm{s}}) &= \ln\left[\prod_{q\in\mathbb{O}} \mathrm{p}_{\boldsymbol{y}}(\boldsymbol{y}[q];\boldsymbol{C_h},\boldsymbol{C_n})\right] \\
&= -BMN\ln\pi - B\ln\det\left[\boldsymbol{S}\boldsymbol{C_h}\boldsymbol{S}^\mathrm{H}+\boldsymbol{C_n}\right] \\
&- B\mathrm{tr}\left[\tilde{\boldsymbol{C}}_{\boldsymbol{y}}\left(\boldsymbol{S}\boldsymbol{C_h}\boldsymbol{S}^\mathrm{H}+\boldsymbol{C_n}\right)^{-1}\right] \qquad (3.23)
\end{aligned}$$

with $\boldsymbol{C_n} = \boldsymbol{I}_N\otimes\boldsymbol{C}_{\boldsymbol{n},\mathrm{s}}$ and sufficient statistic

$$\tilde{\boldsymbol{C}}_{\boldsymbol{y}} = \frac{1}{B}\sum_{q\in\mathbb{O}} \boldsymbol{y}[q]\boldsymbol{y}[q]^\mathrm{H}. \qquad (3.24)$$

For derivation of the algorithms and to obtain further insights, we also require alternative sufficient statistics $\boldsymbol{t}[q], q\in\mathbb{O}$, and $\boldsymbol{t}_\mathrm{T}[q], q\in\mathbb{O}_\mathrm{T}$, where the former is defined as

$$\boldsymbol{t}[q] = \begin{bmatrix} \hat{\boldsymbol{h}}[q] \\ \bar{\boldsymbol{n}}[q] \end{bmatrix} = \boldsymbol{T}\boldsymbol{y}[q], q \in \mathbb{O}, \tag{3.25}$$

with the invertible transformation (Figure 3.1)

$$\boldsymbol{T} = \begin{bmatrix} \boldsymbol{S}^{\dagger} \\ \boldsymbol{A}^{\mathrm{H}} \end{bmatrix}. \tag{3.26}$$

This transformation is introduced to separate signal and noise subspace. $\hat{\boldsymbol{h}}[q] = \hat{\boldsymbol{h}}_{\mathrm{ML}}[q]$ is the ML estimate (2.31) of $\boldsymbol{h}[q]$ for the model in Section 2.2.2. Defining $\boldsymbol{A} \in \mathbb{C}^{MN \times M\bar{N}}$ such that it spans the space orthogonal to the column space of $\boldsymbol{S}$, i.e., $\boldsymbol{A}^{\mathrm{H}}\boldsymbol{S} = \boldsymbol{0}_{M\bar{N} \times P}$, the signal $\bar{\boldsymbol{n}}[q] = \boldsymbol{A}^{\mathrm{H}}\boldsymbol{y}[q] = \boldsymbol{A}^{\mathrm{H}}\boldsymbol{n}[q] \in \mathbb{C}^{M\bar{N}}$ only contains noise. The columns of $\boldsymbol{A}$ are a basis for the $M\bar{N}$ dimensional noise subspace ($\bar{N} = N - K(L+1)$) and are chosen such that $\boldsymbol{A}^{\mathrm{H}}\boldsymbol{A} = \boldsymbol{I}_{M\bar{N}}$. From the structure of $\boldsymbol{S} = \bar{\boldsymbol{S}}^{\mathrm{T}} \otimes \boldsymbol{I}_M$, it follows $\boldsymbol{A} = \bar{\boldsymbol{A}}^{\mathrm{T}} \otimes \boldsymbol{I}_M$ and $\bar{\boldsymbol{S}}\bar{\boldsymbol{A}}^{\mathrm{H}} = \boldsymbol{0}_{K(L+1) \times \bar{N}}$.

The sufficient statistic for model 1 is defined equivalently by

$$\boldsymbol{t}_{\mathrm{T}}[q] = \begin{bmatrix} \hat{\boldsymbol{h}}_{\mathrm{T}}[q] \\ \bar{\boldsymbol{n}}_{\mathrm{T}}[q] \end{bmatrix} = \boldsymbol{T}_{\mathrm{T}}\boldsymbol{y}_{\mathrm{T}}[q], q \in \mathbb{O}_{\mathrm{T}}, \tag{3.27}$$

where

$$\boldsymbol{T}_{\mathrm{T}} = \begin{bmatrix} \boldsymbol{S}_{\mathrm{T}}^{\dagger} \\ \boldsymbol{A}_{\mathrm{T}}^{\mathrm{H}} \end{bmatrix} = \begin{bmatrix} \boldsymbol{I}_B \otimes \boldsymbol{S}^{\dagger} \\ \boldsymbol{I}_B \otimes \boldsymbol{A}^{\mathrm{H}} \end{bmatrix} \in \mathbb{C}^{BMN \times BMN} \tag{3.28}$$

and its properties, e.g., $\boldsymbol{A}_{\mathrm{T}}^{\mathrm{H}}\boldsymbol{S}_{\mathrm{T}} = \boldsymbol{0}_{BM\bar{N} \times BP}$, follow from above.

The log-likelihood function of $\{\boldsymbol{t}_{\mathrm{T}}[q]\}_{q \in \mathbb{O}_{\mathrm{T}}}$ is

$$\begin{aligned}
\mathrm{L}_{\boldsymbol{t}_{\mathrm{T}}}(\{\boldsymbol{t}_{\mathrm{T}}[q]\}_{q \in \mathbb{O}_{\mathrm{T}}}; \boldsymbol{C}_{\boldsymbol{h}_{\mathrm{T}}}, \boldsymbol{C}_{\boldsymbol{n},\mathrm{s}}) &= \ln\left[\prod_{q \in \mathbb{O}_{\mathrm{T}}} \mathrm{p}_{\boldsymbol{t}_{\mathrm{T}}}(\boldsymbol{t}_{\mathrm{T}}[q]; \boldsymbol{C}_{\boldsymbol{h}_{\mathrm{T}}}, \boldsymbol{C}_{\boldsymbol{n},\mathrm{s}})\right] \\
&= \sum_{q \in \mathbb{O}_{\mathrm{T}}} \left\{\ln \mathrm{p}_{\hat{\boldsymbol{h}}_{\mathrm{T}}}(\hat{\boldsymbol{h}}_{\mathrm{T}}[q]; \boldsymbol{C}_{\boldsymbol{h}_{\mathrm{T}}}, \boldsymbol{C}_{\boldsymbol{n},\mathrm{s}}) + \ln \mathrm{p}_{\bar{\boldsymbol{n}}_{\mathrm{T}}}(\bar{\boldsymbol{n}}_{\mathrm{T}}[q]; \boldsymbol{C}_{\boldsymbol{n},\mathrm{s}})\right\} \\
&= -IBP\ln\pi - I\ln\det\left[\boldsymbol{C}_{\boldsymbol{h}_{\mathrm{T}}} + \boldsymbol{I}_B \otimes (\bar{\boldsymbol{S}}^*\bar{\boldsymbol{S}}^{\mathrm{T}})^{-1} \otimes \boldsymbol{C}_{\boldsymbol{n},\mathrm{s}}\right] \\
&\quad - I\mathrm{tr}\left[\tilde{\boldsymbol{C}}_{\hat{\boldsymbol{h}}_{\mathrm{T}}}\left(\boldsymbol{C}_{\boldsymbol{h}_{\mathrm{T}}} + \boldsymbol{I}_B \otimes (\bar{\boldsymbol{S}}^*\bar{\boldsymbol{S}}^{\mathrm{T}})^{-1} \otimes \boldsymbol{C}_{\boldsymbol{n},\mathrm{s}}\right)^{-1}\right] \\
&\quad - IBM\bar{N}\ln\pi - I\ln\det[\boldsymbol{I}_{B\bar{N}} \otimes \boldsymbol{C}_{\boldsymbol{n},\mathrm{s}}] \\
&\quad - I\mathrm{tr}\left[\tilde{\boldsymbol{C}}_{\bar{\boldsymbol{n}}_{\mathrm{T}}}\left(\boldsymbol{I}_{B\bar{N}} \otimes \boldsymbol{C}_{\boldsymbol{n},\mathrm{s}}^{-1}\right)\right]
\end{aligned} \tag{3.29}$$

with sample covariance matrices

$$\tilde{\boldsymbol{C}}_{\hat{\boldsymbol{h}}_{\mathrm{T}}} = \frac{1}{I} \sum_{q \in \mathbb{O}_{\mathrm{T}}} \hat{\boldsymbol{h}}_{\mathrm{T}}[q] \hat{\boldsymbol{h}}_{\mathrm{T}}[q]^{\mathrm{H}} \tag{3.30}$$

and

$$\tilde{\boldsymbol{C}}_{\bar{\boldsymbol{n}}_{\mathrm{T}}} = \frac{1}{I} \sum_{q \in \mathbb{O}_{\mathrm{T}}} \bar{\boldsymbol{n}}_{\mathrm{T}}[q] \bar{\boldsymbol{n}}_{\mathrm{T}}[q]^{\mathrm{H}}. \tag{3.31}$$

For the second model, we get

$$\mathrm{L}_{\boldsymbol{t}}(\{\boldsymbol{t}[q]\}_{q \in \mathbb{O}}; \boldsymbol{C}_{\boldsymbol{h}}, \boldsymbol{C}_{\boldsymbol{n},\mathrm{s}}) = \ln \left[\prod_{q \in \mathbb{O}} \mathrm{p}_{\boldsymbol{t}}(\boldsymbol{t}[q]; \boldsymbol{C}_{\boldsymbol{h}}, \boldsymbol{C}_{\boldsymbol{n},\mathrm{s}}) \right] \tag{3.32}$$

$$\begin{aligned}
&= \sum_{q \in \mathbb{O}} \left\{ \ln \mathrm{p}_{\hat{\boldsymbol{h}}}\left(\hat{\boldsymbol{h}}[q]; \boldsymbol{C}_{\boldsymbol{h}}, \boldsymbol{C}_{\boldsymbol{n},\mathrm{s}}\right) + \ln \mathrm{p}_{\bar{\boldsymbol{n}}}(\bar{\boldsymbol{n}}[q]; \boldsymbol{C}_{\boldsymbol{n},\mathrm{s}}) \right\} \\
&= -BP \ln \pi - B \ln \det \left[\boldsymbol{C}_{\boldsymbol{h}} + (\bar{\boldsymbol{S}}^{*} \bar{\boldsymbol{S}}^{\mathrm{T}})^{-1} \otimes \boldsymbol{C}_{\boldsymbol{n},\mathrm{s}} \right] \\
&- B \operatorname{tr} \left[\tilde{\boldsymbol{C}}_{\hat{\boldsymbol{h}}} \left(\boldsymbol{C}_{\boldsymbol{h}} + (\bar{\boldsymbol{S}}^{*} \bar{\boldsymbol{S}}^{\mathrm{T}})^{-1} \otimes \boldsymbol{C}_{\boldsymbol{n},\mathrm{s}} \right)^{-1} \right] \\
&- BM\bar{N} \ln \pi - B \ln \det [\boldsymbol{I}_{\bar{N}} \otimes \boldsymbol{C}_{\boldsymbol{n},\mathrm{s}}] \\
&- B \operatorname{tr} \left[\tilde{\boldsymbol{C}}_{\bar{\boldsymbol{n}}} \left(\boldsymbol{I}_{\bar{N}} \otimes \boldsymbol{C}_{\boldsymbol{n},\mathrm{s}}^{-1} \right) \right]
\end{aligned} \tag{3.33}$$

with

$$\tilde{\boldsymbol{C}}_{\hat{\boldsymbol{h}}} = \frac{1}{B} \sum_{q \in \mathbb{O}} \hat{\boldsymbol{h}}[q] \hat{\boldsymbol{h}}[q]^{\mathrm{H}} \tag{3.34}$$

and

$$\tilde{\boldsymbol{C}}_{\bar{\boldsymbol{n}}} = \frac{1}{B} \sum_{q \in \mathbb{O}} \bar{\boldsymbol{n}}[q] \bar{\boldsymbol{n}}[q]^{\mathrm{H}}. \tag{3.35}$$

3.3 Maximum Likelihood Estimation

The problem of estimating the channel and noise covariance matrix can be formulated in the framework of estimation of a structured covariance matrix reviewed in Section 3.1: The covariance matrix of the observations $\{\boldsymbol{y}_{\mathrm{T}}[q]\}_{q=1}^{I}$ is structured and uniquely parameterized by the channel covariance matrix $\boldsymbol{C}_{\boldsymbol{h}_{\mathrm{T}}}$ and noise covariance matrix $\boldsymbol{C}_{\boldsymbol{n},\mathrm{s}}$ as $\boldsymbol{C}_{\boldsymbol{y}_{\mathrm{T}}} = \boldsymbol{S}_{\mathrm{T}} \boldsymbol{C}_{\boldsymbol{h}_{\mathrm{T}}} \boldsymbol{S}_{\mathrm{T}}^{\mathrm{H}} + \boldsymbol{I}_{N} \otimes \boldsymbol{C}_{\boldsymbol{n},\mathrm{s}}$ (3.21) (Model 1). Our goal is not necessarily a more accurate estimation of $\boldsymbol{C}_{\boldsymbol{y}_{\mathrm{T}}}$, but we are mainly interested in its parameters $\boldsymbol{C}_{\boldsymbol{h}_{\mathrm{T}}}$ and $\boldsymbol{C}_{\boldsymbol{n},\mathrm{s}}$. The corresponding ML problem is

$$\begin{aligned}\{\hat{\boldsymbol{C}}_{\boldsymbol{h}_\mathrm{T}}^{\mathrm{ML}}, \hat{\boldsymbol{C}}_{\boldsymbol{n},\mathrm{s}}^{\mathrm{ML}}\} &= \underset{\boldsymbol{C}_{\boldsymbol{h}_\mathrm{T}} \in \mathbb{T}_{+,0}^{P,B}, \boldsymbol{C}_{\boldsymbol{n},\mathrm{s}} \in \mathbb{S}_{+,0}^{M}}{\operatorname{argmax}} \quad \mathrm{L}_{\boldsymbol{y}_\mathrm{T}}(\{\boldsymbol{y}_\mathrm{T}[q]\}_{q\in\mathbb{O}_\mathrm{T}}; \boldsymbol{C}_{\boldsymbol{h}_\mathrm{T}}, \boldsymbol{C}_{\boldsymbol{n},\mathrm{s}}) \\ &= \underset{\boldsymbol{C}_{\boldsymbol{h}_\mathrm{T}} \in \mathbb{T}_{+,0}^{P,B}, \boldsymbol{C}_{\boldsymbol{n},\mathrm{s}} \in \mathbb{S}_{+,0}^{M}}{\operatorname{argmax}} \quad \mathrm{L}_{\boldsymbol{t}_\mathrm{T}}(\{\boldsymbol{t}_\mathrm{T}[q]\}_{q\in\mathbb{O}_\mathrm{T}}; \boldsymbol{C}_{\boldsymbol{h}_\mathrm{T}}, \boldsymbol{C}_{\boldsymbol{n},\mathrm{s}}) \end{aligned} \qquad (3.36)$$

with the log-likelihood functions of the direct observations $\{\boldsymbol{y}_\mathrm{T}[q]\}_{q\in\mathbb{O}_\mathrm{T}}$ and their transformation $\{\boldsymbol{t}_\mathrm{T}[q]\}_{q\in\mathbb{O}_\mathrm{T}}$ (3.27) given in (3.21) and (3.29).

For model 2, the ML problem is based on (3.23) and (3.33) and reads

$$\begin{aligned}\{\hat{\boldsymbol{C}}_{\boldsymbol{h}}^{\mathrm{ML}}, \hat{\boldsymbol{C}}_{\boldsymbol{n},\mathrm{s}}^{\mathrm{ML}}\} &= \underset{\boldsymbol{C}_{\boldsymbol{h}} \in \mathbb{S}_{+,0}^{P}, \boldsymbol{C}_{\boldsymbol{n},\mathrm{s}} \in \mathbb{S}_{+,0}^{M}}{\operatorname{argmax}} \quad \mathrm{L}_{\boldsymbol{y}}(\{\boldsymbol{y}[q]\}_{q\in\mathbb{O}}; \boldsymbol{C}_{\boldsymbol{h}}, \boldsymbol{C}_{\boldsymbol{n},\mathrm{s}}) \\ &= \underset{\boldsymbol{C}_{\boldsymbol{h}} \in \mathbb{S}_{+,0}^{P}, \boldsymbol{C}_{\boldsymbol{n},\mathrm{s}} \in \mathbb{S}_{+,0}^{M}}{\operatorname{argmax}} \quad \mathrm{L}_{\boldsymbol{t}}(\{\boldsymbol{t}[q]\}_{q\in\mathbb{O}}; \boldsymbol{C}_{\boldsymbol{h}}, \boldsymbol{C}_{\boldsymbol{n},\mathrm{s}}). \end{aligned} \qquad (3.37)$$

From the discussion in Section 3.1.1 (the necessary conditions (3.5) in particular), it seems that an *explicit* solution of these optimization problems is impossible. Alternatives are the numerical solution of the system of nonlinear equations (3.5) [5] or a numerical and iterative optimization of the unconstrained optimization problems [124]. Numerical optimization techniques create a sequence of optimization variables corresponding to monotonically increasing values of the log-likelihood function.

3.3.1 Application of the Space-Alternating Generalized Expectation Maximization Algorithm

One important iterative method is the *expectation maximization* (EM) algorithm [142]: It ensures convergence to a local maximum of the log-likelihood function. Typically, its formulation is intuitive, but this comes at the price of a relatively slow convergence [68]. It relies on the definition of a complete data space, whose complete knowledge would significantly simplify the optimization problem. The complete data space is not fully accessible. Only an incomplete data is available, which is a non-invertible (deterministic) function of the complete data space. Thus, in the expectation step (E-step) of the EM algorithm a conditional mean (CM) estimate of the complete data log-likelihood function is performed followed by the maximization step (M-step) maximizing the estimated log-likelihood function.

A generalization of the EM algorithm is the *space-alternating generalized expectation maximization* (SAGE) algorithm [68, 142], which aims at an improved convergence rate. Its key idea is the definition of (multiple) hidden-data spaces generalizing the concept of complete data. The observation may be a *stochastic* function of a hidden-data space. The iterative procedure chooses a subset of the parameters to be estimated and the corresponding hidden-data spaces for every iteration and performs an E- and M-Step

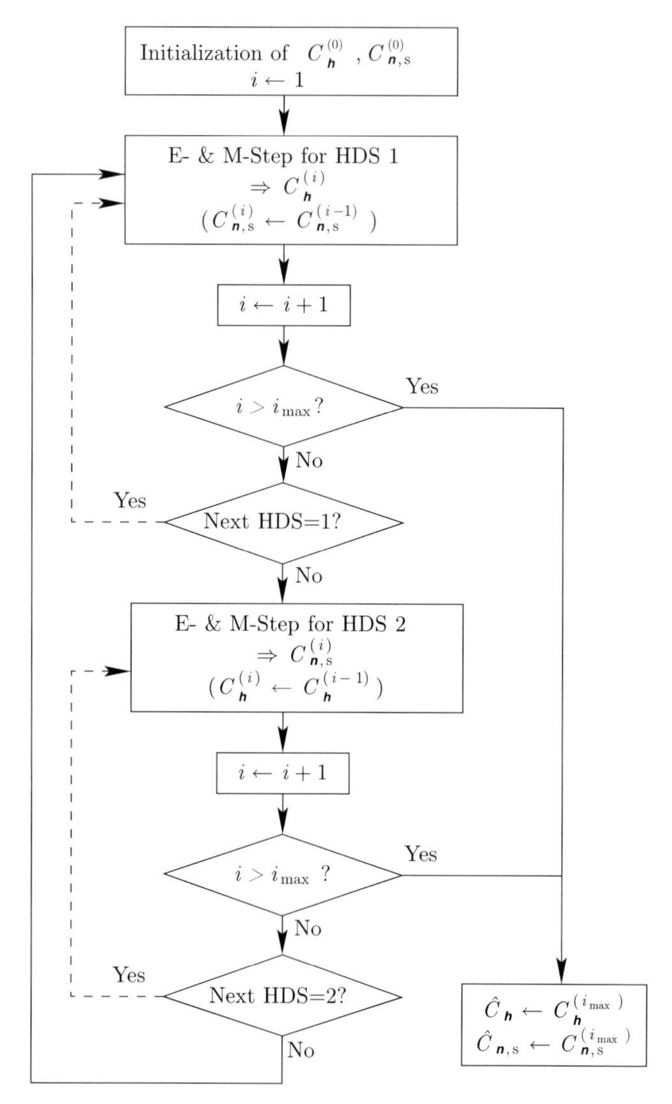

Fig. 3.2 SAGE iterations selecting a hidden data space (HDS) in every iteration for Model 2 (similarly for model 1).

for the chosen hidden-data. We are completely free in choosing a suitable sequence of hidden-data spaces. This can improve convergence rate. Generating a monotonically increasing sequence of log-likelihood values and estimates, it converges asymptotically to a local maximum of the log-likelihood function.

First we have to define suitable hidden data spaces for our estimation problems, which result in simple M-Steps. We start with the following *Gedanken-*

experiment for model 2:[10] If $\boldsymbol{h}[q]$ and $\boldsymbol{n}[q]$ were observable for $q \in \mathbb{O}$, the corresponding ML problems would be

$$\max_{\boldsymbol{C}_{\boldsymbol{h}} \in \mathbb{S}_{+,0}^{P}} \prod_{q \in \mathbb{O}} \mathrm{p}_{\boldsymbol{h}}(\boldsymbol{h}[q]; \boldsymbol{C}_{\boldsymbol{h}}) \tag{3.38}$$

and

$$\max_{\boldsymbol{C}_{\boldsymbol{n},\mathrm{s}} \in \mathbb{S}_{+,0}^{M}} \prod_{q \in \mathbb{O}} \prod_{n=1}^{B} \mathrm{p}_{\boldsymbol{n}[q,n]}(\boldsymbol{n}[q,n]; \boldsymbol{C}_{\boldsymbol{n},\mathrm{s}}) \tag{3.39}$$

with $\boldsymbol{n}[q] = [\boldsymbol{n}[q,1]^{\mathrm{T}}, \boldsymbol{n}[q,2]^{\mathrm{T}}, \ldots, \boldsymbol{n}[q,N]^{\mathrm{T}}]^{\mathrm{T}}$ (3.18). Due to the complex Gaussian pdf the solution to both problems is simple and given by the sample covariance matrices of the corresponding (hypothetical) observations

$$\tilde{\boldsymbol{C}}_{\boldsymbol{h}} = \frac{1}{B} \sum_{q \in \mathbb{O}} \boldsymbol{h}[q]\boldsymbol{h}[q]^{\mathrm{H}}, \tag{3.40}$$

$$\tilde{\boldsymbol{C}}_{\boldsymbol{n},\mathrm{s}} = \frac{1}{BN} \sum_{q \in \mathbb{O}} \sum_{n=1}^{N} \boldsymbol{n}[q,n]\boldsymbol{n}[q,n]^{\mathrm{H}}, \tag{3.41}$$

or $\tilde{c}_n = \frac{1}{BNM} \sum_{q \in \mathbb{O}} \sum_{n=1}^{N} \|\boldsymbol{n}[q,n]\|_2^2$ for spatially uncorrelated noise.

We define the following two hidden-data spaces (HDS) for *model 2*:

- **HDS 1:** The random vector $\boldsymbol{h}[q]$ is an admissible HDS w.r.t. $\boldsymbol{C}_{\boldsymbol{h}}$ because $\mathrm{p}_{\boldsymbol{y},\boldsymbol{h}}(\boldsymbol{y}[q], \boldsymbol{h}[q]; \boldsymbol{C}_{\boldsymbol{h}}, \boldsymbol{C}_{\boldsymbol{n},\mathrm{s}}) = \mathrm{p}_{\boldsymbol{y}|\boldsymbol{h}[q]}(\boldsymbol{y}[q]|\boldsymbol{h}[q]; \boldsymbol{C}_{\boldsymbol{n},\mathrm{s}})\, \mathrm{p}_{\boldsymbol{h}}(\boldsymbol{h}[q]; \boldsymbol{C}_{\boldsymbol{h}})$, i.e., the conditional distribution is independent of $\boldsymbol{C}_{\boldsymbol{h}}$ [68].
- **HDS 2:** The random vector $\boldsymbol{n}[q]$ is an admissible HDS w.r.t. $\boldsymbol{C}_{\boldsymbol{n},\mathrm{s}}$ (or c_n) because $\mathrm{p}_{\boldsymbol{y},\boldsymbol{n}}(\boldsymbol{y}[q], \boldsymbol{n}[q]; \boldsymbol{C}_{\boldsymbol{h}}, \boldsymbol{C}_{\boldsymbol{n},\mathrm{s}}) = \mathrm{p}_{\boldsymbol{y}|\boldsymbol{n}[q]}(\boldsymbol{y}[q]|\boldsymbol{n}[q]; \boldsymbol{C}_{\boldsymbol{h}})\, \mathrm{p}_{\boldsymbol{n}}(\boldsymbol{n}[q]; \boldsymbol{C}_{\boldsymbol{n},\mathrm{s}})$, i.e., the conditional distribution is independent of $\boldsymbol{C}_{\boldsymbol{n},\mathrm{s}}$ [68].

For *model 1*, we will slightly extend the definition of HDS 1 in Section 3.3.3 to be able to incorporate the block-Toeplitz structure of $\boldsymbol{C}_{\boldsymbol{h}_{\mathrm{T}}}$, but the general algorithm introduced below is still applicable.

We start with an initialization $\boldsymbol{C}_{\boldsymbol{h}}^{(0)}$ and $\boldsymbol{C}_{\boldsymbol{n},\mathrm{s}}^{(0)}$. Every iteration of the SAGE algorithm to solve the ML problem (3.37) (and (3.36) for model 1) consists of three steps (Figure 3.2):

1. Choose a HDS for iteration $i+1$. We propose three alternatives:

   a. Start with HDS 1 and then alternate between HDS 1 and HDS 2. Alternating with period $N_{\mathrm{iter}} = 1$ may result in an increased estimation error for $\boldsymbol{C}_{\boldsymbol{n},\mathrm{s}}^{(i)}$, since the algorithm cannot distinguish sufficiently between the

[10] The ML problem for model 2 is simpler than for model 1 as $\boldsymbol{C}_{\boldsymbol{h}}$ does not have any special structure.

correlations in $\boldsymbol{y}[q]$ origining from the signal and noise, respectively, and is trapped in a local maximum.

b. Choose HDS 1 in the first N_{iter} iterations, then switch to HDS 2 for N_{iter} iterations. Now alternate with this period. This has the advantage that the estimate $\boldsymbol{C}_{\boldsymbol{h}}^{(N_{\text{iter}})}$ of $\boldsymbol{C}_{\boldsymbol{h}}$ has already improved considerably before we start improving $\boldsymbol{C}_{\boldsymbol{n},\text{s}}^{(0)}$. The resulting local maximum of the likelihood function corresponds to a better estimation accuracy.

c. Choose HDS 1 and HDS 2 simultaneously, which is similar to the original EM algorithm as we do not alternate between the HDSs. The convergence is slower than for the first variation.

2. *E-step:* Perform the CM estimate of the log-likelihood for the selected HDS given $\boldsymbol{C}_{\boldsymbol{h}}^{(i)}$ and $\boldsymbol{C}_{\boldsymbol{n},\text{s}}^{(i)}$ from the previous iteration.
3. *M-step:* Maximize the estimated log-likelihood w.r.t. $\boldsymbol{C}_{\boldsymbol{h}}$ for HDS 1 and $\boldsymbol{C}_{\boldsymbol{n},\text{s}}$ for HDS 2.

The algorithm is shown in Figure 3.2, where we choose a maximum number of iterations $i_{\max}$ to have a fixed complexity. In the following subsections we derive the E- and M-step for both hidden data spaces explicitly: We start with HDS 2 for the noise covariance matrix $\boldsymbol{C}_{\boldsymbol{n},\text{s}}$ or variance $c_{\boldsymbol{n}}$, respectively. For problem (3.36) and model 1, we give the iterative solution in Section 3.3.3. As a special case, we treat model 1 with $P = 1$ in Section 3.3.5. In Section 3.3.4 we present both steps for estimating the channel correlations based on HDS 1, model 2, and problem (3.37).

3.3.2 Estimation of the Noise Covariance Matrix

The HDS 2 is defined as $\boldsymbol{n}_{\text{T}}[q]$ and $\boldsymbol{n}[q]$ for model 1 and model 2, respectively. We derive the E- and M-step starting with the more general model 1.

Model 1

The log-likelihood function of the HDS 2 is

$$\begin{aligned} \mathrm{L}_{\boldsymbol{n}_{\text{T}}}(\{\boldsymbol{n}_{\text{T}}[q]\}_{q\in\mathbb{O}_{\text{T}}};\boldsymbol{C}_{\boldsymbol{n},\text{s}}) &= \sum_{q\in\mathbb{O}_{\text{T}}}\sum_{b=1}^{B}\sum_{n=1}^{N}\ln \mathrm{p}_{\boldsymbol{n}}(\boldsymbol{n}[q-b,n];\boldsymbol{C}_{\boldsymbol{n},\text{s}}) \\ &= IBN\left(-M\ln\pi - \ln\det[\boldsymbol{C}_{\boldsymbol{n},\text{s}}] - \operatorname{tr}\left[\tilde{\boldsymbol{C}}_{\boldsymbol{n},\text{s}}\boldsymbol{C}_{\boldsymbol{n},\text{s}}^{-1}\right]\right) \end{aligned} \tag{3.42}$$

with sample covariance matrix

$$\tilde{\boldsymbol{C}}_{\boldsymbol{n},\mathrm{s}} = \frac{1}{IBN} \sum_{q\in\mathbb{O}_\mathrm{T}} \sum_{b=1}^{B} \sum_{n=1}^{N} \boldsymbol{n}[q-b,n]\boldsymbol{n}[q-b,n]^\mathrm{H} \tag{3.43}$$

based on the temporal samples $\boldsymbol{n}[q,n]$ of the noise process, which are defined in (3.18).

The CM estimate of the log-likelihood function (3.42)

$$\mathrm{E}_{\boldsymbol{n}_\mathrm{T}}\left[\mathrm{L}_{\boldsymbol{n}_\mathrm{T}}(\{\boldsymbol{n}_\mathrm{T}[q]\}_{q\in\mathbb{O}_\mathrm{T}};\boldsymbol{C}_{\boldsymbol{n},\mathrm{s}})|\{\boldsymbol{y}_\mathrm{T}[q]\}_{q\in\mathbb{O}_\mathrm{T}};\boldsymbol{C}_{\boldsymbol{h}_\mathrm{T}}^{(i)},\boldsymbol{C}_{\boldsymbol{n},\mathrm{s}}^{(i)}\right] \tag{3.44}$$

forms the *E-step* in a SAGE iteration based HDS 2. It results in the CM estimate of the sample covariance matrix (3.43)

$$\mathrm{E}_{\boldsymbol{n}}\left[\tilde{\boldsymbol{C}}_{\boldsymbol{n},\mathrm{s}}|\{\boldsymbol{y}_\mathrm{T}[q]\}_{q\in\mathbb{O}_\mathrm{T}};\boldsymbol{C}_{\boldsymbol{h}_\mathrm{T}}^{(i)},\boldsymbol{C}_{\boldsymbol{n},\mathrm{s}}^{(i)}\right], \tag{3.45}$$

which now substitutes $\tilde{\boldsymbol{C}}_{\boldsymbol{n},\mathrm{s}}$ in (3.42). As $\boldsymbol{n}[q-b,n] = \boldsymbol{J}_{b,n}\boldsymbol{n}_\mathrm{T}[q]$ with selection matrix $\boldsymbol{J}_{b,n} = \boldsymbol{e}_{(b-1)N+n}^\mathrm{T} \otimes \boldsymbol{I}_M \in \{0,1\}^{M\times BMN}$ and the definition of $\mathbb{O}_\mathrm{T}$ in Section 3.2, we obtain

$$\mathrm{E}_{\boldsymbol{n}}\left[\boldsymbol{n}[q-b,n]\boldsymbol{n}[q-b,n]^\mathrm{H}|\boldsymbol{y}_\mathrm{T}[q];\boldsymbol{C}_{\boldsymbol{h}_\mathrm{T}}^{(i)},\boldsymbol{C}_{\boldsymbol{n},\mathrm{s}}^{(i)}\right] = \boldsymbol{J}_{b,n}\boldsymbol{R}_{\boldsymbol{n}_\mathrm{T}|\boldsymbol{y}_\mathrm{T}[q]}^{(i)}\boldsymbol{J}_{b,n}^\mathrm{T} \tag{3.46}$$

for every term in (3.43). With (A.2) and (A.3) the conditional correlation matrix reads

$$\boldsymbol{R}_{\boldsymbol{n}_\mathrm{T}|\boldsymbol{y}_\mathrm{T}[q]}^{(i)} = \mathrm{E}_{\boldsymbol{n}_\mathrm{T}}\left[\boldsymbol{n}_\mathrm{T}[q]\boldsymbol{n}_\mathrm{T}[q]^\mathrm{H}|\boldsymbol{y}_\mathrm{T}[q];\boldsymbol{C}_{\boldsymbol{h}_\mathrm{T}}^{(i)},\boldsymbol{C}_{\boldsymbol{n},\mathrm{s}}^{(i)}\right] = \hat{\boldsymbol{n}}_\mathrm{T}^{(i)}[q]\hat{\boldsymbol{n}}_\mathrm{T}^{(i)}[q]^\mathrm{H} + \boldsymbol{C}_{\boldsymbol{n}_\mathrm{T}|\boldsymbol{y}_\mathrm{T}[q]}^{(i)}, \tag{3.47}$$

where

$$\begin{aligned}\hat{\boldsymbol{n}}_\mathrm{T}^{(i)}[q] &= \mathrm{E}_{\boldsymbol{n}_\mathrm{T}}\left[\boldsymbol{n}_\mathrm{T}[q]|\boldsymbol{y}_\mathrm{T}[q];\boldsymbol{C}_{\boldsymbol{h}_\mathrm{T}}^{(i)},\boldsymbol{C}_{\boldsymbol{n},\mathrm{s}}^{(i)}\right] \\ &= \boldsymbol{C}_{\boldsymbol{n}_\mathrm{T}\boldsymbol{y}_\mathrm{T}}^{(i)}\boldsymbol{C}_{\boldsymbol{y}_\mathrm{T}}^{(i),-1}\boldsymbol{y}_\mathrm{T}[q] = \boldsymbol{W}_{\boldsymbol{n}_\mathrm{T}}^{(i)}\boldsymbol{y}_\mathrm{T}[q]\end{aligned} \tag{3.48}$$

is the CM estimate of the noise $\boldsymbol{n}_\mathrm{T}[q]$ based on the covariance matrices from the previous iteration i. With the definitions in Section 3.2 we have

$$\begin{aligned}\boldsymbol{C}_{\boldsymbol{y}_\mathrm{T}}^{(i)} &= \boldsymbol{S}_\mathrm{T}\boldsymbol{C}_{\boldsymbol{h}_\mathrm{T}}^{(i)}\boldsymbol{S}_\mathrm{T}^\mathrm{H} + \boldsymbol{I}_{BN} \otimes \boldsymbol{C}_{\boldsymbol{n},\mathrm{s}}^{(i)} \\ \boldsymbol{C}_{\boldsymbol{n}_\mathrm{T}\boldsymbol{y}_\mathrm{T}}^{(i)} &= \boldsymbol{C}_{\boldsymbol{n}_\mathrm{T}}^{(i)} = \boldsymbol{I}_{BN} \otimes \boldsymbol{C}_{\boldsymbol{n},\mathrm{s}}^{(i)}.\end{aligned}$$

The conditional covariance matrix is (cf. (A.3))

$$\boldsymbol{C}_{\boldsymbol{n}_\mathrm{T}|\boldsymbol{y}_\mathrm{T}[q]}^{(i)} = \boldsymbol{C}_{\boldsymbol{n}_\mathrm{T}}^{(i)} - \boldsymbol{W}_{\boldsymbol{n}_\mathrm{T}}^{(i)}\boldsymbol{C}_{\boldsymbol{n}_\mathrm{T}\boldsymbol{y}_\mathrm{T}}^{(i),\mathrm{H}} = \boldsymbol{C}_{\boldsymbol{n}_\mathrm{T}}^{(i)} - \boldsymbol{C}_{\boldsymbol{n}_\mathrm{T}}^{(i)}\boldsymbol{C}_{\boldsymbol{y}_\mathrm{T}}^{(i),-1}\boldsymbol{C}_{\boldsymbol{n}_\mathrm{T}}^{(i)}. \tag{3.49}$$

Maximization of the CM estimate of the log-likelihood for the HDS 2 in (3.44) leads to the improved estimate of $\boldsymbol{C}_{\boldsymbol{n},\mathrm{s}}$ in step $i+1$, which is the CM estimate of the sample covariance matrix (3.45)

$$\boxed{\begin{aligned} \boldsymbol{C}_{\boldsymbol{n},\mathrm{s}}^{(i+1)} &= \frac{1}{IBN} \sum_{q\in\mathbb{O}_\mathrm{T}} \sum_{b=1}^{B} \sum_{n=1}^{N} \boldsymbol{J}_{b,n} \boldsymbol{R}_{\boldsymbol{n}_\mathrm{T}|\boldsymbol{y}_\mathrm{T}[q]}^{(i)} \boldsymbol{J}_{b,n}^\mathrm{T} \\ &= \frac{1}{BN} \sum_{b=1}^{B} \sum_{n=1}^{N} \boldsymbol{J}_{b,n} \left(\boldsymbol{W}_{\boldsymbol{n}_\mathrm{T}}^{(i)} \tilde{\boldsymbol{C}}_{\boldsymbol{y}_\mathrm{T}} \boldsymbol{W}_{\boldsymbol{n}_\mathrm{T}}^{(i),\mathrm{H}} + \boldsymbol{C}_{\boldsymbol{n}_\mathrm{T}|\boldsymbol{y}_\mathrm{T}[q]}^{(i)} \right) \boldsymbol{J}_{b,n}^\mathrm{T} \end{aligned}} \tag{3.50}$$

with sample covariance matrix of the observations $\tilde{\boldsymbol{C}}_{\boldsymbol{y}_\mathrm{T}}$ (3.22) and $\boldsymbol{W}_{\boldsymbol{n}_\mathrm{T}}^{(i)}$ from (3.48).

Assuming spatially white noise the estimate for the noise variance is

$$\begin{aligned} c_n^{(i+1)} &= \frac{1}{M} \mathrm{E}_{\boldsymbol{n}}\left[\mathrm{tr}\left[\tilde{\boldsymbol{C}}_{\boldsymbol{n},\mathrm{s}}\right] |\{\boldsymbol{y}_\mathrm{T}[q]\}_{q\in\mathbb{O}_\mathrm{T}}; \boldsymbol{C}_{\boldsymbol{h}_\mathrm{T}}^{(i)}, c_n^{(i)} \right] \\ &= \frac{1}{IBMN} \sum_{q\in\mathbb{O}_\mathrm{T}} \mathrm{tr}\left[\boldsymbol{R}_{\boldsymbol{n}_\mathrm{T}|\boldsymbol{y}_\mathrm{T}[q]}^{(i)} \right] \\ &= \frac{1}{BMN} \mathrm{tr}\left[\boldsymbol{W}_{\boldsymbol{n}_\mathrm{T}}^{(i)} \tilde{\boldsymbol{C}}_{\boldsymbol{y}_\mathrm{T}} \boldsymbol{W}_{\boldsymbol{n}_\mathrm{T}}^{(i),\mathrm{H}} + \boldsymbol{C}_{\boldsymbol{n}_\mathrm{T}|\boldsymbol{y}_\mathrm{T}[q]}^{(i)} \right] \end{aligned} \tag{3.51}$$

because $\mathrm{tr}[\tilde{\boldsymbol{C}}_{\boldsymbol{n},\mathrm{s}}] = \frac{1}{IBN} \sum_{q\in\mathbb{O}_\mathrm{T}} \sum_{b=1}^{B} \|\boldsymbol{n}[q-b]\|_2^2$.

The final estimates are given after iteration i_max of the SAGE algorithm

$$\begin{aligned} \hat{\boldsymbol{C}}_{\boldsymbol{n},\mathrm{s}} &= \boldsymbol{C}_{\boldsymbol{n},\mathrm{s}}^{(i_\mathrm{max})} \\ \hat{c}_n &= c_n^{(i_\mathrm{max})}. \end{aligned} \tag{3.52}$$

Model 2

Model 2 is a special case of model 1 for $I = 1$ and $\boldsymbol{C}_{\boldsymbol{h}_\mathrm{T}} = \boldsymbol{I}_B \otimes \boldsymbol{C}_{\boldsymbol{h}}$. The log-likelihood function of HDS 2 is

$$\begin{aligned} \mathrm{L}_{\boldsymbol{n}}(\{\boldsymbol{n}[q]\}_{q\in\mathbb{O}}; \boldsymbol{C}_{\boldsymbol{n},\mathrm{s}}) &= \sum_{q\in\mathbb{O}} \sum_{n=1}^{N} \ln \mathrm{p}_{\boldsymbol{n}}(\boldsymbol{n}[q,n]; \boldsymbol{C}_{\boldsymbol{n},\mathrm{s}}) \\ &= -BNM \ln \pi - BN \ln \det[\boldsymbol{C}_{\boldsymbol{n},\mathrm{s}}] - BN \mathrm{tr}\left[\tilde{\boldsymbol{C}}_{\boldsymbol{n},\mathrm{s}} \boldsymbol{C}_{\boldsymbol{n},\mathrm{s}}^{-1} \right] \end{aligned} \tag{3.53}$$

with sample covariance matrix

$$\tilde{\boldsymbol{C}}_{\boldsymbol{n},\mathrm{s}} = \frac{1}{BN} \sum_{q\in\mathbb{O}} \sum_{n=1}^{N} \boldsymbol{n}[q,n] \boldsymbol{n}[q,n]^\mathrm{H}. \tag{3.54}$$

Maximization of the CM estimate from the *E-step*

$$\mathrm{E}_{\boldsymbol{n}}\left[\mathrm{L}_{\boldsymbol{n}}(\{\boldsymbol{n}[q]\}_{q\in\mathbb{O}}; \boldsymbol{C}_{\boldsymbol{n},\mathrm{s}}) | \{\boldsymbol{y}[q]\}_{q\in\mathbb{O}}; \boldsymbol{C}_{\boldsymbol{h}}^{(i)}, \boldsymbol{C}_{\boldsymbol{n},\mathrm{s}}^{(i)} \right] \tag{3.55}$$

leads to the estimates in iteration $i+1$

$$\boldsymbol{C}_{\boldsymbol{n},\mathrm{s}}^{(i+1)} = \mathrm{E}_{\boldsymbol{n}}\left[\tilde{\boldsymbol{C}}_{\boldsymbol{n},\mathrm{s}} | \{\boldsymbol{y}[q]\}_{q\in\mathbb{O}}; \boldsymbol{C}_{\boldsymbol{h}}^{(i)}, \boldsymbol{C}_{\boldsymbol{n},\mathrm{s}}^{(i)}\right] = \frac{1}{BN}\sum_{q\in\mathbb{O}}\sum_{n=1}^{N} \boldsymbol{J}_n \boldsymbol{R}_{\boldsymbol{n}|\boldsymbol{y}[q]}^{(i)} \boldsymbol{J}_n^{\mathrm{T}}, \tag{3.56}$$

$$c_n^{(i+1)} = \frac{1}{M}\mathrm{E}_{\boldsymbol{n}}\left[\mathrm{tr}\left[\tilde{\boldsymbol{C}}_{\boldsymbol{n},\mathrm{s}}\right] | \{\boldsymbol{y}[q]\}_{q\in\mathbb{O}}; \boldsymbol{C}_{\boldsymbol{h}}^{(i)}, c_n^{(i)}\right] = \frac{1}{BMN}\sum_{q\in\mathbb{O}} \mathrm{tr}\left[\boldsymbol{R}_{\boldsymbol{n}|\boldsymbol{y}[q]}^{(i)}\right] \tag{3.57}$$

with $\boldsymbol{n}[q,n] = \boldsymbol{J}_n\boldsymbol{n}[q]$ and $\boldsymbol{J}_n = \boldsymbol{e}_n^{\mathrm{T}} \otimes \boldsymbol{I}_M \in \{0,1\}^{M\times MN}$. The conditional correlation matrix based on the previous iteration i reads

$$\boldsymbol{R}_{\boldsymbol{n}|\boldsymbol{y}[q]}^{(i)} = \mathrm{E}_{\boldsymbol{n}}\left[\boldsymbol{n}[q]\boldsymbol{n}[q]^{\mathrm{H}} | \boldsymbol{y}[q]; \boldsymbol{C}_{\boldsymbol{h}}^{(i)}, \boldsymbol{C}_{\boldsymbol{n},\mathrm{s}}^{(i)}\right] = \hat{\boldsymbol{n}}^{(i)}[q]\hat{\boldsymbol{n}}^{(i)}[q]^{\mathrm{H}} + \boldsymbol{C}_{\boldsymbol{n}|\boldsymbol{y}[q]}^{(i)} \tag{3.58}$$

with

$$\begin{aligned}\hat{\boldsymbol{n}}^{(i)}[q] &= \mathrm{E}_{\boldsymbol{n}}\left[\boldsymbol{n}[q] | \boldsymbol{y}[q]; \boldsymbol{C}_{\boldsymbol{h}}^{(i)}, \boldsymbol{C}_{\boldsymbol{n},\mathrm{s}}^{(i)}\right] \\ &= \boldsymbol{C}_{\boldsymbol{ny}}^{(i)}\boldsymbol{C}_{\boldsymbol{y}}^{(i),-1}\boldsymbol{y}[q] = \boldsymbol{W}_{\boldsymbol{n}}^{(i)}\boldsymbol{y}[q] &(3.59)\\ \boldsymbol{C}_{\boldsymbol{n}|\boldsymbol{y}[q]}^{(i)} &= \boldsymbol{C}_{\boldsymbol{n}}^{(i)} - \boldsymbol{W}_{\boldsymbol{n}}^{(i)}\boldsymbol{C}_{\boldsymbol{ny}}^{(i),\mathrm{H}} = \boldsymbol{C}_{\boldsymbol{n}}^{(i)} - \boldsymbol{C}_{\boldsymbol{n}}^{(i)}\boldsymbol{C}_{\boldsymbol{y}}^{(i),-1}\boldsymbol{C}_{\boldsymbol{n}}^{(i)} &(3.60)\end{aligned}$$

and

$$\begin{aligned}\boldsymbol{C}_{\boldsymbol{n}}^{(i)} &= \boldsymbol{I}_N \otimes \boldsymbol{C}_{\boldsymbol{n},\mathrm{s}}^{(i)} \\ \boldsymbol{C}_{\boldsymbol{ny}}^{(i)} &= \boldsymbol{C}_{\boldsymbol{n}}^{(i)} \\ \boldsymbol{C}_{\boldsymbol{y}}^{(i)} &= \boldsymbol{S}\boldsymbol{C}_{\boldsymbol{h}}^{(i)}\boldsymbol{S}^{\mathrm{H}} + \boldsymbol{C}_{\boldsymbol{n}}^{(i)}.\end{aligned}$$

In summary the improved estimates in iteration $i+1$ read

$$\boldsymbol{C}_{\boldsymbol{n},\mathrm{s}}^{(i+1)} = \frac{1}{N}\sum_{n=1}^{N} \boldsymbol{J}_n\left(\boldsymbol{W}_{\boldsymbol{n}}^{(i)}\tilde{\boldsymbol{C}}_{\boldsymbol{y}}\boldsymbol{W}_{\boldsymbol{n}}^{(i),\mathrm{H}} + \boldsymbol{C}_{\boldsymbol{n}|\boldsymbol{y}[q]}^{(i)}\right)\boldsymbol{J}_n^{\mathrm{T}}, \tag{3.61}$$

$$c_n^{(i+1)} = \frac{1}{MN}\mathrm{tr}\left[\boldsymbol{W}_{\boldsymbol{n}}^{(i)}\tilde{\boldsymbol{C}}_{\boldsymbol{y}}\boldsymbol{W}_{\boldsymbol{n}}^{(i),\mathrm{H}} + \boldsymbol{C}_{\boldsymbol{n}|\boldsymbol{y}[q]}^{(i)}\right] \tag{3.62}$$

with sample covariance matrix $\tilde{\boldsymbol{C}}_{\boldsymbol{y}}$ (3.24).

Initialization

We propose to initialize the SAGE algorithm with the ML estimate of the noise covariance matrix $\boldsymbol{C}_{\boldsymbol{n},\mathrm{s}}$ based on the observations in the noise subspace $\{\bar{\boldsymbol{n}}_{\mathrm{T}}[q]\}_{q\in\mathbb{O}_{\mathrm{T}}}$ (3.27). With (3.29) this optimization problem reads

$$\begin{aligned} \boldsymbol{C}_{\boldsymbol{n},\mathrm{s}}^{(0)} &= \operatorname*{argmax}_{\boldsymbol{C}_{\boldsymbol{n},\mathrm{s}} \in \mathbb{S}_{+,0}^{M}} \sum_{q \in \mathbb{O}_{\mathrm{T}}} \sum_{b=1}^{B} \ln \mathrm{p}_{\bar{\boldsymbol{n}}}(\bar{\boldsymbol{n}}[q-b]; \boldsymbol{C}_{\boldsymbol{n},\mathrm{s}}) \\ &= \operatorname*{argmax}_{\boldsymbol{C}_{\boldsymbol{n},\mathrm{s}} \in \mathbb{S}_{+,0}^{M}} IB\bar{N} \left(-\ln \det[\boldsymbol{C}_{\boldsymbol{n},\mathrm{s}}] - \operatorname{tr}\left[\tilde{\boldsymbol{C}}_{\bar{\boldsymbol{n}},\mathrm{s}} \boldsymbol{C}_{\boldsymbol{n},\mathrm{s}}^{-1} \right] \right), \end{aligned} \tag{3.63}$$

where we used $\det[\boldsymbol{I}_{B\bar{N}} \otimes \boldsymbol{C}_{\boldsymbol{n},\mathrm{s}}] = (\det \boldsymbol{C}_{\boldsymbol{n},\mathrm{s}})^{B\bar{N}}$ (A.13) in the last step. Because $\boldsymbol{C}_{\boldsymbol{n},\mathrm{s}} \in \mathbb{S}_{+,0}^{M}$, its maximization yields the sample covariance matrix of the noise in the noise subspace

$$\boxed{\begin{aligned} \boldsymbol{C}_{\boldsymbol{n},\mathrm{s}}^{(0)} = \tilde{\boldsymbol{C}}_{\bar{\boldsymbol{n}},\mathrm{s}} &= \frac{1}{IB\bar{N}} \sum_{q \in \mathbb{O}_{\mathrm{T}}} \sum_{b=1}^{B} \sum_{n=1}^{\bar{N}} \bar{\boldsymbol{n}}[q-b,n] \bar{\boldsymbol{n}}[q-b,n]^{\mathrm{H}} \\ &= \frac{1}{IB\bar{N}} \sum_{q \in \mathbb{O}_{\mathrm{T}}} \sum_{b=1}^{B} \boldsymbol{Y}[q-b] \bar{\boldsymbol{A}}^{\mathrm{H}} \bar{\boldsymbol{A}} \boldsymbol{Y}[q-b]^{\mathrm{H}} \\ &= \frac{1}{IB\bar{N}} \sum_{q \in \mathbb{O}_{\mathrm{T}}} \sum_{b=1}^{B} (\boldsymbol{Y}[q-b] - \hat{\boldsymbol{H}}[q-b]\bar{\boldsymbol{S}})(\boldsymbol{Y}[q-b] - \hat{\boldsymbol{H}}[q-b]\bar{\boldsymbol{S}})^{\mathrm{H}}. \end{aligned}} \tag{3.64}$$

Here, we define $\bar{\boldsymbol{n}}[q,n]$ by $\bar{\boldsymbol{n}}[q] = [\bar{\boldsymbol{n}}[q,1]^{\mathrm{T}}, \bar{\boldsymbol{n}}[q,2]^{\mathrm{T}}, \ldots, \bar{\boldsymbol{n}}[q,\bar{N}]^{\mathrm{T}}]^{\mathrm{T}}$ (similar to (3.18)) and the ML channel estimate $\hat{\boldsymbol{H}}[q] = \hat{\boldsymbol{H}}_{\mathrm{ML}}[q] = \boldsymbol{Y}[q]\bar{\boldsymbol{S}}^{\dagger}$ from (2.31) with $\hat{\boldsymbol{H}}_{\mathrm{ML}} = \mathbf{unvec}\left[\hat{\boldsymbol{h}}_{\mathrm{ML}}\right]$.[11] $\bar{\boldsymbol{P}}^{\perp} = \bar{\boldsymbol{A}}^{\mathrm{H}} \bar{\boldsymbol{A}} = \boldsymbol{I}_N - \bar{\boldsymbol{S}}^{\dagger} \bar{\boldsymbol{S}}$ is the projector on the noise subspace with $\bar{\boldsymbol{S}}^{\dagger} = \bar{\boldsymbol{S}}^{\mathrm{H}} (\bar{\boldsymbol{S}} \bar{\boldsymbol{S}}^{\mathrm{H}})^{-1}$.

This estimate is unbiased in contrast to the estimate of the noise covariance matrix in (2.32). Note that the difference is due to the scaling by $\bar{N}$ instead of N.

For spatially white noise, the ML estimation problem is (cf. (3.29))

$$\max_{c_{\boldsymbol{n}} \in \mathbb{R}_{+,0}} \sum_{q \in \mathbb{O}_{\mathrm{T}}} \sum_{b=1}^{B} \ln \mathrm{p}_{\bar{\boldsymbol{n}}}(\bar{\boldsymbol{n}}[q-b]; \boldsymbol{C}_{\boldsymbol{n},\mathrm{s}} = c_{\boldsymbol{n}} \boldsymbol{I}_M) \, . \tag{3.65}$$

It is solved by the unbiased estimate of the variance

$$c_{\boldsymbol{n}}^{(0)} = \tilde{c}_{\boldsymbol{n}} = \frac{1}{IB\bar{N}M} \sum_{q \in \mathbb{O}_{\mathrm{T}}} \sum_{b=1}^{B} \sum_{n=1}^{\bar{N}} \|\bar{\boldsymbol{n}}[q-b,n]\|_2^2 \tag{3.66}$$

which can be written as

[11] Additionally, the relations $\bar{\boldsymbol{n}}[q] = \boldsymbol{A}^{\mathrm{H}} \boldsymbol{y}[q] = (\bar{\boldsymbol{A}}^{*} \otimes \boldsymbol{I}_M) \boldsymbol{y}[q]$ and $\bar{\boldsymbol{n}}[q] = \mathbf{vec}[\bar{\boldsymbol{N}}[q]]$, $\boldsymbol{y}[q] = \mathbf{vec}[\boldsymbol{Y}[q]]$ together with (A.15) are required (Section 3.2).

$$
\begin{aligned}
c_n^{(0)} = \tilde{c}_n &= \frac{1}{IB\bar{N}M} \sum_{q \in \mathbb{O}_\mathrm{T}} \sum_{b=1}^{B} \| \boldsymbol{Y}[q-b] \bar{\boldsymbol{A}}^{\mathrm{H}} \|_\mathrm{F}^2 \\
&= \frac{1}{IB\bar{N}M} \sum_{q \in \mathbb{O}_\mathrm{T}} \sum_{b=1}^{B} \| \boldsymbol{Y}[q-b] - \hat{\boldsymbol{H}}[q-b] \bar{\boldsymbol{S}} \|_\mathrm{F}^2 .
\end{aligned} \tag{3.67}
$$

Interpretation

Consider the updates (3.51) and (3.62) for estimation of the noise variance c_n. How do they change the initialization from (3.67), which is based on the noise subspace only?

The estimate in iteration $i+1$ (3.51) can be written as

$$
c_n^{(i+1)} = \frac{1}{MN} \left(M\bar{N}\tilde{c}_n + P c_n^{(i)} \left(1 - \frac{\beta c_n^{(i)}}{PB} \right) \right), \tag{3.68}
$$

where

$$
\beta = \mathrm{tr}\left[\left(\boldsymbol{S}_\mathrm{T}^\mathrm{H} \boldsymbol{S}_\mathrm{T} \right)^{-1} \left(\boldsymbol{C}_{\hat{\boldsymbol{h}}_\mathrm{T}}^{(i),-1} - \boldsymbol{C}_{\hat{\boldsymbol{h}}_\mathrm{T}}^{(i),-1} \tilde{\boldsymbol{C}}_{\hat{\boldsymbol{h}}_\mathrm{T}} \boldsymbol{C}_{\hat{\boldsymbol{h}}_\mathrm{T}}^{(i),-1} \right) \right] > 0 \tag{3.69}
$$

is the trace of a generalized Schur complement [132]. We observe that the observations of the noise in the signal subspace are taken into account in order to improve the estimate of the noise variance (similarly for general $\boldsymbol{C}_{\boldsymbol{n},\mathrm{s}}$).[12]

The last inequality in $\beta < \mathrm{tr}\left[\left(\boldsymbol{S}_\mathrm{T}^\mathrm{H} \boldsymbol{S}_\mathrm{T} \right)^{-1} \boldsymbol{C}_{\hat{\boldsymbol{h}}_\mathrm{T}}^{(i),-1} \right] \leq PB/c_n^{(i)}$ follows from $\boldsymbol{C}_{\hat{\boldsymbol{h}}_\mathrm{T}}^{(i)} = \boldsymbol{C}_{\boldsymbol{h}_\mathrm{T}}^{(i)} + c_n^{(i)} (\boldsymbol{S}_\mathrm{T}^\mathrm{H} \boldsymbol{S}_\mathrm{T})^{-1}$ omitting $\boldsymbol{C}_{\boldsymbol{h}_\mathrm{T}}^{(i)}$. Together with $\beta > 0$ this relation yields a bound on the weight for $c_n^{(i)}$ in (3.68): $0 < 1 - \beta c_n^{(i)}/(PB) < 1$. Together with $MN = M\bar{N} + P$ (by definition) and interpreting (3.68) as weighted mean we conclude that

$$
c_n^{(i+1)} < \tilde{c}_n, \; i > 0, \tag{3.70}
$$

with initialization $c_n^{(0)} = \tilde{c}_n$ from (3.67).

For a sufficiently large noise subspace, $\tilde{c}_n$ is already a very accurate estimate. Numerical evaluations show that the SAGE algorithm tends to introduce a bias for a sufficiently large size of the noise subspace and for this initialization. In this case the MSE of the estimate $c_n^{(i)}$ for the noise variance c_n is increased compared to the initialization.

[12] The positive semidefinite Least-Squares approaches in Section 3.5.3 show a similar behavior, but they take into account only a few dimensions of the signal subspace.

3.3.3 Estimation of Channel Correlations in the Time, Delay, and Space Dimensions

The channel covariance matrix $\boldsymbol{C}_{\boldsymbol{h}_{\mathrm{T}}}$ of all channel parameters corresponding to B samples in time, M samples in space, K transmitters, and for a channel order L has block Toeplitz structure $\mathbb{T}_{+,0}^{P,B}$ (Model 1 in Section 3.2). Fuhrmann et al.[80] proposed the EM algorithm for the case that the parameters are observed directly (without noise). This is an extension of their algorithm for ML estimation of a covariance matrix with Toeplitz structure [81, 149]. The idea for this algorithm stems from the observation that the ML problem for a positive definite covariance matrix with Toeplitz structure is very likely to be ill-posed, i.e., yields a singular estimate, as presented in Theorems 3.2 and 3.3. They also showed that a useful regularization is the restriction to covariance matrices with positive definite circulant extension of length $\bar{B}$, which results in a positive definite estimate with probability one (Section 3.1.2). A Toeplitz covariance matrix with circulant extension is the upper left block of a circulant covariance matrix, which describes the autocovariance function of a (wide-sense) periodic random sequence with period $\bar{B}$. Moreover, the ML estimation of a circulant covariance matrix can be obtained explicitly [46] (see Section 3.1.1). Based on these two results Fuhrmann et al.choose a full period of the assumed periodic random process as complete data space and solve the optimization problem iteratively based on the EM algorithm. In [80] this is extended to block Toeplitz structure with the full period of a multivariate periodic random sequence as complete data.

We follow their approach and choose a hidden data space for $\boldsymbol{C}_{\boldsymbol{h}_{\mathrm{T}}}$ which extends $\boldsymbol{h}_{\mathrm{T}} \in \mathbb{C}^{BP}$ to a full period $\bar{B}$ of a P-variate periodic random sequence. This is equivalent to the assumption of a $\boldsymbol{C}_{\boldsymbol{h}_{\mathrm{T}}}$ with positive definite block circulant extension $\mathbb{T}_{\mathrm{ext}+,0}^{P,B}$. Necessarily we have $\bar{B} \geq 2B-1$.[13] Formally, the hidden data space is the random vector

$$\boldsymbol{z}[q] = \begin{bmatrix} \boldsymbol{h}_{\mathrm{T}}[q] \\ \boldsymbol{h}_{\mathrm{ext}}[q] \end{bmatrix} \in \mathbb{C}^{\bar{B}P}, \tag{3.71}$$

where $\boldsymbol{h}_{\mathrm{ext}}[q] \in \mathbb{C}^{(\bar{B}-B)P}$ contains the remaining $\bar{B}-B$ samples of the P-variate random sequence. Its block circulant covariance matrix reads

$$\boldsymbol{C}_{\boldsymbol{z}} = (\boldsymbol{V} \otimes \boldsymbol{I}_P)\boldsymbol{C}_{\boldsymbol{a}}(\boldsymbol{V}^{\mathrm{H}} \otimes \boldsymbol{I}_P), \tag{3.72}$$

where the block diagonal matrix

---

[13] To ensure an asymptotically unbiased estimate, $\bar{B}$ has to be chosen such that the true $\boldsymbol{C}_{\boldsymbol{h}_{\mathrm{T}}}$ has a positive definite circulant extension of size $\bar{B}$. In case $P=1$, sufficient lower bounds on $\bar{B}$ are given by [47, 155], which depend on the minimum eigenvalue of $\boldsymbol{C}_{\boldsymbol{h}_{\mathrm{T}}}$. If the observation size I and B are small, unbiasedness is not important, as the error is determined by the small sample size. Thus, we can choose the minimum $\bar{B} = 2B-1$.

$$\boldsymbol{C}_{\boldsymbol{a}} = \begin{bmatrix} \boldsymbol{C}_{\boldsymbol{a}[1]} & & \\ & \ddots & \\ & & \boldsymbol{C}_{\boldsymbol{a}[\bar{B}]} \end{bmatrix} \tag{3.73}$$

is the covariance matrix of $\boldsymbol{a}[q] = [\boldsymbol{a}[q,1]^{\mathrm{T}}, \boldsymbol{a}[q,2]^{\mathrm{T}}, \ldots, \boldsymbol{a}[q,\bar{B}]^{\mathrm{T}}]^{\mathrm{T}}$, $\boldsymbol{a}[q,n] \in \mathbb{C}^P$ is given by $\boldsymbol{z}[q] = (\boldsymbol{V} \otimes \boldsymbol{I}_P)\boldsymbol{a}[q]$, and $\boldsymbol{C}_{\boldsymbol{a}[n]} = \mathrm{E}[\boldsymbol{a}[q,n]\boldsymbol{a}[q,n]^{\mathrm{H}}]$. For $\boldsymbol{V}_B \in \mathbb{C}^{B\times\bar{B}}$, we define the partitioning of the unitary matrix $[\boldsymbol{V}]_{k,\ell} = \frac{1}{\sqrt{\bar{B}}}\exp(\mathrm{j}2\pi(k-1)(\ell-1)/\bar{B}), k,\ell = 1,2,\ldots,\bar{B}$,

$$\boldsymbol{V} = \begin{bmatrix} \boldsymbol{V}_B \\ \bar{\boldsymbol{V}}_B \end{bmatrix}. \tag{3.74}$$

The upper left $BP \times BP$ block of $\boldsymbol{C}_{\boldsymbol{z}}$ is the positive definite block Toeplitz matrix

$$\boldsymbol{C}_{\boldsymbol{h}_{\mathrm{T}}} = (\boldsymbol{V}_B \otimes \boldsymbol{I}_P)\boldsymbol{C}_{\boldsymbol{a}}(\boldsymbol{V}_B^{\mathrm{H}} \otimes \boldsymbol{I}_P) \in \mathbb{T}_{\mathrm{ext}+,0}^{P,B}, \tag{3.75}$$

which we would like to estimate.

The log-likelihood of the hidden data space (3.71) is

$$\mathrm{L}_{\boldsymbol{z}}(\{\boldsymbol{z}[q]\}_{q\in\mathbb{O}_{\mathrm{T}}};\boldsymbol{C}_{\boldsymbol{z}}) = -I\bar{B}P\ln\pi - I\ln\det[\boldsymbol{C}_{\boldsymbol{z}}] - I\mathrm{tr}\left[\tilde{\boldsymbol{C}}_{\boldsymbol{z}}\boldsymbol{C}_{\boldsymbol{z}}^{-1}\right] \tag{3.76}$$

with (hypothetical) sample covariance matrix of the hidden data space $\tilde{\boldsymbol{C}}_{\boldsymbol{z}} = \frac{1}{I}\sum_{q\in\mathbb{O}_{\mathrm{T}}} \boldsymbol{z}[q]\boldsymbol{z}[q]^{\mathrm{H}}$. It can be simplified to

$$\begin{aligned}\mathrm{L}_{\boldsymbol{z}}(\{\boldsymbol{z}[q]\}_{q\in\mathbb{O}_{\mathrm{T}}};\boldsymbol{C}_{\boldsymbol{z}}) &= -I\bar{B}P\ln\pi - I\sum_{n=1}^{\bar{B}}\ln\det\left[\boldsymbol{C}_{\boldsymbol{a}[n]}\right] - I\sum_{n=1}^{\bar{B}}\mathrm{tr}\left[\tilde{\boldsymbol{C}}_{\boldsymbol{a}[n]}\boldsymbol{C}_{\boldsymbol{a}[n]}^{-1}\right] \\ &= \sum_{n=1}^{\bar{B}}\mathrm{L}_{\boldsymbol{a}[n]}(\{\boldsymbol{a}[q,n]\}_{q\in\mathbb{O}_{\mathrm{T}}};\boldsymbol{C}_{\boldsymbol{a}[n]})\end{aligned} \tag{3.77}$$

exploiting the block circulant structure of $\boldsymbol{C}_{\boldsymbol{z}}$, which yields $\ln\det[\boldsymbol{C}_{\boldsymbol{z}}] = \ln\det[\boldsymbol{C}_{\boldsymbol{a}}] = \sum_{n=1}^{\bar{B}}\ln\det[\boldsymbol{C}_{\boldsymbol{a}[n]}]$ and

$$\mathrm{tr}\left[\tilde{\boldsymbol{C}}_{\boldsymbol{z}}\boldsymbol{C}_{\boldsymbol{z}}^{-1}\right] = \mathrm{tr}\left[\tilde{\boldsymbol{C}}_{\boldsymbol{a}}\boldsymbol{C}_{\boldsymbol{a}}^{-1}\right] = \sum_{n=1}^{\bar{B}}\mathrm{tr}\left[\tilde{\boldsymbol{C}}_{\boldsymbol{a}[n]}\boldsymbol{C}_{\boldsymbol{a}[n]}^{-1}\right] \tag{3.78}$$

with $\tilde{\boldsymbol{C}}_{\boldsymbol{a}[n]} = \boldsymbol{J}'_n\tilde{\boldsymbol{C}}_{\boldsymbol{a}}\boldsymbol{J}_n'^{\mathrm{T}}$ and $\boldsymbol{J}'_n = [\boldsymbol{0}_{P\times P(n-1)}, \boldsymbol{I}_P, \boldsymbol{0}_{P\times P(\bar{B}-n)}] \in \{0,1\}^{P\times P\bar{B}}$. Note that the log-likelihood function (3.77) of $\{\boldsymbol{z}[q]\}_{q\in\mathbb{O}_{\mathrm{T}}}$ is equivalent to the log-likelihood function of the observations $\{\boldsymbol{a}[q,n]\}_{q\in\mathbb{O}_{\mathrm{T}}}, n = 1,2,\ldots,\bar{B}$, of $\bar{B}$ statistically independent random vectors $\boldsymbol{a}[q,n]$.

E- and M-Step

The CM estimate of the likelihood function (3.76) based on estimates from the previous iteration i is

$$\begin{aligned}&\mathrm{E}_{\boldsymbol{z}}\left[\mathrm{L}_{\boldsymbol{z}}(\{\boldsymbol{z}[q]\}_{q\in\mathbb{O}_\mathrm{T}};\boldsymbol{C}_{\boldsymbol{z}})|\{\boldsymbol{t}_\mathrm{T}[q]\}_{q\in\mathbb{O}_\mathrm{T}};\boldsymbol{C}_{\boldsymbol{z}}^{(i)},\boldsymbol{C}_{\boldsymbol{n},\mathrm{s}}^{(i)}\right]=\\&-I\bar{B}P\ln\pi-I\ln\det[\boldsymbol{C}_{\boldsymbol{z}}]-I\mathrm{tr}\left[\mathrm{E}_{\boldsymbol{z}}\left[\tilde{\boldsymbol{C}}_{\boldsymbol{z}}|\{\hat{\boldsymbol{h}}_\mathrm{T}[q]\}_{q\in\mathbb{O}_\mathrm{T}};\boldsymbol{C}_{\boldsymbol{z}}^{(i)},\boldsymbol{C}_{\boldsymbol{n},\mathrm{s}}^{(i)}\right]\boldsymbol{C}_{\boldsymbol{z}}^{-1}\right].\end{aligned}\tag{3.79}$$

With (A.2) and (A.3) the CM estimate of the sample covariance matrix $\tilde{\boldsymbol{C}}_{\boldsymbol{z}}$ of the HDS can be written as

$$\mathrm{E}_{\boldsymbol{z}}\left[\tilde{\boldsymbol{C}}_{\boldsymbol{z}}|\{\hat{\boldsymbol{h}}_\mathrm{T}[q]\}_{q\in\mathbb{O}_\mathrm{T}};\boldsymbol{C}_{\boldsymbol{z}}^{(i)},\boldsymbol{C}_{\boldsymbol{n},\mathrm{s}}^{(i)}\right]=\boldsymbol{W}_{\boldsymbol{z}}^{(i)}\tilde{\boldsymbol{C}}_{\hat{\boldsymbol{h}}_\mathrm{T}}\boldsymbol{W}_{\boldsymbol{z}}^{(i),\mathrm{H}}+\boldsymbol{C}_{\boldsymbol{z}|\hat{\boldsymbol{h}}_\mathrm{T}[q]}^{(i)}\tag{3.80}$$

with sample covariance matrix of the observations $\tilde{\boldsymbol{C}}_{\hat{\boldsymbol{h}}_\mathrm{T}}=\frac{1}{I}\sum_{q\in\mathbb{O}_\mathrm{T}}\hat{\boldsymbol{h}}_\mathrm{T}[q]\hat{\boldsymbol{h}}_\mathrm{T}[q]^\mathrm{H}$, conditional mean

$$\begin{aligned}\hat{\boldsymbol{z}}^{(i)}[q]&=\mathrm{E}_{\boldsymbol{z}}\left[\boldsymbol{z}[q]|\boldsymbol{t}_\mathrm{T}[q];\boldsymbol{C}_{\boldsymbol{z}}^{(i)},\boldsymbol{C}_{\boldsymbol{n},\mathrm{s}}^{(i)}\right]\\&=\mathrm{E}_{\boldsymbol{z}}\left[\boldsymbol{z}[q]|\hat{\boldsymbol{h}}_\mathrm{T}[q];\boldsymbol{C}_{\boldsymbol{z}}^{(i)},\boldsymbol{C}_{\boldsymbol{n},\mathrm{s}}^{(i)}\right]&&(3.81)\\&=\underbrace{\boldsymbol{C}_{\boldsymbol{z}}^{(i)}\begin{bmatrix}\boldsymbol{I}_{BP}\\\boldsymbol{0}_{P(\bar{B}-B)\times BP}\end{bmatrix}\boldsymbol{C}_{\hat{\boldsymbol{h}}_\mathrm{T}}^{(i),-1}}_{\boldsymbol{W}_{\boldsymbol{z}}^{(i)}}\hat{\boldsymbol{h}}_\mathrm{T}[q],&&(3.82)\end{aligned}$$

$\boldsymbol{C}_{\hat{\boldsymbol{h}}_\mathrm{T}}^{(i)}=\boldsymbol{C}_{\boldsymbol{h}_\mathrm{T}}^{(i)}+\boldsymbol{I}_B\otimes(\bar{\boldsymbol{S}}^*\bar{\boldsymbol{S}}^\mathrm{T})^{-1}\otimes\boldsymbol{C}_{\boldsymbol{n},\mathrm{s}}^{(i)}$, and conditional covariance matrix

$$\boldsymbol{C}_{\boldsymbol{z}|\hat{\boldsymbol{h}}_\mathrm{T}[q]}^{(i)}=\boldsymbol{C}_{\boldsymbol{z}}^{(i)}-\boldsymbol{W}_{\boldsymbol{z}}^{(i)}[\boldsymbol{I}_{BP},\boldsymbol{0}_{P(\bar{B}-B)\times BP}]\boldsymbol{C}_{\boldsymbol{z}}^{(i)}.\tag{3.83}$$

It follows from the definition of $\boldsymbol{t}_\mathrm{T}[q]$ (3.27) that $\hat{\boldsymbol{h}}_\mathrm{T}[q]$ and $\bar{\boldsymbol{n}}_\mathrm{T}[q]$ are statistically independent and the conditioning on $\boldsymbol{t}_\mathrm{T}[q]$ in (3.79) reduces to $\hat{\boldsymbol{h}}_\mathrm{T}[q]$ in (3.80) and (3.81).

We can rewrite (3.79) similarly to (3.77), which corresponds to the estimation of $\bar{B}$ unstructured positive definite covariance matrices $\boldsymbol{C}_{\boldsymbol{a}[n]}, n=1,2,\ldots,\bar{B}$. Therefore, maximization of (3.79) results in the estimate of the blocks

$$\boldsymbol{C}_{\boldsymbol{a}[n]}^{(i+1)}=\boldsymbol{J}_n'(\boldsymbol{V}^\mathrm{H}\otimes\boldsymbol{I}_P)\left(\boldsymbol{W}_{\boldsymbol{z}}^{(i)}\tilde{\boldsymbol{C}}_{\hat{\boldsymbol{h}}_\mathrm{T}}\boldsymbol{W}_{\boldsymbol{z}}^{(i),\mathrm{H}}+\boldsymbol{C}_{\boldsymbol{z}|\hat{\boldsymbol{h}}_\mathrm{T}[q]}^{(i)}\right)(\boldsymbol{V}\otimes\boldsymbol{I}_P)\boldsymbol{J}_n'^{,\mathrm{T}}\tag{3.84}$$

on the diagonal of $\boldsymbol{C}_{\boldsymbol{a}}^{(i+1)}=(\boldsymbol{V}^\mathrm{H}\otimes\boldsymbol{I}_P)\boldsymbol{C}_{\boldsymbol{z}}^{(i+1)}(\boldsymbol{V}\otimes\boldsymbol{I}_P)$ in iteration $i+1$. The upper left block of $\boldsymbol{C}_{\boldsymbol{z}}^{(i+1)}$ is the estimate of interest to us:

$$\boxed{\begin{aligned}\boldsymbol{C}_{\boldsymbol{h}_\mathrm{T}}^{(i+1)} &= (\boldsymbol{V}_B \otimes \boldsymbol{I}_P)\boldsymbol{C}_{\boldsymbol{a}}^{(i+1)}(\boldsymbol{V}_B^\mathrm{H} \otimes \boldsymbol{I}_P) \\ &= [\boldsymbol{I}_{BP}, \boldsymbol{0}_{BP\times P(\bar{B}-B)}]\boldsymbol{C}_{\boldsymbol{z}}^{(i+1)}[\boldsymbol{I}_{BP}, \boldsymbol{0}_{BP\times P(\bar{B}-B)}]^\mathrm{T} \in \mathbb{T}_{\mathrm{ext}+,0}^{P,B}.\end{aligned}} \tag{3.85}$$

After $i_{\max}$ iterations the algorithm is stopped and our final estimate is $\hat{\boldsymbol{C}}_{\boldsymbol{h}_\mathrm{T}} = \boldsymbol{C}_{\boldsymbol{h}_\mathrm{T}}^{(i_{\max})}$.

Initialization

The SAGE algorithm requires an initialization $\boldsymbol{C}_{\boldsymbol{z}}^{(0)}$ of $\boldsymbol{C}_{\boldsymbol{z}}$. We start with an estimate of $\boldsymbol{C}_{\boldsymbol{h}_\mathrm{T}}$ (for the choice of $\mathbb{O}_\mathrm{T}$ in Section 3.2)

$$\boxed{\boldsymbol{C}_{\boldsymbol{h}_\mathrm{T}}^{(0)} = \frac{1}{IB}\sum_{q=1}^{I} \hat{\boldsymbol{\mathcal{H}}}[q]\hat{\boldsymbol{\mathcal{H}}}[q]^\mathrm{H} \in \mathbb{T}_{+,0}^{P,B},} \tag{3.86}$$

where

$$\hat{\boldsymbol{\mathcal{H}}}[q] = \begin{bmatrix} \hat{\boldsymbol{h}}[q-1] \ldots \hat{\boldsymbol{h}}[q-B] & & \\ \ddots & \ddots & \\ & \hat{\boldsymbol{h}}[q-1] \ldots \hat{\boldsymbol{h}}[q-B] \end{bmatrix} \in \mathbb{C}^{BP\times 2B-1} \tag{3.87}$$

is block Toeplitz. $\boldsymbol{C}_{\boldsymbol{h}_\mathrm{T}}^{(0)}$ is also block Toeplitz. Its blocks are given by the estimates of $\boldsymbol{C}_{\boldsymbol{h}}[\ell]$

$$\boldsymbol{C}_{\boldsymbol{h}}^{(0)}[\ell] = \frac{1}{IB}\sum_{q=1}^{I}\sum_{i=1}^{B-\ell} \hat{\boldsymbol{h}}[q-i]\hat{\boldsymbol{h}}[q-i-\ell]^\mathrm{H}, \ell = 0, 1, \ldots, B-1, \tag{3.88}$$

which is the multivariate extension of the standard biased estimator [203, p. 24]. It is biased due to the normalization with IB instead of $I(B-\ell)$ in (3.88) and the noise in $\hat{\boldsymbol{h}}[q]$. But from the construction in (3.86) we observe that $\boldsymbol{C}_{\boldsymbol{h}_\mathrm{T}}^{(0)}$ is positive semidefinite. For $I = 1$ and $P > 1$, it is always singular because $BP > 2B-1$. Therefore, we choose $I > 1$ to obtain a numerically stable algorithm.[14]

But (3.86) defines only the upper left block of $\boldsymbol{C}_{\boldsymbol{z}}^{(0)}$. We may choose $\boldsymbol{C}_{\boldsymbol{z}}^{(0)}$ to be block circulant with first block-row

$$[\boldsymbol{C}_{\boldsymbol{h}}^{(0)}[0], \boldsymbol{C}_{\boldsymbol{h}}^{(0)}[1], \ldots, \boldsymbol{C}_{\boldsymbol{h}}^{(0)}[B-1], \boldsymbol{0}_{P\times P(\bar{B}-2B+1)}, \boldsymbol{C}_{\boldsymbol{h}}^{(0)}[B-1]^\mathrm{H}, \ldots, \boldsymbol{C}_{\boldsymbol{h}}^{(0)}[1]^\mathrm{H}], \tag{3.89}$$

[14] Assuming a full column rank of $\hat{\boldsymbol{\mathcal{H}}}[q]$, a necessary condition for positive definite $\boldsymbol{C}_{\boldsymbol{h}_\mathrm{T}}^{(0)}$ is $I \geq P/(2-1/B)$ (at least $I \geq P/2$). But non-singularity of $\boldsymbol{C}_{\boldsymbol{h}_\mathrm{T}}^{(0)}$ is not required by the algorithm.

which is not necessarily positive semidefinite. This is only a problem if it results in an indefinite estimate $\boldsymbol{C}_{\boldsymbol{z}}^{(i+1)}$ (3.84) due to an indefinite $\boldsymbol{C}^{(i)}_{\boldsymbol{z}|\hat{\boldsymbol{h}}_{\mathrm{T}}[q]}$ in (3.83). For $\bar{B} = 2B-1$, the block circulant extension is unique. We now show that it is positive semidefinite which yields a positive semidefinite $\boldsymbol{C}_{\boldsymbol{z}}^{(i+1)}$ (3.84).

Define the block circulant $P(2B-1) \times (2B-1)$ matrix

$$\hat{\bar{\boldsymbol{\mathcal{H}}}}[q] = \begin{bmatrix} \hat{\boldsymbol{\mathcal{H}}}[q] \\ \hat{\boldsymbol{\mathcal{H}}}_{\mathrm{ext}}[q] \end{bmatrix} \tag{3.90}$$

with $P(B-1) \times (2B-1)$ block circulant matrix

$$\hat{\boldsymbol{\mathcal{H}}}_{\mathrm{ext}}[q] = \begin{bmatrix} \hat{\boldsymbol{h}}[q-B] & \boldsymbol{0}_{P\times 1} & \dots & \boldsymbol{0}_{P\times 1} & \hat{\boldsymbol{h}}[q-1] & \dots & \hat{\boldsymbol{h}}[q-B+1] \\ \vdots & \ddots & & & & \ddots & \vdots \\ \hat{\boldsymbol{h}}[q-2] & \dots & \hat{\boldsymbol{h}}[q-B] & \boldsymbol{0}_{P\times 1} & \dots & \boldsymbol{0}_{P\times 1} & \hat{\boldsymbol{h}}[q-1] \end{bmatrix}. \tag{3.91}$$

The unique block circulant extension of $\boldsymbol{C}_{\boldsymbol{h}}^{(0)}[\ell]$ (3.86) is

$$\boxed{\boldsymbol{C}_{\boldsymbol{z}}^{(0)} = \frac{1}{IB} \sum_{q=1}^{I} \hat{\bar{\boldsymbol{\mathcal{H}}}}[q] \hat{\bar{\boldsymbol{\mathcal{H}}}}[q]^{\mathrm{H}} \in \mathbb{S}_{+,0}^{(2B-1)P},} \tag{3.92}$$

because $\bar{B} = 2B-1$ and $\hat{\boldsymbol{\mathcal{H}}}[q]\hat{\boldsymbol{\mathcal{H}}}[q]^{\mathrm{H}}$ is the upper left block of $\hat{\bar{\boldsymbol{\mathcal{H}}}}[q]\hat{\bar{\boldsymbol{\mathcal{H}}}}[q]^{\mathrm{H}}$. Due to its definition in (3.92) it is positive semidefinite.[15] This results in the following theorem.

Theorem 3.5. *For* $\bar{B} = 2B-1$, $\boldsymbol{C}_{\boldsymbol{h}_{\mathrm{T}}}^{(0)}$ (3.86) *has a unique positive semidefinite block-circulant extension* $\boldsymbol{C}_{\boldsymbol{z}}^{(0)}$ *given by* (3.90) *and* (3.92).

Therefore, choosing $\bar{B} = 2B-1$ we have a reliable initialization of our iterative algorithm. Its disadvantage is that the minimum $\bar{B}$ is chosen, which may be too small to be able to represent the true covariance matrix as a block Toeplitz matrix with block circulant extension (see [47] for Toeplitz structure). But the advantage is a reduced complexity and a simple initialization procedure, which is not as complicated as the constructive proofs in [47, 155] suggest.

[15] If $\hat{\bar{\boldsymbol{\mathcal{H}}}}[q], q \in \mathbb{O}_{\mathrm{T}}$, have full column rank (see [214]), then $I \geq P$ is necessary for non-singularity of $\boldsymbol{C}_{\boldsymbol{z}}^{(0)}$.

3.3.4 Estimation of Channel Correlations in the Delay and Space Dimensions

Estimation of $\boldsymbol{C_h}$ for model 2 in Section 3.2 is a special case of the derivations in Section 3.3.3: It is obtained for $I = 1$ and $\boldsymbol{C}_{\boldsymbol{h}_\mathrm{T}} = \boldsymbol{I}_B \otimes \boldsymbol{C_h}$ (temporally uncorrelated multivariate random sequence). Similar results are given by Dogandžić [60] based on the "expectation-conditional maximization either algorithm" [142] for non-zero mean of $\boldsymbol{h}[q]$ and $\boldsymbol{C}_{\boldsymbol{n},\mathrm{s}} = c_n \boldsymbol{I}_M$, which uses a different estimation procedure for c_n than ours given in Section 3.3.2. The performance is compared numerically in Section 3.6.

The following results were presented first in [220]. Here, we derive them as a special case of Section 3.3.3. The conditional mean estimate of the log-likelihood function for the HDS $\boldsymbol{h}[q]$ is

$$\mathrm{E}_{\boldsymbol{h}}\left[\mathrm{L}_{\boldsymbol{h}}(\{\boldsymbol{h}[q]\}_{q\in\mathbb{O}};\boldsymbol{C_h})|\{\hat{\boldsymbol{h}}[q]\}_{q\in\mathbb{O}};\boldsymbol{C}_{\boldsymbol{h}}^{(i)},\boldsymbol{C}_{\boldsymbol{n},\mathrm{s}}^{(i)}\right]. \tag{3.93}$$

Its maximization w.r.t. $\boldsymbol{C_h} \in \mathbb{S}_{+,0}^P$ in iteration $i+1$ yields the conditional mean estimate of the sample covariance matrix for the HDS $\tilde{\boldsymbol{C}}_{\boldsymbol{h}} = \frac{1}{B}\sum_{q=1}^{B} \boldsymbol{h}[q]\boldsymbol{h}[q]^\mathrm{H}$

$$\boxed{\begin{aligned}\boldsymbol{C}_{\boldsymbol{h}}^{(i+1)} &= \mathrm{E}_{\boldsymbol{h}}\left[\tilde{\boldsymbol{C}}_{\boldsymbol{h}}|\{\hat{\boldsymbol{h}}[q]\}_{q\in\mathbb{O}};\boldsymbol{C}_{\boldsymbol{h}}^{(i)},\boldsymbol{C}_{\boldsymbol{n},\mathrm{s}}^{(i)}\right]\\ &= \boldsymbol{W}_{\boldsymbol{h}}^{(i)}\tilde{\boldsymbol{C}}_{\hat{\boldsymbol{h}}}\boldsymbol{W}_{\boldsymbol{h}}^{(i),\mathrm{H}} + \boldsymbol{C}_{\boldsymbol{h}|\hat{\boldsymbol{h}}[q]}^{(i)}.\end{aligned}} \tag{3.94}$$

Due to the definition of $\boldsymbol{t}[q]$ (3.25) it depends only on the ML estimates $\hat{\boldsymbol{h}}[q]$ of $\boldsymbol{h}[q]$ via their sample covariance matrix $\tilde{\boldsymbol{C}}_{\hat{\boldsymbol{h}}} = \frac{1}{B}\sum_{q=1}^{B}\hat{\boldsymbol{h}}[q]\hat{\boldsymbol{h}}[q]^\mathrm{H}$. The CM estimator of $\boldsymbol{h}[q]$ and its conditional covariance matrix are defined as (see (A.3))

$$\begin{aligned}\boldsymbol{W}_{\boldsymbol{h}}^{(i)} &= \boldsymbol{C}_{\boldsymbol{h}}^{(i)}\left(\boldsymbol{C}_{\boldsymbol{h}}^{(i)} + (\bar{\boldsymbol{S}}^*\bar{\boldsymbol{S}}^\mathrm{T})^{-1}\otimes\boldsymbol{C}_{\boldsymbol{n},\mathrm{s}}^{(i)}\right)^{-1}\\ \boldsymbol{C}_{\boldsymbol{h}|\hat{\boldsymbol{h}}[q]}^{(i)} &= \boldsymbol{C}_{\boldsymbol{h}}^{(i)} - \boldsymbol{W}_{\boldsymbol{h}}^{(i)}\boldsymbol{C}_{\boldsymbol{h}}^{(i)}\end{aligned} \tag{3.95}$$

with $\boldsymbol{C}_{\boldsymbol{n}}^{(i)} = \boldsymbol{I}_N \otimes \boldsymbol{C}_{\boldsymbol{n},\mathrm{s}}^{(i)}$ based on (3.61) or $\boldsymbol{C}_{\boldsymbol{n}}^{(i)} = c_n^{(i)}\boldsymbol{I}_{MN}$ from (3.62). Because $\boldsymbol{C}_{\boldsymbol{h}|\hat{\boldsymbol{h}}[q]}^{(i)}$ is positive semidefinite, we ensure a positive semidefinite estimate $\boldsymbol{C}_{\boldsymbol{h}}^{(i+1)}$ in every iteration (assuming a positive semidefinite initialization).

Initialization

The iterations are initialized with the sample mean estimate of the covariance matrix $\boldsymbol{C_h}$

$$\boldsymbol{C}_{\boldsymbol{h}}^{(0)} = \tilde{\boldsymbol{C}}_{\hat{\boldsymbol{h}}} = \frac{1}{B}\sum_{q=1}^{B}\hat{\boldsymbol{h}}[q]\hat{\boldsymbol{h}}[q]^{\mathrm{H}}, \tag{3.96}$$

which is biased and positive semidefinite (non-singular, if $B \geq P$). In Section 3.5.1 it is derived based on least-squares approximation. Its bias stems from the noise in $\hat{\boldsymbol{h}}[q]$. Thus, the only task of SAGE is to remove this bias from $\boldsymbol{C}_{\boldsymbol{h}}^{(0)}$ with positive definite estimates $\boldsymbol{C}_{\boldsymbol{h}}^{(i+1)}$.

If the SAGE algorithm is initialized with $\boldsymbol{C}_{\boldsymbol{n},\mathrm{s}}^{(0)} = \mathbf{0}_{M\times M}$ and the first iteration is based on HDS 1, then we obtain our initialization (3.96) in the first iteration.

An alternative initialization is the estimate from the positive semidefinite Least-Squares approach in Section 3.5.3.

3.3.5 Estimation of Temporal Channel Correlations

An interesting special case of Section 3.3.3 is $P = 1$. The ML problem (3.21) reduces to P independent ML problem of this type ($P = 1$), if the elements of $\boldsymbol{h}[q]$ are statistically independent, i.e., we assume a diagonal covariance matrix $\boldsymbol{C}_{\boldsymbol{h}}$. This is also of interest for prediction of a multivariate random sequence with the sub-optimum predictor (2.55), which requires only knowledge of $B+1$ ($Q = B$) samples of the autocovariance sequence for every parameter.

Thus, $\boldsymbol{C}_{\boldsymbol{h}_{\mathrm{T}}}$ has Toeplitz structure with first row $[c_h[0], c_h[1], \ldots, c_h[B-1]]$ in this case and the results from Section 3.3.3 specialize to the EM algorithm which is given in [149] for known noise variance c_n. The HDS is a full period of $\bar{B}$-periodic random sequence in $\boldsymbol{z}[q] \in \mathbb{C}^{\bar{B}}$ with circulant covariance matrix $\boldsymbol{C}_{\boldsymbol{z}}$, which leads to a diagonal $\boldsymbol{C}_{\boldsymbol{a}} = \mathbf{diag}[c_{a[1]}, c_{a[2]}, \ldots, c_{a[\bar{B}]}]$ with $\boldsymbol{a}[q] = \boldsymbol{V}^{\mathrm{H}}\boldsymbol{z}[q]$, $\boldsymbol{a}[q] = [a[q,1], a[q,2], \ldots, a[q,\bar{B}]]^{\mathrm{T}}$, and $c_{a[n]} = \mathrm{E}[|a[q,n]|^2]$. From (3.85) and (3.84) the E- and M-Step for $P = 1$ read

$$\begin{aligned}
c_{a[n]}^{(i+1)} &= \left[\boldsymbol{V}^{\mathrm{H}}\left(\boldsymbol{W}_{\boldsymbol{z}}^{(i)}\tilde{\boldsymbol{C}}_{\hat{\boldsymbol{h}}_{\mathrm{T}}}\boldsymbol{W}_{\boldsymbol{z}}^{(i),\mathrm{H}} + \boldsymbol{C}_{\boldsymbol{z}|\hat{\boldsymbol{h}}_{\mathrm{T}}[q]}^{(i)}\right)\boldsymbol{V}\right]_{n,n} \\
\boldsymbol{C}_{\boldsymbol{a}}^{(i+1)} &= \mathbf{diag}\left[c_{a[1]}^{(i+1)}, c_{a[2]}^{(i+1)}, \ldots, c_{a[\bar{B}]}^{(i+1)}\right] \in \mathbb{D}_{+,0}^{\bar{B}} \\
\boldsymbol{C}_{\boldsymbol{z}}^{(i+1)} &= \boldsymbol{V}\boldsymbol{C}_{\boldsymbol{a}}^{(i+1)}\boldsymbol{V}^{\mathrm{H}} \in \mathbb{T}_{\mathrm{c}+,0}^{\bar{B}} \\
\boldsymbol{C}_{\boldsymbol{h}_{\mathrm{T}}}^{(i+1)} &= [\boldsymbol{I}_B, \mathbf{0}_{B\times(\bar{B}-B)}]\boldsymbol{C}_{\boldsymbol{z}}^{(i+1)}[\boldsymbol{I}_B, \mathbf{0}_{B\times(\bar{B}-B)}]^{\mathrm{T}} \in \mathbb{T}_{\mathrm{ext}+,0}^{B}
\end{aligned} \tag{3.97}$$

with $\boldsymbol{W}_{\boldsymbol{z}}^{(i)}$ and $\boldsymbol{C}_{\boldsymbol{z}|\hat{\boldsymbol{h}}_{\mathrm{T}}[q]}^{(i)}$ as in (3.82) and (3.83).

Initialization

As initialization for $\boldsymbol{C}_{\boldsymbol{h}_{\mathrm{T}}}^{(0)}$ we choose (3.86) for $P = 1$ and $I = 1$, which is a positive definite and biased estimate [169, p. 153]. For $\bar{B} = 2B - 1$, we obtain the unique positive definite circulant extension $\boldsymbol{C}_{\boldsymbol{z}}^{(0)}$ based on Theorem 3.5.[16]

3.3.6 Extensions and Computational Complexity

In Section 2.1 we discussed different restrictions on the structure of $\boldsymbol{C}_{\boldsymbol{h}_{\mathrm{T}}}$. As one example, we now choose

$$\boldsymbol{C}_{\boldsymbol{h}_{\mathrm{T}}} = \sum_{k=1}^{K} \boldsymbol{C}_{\mathrm{T},k} \otimes \boldsymbol{e}_k \boldsymbol{e}_k^{\mathrm{T}} \otimes \boldsymbol{C}_{\boldsymbol{h}_k} \subseteq \mathbb{S}_{+,0}^{BMK} \tag{3.98}$$

for $L = 0$ (frequency flat channel) and $Q = B$ in (2.11).[17] The covariance matrix is a nonlinear function in $\boldsymbol{C}_{\mathrm{T},k} \in \mathbb{T}_{+,0}^{B}$ with Toeplitz structure and $\boldsymbol{C}_{\boldsymbol{h}_k} \in \mathbb{S}_{+,0}^{M}$. We would like to estimate $\boldsymbol{C}_{\boldsymbol{h}_{\mathrm{T}}}$ with this constraint on its structure.

The advantage of this structure for $\boldsymbol{C}_{\boldsymbol{h}_{\mathrm{T}}}$ is a reduction of the number of unknown (real-valued) parameters from $(2B-1)M^2K^2$ to $K(2B-1+M^2)$ and a reduction of computational complexity.

Two assumptions yield this structure (compare Section 2.1):

- $\boldsymbol{h}_k[q]$ in $\boldsymbol{h}[q] = [\boldsymbol{h}_1[q]^{\mathrm{T}}, \boldsymbol{h}_2[q]^{\mathrm{T}}, \ldots, \boldsymbol{h}_K[q]^{\mathrm{T}}]^{\mathrm{T}}$ are assumed to be statistically independent for different k, i.e., $\mathrm{E}[\boldsymbol{h}_k[q]\boldsymbol{h}_{k'}[q]^{\mathrm{H}}] = \boldsymbol{C}_{\boldsymbol{h}_k}\delta[k-k']$ ($\boldsymbol{C}_{\boldsymbol{h}}$ is block-diagonal).
- The covariance matrix of $\boldsymbol{h}_{\mathrm{T},k}[q] = [\boldsymbol{h}_k[q-1]^{\mathrm{T}}, \boldsymbol{h}_k[q-2]^{\mathrm{T}}, \ldots, \boldsymbol{h}_k[q-B]^{\mathrm{T}}]^{\mathrm{T}} \in \mathbb{C}^{BM}$ has Kronecker structure, i.e., $\boldsymbol{C}_{\boldsymbol{h}_{\mathrm{T},k}} = \boldsymbol{C}_{\mathrm{T},k} \otimes \boldsymbol{C}_{\boldsymbol{h}_k}$.

We can extend the definition of the HDS 1 and introduce K different HDSs: $\boldsymbol{z}_k[q] = [\boldsymbol{h}_{\mathrm{T},k}[q]^{\mathrm{T}}, \boldsymbol{h}_{\mathrm{ext},k}[q]^{\mathrm{T}}]^{\mathrm{T}} \in \mathbb{C}^{\bar{B}M}, k \in \{1, 2, \ldots, K\}$ with $\boldsymbol{h}_{\mathrm{ext},k}[q] \in \mathbb{C}^{M(\bar{B}-B)}$ are the HDSs w.r.t. $\boldsymbol{C}_{\mathrm{T},k}$ and $\boldsymbol{C}_{\boldsymbol{h}_k}$ (compare (3.71)). Restricting $\boldsymbol{z}_k$ to be a full period of a $\bar{B}$-periodic M-variate random sequence we get the following stochastic model of $\boldsymbol{z}_k$: The covariance matrix of $\boldsymbol{z}_k[q]$ is $\boldsymbol{C}_{\boldsymbol{z}_k} = \boldsymbol{C}_{\mathrm{c},k} \otimes \boldsymbol{C}_{\boldsymbol{h}_k}$, where $\boldsymbol{C}_{\mathrm{c},k} \in \mathbb{T}_{\mathrm{c}+,0}^{\bar{B}}$ is circulant and positive semidefinite with upper left block $\boldsymbol{C}_{\mathrm{T},k}$. Thus, we restrict $\boldsymbol{C}_{\mathrm{T},k}$ to have a positive semidefinite circulant extension in analogy to Sections 3.3.3 and 3.3.5.

To include a *Kronecker* structure in the M-step is difficult due to the nonlinearity. A straightforward solution would be based on alternating variables:

[16] In [47] it is proposed to choose $\bar{B}$ as an integer multiple of B, which ensures existence of a positive semidefinite circulant extension.

[17] Estimation with other constraints on the structure of $\boldsymbol{C}_{\boldsymbol{h}}$ — as presented in Section 2.1 — can be treated in a similar fashion.

Choice of HDS	Choice of Model	Complexity (full)	Complexity (incl. structure)
HDS 1	Model 1	$B^3N^3M^3$	B^3P^3
"	Model 1 (Kronecker)	$KM^3N^3 + KB^3$	$KP^3 + KB^2$
"	Model 2	M^3N^3	P^3
"	Model 1 ($P = 1$)	B^3	B^2
HDS 2	Model 1	$B^3N^3M^3$	B^3P^3
"	Model 2	M^3N^3	P^3

Table 3.1 Order of computational complexity of SAGE per iteration for an update of $\boldsymbol{C}_{\boldsymbol{h}_{\mathrm{T}}}$ (HDS 1) and $\boldsymbol{C}_{\boldsymbol{n},\mathrm{s}}$ (HDS 2).

First, optimize the estimated log-likelihood of $\boldsymbol{z}_k[q]$ (from E-step) w.r.t. $\boldsymbol{C}_{\boldsymbol{h}_k}$ for all K receivers keeping $\boldsymbol{C}_{\mathrm{c},k}$ fixed; in a second step maximize the estimated log-likelihood of $\boldsymbol{z}_k[q]$ w.r.t. $\boldsymbol{C}_{\mathrm{c},k}$ for fixed $\boldsymbol{C}_{\boldsymbol{h}_k}$ from the previous iteration. Simulations show that alternating between both covariance matrices does not improve the estimate sufficiently to justify the effort. Therefore, we propose to maximize the estimated log-likelihood of $\boldsymbol{z}_k[q]$ w.r.t. $\boldsymbol{C}_{\boldsymbol{h}_k}$ assuming $\boldsymbol{C}_{\mathrm{c},k} = \boldsymbol{I}_{\bar{B}}$ and w.r.t. $\boldsymbol{C}_{\mathrm{c},k}$ assuming $\boldsymbol{C}_{\boldsymbol{h}_k} = \boldsymbol{I}_M$. Before both estimates are combined as in (3.98), $\hat{\boldsymbol{C}}_{\mathrm{T},k}$ is scaled to have ones on its diagonal.[18] The details of the iterative maximization with SAGE for different k and HDS 1 or 2 follow the argumentation in Sections 3.3.4 and 3.3.5.

To judge the *computational complexity* of SAGE algorithm for different assumptions on $\boldsymbol{C}_{\boldsymbol{h}_{\mathrm{T}}}$, we give the order of complexity per iteration in Table 3.1: The order is determined by the system of linear equations to compute $\boldsymbol{W}_{\boldsymbol{z}}^{(i)}$ (3.82), $\boldsymbol{W}_{\boldsymbol{h}}^{(i)}$ (3.95), and $\boldsymbol{W}_{\boldsymbol{n}}^{(i)}$ (3.48). When the structure of these systems of equations is not exploited, the (full) complexity is given by the third column in Table 3.1. The structure can be exploited based on the Matrix Inversion Lemma (A.19) or working with the sufficient statistic $\boldsymbol{t}_{\mathrm{T}}[q]$. Moreover, for $P = 1$, we have a Toeplitz matrix $\boldsymbol{C}_{\hat{\boldsymbol{h}}_{\mathrm{T}}}^{(i)}$, which results in a reduction from B^3 to B^2 if exploited. Application of *both* strategies results in the order of complexity given in the last column of Table 3.1.

We conclude that the complexity of estimating $\boldsymbol{C}_{\boldsymbol{h}_{\mathrm{T}}}$ assuming only block Toeplitz structure is large, but a significant reduction is possible for the model in (3.98). In Section 3.6, we observe that iterations with HDS 2 do not lead to improved performance; thus, the computationally complexity associated with HDS 2 can be avoided, if we use the initialization (3.64).

[18] The parameterization in (3.98) is unique up to a real-valued scaling.

3.4 Completion of Partial Band-Limited Autocovariance Sequences

In Section 3.3.5 we considered estimation of B consecutive samples of the autocovariance sequence $c_h[\ell] = \mathrm{E}[h[q]h[q-\ell]^*]$ with $\ell \in \{0, 1, \ldots, B-1\}$ for a univariate random sequence $h[q]$. For known second order moment of the noise, it is based on noisy observations (estimates) of $h[q]$

$$\tilde{h}[q] = h[q] + e[q]. \tag{3.99}$$

All B observations were collected in $\tilde{\boldsymbol{h}}_{\mathrm{T}}[q] = [\tilde{h}[q-\ell_1], \tilde{h}[q-\ell_2], \ldots, \tilde{h}[q-\ell_B]]^{\mathrm{T}}$ (denoted as $\hat{\boldsymbol{h}}_{\mathrm{T}}[q]$ in Section 3.3), where we assumed that $\ell_i = i$, i.e., equidistant and consecutive samples were available.

Let us consider the case that $\ell_i = 2i - 1$ as an example, i.e., equidistant samples with period $N_{\mathrm{P}} = 2$ are available. Moreover, for this choice of ℓ_i, we are given I statistically independent observations $\tilde{\boldsymbol{h}}_{\mathrm{T}}[q]$ with $q \in \mathbb{O}_{\mathrm{T}} = \{1, 2B+1, \ldots, (I-1)2B+1\}$. The corresponding likelihood function depends on

$$[c_h[0], c_h[2], \ldots, c_h[2(B-1)]]^{\mathrm{T}} \in \mathbb{C}^B \tag{3.100}$$

and the variance c_e of $e[q]$. We are interested in $c_h[\ell]$ for $\ell \in \{0, 1, \ldots, 2B-1\}$ but from the maximum likelihood problem we can only infer $c_h[\ell]$ for $\ell \in \{0, 2, \ldots, 2(B-1)\}$. Therefore, the problem is ill-posed.

Assuming that $c_h[\ell]$ is a *band-limited sequence* with maximum frequency $f_{\max} \leq 0.25$, the condition of the sampling theorem [170] is satisfied. In principle, it would be possible to reconstruct $c_h[\ell]$ for odd values of ℓ from its samples for even ℓ, if the decimated sequence of infinite length was available. We would only have to ensure that the interpolated sequence remains positive semidefinite.

This problem is referred to as "missing data", "gapped data", or "missing observations" in the literature, for example, in spectral analysis or ARMA parameter estimation [203, 180]. Different approaches are possible:

- Interpolate the observations $\tilde{h}[q]$ and estimate the auto-covariance sequence based on the completed sequence. This method is applied in spectral analysis [203, p. 259]. If we apply a ML approach to the interpolated observations, we have to model the resulting error sequence which now also depends on the autocovariance sequence $c_h[\ell]$. This complicates the problem, if we do not ignore this effect.
- Apply the SAGE algorithm from Section 3.3 with the assumption that only even samples for the first $2B$ samples of $\tilde{h}[q-\ell]$ are given. Modeling the random sequence as periodic with period $\bar{B} \geq 4B-1$ and a HDS given by the full period (including the samples for odd ℓ), we can proceed iteratively. We only require a good initialization of $c_h[\ell]$ which is difficult to obtain because more than half of the samples in one period $\bar{B}$ are missing.

- Define the positive semidefinite $2B \times 2B$ Toeplitz matrix $\boldsymbol{C}_\mathrm{T} \in \mathbb{T}_{+,0}^{2B}$ with first row $[\hat{c}_h[0], \hat{c}_h[1], \ldots, \hat{c}_h[2B-1]]$. Given estimates $\hat{c}_h[\ell], \ell \in \{0, 2, \ldots, 2(B-1)\}$, e.g., from the SAGE algorithm in Section 3.3.5, this matrix is not known completely. We would like to find a completion of this matrix. The existence of a completion for *every* given estimate $\hat{c}_h[\ell], \ell \in \{0, 2, \ldots, 2(B-1)\}$, has to be ensured which we discuss below. Generally, the choice of the unknown elements to obtain a positive semidefinite Toeplitz matrix $\boldsymbol{C}_\mathrm{T}$ is not unique.
  In [131] the completion of this matrix is discussed which maximizes the entropy $\ln\det[\boldsymbol{C}_\mathrm{T}]$ (see also the references in [130]). The choice of the entropy is only one possible cost function and, depending on the application and context, other cost functions may be selected to choose a completion from the convex set of possible completions. No optimization has been proposed so far, which includes the constraint on the total covariance sequence, i.e., the extension of $c_h[\ell]$ for $\ell \in \{-(2B-1), \ldots, 2B+1\}$ to $\ell \in \mathbb{Z}$, to be *band-limited* with maximum frequency $f_\mathrm{max} \leq 0.25$.

Completion of Perfectly Known Partial Band-Limited Autocovariance Sequences

Given $c_h[\ell]$ for the periodic pattern $\ell = 2i$ with $i \in \{0, 1, \ldots, B-1\}$, find a completion of this sequence for $\ell \in \{1, 3, \ldots, 2B-1\}$ which satisfies the following two conditions on $c_h[\ell], \ell = 0, 1, \ldots, 2B-1$:

1. It has a positive semidefinite extension on $\mathbb{Z}$, i.e., its Fourier transform is real and positive.
2. An extension on $\mathbb{Z}$ exists which is band-limited with maximum frequency $f_\mathrm{max} \leq 0.25$.[19]

This corresponds to interpolation of the given decimated sequence $c_h[\ell]$ at $\ell \in \{1, 3, \ldots, 2B-1\}$.

The required background to solve this problem is given in Appendix B. The characterization of the corresponding generalized band-limited trigonometric moment problem in Theorem B.4 gives the necessary and sufficient conditions required for the existence of a solution: The Toeplitz matrices $\bar{\boldsymbol{C}}_\mathrm{T}$ with first row $[c_h[0], c_h[2], \ldots, c_h[2(B-1)]]$ and $\bar{\boldsymbol{C}}_\mathrm{T}'$ with first row $[c_h'[0], c_h'[2], \ldots, c_h'[2B-4]]$ for $c_h'[2i] = c_h[2(i-1)] - 2\cos(2\pi f_\mathrm{max}')c_h[2i] + c_h[2(i+1)]$ and $f_\mathrm{max}' = 2f_\mathrm{max}$ are required to be positive semidefinite. Generally, neither of both matrices is singular and the band-limited positive semidefinite extension is not unique.

First, we assume that both conditions are satisfied as $c_h[2i]$ is taken from a autocovariance sequence band-limited to f_max. We are only interested in the completion on the interval $\{0, 1, \ldots, 2B-1\}$. This corresponds to

[19] The Fourier transform of the extension is non-zero only on the interval $[-2\pi f_\mathrm{max}, 2\pi f_\mathrm{max}]$.

a *matrix completion problem* for the Toeplitz covariance matrix $\boldsymbol{C}_{\mathrm{T}}$ with first row $[c_h[0], c_h[1], \ldots, c_h[2B-1]]$ given $c_h[\ell]$ for the periodic pattern above. Admissible completions are $\{c_h[\ell]\}_{\ell=0}^{2B-1}$ which have a band-limited positive semidefinite extension on $\mathbb{Z}$. This convex set of admissible extensions is characterized by Theorem B.3, where $\boldsymbol{C}'_{\mathrm{T}}$ is Toeplitz with first row $[c_h[0], c_h[1], \ldots, c_h[2B-2]]$ and $c'_h[\ell] = c_h[\ell-1] - 2\cos(2\pi f_{\max}) c_h[\ell] + c_h[\ell+1]$.

As an *application* we choose univariate MMSE prediction (Section 2.3.1) of the random variable $h[q]$ based on $\tilde{\boldsymbol{h}}_{\mathrm{T}}[q] = [\tilde{h}[q-\ell_1], \tilde{h}[q-\ell_2], \ldots, \tilde{h}[q-\ell_B]]^{\mathrm{T}}$ for $\ell_i = 2i-1, i \in \{1, 2, \ldots, B\}$,

$$\hat{h}[q] = \boldsymbol{w}^{\mathrm{T}} \tilde{\boldsymbol{h}}_{\mathrm{T}}[q] \tag{3.101}$$

with $\boldsymbol{w}^{\mathrm{T}} = \boldsymbol{c}_{\boldsymbol{h}_{\mathrm{T}} h}^{\mathrm{H}} \boldsymbol{C}_{\tilde{\boldsymbol{h}}_{\mathrm{T}}}^{-1}$, where $\boldsymbol{c}_{\boldsymbol{h}_{\mathrm{T}} h} = [c_h[1], c_h[3], \ldots, c_h[2B-1]]^{\mathrm{T}}$, $\boldsymbol{C}_{\tilde{\boldsymbol{h}}_{\mathrm{T}}} = \boldsymbol{C}_{\boldsymbol{h}_{\mathrm{T}}} + c_{\mathrm{e}} \boldsymbol{I}_B$, and $\boldsymbol{C}_{\boldsymbol{h}_{\mathrm{T}}}$ is Toeplitz with first row $[c_h[0], c_h[2], \ldots, c_h[2(B-1)]]$. Therefore, $\boldsymbol{C}_{\tilde{\boldsymbol{h}}_{\mathrm{T}}}$ is given and $\boldsymbol{c}_{\boldsymbol{h}_{\mathrm{T}} h}$ contains the unknown odd samples of $c_h[\ell]$.

The *minimax prediction* problem (2.109) for the new *uncertainty set* reads

$$\min_{\boldsymbol{w}} \max_{\boldsymbol{c}_{\boldsymbol{h}_{\mathrm{T}} h}} \mathrm{E}\left[\left|h[q] - \boldsymbol{w}^{\mathrm{T}} \tilde{\boldsymbol{h}}_{\mathrm{T}}[q]\right|^2\right] \quad \text{s.t.} \quad \boldsymbol{C}_{\mathrm{T}} \succeq 0, \boldsymbol{C}'_{\mathrm{T}} \succeq 0. \tag{3.102}$$

Due to the minimax Theorem 2.2, it is equivalent to the maxmin problem

$$\max_{\boldsymbol{c}_{\boldsymbol{h}_{\mathrm{T}} h} \in \mathbb{C}^B} c_h[0] - \boldsymbol{c}_{\boldsymbol{h}_{\mathrm{T}} h}^{\mathrm{H}} \boldsymbol{C}_{\tilde{\boldsymbol{h}}_{\mathrm{T}}}^{-1} \boldsymbol{c}_{\boldsymbol{h}_{\mathrm{T}} h} \quad \text{s.t.} \quad \boldsymbol{C}_{\mathrm{T}} \succeq 0, \boldsymbol{C}'_{\mathrm{T}} \succeq 0 \tag{3.103}$$

with the expression for the MMSE parameterized in the unknown $\boldsymbol{c}_{\boldsymbol{h}_{\mathrm{T}} h}$. For this uncertainty set, the minimax MSE solution is equivalent to the *minimum* (weighted) *norm completion* of the covariance matrix $\boldsymbol{C}_{\mathrm{T}}$

$$\boxed{\min_{\boldsymbol{c}_{\boldsymbol{h}_{\mathrm{T}} h}} \|\boldsymbol{c}_{\boldsymbol{h}_{\mathrm{T}} h}\|_{\boldsymbol{C}_{\tilde{\boldsymbol{h}}_{\mathrm{T}}}^{-1}} \quad \text{s.t.} \quad \boldsymbol{C}_{\mathrm{T}} \succeq 0, \boldsymbol{C}'_{\mathrm{T}} \succeq 0.} \tag{3.104}$$

According to Example B.2 in Section B.2 the trivial solution $\boldsymbol{c}_{\boldsymbol{h}_{\mathrm{T}} h} = \boldsymbol{0}_{B \times 1}$ is avoided if $f_{\max} < 0.25$.

Problem (3.104) can be easily transformed to an equivalent formulation involving a linear cost function and constraints describing symmetric cones

$$\min_{\boldsymbol{c}_{\boldsymbol{h}_{\mathrm{T}} h}, t} t \quad \text{s.t.} \quad \boldsymbol{C}_{\mathrm{T}} \succeq 0, \boldsymbol{C}'_{\mathrm{T}} \succeq 0, \|\boldsymbol{c}_{\boldsymbol{h}_{\mathrm{T}} h}\|_{\boldsymbol{C}_{\tilde{\boldsymbol{h}}_{\mathrm{T}}}^{-1}} \leq t \tag{3.105}$$

which can be solved numerically using interior point methods, e.g., SeDuMi [207]. The Toeplitz structure of the matrices involved in (3.105) can be exploited to reduce the numerical complexity by one order (compare [3]).

Example 3.4. Consider $c_h[\ell] = \mathrm{J}_0(2\pi f_{\max} \ell)$, which is perfectly known for $\ell \in \{0, 2, \ldots, 2B-2\}$ with $B = 5$. We perform linear prediction as defined in (3.101). The completion of the covariance sequence for $l \in \{1, 3, \ldots, 2B-1\}$ is chosen according to the minimax MSE (3.102) or equivalently the minimum

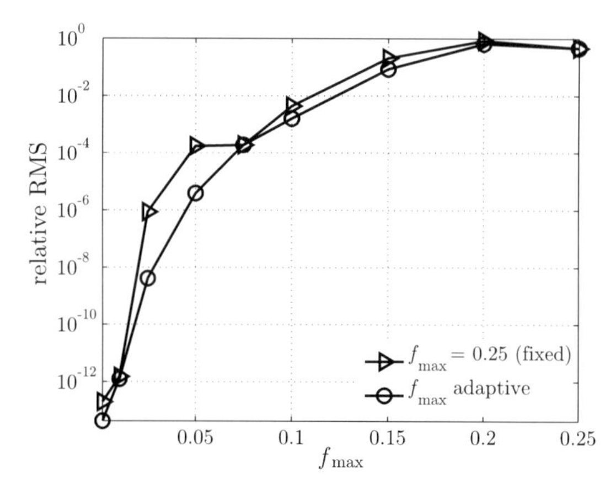

Fig. 3.3 Relative root mean square (RMS) error (3.106) of completed partial band-limited autocovariance sequence: Minimum norm completion with fixed $f_{\max} = 0.25$ and true $f_{\max}$ (adaptive).

norm criterion (3.104). We consider two versions of the algorithm: An upper bound for $f_{\max}$ according to the sampling theorem, i.e., $f_{\max} = 0.25$, or the minimum true $f_{\max}$ is given. The knowledge of $f_{\max}$ is considered in (3.105) via $\boldsymbol{C}'_{\mathrm{T}}$. The relative root mean square (RMS) error

$$\sqrt{\frac{\sum_{i=1}^{B} |c_h[2i-1] - \hat{c}_h[2i-1]|^2}{\sum_{i=1}^{B} |c_h[2i-1]|^2}} \tag{3.106}$$

for the completion $\hat{c}_h[\ell]$ is shown in Figure 3.3. For $f_{\max} \leq 0.1$, the relative RMS error is below 10^{-3}. As the performance of both versions is very similar, the completion based on a fixed $f_{\max} = 0.25$ should be preferred: It does not require estimation of $f_{\max}$, which is often also based on the (decimated) autocovariance sequence $c_h[2i]$ [211]. Obviously, constraining the partial autocovariance sequence to have a band-limited extension is a rather accurate characterization. Furthermore, its interpolation accuracy shows that most of the information on $c_h[\ell]$ for $\ell \in \{0, 1, \ldots, 2B-1\}$ is concentrated in this interval.

The MSE for prediction of $h[q]$ based on the completion with true $f_{\max}$ (adaptive) and fixed $f_{\max} = 0.25$ in Figure 3.4 prove the effectiveness of our approach. The upper bound given by the worst case MSE (3.103) is only shown for adaptive $f_{\max}$; for fixed $f_{\max} = 0.25$ it is almost identical to the actual MSE given in Figure 3.4. Both upper bounds are rather tight w.r.t. the MSE for perfect knowledge of $\{c_h[\ell]\}_{\ell=0}^{2B-1}$. □

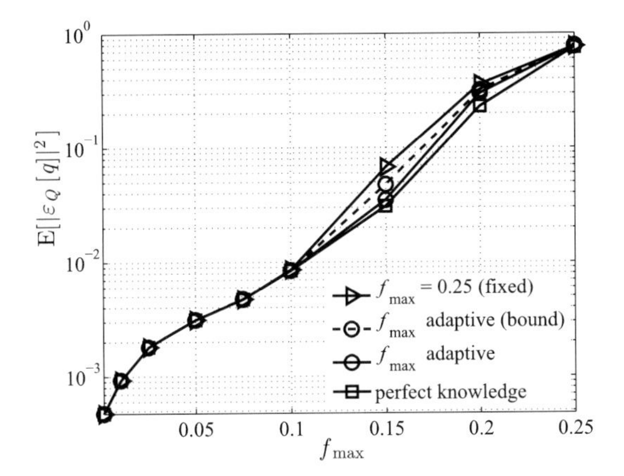

Fig. 3.4 MSE $\mathrm{E}[|\varepsilon_Q[q]|^2]$ for minimax MSE prediction with filter order $Q = B = 5$ based on the completed partial band-limited autocovariance sequence.

Completion of Estimated Partial Band-Limited Autocovariance Sequences

If the true $c_h[\ell]$ for $\ell \in \{0, 2, \ldots, 2(B-1)\}$ is not available, we propose the following two-stage procedure:

- Estimate $c_h[\ell]$ for $\ell \in \{0, 2, \ldots, 2(B-1)\}$ based on the SAGE algorithm introduced in Section 3.3.5.
- Solve the optimization problem (3.102) or (3.104) for $\hat{c}_h[\ell], \ell \in \{0, 2, \ldots, 2(B-1)\}$, to construct an estimate for $c_h[\ell]$ for $\ell \in \{0, 1, \ldots, 2B-1\}$, which results in a positive semidefinite Toeplitz matrix with first row $\hat{c}_h[\ell], \ell \in \{0, 1, \ldots, 2B-1\}$. If the completion is based on the side information $f_{\max} \leq 0.25$, i.e., we choose $f_{\max} = 0.25$, a solution to (3.104) always exists (Theorem B.4).
  But if we assume an $f_{\max} < 0.25$, a completion of $\hat{c}_h[\ell]$ for $\ell \in \{0, 2, \ldots, 2(B-1)\}$ such that it has positive definite and band-limited extension on $\mathbb{Z}$ may not exist. In this case we first project $\hat{\boldsymbol{c}} = [\hat{c}_h[0], \hat{c}_h[2], \ldots, \hat{c}_h[2(B-1)]]^{\mathrm{T}}$ on the space of positive semidefinite sequences band-limited to $f_{\max}$ solving the optimization problem

$$\min_{\boldsymbol{c}} \|\boldsymbol{c} - \hat{\boldsymbol{c}}\|_2 \quad \text{s.t.} \quad \bar{\boldsymbol{C}}_{\mathrm{T}} \succeq 0, \bar{\boldsymbol{C}}_{\mathrm{T}}' \succeq 0 \tag{3.107}$$

  with minimizer $\hat{\boldsymbol{c}}^{\mathrm{LS}}$. The constraints result from Theorem B.4, where $\bar{\boldsymbol{C}}_{\mathrm{T}}$ is Toeplitz with first row $[\hat{c}_h[0], \hat{c}_h[2], \ldots, \hat{c}_h[2(B-1)]$ and $\bar{\boldsymbol{C}}_{\mathrm{T}}'$ is Toeplitz with first row $[\hat{c}_h'[0], \hat{c}_h'[2], \ldots, \hat{c}_h'[2B-4]$ for $\hat{c}_h'[2i] = \hat{c}_h[2(i-1)] - 2\cos(2\pi f_{\max}')\hat{c}_h[2i] + \hat{c}_h[2(i+1)]$ and $f_{\max}' = 2f_{\max}$. This problem can be transformed into a cone program with second order and positive semidefinite cone constraints.

Problem (3.102) with $\hat{\boldsymbol{c}}$ or $\hat{\boldsymbol{c}}^{\mathrm{LS}}$ is no robust optimization with guaranteed maximum MSE anymore, because $\hat{c}_h[\ell]$ are treated as error-free in this formulation.[20] As in (3.104) it has to be understood as a minimum norm completion, where the weighting is determined by the application.

Extensions to Other Patterns of Partial Autocovariance Sequences

An extension of this completion algorithm (3.104) to other patterns $\mathcal{P}$ of partial covariance sequences is only possible, if $c_h[\ell]$ or $\hat{c}_h[\ell]$ are given for a periodic pattern $\ell = N_{\mathrm{P}}i$ with $i \in \{0, 1, \ldots, B-1\}$ as stated by Theorem B.4. For irregular (non-periodic) patterns of the partial covariance sequence, a completion may not exist for *all* admissible values of $c_h[\ell]$ or $\hat{c}_h[\ell]$ with $\ell \in \mathcal{P}$ (see the example in [131] for $f_{\max} = 1/2$).

Thus, for an *irregular* spacing of the (noisy) channel observations, *prediction* based on an estimated autocovariance sequence is problematic: Estimation of the autocovariance sequence from irregular samples is difficult *and* a completion does not exist in general, i.e., for all estimated partial autocovariance sequences forming partial positive semidefinite Toeplitz covariance matrix (Appendix B). In this case, we recommend prediction based on a worst case autocovariance sequence as introduced in Section 2.4.3.2.

For a periodic spacing, our results from above show that a band-limited completion is possible. The proposed algorithm is still rather complex, but illustrates the capability of this approach, which we explore further in Section 3.6.

3.5 Least-Squares Approaches

In Section 3.3 we propose the SAGE algorithm to solve the ML problem (3.37) for estimation of $\boldsymbol{C_h}$ and $\boldsymbol{C}_{\boldsymbol{n},\mathrm{s}}$ (Model 2) iteratively. Of course, this ML problem inherently yields an estimate of $\boldsymbol{C_y}$ taking into account its linear structure, i.e., $\boldsymbol{C_y}$ is restricted to the set

$$\mathcal{S} = \left\{ \boldsymbol{C_y} | \boldsymbol{C_y} = \boldsymbol{S}\boldsymbol{C_h}\boldsymbol{S}^{\mathrm{H}} + \boldsymbol{I}_N \otimes \boldsymbol{C}_{\boldsymbol{n},\mathrm{s}}, \boldsymbol{C_h} \succeq 0, \boldsymbol{C}_{\boldsymbol{n},\mathrm{s}} \succeq 0 \right\} \tag{3.108}$$

$$= \left\{ \boldsymbol{C_y} | \boldsymbol{C_y} = \boldsymbol{S}\left(\boldsymbol{C_h} + (\boldsymbol{S}^{\mathrm{H}}\boldsymbol{S})^{-1}(\boldsymbol{I}_N \otimes \boldsymbol{C}_{\boldsymbol{n},\mathrm{s}})\right)\boldsymbol{S}^{\mathrm{H}} + \boldsymbol{P}^{\perp}(\boldsymbol{I}_N \otimes \boldsymbol{C}_{\boldsymbol{n},\mathrm{s}}), \right.$$
$$\left. \boldsymbol{C_h} \succeq 0, \boldsymbol{C}_{\boldsymbol{n},\mathrm{s}} \succeq 0 \right\}. \tag{3.109}$$

The second description of $\mathcal{S}$ is based on the decomposition in signal and noise subspace with the orthogonal projector on the $M\bar{N}$-dimensional noise subspace $\boldsymbol{P}^{\perp} = \boldsymbol{I}_{MN} - \boldsymbol{S}\boldsymbol{S}^{\dagger} = \bar{\boldsymbol{P}}^{\perp,\mathrm{T}} \otimes \boldsymbol{I}_M$, $\bar{\boldsymbol{P}}^{\perp} = \boldsymbol{I}_N - \bar{\boldsymbol{S}}^{\dagger}\bar{\boldsymbol{S}}$.

---

[20] Introducing a convex uncertainty set for $\hat{c}_h[\ell]$ would turn it into a robust minimax optimization problem again.

For the following argumentation, we write the ML problem (3.37) as a maximization w.r.t. $\boldsymbol{C}_{\boldsymbol{y}}$

$$\max_{\boldsymbol{C}_{\boldsymbol{y}} \in \mathcal{S}} \mathrm{L}_{\boldsymbol{y}}(\{\boldsymbol{y}[q]\}_{q\in\mathbb{O}}; \boldsymbol{C}_{\boldsymbol{y}}) \tag{3.110}$$

with log-likelihood function

$$\mathrm{L}_{\boldsymbol{y}}(\{\boldsymbol{y}[q]\}_{q\in\mathbb{O}}; \boldsymbol{C}_{\boldsymbol{y}}) = -BMN \ln \pi - B \ln \det[\boldsymbol{C}_{\boldsymbol{y}}] - B \mathrm{tr}\left[\tilde{\boldsymbol{C}}_{\boldsymbol{y}} \boldsymbol{C}_{\boldsymbol{y}}^{-1}\right]. \tag{3.111}$$

The iterative solution of this problem is rather complex (Section 3.3.4 and 3.3.6). To reduce the numerical complexity of ML estimation, Stoica and Söderström introduced the *extended invariance principle* (EXIP) [204] (A brief review is given in Section 3.1.1). It yields an algorithm with the following steps:

1. The set $\mathcal{S}$ for $\boldsymbol{C}_{\boldsymbol{y}}$ is extended to $\mathcal{S}_{\mathrm{ext}}$ such that the ML problem (3.110) has a simple solution.[21] In general the ML estimate based on the extended set is not an element of $\mathcal{S}$ with probability one.[22]
2. An estimate in $\mathcal{S}$ is obtained from a *weighted Least-Squares* (LS) approximation of the ML estimate of the parameters for the extended set $\mathcal{S}_{\mathrm{ext}}$.[23]

This estimate is asymptotically (in B) equivalent to the ML estimate from (3.110) if the weighting in the LS approximation is chosen as the Hessian matrix of the cost function (3.111). The Hessian matrix is evaluated based on a consistent estimate of the parameters for the extended set.

We choose two natural extensions of $\mathcal{S}$. The first is based on the decomposition of $\boldsymbol{C}_{\boldsymbol{y}}$ in the covariance matrix of the signal and noise subspace (3.109)

$$\boxed{\mathcal{S}_{\mathrm{ext1}} = \left\{\boldsymbol{C}_{\boldsymbol{y}} | \boldsymbol{C}_{\boldsymbol{y}} = \boldsymbol{S}\boldsymbol{C}_{\hat{\boldsymbol{h}}}\boldsymbol{S}^{\mathrm{H}} + \boldsymbol{P}^{\perp}(\boldsymbol{I}_N \otimes \boldsymbol{C}_{\boldsymbol{n},\mathrm{s}}), \boldsymbol{C}_{\hat{\boldsymbol{h}}} \succeq 0, \boldsymbol{C}_{\boldsymbol{n},\mathrm{s}} \succeq 0\right\},} \tag{3.112}$$

where $\boldsymbol{C}_{\hat{\boldsymbol{h}}} = \mathrm{E}[\hat{\boldsymbol{h}}[q]\hat{\boldsymbol{h}}[q]^{\mathrm{H}}]$ is the covariance of the ML channel estimate (3.25) (see also (2.31)). The second choice does not assume any structure of $\boldsymbol{C}_{\boldsymbol{y}}$ and is the set of all positive semidefinite $MN \times MN$ matrices

$$\boxed{\mathcal{S}_{\mathrm{ext2}} = \{\boldsymbol{C}_{\boldsymbol{y}} | \boldsymbol{C}_{\boldsymbol{y}} \succeq 0\} = \mathbb{S}_{+,0}^{MN}.} \tag{3.113}$$

The relation of the extended sets to $\mathcal{S}$ is $\mathcal{S} \subseteq \mathcal{S}_{\mathrm{ext1}} \subseteq \mathcal{S}_{\mathrm{ext2}}$.

Application of the EXIP yields the following optimization problems:

---

[21] $\mathcal{S}$ is a subset of the extended set.

[22] If it is an element of $\mathcal{S}$, it is also ML estimate based on $\mathcal{S}$ due to the ML invariance principle [124].

[23] This corresponds to a weighted LS solution in $\mathcal{S}$ of the likelihood equation corresponding to the extended set. The likelihood equation is the gradient of the log-likelihood function (e.g., (3.2)) set to zero, which is a necessary condition for a ML solution (compare with (3.4)).

- The ML estimates of the parameters $\boldsymbol{C}_{\hat{\boldsymbol{h}}}$ and $\boldsymbol{C}_{\boldsymbol{n},\mathrm{s}}$ for $\boldsymbol{C}_{\boldsymbol{y}}$ constrained to $\mathcal{S}_{\mathrm{ext1}}$ are (Step 1)

$$\begin{aligned}\tilde{\boldsymbol{C}}_{\hat{\boldsymbol{h}}} &= \frac{1}{B}\sum_{q\in\mathbb{O}} \hat{\boldsymbol{h}}[q]\hat{\boldsymbol{h}}[q]^{\mathrm{H}} = \boldsymbol{S}^{\dagger}\tilde{\boldsymbol{C}}_{\boldsymbol{y}}\boldsymbol{S}^{\dagger,\mathrm{H}} \\ \tilde{\boldsymbol{C}}_{\boldsymbol{n},\mathrm{s}} &= \frac{1}{B\bar{N}}\sum_{q\in\mathbb{O}} \boldsymbol{Y}[q]\bar{\boldsymbol{P}}^{\perp}\boldsymbol{Y}[q]^{\mathrm{H}}\end{aligned} \tag{3.114}$$

with $\bar{N} = N - K(L+1)$ and $\boldsymbol{Y}[q] = [\boldsymbol{y}[q,1], \boldsymbol{y}[q,2], \ldots, \boldsymbol{y}[q,N]]$ ($\boldsymbol{y}[q,n]$ defined equivalently to (3.18)). The corresponding weighted LS approximation of these parameters reads (Step 2)

$$\min_{\substack{\boldsymbol{C}_{\boldsymbol{h}}\in\mathbb{C}^{P\times P}\\ \boldsymbol{C}_{\boldsymbol{n},\mathrm{s}}\in\mathbb{C}^{M\times M}}} \|\tilde{\boldsymbol{C}}_{\hat{\boldsymbol{h}}} - \boldsymbol{C}_{\boldsymbol{h}} - (\boldsymbol{S}^{\mathrm{H}}\boldsymbol{S})^{-1}(\boldsymbol{I}_N \otimes \boldsymbol{C}_{\boldsymbol{n},\mathrm{s}})\|_{\mathrm{F},\boldsymbol{W}_1}^2 + \|\tilde{\boldsymbol{C}}_{\boldsymbol{n},\mathrm{s}} - \boldsymbol{C}_{\boldsymbol{n},\mathrm{s}}\|_{\mathrm{F},\boldsymbol{W}_2}^2 \tag{3.115}$$

with weights $\boldsymbol{W}_1 = \sqrt{B}\tilde{\boldsymbol{C}}_{\hat{\boldsymbol{h}}}^{-1}$ and $\boldsymbol{W}_2 = \sqrt{B\bar{N}}\tilde{\boldsymbol{C}}_{\boldsymbol{n},\mathrm{s}}^{-1}$. The weights are related to the inverse error covariance matrix of the estimates (3.114) [133] which is based on consistent estimates of the involved covariance matrices. The weighted Frobenius norm is defined as $\|\boldsymbol{A}\|_{\mathrm{F},\boldsymbol{W}}^2 = \mathrm{tr}[\boldsymbol{A}\boldsymbol{W}\boldsymbol{A}\boldsymbol{W}]$ with $\boldsymbol{A} = \boldsymbol{A}^{\mathrm{H}}$ and $\boldsymbol{W} \succeq 0$. For white noise $\boldsymbol{C}_{\boldsymbol{n},\mathrm{s}} = c_n\boldsymbol{I}_M$, we obtain

$$\min_{\boldsymbol{C}_{\boldsymbol{h}}\in\mathbb{C}^{P\times P}, c_n\in\mathbb{R}} \|\tilde{\boldsymbol{C}}_{\hat{\boldsymbol{h}}} - \boldsymbol{C}_{\boldsymbol{h}} - (\boldsymbol{S}^{\mathrm{H}}\boldsymbol{S})^{-1}c_n\|_{\mathrm{F},\boldsymbol{W}_1}^2 + (\tilde{c}_n - c_n)^2 w_2^2, \tag{3.116}$$

where $\tilde{c}_n$ is an estimate of c_n based on the noise subspace (compare (3.29) and (3.67))

$$\tilde{c}_n = \frac{1}{B\bar{N}M}\sum_{q\in\mathbb{O}} \|\boldsymbol{P}^{\perp}\boldsymbol{y}[q]\|_2^2. \tag{3.117}$$

The new weight $w_2 = \sqrt{BM\bar{N}}\,\tilde{c}_n^{-1}$ is the inverse of the square root of the error covariance of $\tilde{c}_n$ which depends on the unknown c_n; it is evaluated for its consistent estimate $\tilde{c}_n$ which is sufficient according to the EXIP.
- The ML estimate of $\boldsymbol{C}_{\boldsymbol{y}}$ for the extended set $\mathcal{S}_{\mathrm{ext2}}$ (3.113) is the sample covariance matrix (Step 1)

$$\tilde{\boldsymbol{C}}_{\boldsymbol{y}} = \frac{1}{B}\sum_{q\in\mathbb{O}} \boldsymbol{y}[q]\boldsymbol{y}[q]^{\mathrm{H}}. \tag{3.118}$$

Its weighted LS approximation reads (Step 2)

$$\min_{\boldsymbol{C}_{\boldsymbol{h}}\in\mathbb{C}^{P\times P}, \boldsymbol{C}_{\boldsymbol{n},\mathrm{s}}\in\mathbb{C}^{M\times M}} \|\tilde{\boldsymbol{C}}_{\boldsymbol{y}} - \boldsymbol{S}\boldsymbol{C}_{\boldsymbol{h}}\boldsymbol{S}^{\mathrm{H}} - \boldsymbol{I}_N \otimes \boldsymbol{C}_{\boldsymbol{n},\mathrm{s}}\|_{\mathrm{F},\boldsymbol{W}}^2 \tag{3.119}$$

and for spatially white noise

$$\min_{\boldsymbol{C}_{\boldsymbol{h}} \in \mathbb{C}^{P \times P}, c_n \in \mathbb{R}} \|\tilde{\boldsymbol{C}}_{\boldsymbol{y}} - \boldsymbol{S}\boldsymbol{C}_{\boldsymbol{h}}\boldsymbol{S}^{\mathrm{H}} - c_n \boldsymbol{I}_{MN}\|_{\mathrm{F},\boldsymbol{W}}^2. \tag{3.120}$$

The weighting matrix is $\boldsymbol{W} = \sqrt{B}\tilde{\boldsymbol{C}}_{\boldsymbol{y}}^{-1}$ [133] based on the consistent estimate $\tilde{\boldsymbol{C}}_{\boldsymbol{y}}$ of $\boldsymbol{C}_{\boldsymbol{y}}$.

All weighted LS problems resulting from the EXIP yield estimates of $\boldsymbol{C}_{\boldsymbol{h}}$ and $\boldsymbol{C}_{\boldsymbol{n},\mathrm{s}}$ (or c_n) which are linear in $\tilde{\boldsymbol{C}}_{\hat{\boldsymbol{h}}}$, $\tilde{\boldsymbol{C}}_{\boldsymbol{n},\mathrm{s}}$, or $\tilde{\boldsymbol{C}}_{\boldsymbol{y}}$. As this approach is only asymptotically equivalent to ML, the estimates may be indefinite for small B. Moreover, for small B the weighting matrices are not defined because the corresponding matrices are not invertible. For example, $\tilde{\boldsymbol{C}}_{\boldsymbol{y}}$ is invertible only if $B \geq MN$. But MN, e.g., $M = 4$ and $N = 32$, can be larger than the maximum number of statistically independent observations $\boldsymbol{y}[q]$ available within the stationarity time of the channel parameters $\boldsymbol{h}[q]$.

Therefore, we consider unweighted LS approximation with $\boldsymbol{W}_1 = \boldsymbol{I}_P$, $\boldsymbol{W}_2 = \boldsymbol{I}_M$, $w_2 = 1$, and $\boldsymbol{W} = \boldsymbol{I}_{MN}$ in the sequel. First, we derive solutions for LS approximations assuming $\boldsymbol{C}_{\boldsymbol{n},\mathrm{s}} = c_n \boldsymbol{I}_M$, which we generalize in Section 3.5.4.

3.5.1 A Heuristic Approach

Before we treat the LS problems from above, we introduce a simple heuristic estimator of $\boldsymbol{C}_{\boldsymbol{h}}$ which provides further insights in the improvements of the solutions to follow.

In [56] we proposed the sample covariance matrix $\tilde{\boldsymbol{C}}_{\hat{\boldsymbol{h}}}$ (3.114) of the ML channel estimates $\hat{\boldsymbol{h}}[q] = \boldsymbol{S}^{\dagger}\boldsymbol{y}[q]$ (2.31) as an estimate of $\boldsymbol{C}_{\boldsymbol{h}}$:

$$\boxed{\hat{\boldsymbol{C}}_{\boldsymbol{h}}^{\mathrm{Heur}} = \boldsymbol{S}^{\dagger}\tilde{\boldsymbol{C}}_{\boldsymbol{y}}\boldsymbol{S}^{\dagger,\mathrm{H}} = \frac{1}{B}\sum_{q=1}^{B} \hat{\boldsymbol{h}}[q]\hat{\boldsymbol{h}}[q]^{\mathrm{H}}.} \tag{3.121}$$

It is the solution of the LS approximation problem

$$\min_{\boldsymbol{C}_{\boldsymbol{h}}} \|\tilde{\boldsymbol{C}}_{\boldsymbol{y}} - \boldsymbol{S}\boldsymbol{C}_{\boldsymbol{h}}\boldsymbol{S}^{\mathrm{H}}\|_{\mathrm{F}}^2. \tag{3.122}$$

Implicitly the noise variance is assumed small and neglected. Therefore, $\hat{\boldsymbol{C}}_{\boldsymbol{h}}^{\mathrm{Heur}}$ is biased with bias given by the last term of

$$\boxed{\mathrm{E}\left[\hat{\boldsymbol{C}}_{\boldsymbol{h}}^{\mathrm{Heur}}\right] = \boldsymbol{C}_{\boldsymbol{h}} + c_n(\boldsymbol{S}^{\mathrm{H}}\boldsymbol{S})^{-1}.} \tag{3.123}$$

This estimate is always positive semidefinite.[24]

3.5.2 Unconstrained Least-Squares

The LS approximation problem (3.116) derived from the EXIP based on $\mathcal{S}_{\text{ext1}}$ (Step 2) is

$$(\hat{\boldsymbol{C}}_{\boldsymbol{h}}^{\text{LS}}, \hat{c}_n^{\text{LS}}) = \underset{\boldsymbol{C}_{\boldsymbol{h}} \in \mathbb{C}^{P \times P}, c_n \in \mathbb{R}}{\operatorname{argmin}} \|\tilde{\boldsymbol{C}}_{\hat{\boldsymbol{h}}} - \boldsymbol{C}_{\boldsymbol{h}} - (\boldsymbol{S}^{\text{H}}\boldsymbol{S})^{-1} c_n\|_{\text{F}}^2 + (\tilde{c}_n - c_n)^2. \tag{3.124}$$

With the obvious solution

$$\boxed{\begin{aligned} \hat{c}_n^{\text{LS}} &= \frac{1}{M\bar{N}} \operatorname{trace}\left[\tilde{\boldsymbol{C}}_{\boldsymbol{y}} - \boldsymbol{P}\tilde{\boldsymbol{C}}_{\boldsymbol{y}}\boldsymbol{P}\right] \\ &= \frac{1}{M\bar{N}} \operatorname{trace}\left[\boldsymbol{P}^{\perp}\tilde{\boldsymbol{C}}_{\boldsymbol{y}}\boldsymbol{P}^{\perp}\right] \\ \hat{\boldsymbol{C}}_{\boldsymbol{h}}^{\text{LS}} &= \boldsymbol{S}^{\dagger}\tilde{\boldsymbol{C}}_{\boldsymbol{y}}\boldsymbol{S}^{\dagger,\text{H}} - (\boldsymbol{S}^{\text{H}}\boldsymbol{S})^{-1}\hat{c}_n^{\text{LS}} \end{aligned}} \tag{3.125}$$

the cost function (3.124) is zero.

The second LS problem (3.120) which is based on the second parameterization $\mathcal{S}_{\text{ext2}}$ results in [103, 41, 54]

$$\min_{\boldsymbol{C}_{\boldsymbol{h}}, c_n} \|\tilde{\boldsymbol{C}}_{\boldsymbol{y}} - \boldsymbol{S}\boldsymbol{C}_{\boldsymbol{h}}\boldsymbol{S}^{\text{H}} - c_n \boldsymbol{I}_{MN}\|_{\text{F}}^2. \tag{3.126}$$

Introducing the projectors $\boldsymbol{P} = \boldsymbol{S}\boldsymbol{S}^{\dagger}$ on the signal subspace and $\boldsymbol{P}^{\perp} = (\boldsymbol{I}_{MN} - \boldsymbol{S}\boldsymbol{S}^{\dagger})$ on the noise subspace together with $\tilde{\boldsymbol{C}}_{\boldsymbol{y}} = (\boldsymbol{P}^{\perp} + \boldsymbol{P})\tilde{\boldsymbol{C}}_{\boldsymbol{y}}(\boldsymbol{P}^{\perp} + \boldsymbol{P})$ and $\boldsymbol{I}_{MN} = \boldsymbol{P} + \boldsymbol{P}^{\perp}$ we can decompose this cost function in an approximation error in the signal and noise subspace[25]

$$\begin{aligned} &\|\tilde{\boldsymbol{C}}_{\boldsymbol{y}} - \boldsymbol{S}\boldsymbol{C}_{\boldsymbol{h}}\boldsymbol{S}^{\text{H}} - c_n \boldsymbol{I}_{MN}\|_{\text{F}}^2 = \\ &\qquad \|\boldsymbol{P}\tilde{\boldsymbol{C}}_{\boldsymbol{y}}\boldsymbol{P} - \boldsymbol{S}\boldsymbol{C}_{\boldsymbol{h}}\boldsymbol{S}^{\text{H}} - c_n \boldsymbol{P}\|_{\text{F}}^2 + \|\boldsymbol{P}^{\perp}\tilde{\boldsymbol{C}}_{\boldsymbol{y}}\boldsymbol{P}^{\perp} - c_n \boldsymbol{P}^{\perp}\|_{\text{F}}^2. \end{aligned} \tag{3.127}$$

Choosing $\boldsymbol{C}_{\boldsymbol{h}} = \boldsymbol{S}^{\dagger}\tilde{\boldsymbol{C}}_{\boldsymbol{y}}\boldsymbol{S}^{\dagger,\text{H}} - (\boldsymbol{S}^{\text{H}}\boldsymbol{S})^{-1} c_n$ the approximation error in the signal subspace is zero and minimization of the error in the noise subspace

$$\min_{c_n} \|\boldsymbol{P}^{\perp}\tilde{\boldsymbol{C}}_{\boldsymbol{y}}\boldsymbol{P}^{\perp} - c_n \boldsymbol{P}^{\perp}\|_{\text{F}}^2, \tag{3.128}$$

yields $\hat{c}_n^{\text{LS}}$.

---

[24] Note that the SAGE algorithm (Section 3.3.4) provides $\hat{\boldsymbol{C}}_{\boldsymbol{h}}^{\text{Heur}}$ in the first iteration when initialized with $c_n^{(0)} = 0$.

[25] Note that $\|\boldsymbol{A} + \boldsymbol{B}\|_{\text{F}}^2 = \|\boldsymbol{A}\|_{\text{F}}^2 + \|\boldsymbol{B}\|_{\text{F}}^2$ for two matrices $\boldsymbol{A}$ and $\boldsymbol{B}$ with inner product $\operatorname{tr}[\boldsymbol{A}^{\text{H}}\boldsymbol{B}] = 0$ ("Theorem of Pythagoras").

Both LS problems (3.124) and (3.126) result in the same estimates $\hat{\boldsymbol{C}}_{\boldsymbol{h}}^{\mathrm{LS}}$ and $\hat{c}_n^{\mathrm{LS}}$. The bias of the heuristic estimator $\hat{\boldsymbol{C}}_{\boldsymbol{h}}^{\mathrm{Heur}}$ (3.123) is removed in $\hat{\boldsymbol{C}}_{\boldsymbol{h}}^{\mathrm{LS}}$ [41], which is now unbiased. The estimate $\hat{\boldsymbol{C}}_{\boldsymbol{h}}^{\mathrm{LS}}$ is indefinite with a non-zero probability which is mainly determined by $\boldsymbol{C}_{\boldsymbol{h}}$, B, N, and c_n. Depending on the application, this can lead to a substantial performance degradation.

On the other hand, if $\hat{\boldsymbol{C}}_{\boldsymbol{h}}^{\mathrm{LS}}$ is positive semidefinite, $\hat{\boldsymbol{C}}_{\boldsymbol{h}}^{\mathrm{LS}}$ and $\hat{c}_n^{\mathrm{LS}}$ are the ML estimates: $\hat{\boldsymbol{C}}_{\boldsymbol{h}}^{\mathrm{LS}}$ and $\hat{c}_n^{\mathrm{LS}}$ are the unique positive semidefinite solutions to the likelihood equation (cf. (3.4)). Here, the likelihood equation is $\tilde{\boldsymbol{C}}_{\boldsymbol{y}} = \boldsymbol{S}\boldsymbol{C}_{\boldsymbol{h}}\boldsymbol{S}^{\mathrm{H}} + c_n \boldsymbol{I}_{MN}$ and the approximation error in (3.126) is zero in this case.[26]

3.5.3 Least-Squares with Positive Semidefinite Constraint

Because $\hat{\boldsymbol{C}}_{\boldsymbol{h}}^{\mathrm{LS}}$ (3.125) can be indefinite, we introduce positive semidefinite constraints in both LS problems (3.124) and (3.126). We start with the latter problem because the solution for the first follows from it.

LS Approximation for $\mathcal{S}_{\mathrm{ext2}}$

We extend problem (3.126) to

$$\boxed{(\hat{\boldsymbol{C}}_{\boldsymbol{h}}^{\mathrm{psd2}}, \hat{c}_n^{\mathrm{psd2}}) = \underset{\boldsymbol{C}_{\boldsymbol{h}}, c_n}{\operatorname{argmin}} \|\tilde{\boldsymbol{C}}_{\boldsymbol{y}} - \boldsymbol{S}\boldsymbol{C}_{\boldsymbol{h}}\boldsymbol{S}^{\mathrm{H}} - c_n \boldsymbol{I}_{MN}\|_{\mathrm{F}}^2 \text{ s.t. } \boldsymbol{C}_{\boldsymbol{h}} \succeq 0.} \quad (3.129)$$

To solve it analytically we proceed based on (3.127) and define $\boldsymbol{S}' = \boldsymbol{S}(\boldsymbol{S}^{\mathrm{H}}\boldsymbol{S})^{-1/2}$ with orthonormalized columns, i.e., $\boldsymbol{S}'^{\mathrm{H}}\boldsymbol{S}' = \boldsymbol{I}_P$. Thus, we have $\boldsymbol{P} = \boldsymbol{S}'\boldsymbol{S}'^{\mathrm{H}}$. We rewrite (3.127) using the EVD of

$$\boldsymbol{S}'^{\mathrm{H}}\tilde{\boldsymbol{C}}_{\boldsymbol{y}}\boldsymbol{S}' = \boldsymbol{V}\boldsymbol{\Sigma}\boldsymbol{V}^{\mathrm{H}} \quad (3.130)$$

with $\boldsymbol{\Sigma} = \mathbf{diag}[\sigma_1, \sigma_2, \ldots, \sigma_P]$, $\sigma_i \geq \sigma_{i+1} \geq 0$, and of

$$\boldsymbol{X} = (\boldsymbol{S}^{\mathrm{H}}\boldsymbol{S})^{1/2}\boldsymbol{C}_{\boldsymbol{h}}(\boldsymbol{S}^{\mathrm{H}}\boldsymbol{S})^{1/2} = \boldsymbol{U}_{\mathrm{x}}\boldsymbol{D}_{\mathrm{x}}\boldsymbol{U}_{\mathrm{x}}^{\mathrm{H}} \quad (3.131)$$

with $\boldsymbol{D}_{\mathrm{x}} = \mathbf{diag}[d_1, d_2, \ldots, d_P]$:

[26] In [60] the solution (3.125) is also obtained as a special case of the iterative solution of (3.110) with the "expectation-conditional maximization either algorithm".

$$\begin{aligned}\|\tilde{\boldsymbol{C}}_{\boldsymbol{y}} - \boldsymbol{S}\boldsymbol{C}_{\boldsymbol{h}}\boldsymbol{S}^{\mathrm{H}} - c_n\boldsymbol{I}_{MN}\|_{\mathrm{F}}^2 &= \|\boldsymbol{S}'(\boldsymbol{S}'^{\mathrm{H}}\tilde{\boldsymbol{C}}_{\boldsymbol{y}}\boldsymbol{S}' - \boldsymbol{X} - c_n\boldsymbol{I}_P)\boldsymbol{S}'^{\mathrm{H}}\|_{\mathrm{F}}^2 + \\ &\quad + \|\boldsymbol{P}^{\perp}\tilde{\boldsymbol{C}}_{\boldsymbol{y}}\boldsymbol{P}^{\perp} - c_n\boldsymbol{P}^{\perp}\|_{\mathrm{F}}^2 \\ &= \|\boldsymbol{S}'^{\mathrm{H}}\tilde{\boldsymbol{C}}_{\boldsymbol{y}}\boldsymbol{S}' - \boldsymbol{U}_{\mathrm{x}}\boldsymbol{D}_{\mathrm{x}}\boldsymbol{U}_{\mathrm{x}}^{\mathrm{H}} - c_n\boldsymbol{I}_P\|_{\mathrm{F}}^2 + \|\boldsymbol{P}^{\perp}\tilde{\boldsymbol{C}}_{\boldsymbol{y}}\boldsymbol{P}^{\perp} - c_n\boldsymbol{P}^{\perp}\|_{\mathrm{F}}^2. \quad (3.132)\end{aligned}$$

First, we minimize this expression w.r.t. $\boldsymbol{U}_{\mathrm{x}}$ subject to the constraint $\boldsymbol{U}_{\mathrm{x}}^{\mathrm{H}}\boldsymbol{U}_{\mathrm{x}} = \boldsymbol{I}_P$. The corresponding Lagrangian function with Lagrange parameters $\{m_i\}_{i=1}^P \geq 0$ and $\boldsymbol{M} \in \mathbb{C}^{P\times P}$ is

$$\begin{aligned}\mathrm{F}(\boldsymbol{U}_{\mathrm{x}}, \{d_i\}_{i=1}^P, c_n, \{m_i\}_{i=1}^P, \boldsymbol{M}) &= \|\boldsymbol{S}'^{\mathrm{H}}\tilde{\boldsymbol{C}}_{\boldsymbol{y}}\boldsymbol{S}' - \boldsymbol{U}_{\mathrm{x}}\boldsymbol{D}_{\mathrm{x}}\boldsymbol{U}_{\mathrm{x}}^{\mathrm{H}} - c_n\boldsymbol{I}_P\|_{\mathrm{F}}^2 \\ + \|\boldsymbol{P}^{\perp}\tilde{\boldsymbol{C}}_{\boldsymbol{y}}\boldsymbol{P}^{\perp} - c_n\boldsymbol{P}^{\perp}\|_{\mathrm{F}}^2 &+ \mathrm{tr}\left[\boldsymbol{M}^{\mathrm{T}}(\boldsymbol{U}_{\mathrm{x}}^{\mathrm{H}}\boldsymbol{U}_{\mathrm{x}} - \boldsymbol{I}_P)\right] + \mathrm{tr}\left[\boldsymbol{M}^*(\boldsymbol{U}_{\mathrm{x}}^{\mathrm{H}}\boldsymbol{U}_{\mathrm{x}} - \boldsymbol{I}_P)\right] - \\ &- \sum_{i=1}^P d_i m_i. \quad (3.133)\end{aligned}$$

The Karush-Kuhn-Tucker (KKT) conditions lead to

$$\frac{\partial \mathrm{F}}{\partial \boldsymbol{U}_{\mathrm{x}}^*} = \boldsymbol{U}_{\mathrm{x}}\boldsymbol{M}^{\mathrm{T}} + \boldsymbol{U}_{\mathrm{x}}\boldsymbol{M}^* - 2(\boldsymbol{S}'^{\mathrm{H}}\tilde{\boldsymbol{C}}_{\boldsymbol{y}}\boldsymbol{S}' - c_n\boldsymbol{I}_P)\boldsymbol{U}_{\mathrm{x}}\boldsymbol{D}_{\mathrm{x}} = \boldsymbol{0}_{P\times P}. \quad (3.134)$$

The left hand side of the resulting equation

$$\frac{1}{2}(\boldsymbol{M}^{\mathrm{T}} + \boldsymbol{M}^*) = \boldsymbol{U}_{\mathrm{x}}^{\mathrm{H}}(\boldsymbol{S}'^{\mathrm{H}}\tilde{\boldsymbol{C}}_{\boldsymbol{y}}\boldsymbol{S}' - c_n\boldsymbol{I}_P)\boldsymbol{U}_{\mathrm{x}}\boldsymbol{D}_{\mathrm{x}} \quad (3.135)$$

is Hermitian. If all $d_i > 0$ are distinct, the right hand side is only Hermitian if $\boldsymbol{U}_{\mathrm{x}} = \boldsymbol{V}$. Later we see that $d_i > 0$ are distinct if the eigenvalues σ_i are distinct.

Due to the definition of the Frobenius norm and $\boldsymbol{U}_{\mathrm{x}} = \boldsymbol{V}$, which is the optimum choice of eigenvectors, the cost function is now equivalent to

$$\|\tilde{\boldsymbol{C}}_{\boldsymbol{y}} - \boldsymbol{S}\boldsymbol{C}_{\boldsymbol{h}}\boldsymbol{S}^{\mathrm{H}} - c_n\boldsymbol{I}_{MN}\|_{\mathrm{F}}^2 = \sum_{i=1}^P (\sigma_i - d_i - c_n)^2 + \|\boldsymbol{P}^{\perp}\tilde{\boldsymbol{C}}_{\boldsymbol{y}}\boldsymbol{P}^{\perp} - c_n\boldsymbol{P}^{\perp}\|_{\mathrm{F}}^2. \quad (3.136)$$

Completing the squares in (3.136) and omitting the constant, the optimization problem (3.129) is

$$\min_{\{d_i\}_{i=1}^{MK}, c_n} \sum_{i=1}^P (\sigma_i - d_i - c_n)^2 + M\bar{N}\left(c_n - \hat{c}_n^{\mathrm{LS}}\right)^2 \text{ s.t. } d_i \geq 0. \quad (3.137)$$

Note that $\mathrm{tr}[\boldsymbol{P}^{\perp}] = M\bar{N}$. Assuming distinct σ_i this optimization problem can be solved interpreting its KKT conditions as follows:

1. If the unconstrained LS solution (3.125) is positive semidefinite, i.e., $d_i \geq 0$ for all i, it is the solution to (3.137).

2. The constraints on d_i are either active or inactive corresponding to d_i being zero or positive, i.e., 2^P possibilities should be checked in general. These can be reduced exploiting the order of σ_i, which is our focus in the sequel.
3. As a first step we could set $d_j = 0$ for all j with $\sigma_j - \hat{c}_n^{\mathrm{LS}} < 0$. All indices j with $\sigma_j - \hat{c}_n^{\mathrm{LS}} < 0$ are collected in the set $\mathcal{Z} \subseteq \{1, 2, \ldots, P\}$ with cardinality $Z = |\mathcal{Z}|$. For the remaining indices, we choose $d_i = \sigma_j - c_n$. Thus, the cost function in (3.137) is
$$\sum_{i \in \mathcal{Z}} (\sigma_i - c_n)^2 + M\bar{N} \left(c_n - \hat{c}_n^{\mathrm{LS}}\right)^2 . \tag{3.138}$$
Minimization w.r.t. c_n yields
$$\hat{c}_n^{\mathrm{psd2}} = \frac{1}{M\bar{N} + Z} \left(M\bar{N}\, \hat{c}_n^{\mathrm{LS}} + \sum_{i \in \mathcal{Z}} \sigma_i \right) . \tag{3.139}$$
Because $\sigma_i < \hat{c}_n^{\mathrm{LS}}, i \in \mathcal{Z}$, we have $\hat{c}_n^{\mathrm{psd}} \leq \hat{c}_n^{\mathrm{LS}}$ with strict inequality in case $Z > 0$. Thus, also information about the noise in the signal subspace is considered, i.e., part of the signal subspace is treated as if it contained noise only.[27]
4. But the cost function (3.138) could be reduced further if only fewer d_i are chosen zero. This is possible if we start with the smallest σ_i, i.e., $i = P$, and check whether $\sigma_i - \hat{c}_n^{\mathrm{LS}} < 0$. If negative, we set $d_P = 0$. This decreases $\hat{c}_n^{\mathrm{psd}}$ based on $\mathcal{Z} = \{P\}$. We continue with $j = P - 1$ and check if $\sigma_{P-1} - \hat{c}_n^{\mathrm{psd}} < 0$. $\hat{c}_n^{\mathrm{psd}}$ is recomputed based on (3.139) in every step. As $\hat{c}_n^{\mathrm{psd}}$ decreases, generally fewer $\sigma_i - \hat{c}_n^{\mathrm{psd}}$ are negative than for $\sigma_i - \hat{c}_n^{\mathrm{LS}}$. We continue decreasing i until $\sigma_i - \hat{c}_n^{\mathrm{psd}} \geq 0$, for which we choose $d_i = \sigma_i - \hat{c}_n^{\mathrm{psd}}$.
5. Thus, the number $Z = |\mathcal{Z}|$ of zero d_i is minimized, i.e., fewer terms appear in the cost function (3.138). This decreases the cost function.

These observations yield Algorithm 3.1 which provides a computationally efficient solution to optimization problem (3.137). The estimates $\hat{c}_n^{\mathrm{psd}}$ and $\hat{\boldsymbol{C}}_{\boldsymbol{h}}^{\mathrm{psd}}$ are biased as long as the probability for any $\sigma_i - \hat{c}_n^{\mathrm{LS}} < 0$ is non-zero, which is more likely for small B, large noise variance c_n, and higher correlations in $\boldsymbol{C}_{\boldsymbol{h}}$.

LS Approximation for $\mathcal{S}_{\mathrm{ext1}}$

An alternative approach is the positive semidefinite LS approximation of $\tilde{\boldsymbol{C}}_{\hat{\boldsymbol{h}}}$

[27] The SAGE algorithm for solving the ML problem (3.110) on $\mathcal{S}$ also exploits the noise in the signal space to improve its estimate of c_n (Section 3.3.2).

Algorithm 3.1 Positive semidefinite estimate of channel covariance matrix and noise variance related to the EXIP with $\mathcal{S}_{\text{ext2}}$.

```
    Z = 0, 𝒵 = {}
 2: ĉ_n = ĉ_n^LS
    S' = S(S^H S)^{-1/2}
 4: compute EVD S'^H C̃_y S' = V Σ V^H
    Σ = diag[σ_1, σ_2, ..., σ_P], σ_j ≥ σ_{j+1}, ∀j
 6: for i = P, P − 1, ..., 1 do
      if σ_i − ĉ_n < 0 then
 8:     d_i = 0
        Z ← Z + 1
10:     𝒵 ← 𝒵 ∪ {i}
        ĉ_n = 1/(M N̄ + Z) (M N̄ ĉ_n^LS + Σ_{i∈𝒵} σ_i)
12:   else
        d_i = σ_i − ĉ_n
14:   end if
    end for
16: D = diag[d_1, d_2, ..., d_P]
    ĉ_n^psd2 = ĉ_n
18: Ĉ_h^psd2 = (S^H S)^{-1/2} V D V^H (S^H S)^{-1/2}
```

$$\boxed{\begin{aligned}(\hat{\boldsymbol{C}}_{\boldsymbol{h}}^{\mathrm{psd1}}, \hat{c}_n^{\mathrm{psd1}}) = \underset{\boldsymbol{C}_{\boldsymbol{h}}, c_n}{\operatorname{argmin}} \|\tilde{\boldsymbol{C}}_{\hat{\boldsymbol{h}}} - \boldsymbol{C}_{\boldsymbol{h}} - (\boldsymbol{S}^{\mathrm{H}}\boldsymbol{S})^{-1} c_n\|_{\mathrm{F}}^2 + (\tilde{c}_n - c_n)^2 \\ \text{s.t.} \quad \boldsymbol{C}_{\boldsymbol{h}} \succeq 0.\end{aligned}} \tag{3.140}$$

For $\boldsymbol{S}^{\mathrm{H}}\boldsymbol{S} = N\boldsymbol{I}_P$, a solution can be derived with the same steps as before. We parameterize $\boldsymbol{C}_{\boldsymbol{h}}$ with its eigenvalue decomposition $\boldsymbol{C}_{\boldsymbol{h}} = \boldsymbol{U}\boldsymbol{D}\boldsymbol{U}^{\mathrm{H}}$, where $\boldsymbol{U}^{\mathrm{H}}\boldsymbol{U} = \boldsymbol{I}_P$, $\boldsymbol{D} = \mathbf{diag}[d_1, d_2, \ldots, d_P]$, and $d_i \geq 0$.

The eigenvectors of $\tilde{\boldsymbol{C}}_{\hat{\boldsymbol{h}}}$ and $\boldsymbol{C}_{\boldsymbol{h}}$ can be shown to be identical. The optimization problem for the remaining parameters is

$$\min_{\boldsymbol{C}_{\boldsymbol{h}}, c_n} \sum_{i=1}^{P} (\sigma_i - d_i - c_n/N)^2 + (c_n - \hat{c}_n^{\mathrm{LS}})^2 \quad \text{s.t.} \quad d_i \geq 0. \tag{3.141}$$

It is solved by Algorithm 3.2.

The estimate of c_n is

$$\hat{c}_n^{\mathrm{psd1}} = \frac{1}{1 + Z/N^2}\left(\hat{c}_n^{\mathrm{LS}} + \frac{1}{N}\sum_{i\in\mathcal{Z}} \sigma_i\right) \leq \hat{c}_n^{\mathrm{LS}} \tag{3.142}$$

Algorithm 3.2 Positive semidefinite estimate of channel covariance matrix and noise variance related to the EXIP with $\mathcal{S}_{\text{ext1}}$.

```
    Z = 0, 𝒵 = {}
 2: ĉ_n = ĉ_n^LS
    compute EVD S† C̃_y S†H = V Σ V^H
 4: Σ = diag[σ_1, σ_2, ..., σ_P], σ_j ≥ σ_{j+1} ∀j
    for i = P, P − 1, ..., 1 do
 6:   if σ_i − ĉ_n/N < 0 then
        d_i = 0
 8:     Z ← Z + 1
        𝒵 ← 𝒵 ∪ {i}
10:     ĉ_n = 1/(1+Z/N²) (ĉ_n^LS + 1/N Σ_{i∈𝒵} σ_i)
      else
12:     d_i = σ_i − ĉ_n/N
      end if
14: end for
    D = diag[d_1, d_2, ..., d_P]
16: ĉ_n^psd1 = ĉ_n
    Ĉ_h^psd1 = V D V^H
```

with the set $\mathcal{Z}$ of indices i for which $d_i = 0$. For the same set $\mathcal{Z}$, this estimate is closer to $\hat{c}_n^{\text{LS}}$ than $\hat{c}_n^{\text{psd2}}$ because $\hat{c}_n^{\text{psd2}} \leq \hat{c}_n^{\text{psd1}} \leq \hat{c}_n^{\text{LS}}$. If $\hat{c}_n^{\text{LS}}$ is already accurate, i.e., for sufficiently large B and $\bar{N}$, the bias increases the MSE w.r.t. c_n; but for small $\bar{N}$ and B the bias reduces the estimation error (this is not observed for the selected scenarios in Section 3.6).

A Heuristic Modification

Because $\hat{c}_n^{\text{psd1}}$ and $\hat{c}_n^{\text{psd2}}$ are biased and the their estimation error is typically larger than for the unbiased estimate $\hat{c}_n^{\text{LS}}$, the following heuristic achieves better result, when applied to MMSE channel estimation:

1. Estimate c_n using $\hat{c}_n^{\text{LS}}$ (3.125), which is unbiased and based on the true noise subspace.
2. Solve optimization problem (3.141)

$$\min_{\{d_i\}_{i=1}^{MK}} \sum_{i=1}^{P} (\sigma_i - d_i - \hat{c}_n^{\text{LS}})^2 \quad \text{s.t.} \quad d_i \geq 0, \tag{3.143}$$

   i.e., set all $d_i = 0$ for $\sigma_i - \hat{c}_n^{\text{LS}} < 0$. The solution reads

$$\boxed{\hat{\boldsymbol{C}}_{\boldsymbol{h}}^{\mathrm{psd3}} = \boldsymbol{V}\boldsymbol{D}^{+}\boldsymbol{V}^{\mathrm{H}},} \tag{3.144}$$

where $\boldsymbol{D}^{+}$ performs $\max(0, d_i)$ for all elements in $\boldsymbol{D}$ from the EVD of

$$\hat{\boldsymbol{C}}_{\boldsymbol{h}}^{\mathrm{LS}} = \boldsymbol{V}\boldsymbol{D}\boldsymbol{V}^{\mathrm{H}}. \tag{3.145}$$

This is equivalent to discarding the negative definite part of the estimate $\hat{\boldsymbol{C}}_{\boldsymbol{h}}^{\mathrm{LS}}$ (3.125) similar to [92].

Computational Complexity

The additional complexity for computing the positive definite solution compared to (3.125) results from the EVD of $\boldsymbol{S}'^{\mathrm{H}}\tilde{\boldsymbol{C}}_{\boldsymbol{y}}\boldsymbol{S}'$ and computation of $(\boldsymbol{S}^{\mathrm{H}}\boldsymbol{S})^{1/2}$ and
$(\boldsymbol{S}^{\mathrm{H}}\boldsymbol{S})^{-1/2}$ (only for Algorithm 3.1). For the indefinite least-squares estimate (3.125), a tracking-algorithm of very low-complexity was presented by [141], whereas tracking of eigenvalues and eigenvectors is more difficult and complex.

3.5.4 Generalization to Spatially Correlated Noise

Next we generalize our previous results to a general noise covariance matrix $\boldsymbol{C}_{\boldsymbol{n}} = \boldsymbol{I}_N \otimes \boldsymbol{C}_{\boldsymbol{n},\mathrm{s}}$ and solve (3.115) and (3.119) without weighting. We start with (3.119) and rewrite (3.127) for the more general noise covariance matrix $\boldsymbol{C}_{\boldsymbol{n}} = \boldsymbol{I}_N \otimes \boldsymbol{C}_{\boldsymbol{n},\mathrm{s}}$

$$\begin{aligned}\|\tilde{\boldsymbol{C}}_{\boldsymbol{y}} - \boldsymbol{S}\boldsymbol{C}_{\boldsymbol{h}}\boldsymbol{S}^{\mathrm{H}} - \boldsymbol{I}_N \otimes \boldsymbol{C}_{\boldsymbol{n},\mathrm{s}}\|_{\mathrm{F}}^2 &= \|\boldsymbol{P}\tilde{\boldsymbol{C}}_{\boldsymbol{y}}\boldsymbol{P} - \boldsymbol{S}\boldsymbol{C}_{\boldsymbol{h}}\boldsymbol{S}^{\mathrm{H}} - \boldsymbol{P}(\boldsymbol{I}_N \otimes \boldsymbol{C}_{\boldsymbol{n},\mathrm{s}})\boldsymbol{P}\|_{\mathrm{F}}^2 \\ &+ \|\boldsymbol{P}^{\perp}\tilde{\boldsymbol{C}}_{\boldsymbol{y}}\boldsymbol{P}^{\perp} - \boldsymbol{P}^{\perp}(\boldsymbol{I}_N \otimes \boldsymbol{C}_{\boldsymbol{n},\mathrm{s}})\boldsymbol{P}^{\perp}\|_{\mathrm{F}}^2.\end{aligned} \tag{3.146}$$

As before the first term is zero choosing $\boldsymbol{C}_{\boldsymbol{h}}$ from the space of Hermitian P-dimensional matrices as

$$\boxed{\hat{\boldsymbol{C}}_{\boldsymbol{h}}^{\mathrm{LS}} = \boldsymbol{S}^{\dagger}(\tilde{\boldsymbol{C}}_{\boldsymbol{y}} - \boldsymbol{I}_N \otimes \hat{\boldsymbol{C}}_{\boldsymbol{n},\mathrm{S}}^{\mathrm{LS}})\boldsymbol{S}^{\dagger,\mathrm{H}}} \tag{3.147}$$

given an estimate $\hat{\boldsymbol{C}}_{\boldsymbol{n},\mathrm{S}}^{\mathrm{LS}}$ as described below. The estimate $\hat{\boldsymbol{C}}_{\boldsymbol{h}}^{\mathrm{LS}}$ is unbiased, but indefinite in general. Now, we can minimize the second term in (3.146)

$$\min_{\boldsymbol{C}_{\boldsymbol{n},\mathrm{s}}} \|\boldsymbol{P}^{\perp}\tilde{\boldsymbol{C}}_{\boldsymbol{y}}\boldsymbol{P}^{\perp} - \boldsymbol{P}^{\perp}(\boldsymbol{I}_N \otimes \boldsymbol{C}_{\boldsymbol{n},\mathrm{s}})\boldsymbol{P}^{\perp}\|_{\mathrm{F}}^2. \tag{3.148}$$

This yields the estimate of the noise covariance matrix $\boldsymbol{C}_{\boldsymbol{n},\mathrm{s}}$

$$\hat{\boldsymbol{C}}_{\boldsymbol{n},\mathrm{s}}^{\mathrm{LS}} = \frac{1}{N-K}\sum_{n=1}^{N}(\boldsymbol{e}_n^{\mathrm{T}} \otimes \boldsymbol{I}_M)\boldsymbol{P}^{\perp}\tilde{\boldsymbol{C}}_{\boldsymbol{y}}\boldsymbol{P}^{\perp}(\boldsymbol{e}_n \otimes \boldsymbol{I}_M), \tag{3.149}$$

which is based on the noise subspace. It is equivalent to the sample covariance of the estimated noise

$$\hat{\boldsymbol{n}}[q,n] = (\boldsymbol{e}_n^{\mathrm{T}} \otimes \boldsymbol{I}_M)\boldsymbol{P}^{\perp}\boldsymbol{y}[q] = \boldsymbol{y}[q,n] - \hat{\boldsymbol{H}}[q]\boldsymbol{s}[n] \tag{3.150}$$

with the ML channel estimate (2.31)

$$\hat{\boldsymbol{H}}[q] = \mathbf{unvec}\left[\hat{\boldsymbol{h}}[q]\right] = \mathbf{unvec}\left[\boldsymbol{S}^{\dagger}\boldsymbol{y}[q]\right]. \tag{3.151}$$

This leads to

$$\boxed{\hat{\boldsymbol{C}}_{\boldsymbol{n},\mathrm{s}}^{\mathrm{LS}} = \frac{1}{B\bar{N}}\sum_{q\in\mathbb{O}}^{B}\sum_{n=1}^{N}\hat{\boldsymbol{n}}[q,n]\hat{\boldsymbol{n}}[q,n]^{\mathrm{H}} = \frac{1}{B\bar{N}}\sum_{q\in\mathbb{O}}^{B}\boldsymbol{Y}[q]\bar{\boldsymbol{P}}^{\perp}\boldsymbol{Y}[q]^{\mathrm{H}}} \tag{3.152}$$

as in (3.64) and (3.114). This estimate of the noise covariance matrix is very similar to the well-known ML estimate [158] of the noise covariance matrix, when estimated jointly with the channel $\boldsymbol{h}[q]$ (cf. (2.32)). The difference is in the scaling by $\bar{N} = N - K$ instead of N which yields an unbiased estimate with smaller estimation error in this case.

Optimization problem (3.115) yields the same solution. Again both problems differ if we add a positive semidefinite constraint on $\boldsymbol{C}_{\boldsymbol{h}}$.

Here a simple positive semidefinite solution of $\boldsymbol{C}_{\boldsymbol{h}}$ can be obtained similar to the heuristic introduced at the end of Section 3.5.3: A priori we choose $\hat{\boldsymbol{C}}_{\boldsymbol{n},\mathrm{s}}^{\mathrm{LS}}$ as the estimate of $\boldsymbol{C}_{\boldsymbol{n},\mathrm{s}}$. Then, for (3.115) with positive semidefinite constraint $\hat{\boldsymbol{C}}_{\boldsymbol{h}}$ is given by the positive semidefinite part of $\hat{\boldsymbol{C}}_{\boldsymbol{h}}^{\mathrm{LS}}$ (3.147), i.e., setting the negative eigenvalues to zero.

3.6 Performance Comparison

The estimators for channel and noise covariance matrices which we present in Sections 3.3 and 3.5 differ significantly regarding their computational complexity. But how much do they differ in estimation quality?

Another important question is, how many observations (measured by B and I) are necessary to obtain sufficient MMSE channel estimation and prediction quality and whether this number is available for a typical stationarity time of the wireless channel.

Estimation quality is often measured by the mean square error (MSE) of the estimates. The MSE does not reflect the requirements of a particular application on the estimation accuracy. Therefore, we apply the esti-

Scenario	I	B	M	K	Assumptions on $\boldsymbol{C}_{\boldsymbol{h}_\mathrm{T}}$ or $\boldsymbol{C}_{\boldsymbol{n},\mathrm{s}}$	Results in Figures:
A	1	$[1, 1000]$	8	8	$\boldsymbol{C}_{\boldsymbol{n},\mathrm{s}} = c_n \boldsymbol{I}_M$	3.5, 3.6
	1	100	8	8		3.7
B	1	$[1, 1000]$	8	1	$\boldsymbol{C}_{\boldsymbol{n},\mathrm{s}} \neq c_n \boldsymbol{I}_M$	3.8, 3.10
	1	100	8	1		3.9, 3.11
C	1	$[6, 50]$	1	1	Clarke's psd	3.14
	1	20	1	1	(Figure 2.12)	3.12, 3.13, 3.16
D	1	20	1	1	bandpass psd (Figure 2.12)	3.15, 3.17
E	20	20	4	4	block Toeplitz Clarke's psd $\boldsymbol{C}_{\boldsymbol{n},\mathrm{s}} = c_n \boldsymbol{I}_M$	3.18

Table 3.2 Overview of simulation scenarios in Section 3.6.

mated covariance matrices to MMSE channel estimation (Section 2.2.1) and MMSE prediction (Section 2.3.1). We discuss the performance for estimation of channel correlations in space, time, and (jointly) in space and time domain separately.

An overview of the selected scenarios is given in Table 3.2. In scenarios with multiple antennas we assume a uniform linear array with half (carrier) wavelength spacing. Moreover, we only consider frequency flat channels ($L = 0$); training symbols are chosen from Hadamard sequences with length $N = 16$ ($\boldsymbol{S}^\mathrm{H}\boldsymbol{S} = N\boldsymbol{I}_P$). More details on the models are given below and in Appendix E.

Estimation of Spatial Channel Covariance Matrix

We compare the following *algorithms* for estimation of the spatial channel covariance matrix $\boldsymbol{C}_{\boldsymbol{h}}$ and the noise variance c_n or spatial noise covariance matrix $\boldsymbol{C}_{\boldsymbol{n},\mathrm{s}}$ (the names employed in the legends of the figures are given in quotation marks at the end):

- ML estimation of $\boldsymbol{C}_{\boldsymbol{h}}$ with SAGE is based on (3.94) (Section 3.3.4) with $i_\mathrm{max} = 10$ iterations. As initialization we use the heuristic positive semidefinite LS estimate (3.144). The unbiased estimates (3.64) or (3.67) are employed for the noise covariance matrix $\boldsymbol{C}_{\boldsymbol{n},\mathrm{s}}$ and variance c_n, respectively. ("SAGE")
- This version is identical to "SAGE" except for estimation of $\boldsymbol{C}_{\boldsymbol{n},\mathrm{s}}$ or c_n, which are estimated iteratively based on HDS 2 for the noise (Section 3.3.2) with initialization from (3.64) or (3.67). In $i_\mathrm{max} = 10$ iterations we use $N_\mathrm{iter} = 5$ iterations for the HDS 1, the next 5 iterations with HDS 2. ("SAGE (include HDS for noise)")
- The heuristic estimator of $\boldsymbol{C}_{\boldsymbol{h}}$ is (3.123) and for estimation of $\boldsymbol{C}_{\boldsymbol{n},\mathrm{s}}$ or c_n on we apply (3.125) and (3.152) (identical to (3.64) and (3.67)). ("Heuristic")

- The unbiased LS estimate is (3.125) for $\boldsymbol{C}_{\boldsymbol{n},\mathrm{s}} = c_n \boldsymbol{I}_M$ and (3.147) with (3.152) for spatially correlated noise. ("Unbiased LS")
- The positive semidefinite LS estimator based on $\mathcal{S}_{\mathrm{ext1}}$ and $\boldsymbol{C}_{\boldsymbol{n},\mathrm{s}} = c_n \boldsymbol{I}_M$ is given in Algorithm 3.2. ("Pos. semidef. LS (ext. 1)")
- The positive semidefinite LS estimator based on $\mathcal{S}_{\mathrm{ext2}}$ and $\boldsymbol{C}_{\boldsymbol{n},\mathrm{s}} = c_n \boldsymbol{I}_M$ is given in Algorithm 3.1. ("Pos. semidef. LS (ext. 2)")
- The heuristic positive semidefinite LS approach is (3.144) with (3.152) or $\hat{c}_n$ from (3.125). ("Pos. semidef. LS (heuristic)")
- The solution of the ML problem for $\boldsymbol{C}_{\boldsymbol{n},\mathrm{s}} = c_n \boldsymbol{I}_M$ based on the expectation-conditional maximization either (ECME) algorithm is given by Equation (3.8) in [60]. ("ECME")

As standard performance measure we choose the average MSE of $\hat{c}_n$ normalized to c_n and the average MSE of $\hat{\boldsymbol{C}}_{\boldsymbol{h}}$ normalized to $\mathrm{tr}[\boldsymbol{C}_{\boldsymbol{h}}]$, where the MSEs are averaged over 1000 estimates (approximation of the expectation operator given in the ordinate of the figures). Choosing channel estimation as application, we give the MSE of the channel estimate $\hat{\boldsymbol{h}} = \hat{\boldsymbol{W}}\boldsymbol{y}$, where $\hat{\boldsymbol{W}}$ is the MMSE channel estimator (Section 2.2.1) evaluated for $\hat{\boldsymbol{C}}_{\boldsymbol{h}}$ and $\hat{\boldsymbol{C}}_{\boldsymbol{n},\mathrm{s}}$.[28] It is also averaged over 1000 independent estimates. Normalization with $\mathrm{tr}[\boldsymbol{C}_{\boldsymbol{h}}]$ ensures that the maximum normalized MSE for perfect knowledge of the covariance matrices is one. For $\hat{\boldsymbol{C}}_{\boldsymbol{h}} = \hat{\boldsymbol{U}}\hat{\boldsymbol{\Lambda}}\hat{\boldsymbol{U}}^{\mathrm{H}}$, $\boldsymbol{S}^{\mathrm{H}}\boldsymbol{S} = N\boldsymbol{I}_P$, and $\hat{\boldsymbol{C}}_{\boldsymbol{n},\mathrm{s}} = \hat{c}_n \boldsymbol{I}_M$, the MMSE estimator (2.22) can be written as (compare Section 2.2.4)

$$\hat{\boldsymbol{W}} = \hat{\boldsymbol{U}}\hat{\boldsymbol{\Lambda}}\left(\hat{\boldsymbol{\Lambda}} + \frac{\hat{c}_n}{N}\boldsymbol{I}_P\right)^{-1}\hat{\boldsymbol{U}}^{\mathrm{H}}\boldsymbol{S}^{\mathrm{H}}/N. \tag{3.153}$$

Now, assume that $\hat{\boldsymbol{C}}_{\boldsymbol{h}}$ is indefinite, i.e., some diagonal-elements $\hat{\lambda}_i$ of $\hat{\boldsymbol{\Lambda}}$ are negative. If $\hat{\lambda}_i \approx -\hat{c}_n/N$, the system of equations, whose solution is $\hat{\boldsymbol{W}}$, is ill-conditioned.[29] This leads to a very large norm of $\hat{\boldsymbol{W}}$ and a very large MSE of $\hat{\boldsymbol{W}}$ and of $\hat{\boldsymbol{h}}$: The regularization of the MMSE estimates, which leads to a loading of $\hat{\boldsymbol{\Lambda}}$ with $\hat{c}_n/N > 0$ (3.153) and decreased norm of $\hat{\boldsymbol{W}}$, is based on the assumption $\hat{\boldsymbol{C}}_{\boldsymbol{h}} \succeq 0$; if this is not true, it has the adverse effect.

As performance references, we employ the MSE in $\hat{\boldsymbol{h}}$ with a ML channel estimator $\boldsymbol{W} = \boldsymbol{S}^{\dagger}$ (2.31): It does not rely on $\boldsymbol{C}_{\boldsymbol{h}}$ and the MMSE estimator is employed aiming at a reduced MSE compared to ML channel estimator ("ML channel estimation"). Thus, this reference will give us the minimum number of observations B for acceptable minimum performance. As a lower bound we use the MSE of the MMSE estimator with perfect knowledge of the covariance matrices ("perfect knowledge").

For the simulations, we consider two scenarios with B statistically independent observations:

---

[28] Given ML estimates of the covariance matrices, it is the ML estimate of the MMSE channel estimator according to the ML invariance principle.

[29] This degree of ill-conditioning does *not* lead to numerical problems on a typical floating point architecture.

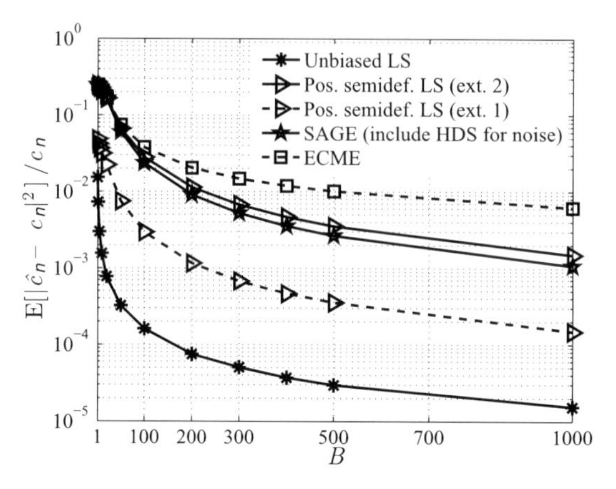

Fig. 3.5 Normalized MSE of noise variance c_n vs. B for Scenario A ($c_n = 1$).

Scenario A: We choose the same parameters as in Section 2.2.4 and Figure 2.6: $M = K = 8$, block-diagonal $\boldsymbol{C}_{\boldsymbol{h}}$ normalized to $\mathrm{tr}[\boldsymbol{C}_{\boldsymbol{h}}] = MK$, Laplace angular power spectrum with spread $\sigma = 5°$ and $\bar{\boldsymbol{\varphi}} = [-45°, -32.1°, -19.3°, -6.4°, 6.4°, 19.3°, 32.1°, 45°]$, and white noise $\boldsymbol{C}_{\boldsymbol{n},\mathrm{s}} = c_n \boldsymbol{I}_M$.

The unbiased estimator for c_n (cf. (3.125) and (3.67)) has the smallest normalized MSE (Figure 3.5). All other estimates attempt to improve $\hat{c}_n$ by exploiting the knowledge about the noise in the signal subspace. The ECME algorithm performs worst. The estimate $\hat{c}_n$ with positive semidefinite LS based on $\mathcal{S}_{\mathrm{ext1}}$ (ext. 1) is considerably better than for $\mathcal{S}_{\mathrm{ext2}}$. The performance of SAGE degrades, when HDS 2 (of the noise) is included in the iterations. Because initialization of SAGE with $\boldsymbol{C}_{\boldsymbol{h}}^{(0)}$ is not good enough, it is not able to take advantage of the additional information about c_n in the signal subspace: The algorithm cannot distinguish sufficiently between noise and signal correlations. Therefore, we consider only the unbiased estimator for the noise covariance matrix in the sequel.

The MSE in $\hat{\boldsymbol{C}}_{\boldsymbol{h}}$ for all estimators (except the heuristic estimator) is approximately identical, because the main difference between the estimators lies in the estimation quality of the small eigenvalues. In [60] the MSE in $\hat{\boldsymbol{C}}_{\boldsymbol{h}}$ of the ECME algorithm was shown to be close to the Cramér-Rao lower bound (sometimes better due to a bias); we do not investigate this relation further, because the MSE in $\hat{\boldsymbol{C}}_{\boldsymbol{h}}$ is similar for all considered estimators. Moreover, it is not a good indicator for the applications considered here, which is shown by the next results.

For $B \leq 200$, the MSE of the MMSE estimator based on unbiased LS is larger than for the ML channel estimator and increases very fast for small B (Figure 3.6): $\hat{\boldsymbol{C}}_{\boldsymbol{h}}$ is indefinite with smallest eigenvalues in the size of $-\hat{c}_n/N$, which leads to a large error in $\hat{\boldsymbol{W}}$ (3.153). For small B, SAGE and the heuristic estimator have a very similar MSE and outperform the ML chan-

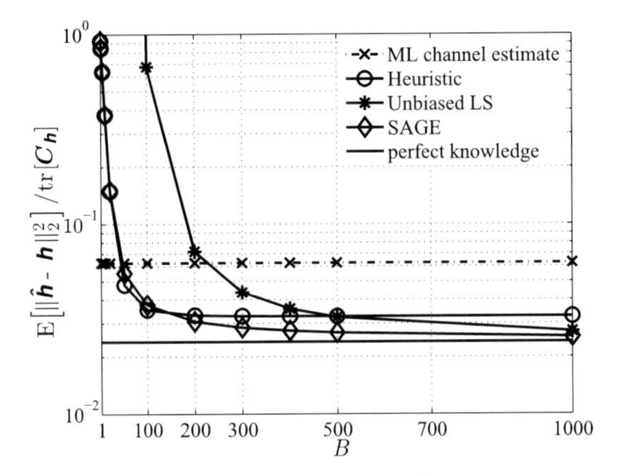

Fig. 3.6 Normalized MSE of channel estimates $\hat{\boldsymbol{h}} = \boldsymbol{W}\boldsymbol{y}$ vs. B for Scenario A ($c_n = 1$).

nel estimator for $B \geq 50$. For large B, SAGE outperforms the heuristic due to the reduced or removed bias. The MMSE performance of all other estimators[30] (positive semidefinite LS, SAGE with HDS for noise, and ECME) is almost identical to SAGE. In Figure 3.6 and 3.7 the heuristic sometimes outperforms SAGE, because its bias serves as loading in the inverse similar to minimax robust channel estimators (Section 2.4.2). Figure 3.7 shows the MSE vs. SNR $= -10\log(c_n)$ for $B = 100$; the degradation in MSE of the unbiased LS is large (Figure 3.7).

A performance of the MMSE estimators, which is worse than for ML channel estimation, can be avoided by a (robust) minimax MSE design as presented in Section 2.4.2: The set size described by $\alpha_{\boldsymbol{h}}^{(2)}$ (2.90) has to be adapted to B, $\boldsymbol{C}_{\boldsymbol{h}}$, and the SNR, but a good adaptation rule is not straightforward.

The number of statistically uncorrelated observations B available depends on the ratio of stationarity time to coherence time: Channel measurements show that this ratio is about 100 in an urban/suburban environment [227].[31] Consequently, the relevant number of observations B is certainly below 200, where the heuristic estimator is close to optimum for uncorrelated noise. This changes for correlated noise.

Scenario B: We choose $M = 8$, $K = 1$, and $\boldsymbol{C}_{\boldsymbol{h}}$ for a Laplace angular power spectrum with spread $\sigma = 5°$ and $\bar{\varphi} = 0°$. The spatial noise correlations are $\boldsymbol{C}_{\boldsymbol{n},\mathrm{s}} = \boldsymbol{C}_{\mathrm{i}} + c_n \boldsymbol{I}_M$, which models white noise with $c_n = 0.01$ additionally to an interferer with mean direction $\bar{\varphi} = 30°$, $30°$ uniform angular power spectrum, and $\mathrm{tr}[\boldsymbol{C}_{\mathrm{i}}] = Mc_{\mathrm{i}}$ (c_{i} is the variance of the interfering signal). Among the LS approaches (besides the simple heuristic) only the unbiased and the

[30] This results are not shown here.

[31] The measure employed by [227] quantifies the time-invariance of the principle eigenmode of $\boldsymbol{C}_{\boldsymbol{h}}$. It is not clear yet, whether these results can be generalized to the whole covariance matrix and to channel estimation.

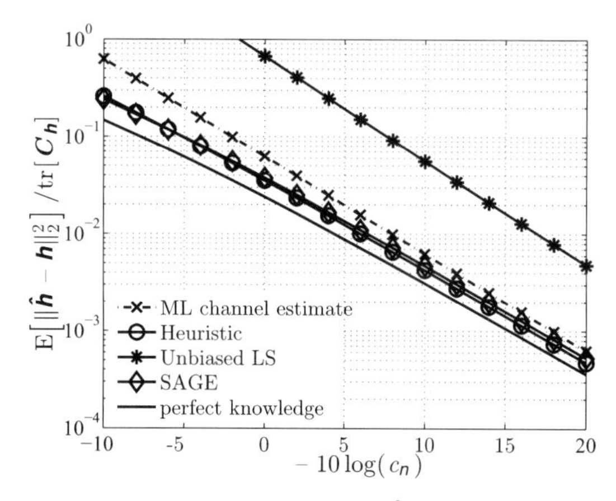

Fig. 3.7 Normalized MSE of channel estimates $\hat{\boldsymbol{h}} = \boldsymbol{W}\boldsymbol{y}$ vs. SNR $= -10\log(c_n)$ for Scenario A ($B = 100$).

heuristic positive semidefinite LS estimators were derived for correlated noise in Section 3.5.4.

In Figures 3.8 and 3.9 the MSE in $\hat{\boldsymbol{C}}_{\boldsymbol{h}}$ is given for different B and c_i, respectively: For high B and small SIR $= -10\log(c_i)$ (large c_i), the improvement of SAGE, whose MSE cannot be distinguished from unbiased LS and positive semidefinite LS approaches, over the heuristic is largest: It removes the bias of the heuristic LS estimator, which is $\mathrm{E}[\hat{\boldsymbol{C}}_{\boldsymbol{h}}^{\mathrm{Heur}}] - \boldsymbol{C}_{\boldsymbol{h}} = \boldsymbol{C}_{\boldsymbol{n},\mathrm{s}}/N$ for $K = 1$.

The MSE in MMSE channel estimation shows a significant performance improvement over the heuristic estimator for small to medium SIR and $B \geq 20$ (Figures 3.10 and 3.11). For $B \geq 5$, all estimators which yield a positive semidefinite $\hat{\boldsymbol{C}}_{\boldsymbol{h}}$ outperform the ML channel estimator. SAGE, which is initialized with the heuristic positive semidefinite LS, only yields a tiny improvement compared to its initialization.

From this observation we conclude that the most important task of the estimators is to remove the bias corresponding to noise correlations from $\hat{\boldsymbol{C}}_{\boldsymbol{h}}$, while maintaining a valid (positive semidefinite) channel covariance matrix. This is already ensured sufficiently by the heuristic positive semidefinite LS estimator, which is much simpler than SAGE.

Estimation and Completion of Autocovariance Sequence

The algorithms for estimating the temporal channel correlations (autocorrelation sequence) from Section 3.3.5 are implemented as follows assuming perfect knowledge of the error variance c_e (compare with models in Section 2.3.1). The two main estimation approaches are:

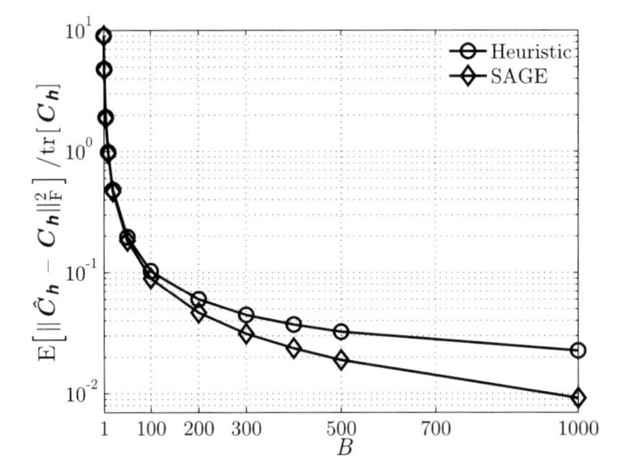

Fig. 3.8 Normalized MSE of spatial covariance matrix vs. B for correlated noise (Scenario B, $c_i = 1$, $c_n = 0.01$).

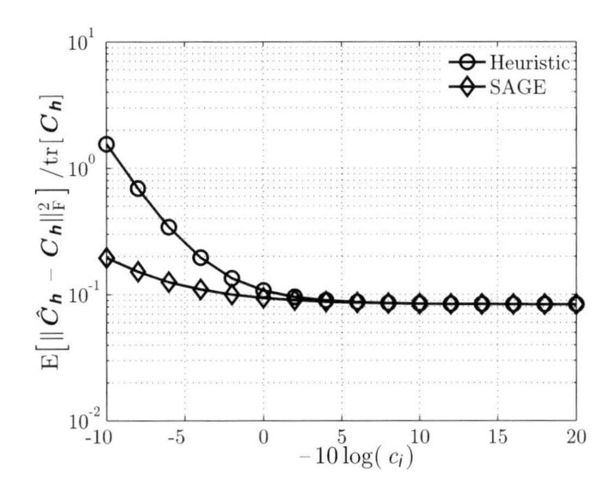

Fig. 3.9 Normalized MSE of spatial covariance matrix vs. SIR $= -10\log(c_i)$ for correlated noise (Scenario B, $B = 100$).

- The biased sample (heuristic) autocovariance estimator (3.86) (for $P = 1$), ("MMSE (Heuristic - Initialization)")
- the SAGE algorithm (3.97) with $\bar{B} = 2B - 1$, initialization by (3.86) (for $P = 1$), and $i_{\max} = 10$ iterations as described in Section 3.3.5. ("MMSE (SAGE)")

For $N_{\mathrm{P}} = 2$, the even samples of the autocovariance sequence are estimated with the biased heuristic or SAGE estimator before completing the necessary odd values based on the minimax problem (3.102). We choose a fixed maximum frequency $f_{\max} = 0.25$ of the autocovariance sequences to be com-

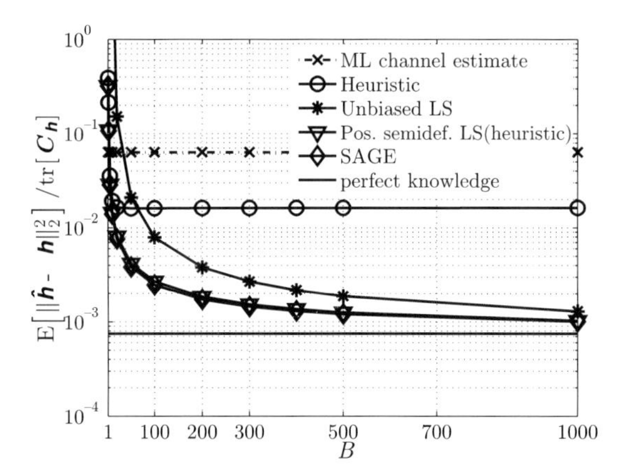

Fig. 3.10 Normalized MSE of channel estimates $\hat{\boldsymbol{h}}$ vs. B for correlated noise (Scenario B, $c_i = 1$).

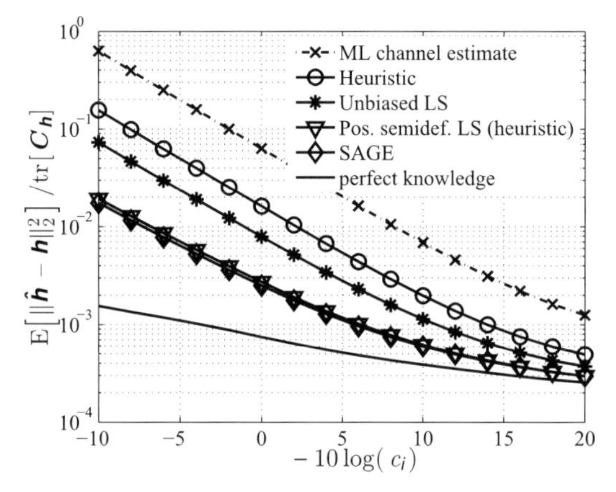

Fig. 3.11 Normalized MSE of channel estimates $\hat{\boldsymbol{h}}$ vs. SIR $= -10\log(c_i)$ for correlated noise (Scenario B, $B = 100$).

pleted. The minimax completion is solved for the second order cone formulation (3.105) with SeDuMi [207].[32]

The estimation accuracy is measured by the average MSE of univariate MMSE prediction $\mathrm{E}[|\varepsilon_Q[q]|^2]$ ($c_h[0] = 1$) with estimation error $\varepsilon_Q[q] = \hat{h}[q] - h[q]$ and $\hat{h}[q] = \boldsymbol{w}^{\mathrm{T}}\boldsymbol{y}_{\mathrm{T}}[q]$ (Section 2.3.1). It is averaged over 1000 estimates. We choose a prediction order $Q = 5$, which requires a minimum of $B = 6$ observations to be able to estimate the necessary $Q + 1 = 6$ samples of the autocovariance sequence (for $N_{\mathrm{P}} = 1$). Here, we do not consider the MSE in

[32] Because we choose $f_{\max} = 0.25$ the optimization problem (3.102) has a solution for all estimates and the projection of the estimates on the set of partial sequences which have a positive semidefinite band-limited (to $f_{\max} = 0.25$) extension (3.107) is not necessary.

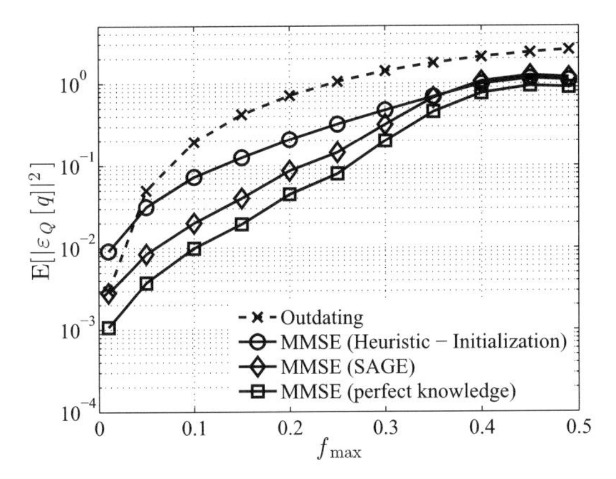

Fig. 3.12 MSE of channel prediction ($Q = 5$) vs. maximum frequency $f_{\max}$ (Scenario C, $c_h[0] = 1$, $c_e = 0.01$, $B = 20$).

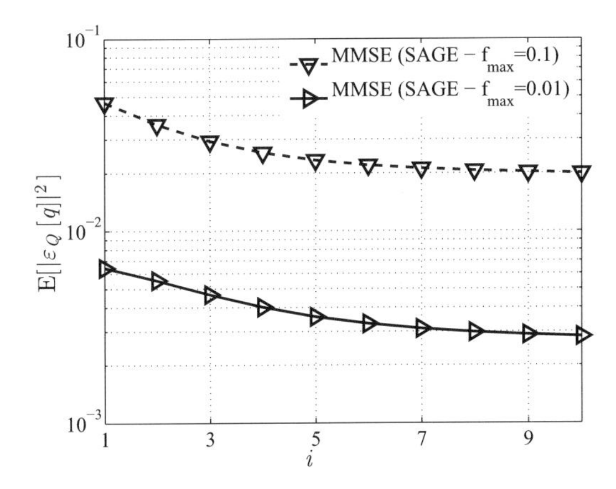

Fig. 3.13 MSE of channel prediction ($Q = 5$) vs. iteration number (Scenario C, $c_h[0] = 1$, $c_e = 0.01$).

the estimated autocovariance sequence, as it is not reduced significantly by SAGE: The gain in MSE is small compared to the gains obtained in MMSE prediction.

As lower bound on the MSE performance we employ the MSE of MMSE prediction with perfect knowledge of the autocovariance sequence and $Q = 5$ ("MMSE (perfect knowledge)"). Simply using the last available observation of the channel $\hat{h}[q-1]$, i.e., $\boldsymbol{w}^{\mathrm{T}} = [1, 0, \ldots, 0]$, serves as an upper limit (a trivial "predictor"), which should be outperformed by any more advanced predictor (denoted as "Outdating" in the figures).

The error variance is chosen as $c_e = 0.01$, which corresponds to $N = 10$ and $c_n = 0.1$ ($c_e = c_n/N$ as in (2.60)).

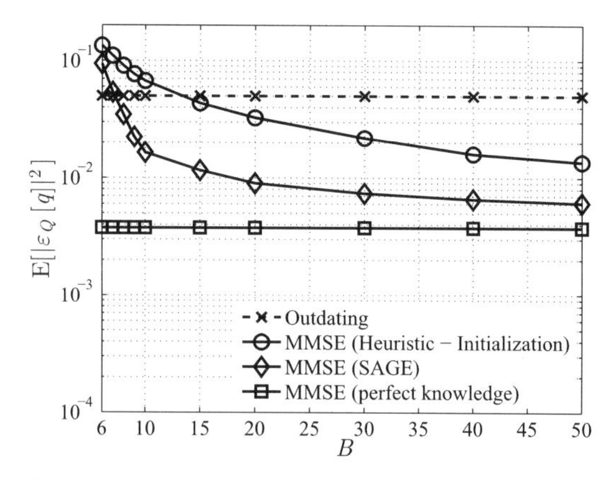

Fig. 3.14 MSE of channel prediction ($Q = 5$) vs. B (Scenario C, $c_h[0] = 1$, $c_e = 0.01$, $f_{\max} = 0.05$).

Scenario C, $N_\mathrm{P} = 1$: Clarke's model with $c_h[0] = 1$ is selected for the power spectral density of the autocovariance sequence (Figure 2.12). MMSE prediction based on estimates from SAGE significantly improves the prediction accuracy over the biased heuristic estimator (Figure 3.12). For $B = 20$, it is already close to the lower bound. The SAGE algorithm converges in 7 iterations (Figure 3.13). Figure 3.14 shows the prediction MSE versus the number of observations B for $f_{\max} = 0.05$: More than $B = 10$ observations should be available to ensure reliable prediction, which outperforms simple "Outdating". A stationarity time of the channel over $B = 10$ observations is sufficient to obtain reliable estimates: The theory and measurements by Matz [140] indicates a stationarity time between 0.7 s and 2 s, which indicates that a sufficient number of (correlated) channel observations is available in wireless communications.

Since Clarke's power spectral density is very close to the rectangular power spectral density, which is the worst-case power spectral density of the minimax predictor, the performance of minimax prediction and prediction with perfect knowledge is very similar (Figure 2.15), but depends on the knowledge of $f_{\max}$.

Scenario D, $N_\mathrm{P} = 1$: Now, we consider the band-pass power spectral density with $c_h[0] = 1$ (Figure 2.12). We also give the minimax robust predictor based on the class of power spectral density band-limited to $f_{\max}$ ($f_{\max}$ perfectly known) as reference. This minimax predictor is too conservative for this scenario (Section 2.4.3.4) in case of moderate too high Doppler frequencies. Note that its performance depends on an (approximate) knowledge of $f_{\max}$.

MMSE prediction based on the heuristic estimator performs worse than the minimax predictor for low $f_{\max}$ (Figure 3.15). But with the SAGE estimator the MSE performance of the predictor follows closely the MSE for prediction with perfect knowledge of the autocovariance sequence.

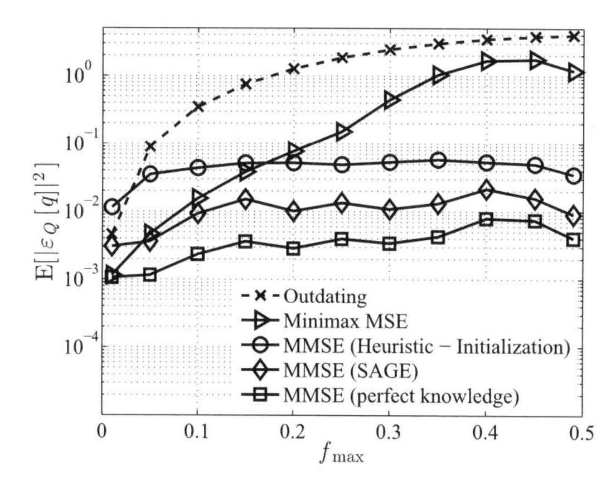

Fig. 3.15 MSE of channel prediction ($Q = 5$) vs. maximum frequency $f_{\max}$ (Scenario D, bandpass power spectral density, $c_h[0] = 1$, $c_e = 0.01$, $B = 20$).

Scenario C and D, $N_{\mathrm{P}} = 2$: Above, we have considered spacing of the observations with period $N_{\mathrm{P}} = 1$. For a spacing with $N_{\mathrm{P}} = 2$, i.e., $\ell_i = N_{\mathrm{P}} i - 1$ in (3.20), the ML problem for estimation of the samples of the autocovariance sequence, which are required for prediction by one sample, is ill-posed. We propose an optimization problem for minimax MSE or minimum norm completion of the missing (odd) samples in Section 3.4. From Figure 3.16 we conclude that prediction quality for $f_{\max} \leq 0.1$ is equivalent to $N_{\mathrm{P}} = 1$ (Figure 3.12). But the complexity of the completion step is significant. For $f_{\max} \geq 0.25$, the MSE of prediction is very close to its maximum of one. And for $f_{\max} \geq 0.2$, the gap between outdating and MMSE prediction with perfect knowledge is small and the SAGE estimator with a completion based on $B = 20$ observations is not accurate enough to provide a gain over "Outdating".

This is different for the band-pass power spectral density (Figure 3.17): Prediction accuracy based on SAGE is superior to all other methods. Only for small $f_{\max}$, minimax prediction is an alternative (given approximate knowledge of $f_{\max}$).

Estimation of Spatial and Temporal Channel Correlations

Finally, we consider estimation of spatial and temporal channel correlations $\boldsymbol{C}_{\boldsymbol{h}_{\mathrm{T}}}$ together with the noise variance ($N = 16$). The noise variance $c_n = 0.1$ ($\boldsymbol{C}_{\boldsymbol{n},\mathrm{s}} = c_n \boldsymbol{I}_M$) is estimated with (3.67). The estimators for $\boldsymbol{C}_{\boldsymbol{h}_{\mathrm{T}}}$ are:

- The heuristic biased estimator of $\boldsymbol{C}_{\boldsymbol{h}_{\mathrm{T}}}$ is given by (3.86) ("MMSE (Heuristic — Initialization)")

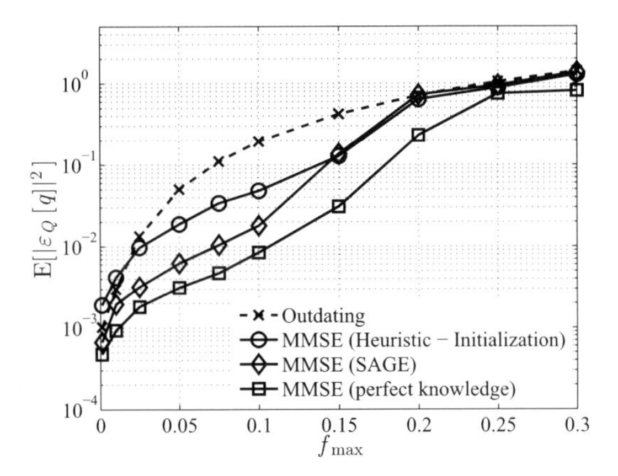

Fig. 3.16 MSE of channel prediction ($Q = 5$) vs. maximum frequency $f_{\max}$ with completion of autocovariance sequence for period $N_{\mathrm{P}} = 2$ of observations (Scenario C, $c_h[0] = 1$, $c_e = 0.01$, $B = 20$).

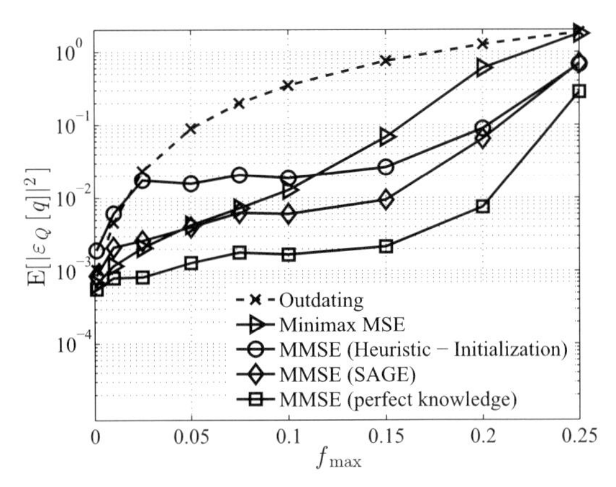

Fig. 3.17 MSE of channel prediction ($Q = 5$) vs. maximum frequency $f_{\max}$ with completion of autocovariance sequence for period $N_{\mathrm{P}} = 2$ of observations (Scenario D, bandpass power spectral density, $c_h[0] = 1$, $c_e = 0.01$, $B = 20$).

- ML estimation is based on the SAGE algorithm (3.85) with $i_{\max} = 10$ iterations, $\bar{B} = 2B - 1$, and the assumption of *block Toeplitz* $\boldsymbol{C}_{\boldsymbol{h}_{\mathrm{T}}}$. ("MMSE (SAGE)")
- The SAGE algorithm is also employed with the same parameters, but assuming the *Kronecker model* for $\boldsymbol{C}_{\boldsymbol{h}_{\mathrm{T}}}$ from (3.98). It estimates temporal and spatial correlations for every transmitter separately and combines them according to (3.98) as described in Section 3.3.6 ("MMSE (SAGE & Kronecker)").

The estimation quality is measured by the average MSE of multivariate MMSE prediction ($Q = 5$ and $N_{\mathrm{P}} = 1$, Section 2.3.1), which is obtained from 200 independent estimates. As upper limit, which should not be exceeded, we

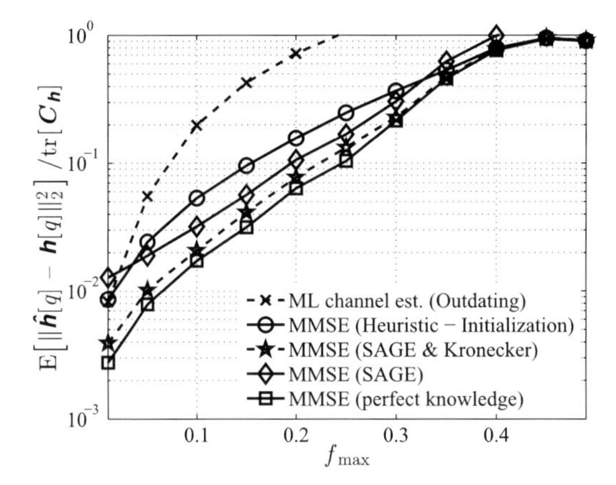

Fig. 3.18 Normalized MSE of multivariate prediction ($Q = 5$) vs. maximum frequency $f_{\max}$ (Scenario E, $c_n = 0.1$).

choose the ML channel estimate based on the last observed training sequence ("ML channel est. (Outdating)"). The lower bound is given by the MSE for MMSE prediction with perfect knowledge of the covariance matrices.

Scenario E: Due to the high computational complexity for estimation of $\boldsymbol{C}_{\boldsymbol{h}_{\mathrm{T}}}$ with block Toeplitz structure, we choose $M = K = 4$. The structure of $\boldsymbol{C}_{\boldsymbol{h}_{\mathrm{T}}}$ is $\boldsymbol{C}_{\boldsymbol{h}_{\mathrm{T}}} = \boldsymbol{C}_{\mathrm{T}} \otimes \boldsymbol{C}_{\boldsymbol{h}}$ (2.12), i.e., identical temporal correlations for all K receivers. The block-diagonal $\boldsymbol{C}_{\boldsymbol{h}}$ is based on a Laplace angular power spectrum with spread $\sigma = 5°$ and mean azimuth angles $\bar{\boldsymbol{\varphi}} = [-45°, -15°, 15°, 45°]$ (normalized to $\mathrm{tr}[\boldsymbol{C}_{\boldsymbol{h}_{\mathrm{T}}}] = MK$). The temporal correlations are described by Clarke's power spectral density. We choose $I = B = 20$, i.e., 400 correlated observations.

Although the number of observations is large, the performance degradation of SAGE with block Toeplitz assumption is significant and the gain over the heuristic biased estimator too small to justify its high computational complexity (Figure 3.18). But the assumption of $\boldsymbol{C}_{\boldsymbol{h}_{\mathrm{T}}}$ with *Kronecker structure* (3.98) improves the performance significantly and reduces the complexity tremendously.

Chapter 4
Linear Precoding with Partial Channel State Information

Introduction to Chapter

A wireless communication link from a transmitter with multiple antennas to multiple receivers, which are decentralized and cannot cooperate, is a (vector) *broadcast channel* (BC) from the point of view of information theory. Such a point-to-multipoint scenario is also referred to as *downlink* in a cellular wireless communication system (Figure 4.1).

If only one transmit antenna was available, the maximum sum capacity would be achieved transmitting to the receiver with maximum signal-to-noise ratio (SNR) in every time slot, i.e., by time division multiple access (TDMA) with maximum throughput scheduling [83, p. 473]. But the additional degrees of freedom resulting from multiple transmit antennas can be exploited to improve the information rate: More than one receiver is served at the same time and the data streams are (approximately) separated spatially (a.k.a. space division multiple access (SDMA)). The sum capacity of the BC cannot be achieved by *linear precoding* (see introduction to Chapter 5), but the restriction to a linear structure $\mathbf{T}_{\mathrm{tx}}(\boldsymbol{d}[n]) = \boldsymbol{P}\boldsymbol{d}[n]$ (Figure 4.1) reduces the computational complexity significantly [106].[1]

The design of a linear precoder (or beamformer) has attracted a lot of attention during the previous ten years; it is impossible to give a complete overview here. We list the main directions of research and refer to survey articles with more detailed information. First we categorize them according to their optimization criterion:

- Signal-to-interference-and-noise ratio (SINR) [19, 185]
- Mean square error (MSE) [116, 108]
- Information rate [83, 183]
- Various ad hoc criteria, e.g., least-squares (see references in [108]), application of receive beamforming to precoding [19, 18].

[1] Linear precoding is also referred to as beamforming, transmit processing, preequalization (for complete CSI), or sometimes predistortion.

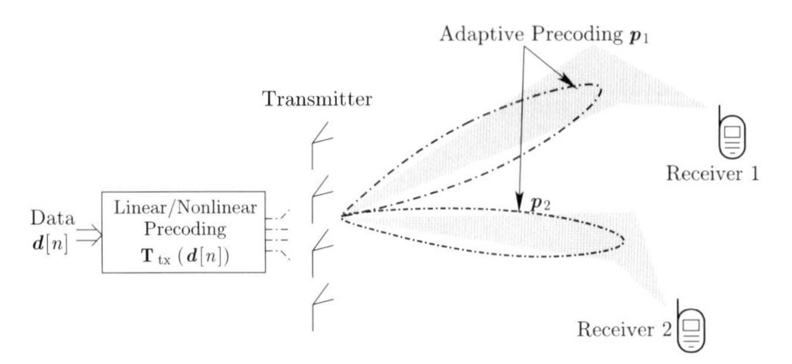

Fig. 4.1 Precoding for the wireless broadcast channel (BC) with multiple transmit antennas and single antenna receivers.

They can also be distinguished by the channel state information (CSI), which is assumed to be available at the transmitter (compare [19]): Complete CSI (C-CSI), partial CSI (P-CSI), and statistical CSI (S-CSI) or channel distribution information (see also Table 4.1 and Section 4.2 for our definitions). For C-CSI, algorithms have been developed based on SINR, MSE, and information rate. The case of P-CSI is mostly addressed for a quantized feedback [107, 188]. Recently, sum rate is optimized in [31]. Robust optimization based on MSE, which applies the Bayesian paradigm (Appendix C) for P-CSI, is proposed in [53, 98, 58]. The standard approach related to SINR for S-CSI is also surveyed in [19, 185]. A first step towards MSE-based optimization for S-CSI which is based on receiver beamforming is introduced in [18]. A similar idea is developed for sum MSE [112, 243] assuming a matched filter receiver. A more systematic derivation for sum MSE is given in [58], which include S-CSI as a special case.

A significant number of algorithms is based on the duality of the BC and the (virtual) multiple access channel (MAC), which is proved for SINR in [185] and for MSE (C-CSI) in [221, 146, 7].

Outline of Chapter

The system model for the broadcast channel and necessary (logical) training channels is given in Section 4.1. After the definition of different degrees of CSI at the transmitter and receiver in Section 4.2, we introduce the performance measures for P-CSI in Section 4.3: The goal is to develop MSE-related and mathematically tractable performance measures; we start from the point of view of information theory to point out the relevance and deficiencies of these practical optimization criteria. Our focus is the conceptual difference between MSE and information rate; moreover, we develop approximations of the BC which result in simple and suboptimum performance measures. We also discuss adaptive linear precoding when only a common training channel is available, i.e., the receivers have incomplete CSI, which is important

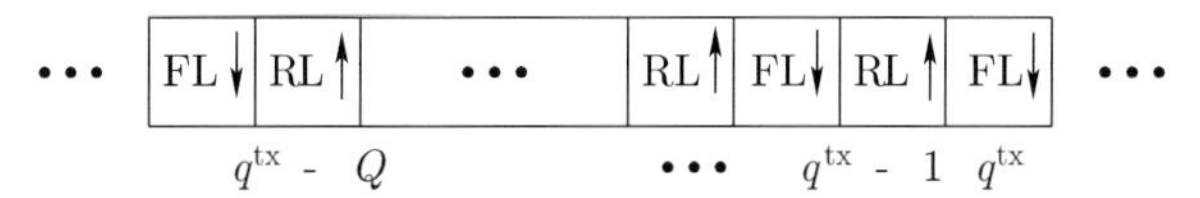

Fig. 4.2 Time division duplex slot structure with alternating allocation of forward link (FL) and reverse link (RL) time slots.

in practice. Algorithms for minimizing the sum MSE are introduced in Section 4.4. They can be generalized to nonlinear precoding, for example. Two dualities between the BC and the (virtual) MAC for P-CSI are developed in Section 4.5. In Section 4.6, we briefly introduce optimization of linear precoding under quality of service constraints, in particular, for systems with a common training channel.

4.1 System Model for the Broadcast Channel

The communication links between a transmitter equipped with M antennas and K single antenna receivers can be organized in a *time division duplex* (TDD) mode. For notational simplicity, we assume that time slots (blocks) with a duration of T_{b} alternate between the forward link (FL) and reverse link (RL), i.e., are assigned with a period of $2T_{\mathrm{b}}$ ($N_{\mathrm{P}} = 2$) in each direction (Figure 4.2). The forward link from the transmitter to all receivers, which is a.k.a. the downlink in a cellular communication system, consists of a (logical) *training channel* with N_{t} training symbols followed by a (logical) data channel with N_{d} data symbols.[2] The *physical channel* is assumed *constant* during one time slot (block) as in Section 2.1.[3]

Reciprocity of forward link and reverse link physical channels $\boldsymbol{h}_k$ (Figure 4.3) is assumed, which allows the transmitter to obtain channel state information (CSI) from observations of a training sequence in the reverse link. The non-trivial aspects of hardware calibration to achieve (perfect) reciprocity of the physical channel, which includes transmitter and receiver hardware, are not considered here (e.g., see [87, 136]).

4.1.1 Forward Link Data Channel

In the forward link the modulated data $d_k[q^{\mathrm{tx}}, n], n \in \{N_{\mathrm{t}} + 1, \ldots, N_{\mathrm{t}} + N_{\mathrm{d}}\}$, for the kth receiver is transmitted in the q^{tx}th time slot. The (linear) precoder $\boldsymbol{P}$ is optimized for this time slot. The frequency flat channels ($L = 0$,

[2] The slot duration is $T_{\mathrm{b}} = (N_{\mathrm{d}} + N_{\mathrm{t}})T_{\mathrm{s}}$ for a symbol period T_{s}.

[3] The relative duration of the data to the training channel determines the throughput penalty due to training.

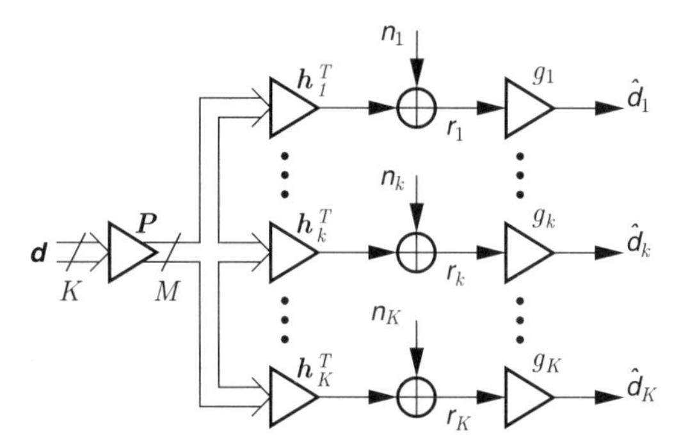

Fig. 4.3 Forward link system model: Linear precoding with M transmit antennas to K non-cooperative single-antenna receivers.

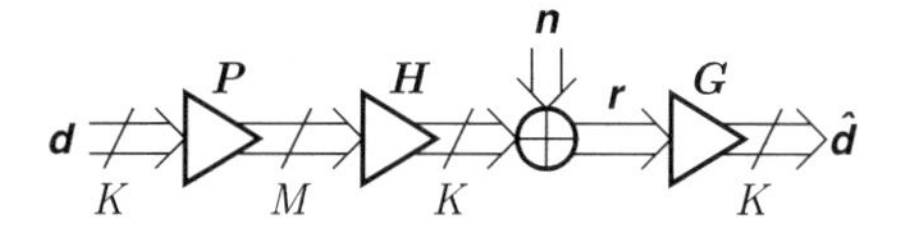

Fig. 4.4 System model for the forward link in matrix-vector notation.

Section 2.1) to all receivers are stacked in

$$\boldsymbol{H} = \begin{bmatrix} \boldsymbol{h}_1^{\mathrm{T}} \\ \boldsymbol{h}_2^{\mathrm{T}} \\ \vdots \\ \boldsymbol{h}_K^{\mathrm{T}} \end{bmatrix} \triangleq \boldsymbol{H}[q^{\mathrm{tx}}] = \begin{bmatrix} \boldsymbol{h}_1[q^{\mathrm{tx}}]^{\mathrm{T}} \\ \boldsymbol{h}_2[q^{\mathrm{tx}}]^{\mathrm{T}} \\ \vdots \\ \boldsymbol{h}_K[q^{\mathrm{tx}}]^{\mathrm{T}} \end{bmatrix} \in \mathbb{C}^{K \times M}, \tag{4.1}$$

where $\boldsymbol{h}_k \in \mathbb{C}^M$ contains the channel coefficients to the kth receiver (in the complex baseband representation). The receive signal of the kth receiver for the nth data symbol ($n \in \{N_\mathrm{t}+1, \ldots, N_\mathrm{t}+N_\mathrm{d}\}$) is

$$r_k[q^{\mathrm{tx}}, n] = \boldsymbol{h}_k[q^{\mathrm{tx}}]^{\mathrm{T}} \boldsymbol{p}_k d_k[q^{\mathrm{tx}}, n] + \boldsymbol{h}_k[q^{\mathrm{tx}}]^{\mathrm{T}} \sum_{\substack{i=1 \\ i \neq k}}^{K} \boldsymbol{p}_i d_i[q^{\mathrm{tx}}, n] + n_k[q^{\mathrm{tx}}, n] \tag{4.2}$$

including additive white Gaussian noise $n_k[q^{\mathrm{tx}}, n]$.

In the sequel, we omit the transmit slot index q^{tx} and symbol index n and write the receive signal based on (4.1) as (Figure 4.3)

$$\boxed{\begin{aligned} r_k &= \boldsymbol{h}_k^{\mathrm{T}} \boldsymbol{p}_k d_k + \boldsymbol{h}_k^{\mathrm{T}} \sum_{\substack{i=1 \\ i \neq k}}^{K} \boldsymbol{p}_i d_i + n_k \\ &= \boldsymbol{h}_k^{\mathrm{T}} \boldsymbol{P} \boldsymbol{d} + n_k. \end{aligned}} \tag{4.3}$$

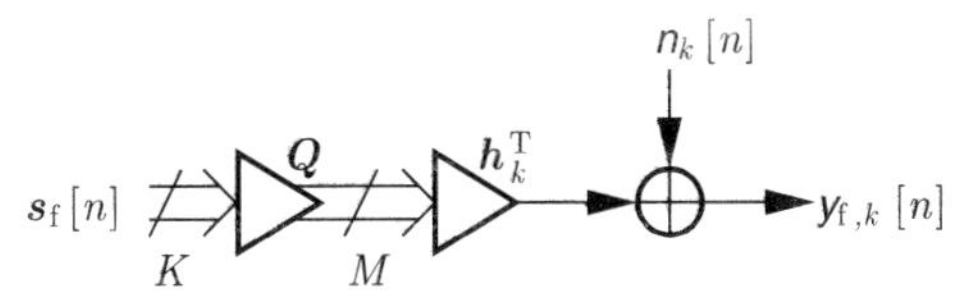

Fig. 4.5 Forward link training channel in time slot q^{tx}.

The linear precoders $\boldsymbol{p}_k$ for all receivers are summarized in $\boldsymbol{P} = [\boldsymbol{p}_1, \boldsymbol{p}_2, \ldots, \boldsymbol{p}_K] \in \mathbb{C}^{M \times K}$. The modulated data symbols d_k are modeled as uncorrelated, zero-mean random variables with variance one, i.e., $\mathrm{E}[\boldsymbol{d}\boldsymbol{d}^{\mathrm{H}}] = \boldsymbol{I}_K$ with $\boldsymbol{d} = [d_1, d_2, \ldots, d_K]^{\mathrm{T}}$. They are chosen from a finite modulation alphabet $\mathbb{D}$ such as QPSK or 16QAM in practice, whereas for more theoretical and principal considerations we choose them as complex Gaussian distributed random variables representing the coded messages. The additive noise is complex Gaussian distributed with variance $c_{n_k} = \mathrm{E}[|n_k|^2]$. Data and noise are assumed to be uncorrelated.

The receivers cannot cooperate. Linear receivers $g_k \in \mathbb{C}$ yield an estimate of the transmitted symbols

$$\hat{d}_k = g_k r_k. \tag{4.4}$$

In matrix-vector notation we have (Figure 4.4)

$$\boxed{\hat{\boldsymbol{d}} = \boldsymbol{G}\boldsymbol{r} = \boldsymbol{G}\boldsymbol{H}\boldsymbol{P}\boldsymbol{d} + \boldsymbol{G}\boldsymbol{n} \in \mathbb{C}^K} \tag{4.5}$$

with receive signal $\boldsymbol{r} = [r_1, r_2, \ldots, r_K]^{\mathrm{T}}$ (equivalently for $\boldsymbol{n}$) and the diagonal receiver and noise covariance matrix

$$\begin{aligned} \boldsymbol{G} &= \mathbf{diag}[g_1, g_2, \ldots, g_K] \quad \text{and} \\ \boldsymbol{C}_{\boldsymbol{n}} &= \mathrm{E}[\boldsymbol{n}\boldsymbol{n}^{\mathrm{H}}] = \mathbf{diag}[c_{n_1}, c_{n_2}, \ldots, c_{n_K}], \end{aligned}$$

respectively.

The precoder $\boldsymbol{P}$ is designed subject to a total power constraint

$$\mathrm{E}[\|\boldsymbol{P}\boldsymbol{d}\|_2^2] = \mathrm{tr}\left[\boldsymbol{P}\boldsymbol{P}^{\mathrm{H}}\right] \le P_{\mathrm{tx}}. \tag{4.6}$$

4.1.2 Forward Link Training Channel

In the forward link, N_t training symbols $\boldsymbol{s}_f[n] = [s_{f,1}[n], s_{f,2}[n], \ldots, s_{f,K}[n]]^{\mathrm{T}} \in \mathbb{B}^K$ are transmitted at $n \in \{1, 2, \ldots, N_t\}$

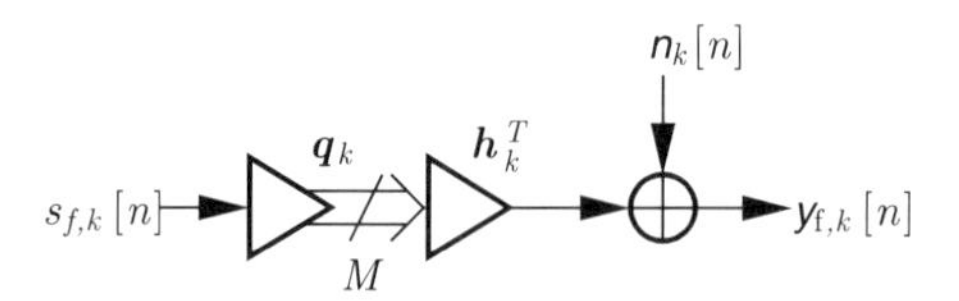

Fig. 4.6 Model of forward link training channel to receiver k for orthogonal training sequences.

(Figure 4.5). For now, we assume one training sequence for each receiver.[4] The training sequences are precoded with $\boldsymbol{Q} = [\boldsymbol{q}_1, \ldots, \boldsymbol{q}_K] \in \mathbb{C}^{M \times K}$. For example, $\boldsymbol{Q}$ can be chosen equal to $\boldsymbol{P}$. The power constraint for the training channel is

$$\boxed{\|\boldsymbol{Q}\|_{\mathrm{F}}^2 \leq P_{\mathrm{tx}}^{\mathrm{t}}.} \tag{4.7}$$

The received training sequence

$$\boldsymbol{y}_{\mathrm{f}}[q^{\mathrm{tx}}, n] = [y_{\mathrm{f},1}[q^{\mathrm{tx}}, n], y_{\mathrm{f},2}[q^{\mathrm{tx}}, n], \ldots, y_{\mathrm{f},K}[q^{\mathrm{tx}}, n]]^{\mathrm{T}}$$

at all receivers is

$$\boldsymbol{y}_{\mathrm{f}}[q^{\mathrm{tx}}, n] = \boldsymbol{H}[q^{\mathrm{tx}}]\boldsymbol{Q}\boldsymbol{s}_{\mathrm{f}}[n] + \boldsymbol{n}[q^{\mathrm{tx}}, n] \in \mathbb{C}^K. \tag{4.8}$$

The noise random process is identical to the forward link data channel in (4.2) and (4.5).

Omitting the current slot index q^{tx} as above the signal at the kth receiver reads

$$y_{\mathrm{f},k}[n] = \boldsymbol{h}_k^{\mathrm{T}}\boldsymbol{Q}\boldsymbol{s}_{\mathrm{f}}[n] + n_k[n]. \tag{4.9}$$

To introduce a model comprising all N_{t} training symbols we define the kth row of $\bar{\boldsymbol{S}}_{\mathrm{f}} \in \mathbb{B}^{K \times N_{\mathrm{t}}}$ as $\boldsymbol{s}_{\mathrm{f},k} = [s_{\mathrm{f},k}[1], s_{\mathrm{f},k}[2], \ldots, s_{\mathrm{f},k}[N_{\mathrm{t}}]]^{\mathrm{T}} \in \mathbb{B}^{N_{\mathrm{t}}}$. With $\boldsymbol{y}_{\mathrm{f},k} = [y_{\mathrm{f},k}[1], y_{\mathrm{f},k}[2], \ldots, y_{\mathrm{f},k}[N_{\mathrm{t}}]]^{\mathrm{T}}$ (equivalently for $\boldsymbol{n}_{\mathrm{f},k}$) we obtain

$$\boldsymbol{y}_{\mathrm{f},k}^{\mathrm{T}} = \boldsymbol{h}_k^{\mathrm{T}}\boldsymbol{Q}\bar{\boldsymbol{S}}_{\mathrm{f}} + \boldsymbol{n}_{\mathrm{f},k}^{\mathrm{T}} \in \mathbb{C}^{N_{\mathrm{t}}}. \tag{4.10}$$

Assuming orthogonal training sequences $\bar{\boldsymbol{S}}_{\mathrm{f}}\bar{\boldsymbol{S}}_{\mathrm{f}}^{\mathrm{H}} = N_{\mathrm{t}}\boldsymbol{I}_K$ the receivers' ML channel estimate is (based on the corresponding model in Figure 4.6)

$$\boxed{\widehat{\boldsymbol{h}_k^{\mathrm{T}}\boldsymbol{q}_k} = \boldsymbol{s}_{\mathrm{f},k}^{\mathrm{H}}\boldsymbol{y}_{\mathrm{f},k}/N_{\mathrm{t}} = \boldsymbol{h}_k^{\mathrm{T}}\boldsymbol{q}_k + e_k} \tag{4.11}$$

with variance $c_{e_k} = c_{n_k}/N_{\mathrm{t}}$ of the estimation error e_k.

---

[4] Due to constraints in a standardized system, the number of training sequences may be limited (compare Section 4.2.2).

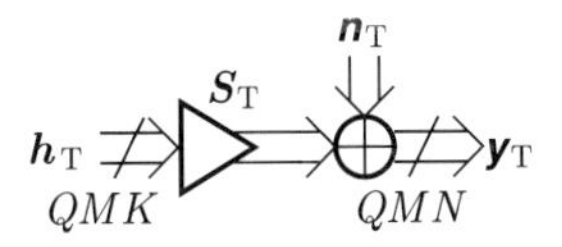

Fig. 4.7 Observations of training sequences from Q previous reverse link slots.

4.1.3 Reverse Link Training Channel

From our assumption of reciprocity of the physical channel described by $\boldsymbol{H}$ (modeling propagation conditions as well as hardware), it follows for the reverse link channel $\boldsymbol{H}^{\mathrm{RL}}[q] = \boldsymbol{H}^{\mathrm{FL}}[q]^{\mathrm{T}} = \boldsymbol{H}[q]^{\mathrm{T}}$. The receive signal of the nth training symbol $\boldsymbol{s}[n] \in \mathbb{B}^K$ from all K transmitters is

$$\boldsymbol{y}[q,n] = \boldsymbol{H}[q]^{\mathrm{T}}\boldsymbol{s}[n] + \boldsymbol{v}[q,n] \in \mathbb{C}^M \tag{4.12}$$

with additive Gaussian noise $\boldsymbol{v}[q,n] \sim \mathcal{N}_\mathrm{c}(\boldsymbol{0}, \boldsymbol{C}_{\boldsymbol{v}})$ (for example, $\boldsymbol{C}_{\boldsymbol{v}} = c_n \boldsymbol{I}_M$).

We have N training symbols per reverse link time slot. The total receive signal $\boldsymbol{y}[q] = [\boldsymbol{y}[q,1]^{\mathrm{T}}, \boldsymbol{y}[q,2]^{\mathrm{T}}, \ldots, \boldsymbol{y}[q,N]^{\mathrm{T}}]^{\mathrm{T}}$ is

$$\boldsymbol{y}[q] = \boldsymbol{S}\boldsymbol{h}[q] + \boldsymbol{v}[q] \in \mathbb{C}^{MN} \tag{4.13}$$

with $\boldsymbol{S} = [\boldsymbol{s}[1], \boldsymbol{s}[2], \ldots, \boldsymbol{s}[N]]^{\mathrm{T}} \otimes \boldsymbol{I}_M$ (as in Section 2.1) and $\boldsymbol{h}[q] = \mathbf{vec}\left[\boldsymbol{H}[q]^{\mathrm{T}}\right]$, where the channels $\boldsymbol{h}_k[q]$ of all K terminals are stacked above each other.

Due to our assumption of time slots which alternate between forward and reverse link (Figure 4.2) at a period of $N_\mathrm{P} = 2$, observations $\boldsymbol{y}[q]$ from the reverse link are available for $q \in \{q^{\mathrm{tx}} - 1, q^{\mathrm{tx}} - 3, \ldots, q^{\mathrm{tx}} - (2Q-1)\}$ (for some Q). All available observations about the forward link channel $\boldsymbol{H}[q^{\mathrm{tx}}]$ at time index q^{tx} are summarized in

$$\boldsymbol{y}_\mathrm{T}[q^{\mathrm{tx}}] = \boldsymbol{S}_\mathrm{T}\boldsymbol{h}_\mathrm{T}[q^{\mathrm{tx}}] + \boldsymbol{v}_\mathrm{T}[q^{\mathrm{tx}}] \in \mathbb{C}^{QMN} \tag{4.14}$$

with $\boldsymbol{y}_\mathrm{T}[q^{\mathrm{tx}}] = [\boldsymbol{y}[q^{\mathrm{tx}}-1]^{\mathrm{T}}, \boldsymbol{y}[q^{\mathrm{tx}}-3]^{\mathrm{T}}, \ldots, \boldsymbol{y}[q^{\mathrm{tx}}-(2Q-1)]^{\mathrm{T}}]^{\mathrm{T}}$. Omitting the transmit slot index we denote it

$$\boxed{\boldsymbol{y}_\mathrm{T} = \boldsymbol{S}_\mathrm{T}\boldsymbol{h}_\mathrm{T} + \boldsymbol{v}_\mathrm{T} \in \mathbb{C}^{QMN}.} \tag{4.15}$$

This serves as our model to describe the transmitter's CSI.

Category of CSI	Knowledge of	Characterization by
Complete CSI (C-CSI)	channel realization	$\boldsymbol{h}$
Partial CSI (P-CSI)	conditional channel PDF	$\mathrm{p}_{\boldsymbol{h}\|\boldsymbol{y}_\mathrm{T}}(\boldsymbol{h}\|\boldsymbol{y}_\mathrm{T})$
Statistical CSI (S-CSI)	channel PDF	$\mathrm{p}_{\boldsymbol{h}}(\boldsymbol{h})$

Table 4.1 Categories of channel state information (CSI) at the transmitter and their definitions.

4.2 Channel State Information at the Transmitter and Receivers

The degree of knowledge about the system's parameters is different at the transmitter and receivers. The most important system parameter is the channel state $\boldsymbol{H}$, i.e., the channel state information (CSI); but also the variances c_{n_k} of the additive noise at the receivers (Figure 4.3) and the assumed signal processing regime at the other side of the communication link $\boldsymbol{P}$ and $\{g_k\}_{k=1}^K$, respectively, are important for optimizing signal processing. We will see below that this information is not identical at the transmitter and receivers in the considered forward link time slot q^{tx}. This asymmetry in system knowledge, e.g., CSI, has a significant impact on the design methodology for the transmitter and receivers: The constrained knowledge has to be incorporated into an optimization problem for the precoding parameters which should be mathematically tractable.

4.2.1 Transmitter

We optimize the transmitter, i.e., $\boldsymbol{P}$, in time slot q^{tx}. Due to the time-variant nature of the channel $\boldsymbol{H}[q]$ and the TDD slot structure (Figure 4.2), only information about previous channel realizations is available.[5] The observations from Q previous reverse link training periods are collected in $\boldsymbol{y}_\mathrm{T}$ (4.15).

The CSI available at the transmitter is described by the probability density function (PDF) $\mathrm{p}_{\boldsymbol{h}|\boldsymbol{y}_\mathrm{T}}(\boldsymbol{h}|\boldsymbol{y}_\mathrm{T})$ of $\boldsymbol{h} = \boldsymbol{h}[q^{\mathrm{tx}}]$ given $\boldsymbol{y}_\mathrm{T}$ (Table 4.1). If we have full access to $\boldsymbol{h}$ via $\boldsymbol{y}_\mathrm{T}$, we refer to it as *complete CSI* (C-CSI) at the transmitter: It is available if $\boldsymbol{C}_{\boldsymbol{v}} = \boldsymbol{0}_{M\times M}$ and the channel is time-invariant, i.e., has a constant autocovariance function. The situation in which $\boldsymbol{y}_\mathrm{T}$ contains only information about $\boldsymbol{h}$ which is incomplete is referred to as *partial CSI* (P-

[5] The assumption of a TDD system and reciprocity simplifies the derivations below. In principal, the derivations are also possible for frequency division duplex system with a (quantized) feedback of the CSI available at the receivers to the transmitter.

CSI). If $\boldsymbol{y}_\mathrm{T}$ is statistically independent of $\boldsymbol{h}$, i.e., only knowledge about the channel's PDF $\mathrm{p}_{\boldsymbol{h}}(\boldsymbol{h}) = \mathrm{p}_{\boldsymbol{h}|\boldsymbol{y}_\mathrm{T}}(\boldsymbol{h}|\boldsymbol{y}_\mathrm{T})$ is available, this is termed *statistical CSI* (S-CSI) or also channel distribution information [83].

The common statistical assumptions about $\boldsymbol{h}_\mathrm{T} \sim \mathcal{N}_\mathrm{c}(\boldsymbol{0}, \boldsymbol{C}_{\boldsymbol{h}_\mathrm{T}})$ and $\boldsymbol{v}_\mathrm{T} \sim \mathcal{N}_\mathrm{c}(\boldsymbol{0}, \boldsymbol{C}_{\boldsymbol{v}_\mathrm{T}})$ yield a complex Gaussian conditional probability distribution $\mathrm{p}_{\boldsymbol{h}|\boldsymbol{y}_\mathrm{T}}(\boldsymbol{h}|\boldsymbol{y}_\mathrm{T})$.[6]

Additionally, we assume that the variances c_{n_k} of the noise processes at the receivers are known to the transmitter.[7]

From its knowledge about the number of training symbols in the forward link and the noise variances at the receivers the transmitter can infer the receivers' degree of CSI. Although the transmitter cannot control the receivers' signal processing capabilities, it can be optimistic and assume, e.g., the Cramér-Rao lower bound for its estimation errors. When employing information theoretic measures to describe the system's performance at the transmitter, we make additional assumptions about the receivers' decoding capabilities.

4.2.2 Receivers

The kth receiver has only access to $\boldsymbol{P}$ and $\boldsymbol{H}$ via the received training sequence in the forward link $\boldsymbol{y}_{\mathrm{f},k}$ (4.10) and the observations of the data r_k (4.3). It is obvious that the variance c_{r_k} of the received data signal can be estimated directly as well as the overall channel $\boldsymbol{h}_k^\mathrm{T}\boldsymbol{q}_k$ (4.11), which consists of the physical channel to receiver k and the precoder for the kth training sequence. With the methods from Chapter 3 also the sum of the noise and interference variance is available.

Therefore, the receivers do not have access to $\boldsymbol{P}$ and $\boldsymbol{H}$ separately, but only to a projection of the corresponding channel $\boldsymbol{h}_k$ on a precoding vector of the training channel $\boldsymbol{q}_k$. If the precoding for the data and training signals are identical, i.e., $\boldsymbol{Q} = \boldsymbol{P}$, we expect to have all relevant information at a receiver for its signal processing, e.g., an MMSE receiver g_k or ML decoding (depending on the conceptual framework in Section 4.3).[8] But these realizability constraints have to be checked when optimizing the receivers.

There are standardized systems, where the number of training sequences S is smaller than the number of receivers. For example, in WCDMA of the UMTS standard [1] different training (pilot) channels are available (see the detailed discussion in [152, 167]):

---

[6] The assumption about the noise is justified by the TDD model; in case of a low rate quantized feedback it is not valid anymore.

[7] In cellular systems with interference from adjacent cells which is not a stationary process, this issue requires further care.

[8] It may be inaccurate due to the finite number N_t of training symbols.

- A *primary common training channel* which is transmitted in the whole cell sector ($S = 1$).
- A *secondary common training channel* which allows for multiple training sequences (its number S is significantly less than the number of receivers).
- One *dedicated training channel* per receiver which is transmitted at a significantly lower power than the primary common training channel.

If only one training sequence ($S = 1$) is available, it is shared by all receivers (common training sequence). For example, it is precoded with a fixed precoder $\boldsymbol{q} \in \mathbb{C}^M$, which may be a "sector-beam" or omnidirectional via the first antenna element ($\boldsymbol{q} = \boldsymbol{e}_1$). Alternatively, a small number $S < K$ of training sequences, e.g., $S = 4$ as in Figure 4.13, is available; they are precoded with a fixed (non-adaptive) precoder $\boldsymbol{Q} = [\boldsymbol{q}_1, \boldsymbol{q}_2, \ldots, \boldsymbol{q}_S]$. For example, a fixed grid-of-beams is chosen where every $\boldsymbol{q}_i$ is an array steering vector (Section E) of the antenna array corresponding to a spatial direction.

With a common training channel the kth receiver can only estimate $\boldsymbol{h}_k^{\mathrm{T}} \boldsymbol{q}_k$ and not $\boldsymbol{h}_k^{\mathrm{T}} \boldsymbol{p}_k$ as required for a coherent demodulation of the data symbols. This *incomplete CSI* presents a limitation for the design of precoding, because in general the transmitter cannot signal further information about $\boldsymbol{p}_k$ to the receivers. Optimization of the precoder $\boldsymbol{P}$ has to incorporate this restriction of the receivers' CSI, which we address in Sections 4.3.3 and 4.6.

4.3 Performance Measures for Partial Channel State Information

When optimizing a precoder based on P-CSI at the transmitter, attention has to be paid to the choice of the optimization criterion which measures the transmission quality over the communication channel to every receiver. The choice of a performance measure is influenced by several aspects, which are partially in conflict:

- It should describe the transmission quality and the quality of service (QoS) constraints of every receiver's data.
- Its mathematical structure should be simple enough to be mathematically tractable.
- The available CSI should be fully exploited, i.e., uncertainties in CSI are modeled and considered.
- It needs to take into account restrictions on the receivers regarding their processing capabilities and channel state information:
  - explicitly by modeling the receivers' processing capabilities or
  - implicitly by creating precoding solutions, which are in accordance with these restrictions.

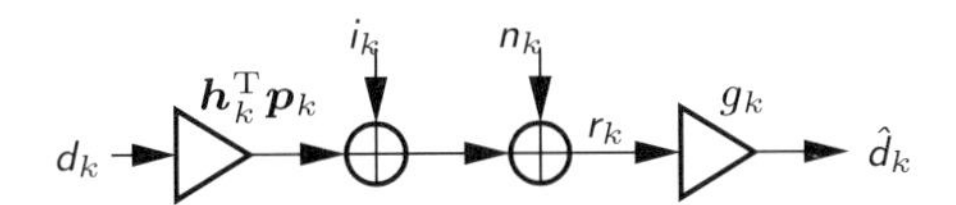

Fig. 4.8 Link model for kth receiver in the broadcast channel (BC) (equivalent to Figure 4.3).

The following argumentation starts with the point of view of *information theory*. The achievable error-free information rate is given by the mutual information between the coded data signal and received signals. Implicitly, it assumes an ML receiver [40] for the transmitted code word of infinite length. The information theoretic performance measures for the transmission quality are difficult to handle analytically for P-CSI. We approximate these measures simplifying the broadcast channel model (Figure 4.8).

A lower bound on the mean information rate is determined by the expected minimum MSE in the estimates $\hat{d}_k$. The expected MSE is a criterion of its own, but this relation shows its relevance for a communication system. It assumes an MMSE receiver g_k, which is optimum from the perspective of *estimation theory*. This less ambitious receiver model may be considered more practical.[9] Also, note that the information rate is independent of the scaling g_k for the receive signal.

In a last step, we include restrictions on the receivers' CSI explicitly. All MSE-based optimization criteria are summarized and compared qualitatively at the end regarding their accuracy-complexity trade-off.

4.3.1 Mean Information Rate, Mean MMSE, and Mean SINR

Complete Channel State Information at the Transmitter

Figure 4.8 gives the link model to the kth receiver, which is equivalent to the broadcast channel in (4.3) and Figure 4.3. The interference from the data $\{d_i\}_{i\neq k}$ designated to the other receivers is summarized in

$$i_k = \sum_{\substack{i=1\\ i\neq k}}^{K} \boldsymbol{h}_k^{\mathrm{T}} \boldsymbol{p}_i d_i, \tag{4.16}$$

which is a zero-mean random variable (d_i are zero-mean) with variance

[9] For example, the MMSE determines also the bit error rate for QAM modulation [160].

$$c_{i_k} = \sum_{\substack{i=1\\i\neq k}}^{K} |\boldsymbol{h}_k^{\mathrm{T}} \boldsymbol{p}_i|^2. \tag{4.17}$$

We assume that d_k are zero-mean complex Gaussian distributed with variance one, for simplicity.[10] Therefore, for known $\boldsymbol{h}_k$ the interference i_k is also complex Gaussian.

Under these constraints and for C-CSI at the transmitter and receivers the *mutual information* $\mathcal{I}(d_k; r_k)$, i.e., the achievable information rate, of the kth receiver's communication link, which is an additive white Gaussian noise (AWGN) channel, is well known and reads

$$\boxed{\mathrm{R}_k(\boldsymbol{P}; \boldsymbol{h}_k) = \mathcal{I}(d_k; r_k) = \log_2[1 + \mathrm{SINR}_k(\boldsymbol{P}; \boldsymbol{h}_k)]} \tag{4.18}$$

with

$$\mathrm{SINR}_k(\boldsymbol{P}; \boldsymbol{h}_k) = \frac{|\boldsymbol{h}_k^{\mathrm{T}} \boldsymbol{p}_k|^2}{\sum\limits_{\substack{i=1\\i\neq k}}^{K} |\boldsymbol{h}_k^{\mathrm{T}} \boldsymbol{p}_i|^2 + c_{n_k}}. \tag{4.19}$$

The maximum *sum rate* (throughput) is a characteristic point of the region of achievable rates, on which we focus in the sequel. The sum rate does not include fairness or quality of service constraints, but measures the maximum achievable error-free data throughput in the system with channel coding. It can be written as

$$\mathrm{R}_{\mathrm{sum}}(\boldsymbol{P}; \boldsymbol{h}) = \sum_{k=1}^{K} \mathrm{R}_k(\boldsymbol{P}; \boldsymbol{h}_k) = \log_2\left[\prod_{k=1}^{K} (1 + \mathrm{SINR}_k(\boldsymbol{P}; \boldsymbol{h}_k))\right] \tag{4.20}$$

$$= -\log_2\left[\prod_{k=1}^{K} \mathrm{MMSE}_k(\boldsymbol{P}; \boldsymbol{h}_k)\right]$$

$$\geq K \log_2[K] - K \log_2\left[\sum_{k=1}^{K} \mathrm{MMSE}_k(\boldsymbol{P}; \boldsymbol{h}_k)\right] \tag{4.21}$$

in terms of

$$\boxed{\mathrm{MMSE}_k(\boldsymbol{P}; \boldsymbol{h}_k) = 1/(1 + \mathrm{SINR}_k(\boldsymbol{P}; \boldsymbol{h}_k))} \tag{4.22}$$

and $\boldsymbol{h} = \mathbf{vec}\left[\boldsymbol{H}^{\mathrm{T}}\right]$. The lower bound in (4.21) based on the sum MMSE follows from the geometric-arithmetic mean inequality

[10] The Gaussian probability distribution is capacity achieving in the broadcast channel with C-CSI at the transmitter, which requires dirty paper coding [232]. Under the restriction of P-CSI it is not known whether this distribution is capacity achieving.

$$\left(\prod_{k=1}^{K} \mathrm{MMSE}_k(\boldsymbol{P};\boldsymbol{h}_k)\right)^{1/K} \leq \sum_{k=1}^{K} \mathrm{MMSE}_k(\boldsymbol{P};\boldsymbol{h}_k)\Big/K. \tag{4.23}$$

The minimum of

$$\begin{aligned}\mathrm{MSE}_k(\boldsymbol{P}, g_k;\boldsymbol{h}_k) &= \mathrm{E}_{r_k,d_k}\left[|g_k r_k - d_k|^2\right] = 1 + |g_k|^2 c_{r_k} - 2\mathrm{Re}\big(g_k \boldsymbol{h}_k^{\mathrm{T}} \boldsymbol{p}_k\big)\\ &= 1 + |g_k|^2 \sum_{i=1}^{K} |\boldsymbol{h}_k^{\mathrm{T}}\boldsymbol{p}_i|^2 + |g_k|^2 c_{n_k} - 2\mathrm{Re}\big(g_k \boldsymbol{h}_k^{\mathrm{T}} \boldsymbol{p}_k\big)\end{aligned} \tag{4.24}$$

with variance of the receive signal $c_{r_k} = |\boldsymbol{h}_k^{\mathrm{T}}\boldsymbol{p}_k|^2 + c_{i_k} + c_{n_k}$ is achieved by[11]

$$g_k = \mathrm{g}_k^{\mathrm{mmse}}(\boldsymbol{P},\boldsymbol{h}_k) = (\boldsymbol{h}_k^{\mathrm{T}}\boldsymbol{p}_k)^*/c_{r_k}, \tag{4.25}$$

which estimates the transmitted (coded) data d_k. It reads

$$\mathrm{MMSE}_k(\boldsymbol{P};\boldsymbol{h}_k) = \min_{g_k} \mathrm{MSE}_k(\boldsymbol{P}, g_k;\boldsymbol{h}_k) = 1 - \frac{|\boldsymbol{h}_k^{\mathrm{T}}\boldsymbol{p}_k|^2}{c_{r_k}}. \tag{4.26}$$

Example 4.1. An algorithm which maximizes (4.20) directly for C-CSI (see also (4.28)) is given in [59].[12] The comparison in [183] shows that other algorithms perform similarly. In Figure 4.9 we compare the sum rate obtained by [59] ("Max. sum rate") with the solution for minimizing the sum MSE, i.e., maximizing the lower bound (4.21), given by [102]. As reference we give the sum rate for time division multiple access (TDMA) with maximum throughput (TP) scheduling where a matched filter $\boldsymbol{p}_k = \boldsymbol{h}_k^*/\|\boldsymbol{h}_k\|_2^2$ is applied for the selected receiver; for $M = 1$, this strategy maximizes sum rate. Fairness is not an issue in all schemes.

The results show that the sum MSE yields a performance close to maximizing the sum rate if $K/M < 1$. For $K = M = 8$, the difference is substantial: Minimizing the MSE does *not* yield the full multiplexing gain [106] $\lim_{P_{\mathrm{tx}}\to\infty} \mathrm{R}_{\mathrm{sum}}(\boldsymbol{P};\boldsymbol{h})/\log_2[P_{\mathrm{tx}}] = \min(M,K)$ at high SNR.

The gain of both linear precoders compared to TDMA (with almost identical rate for $K = 8$ and $K = 6$), which schedules only one receiver at a time, is substantial. □

Partial Channel State Information at the Transmitter

So far our argumentation has been based on the assumption of C-CSI at the transmitter. With P-CSI at the transmitter only the probability density

[11] The LMMSE receiver is realizable for the assumptions in Section 4.2.2.

[12] Note that (4.20) is not concave (or convex) in $\boldsymbol{P}$ and has multiple local maxima.

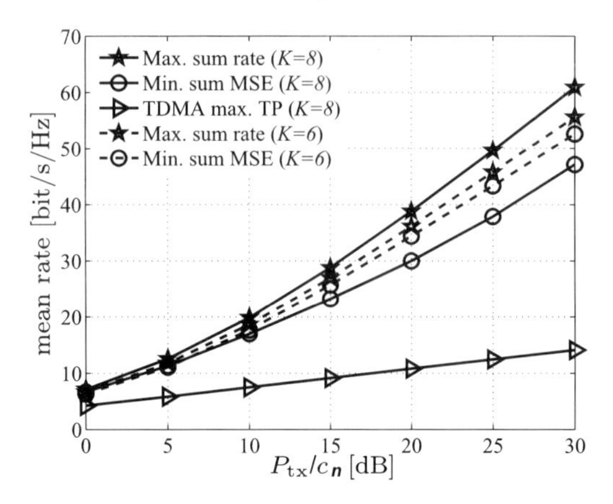

Fig. 4.9 Mean sum rate for linear precoding with C-CSI: Comparison of direct maximization of sum rate [59], minimization of sum MSE (4.69), and TDMA with maximum throughput (TP) scheduling (Scenario 1 with $M = 8$, cf. Section 4.4.4).

$\mathrm{p}_{\boldsymbol{h}|\boldsymbol{y}_\mathrm{T}}(\boldsymbol{h}|\boldsymbol{y}_\mathrm{T})$ (Table 4.1) is available for designing $\boldsymbol{P}$. There are two standard approaches for dealing with this uncertainty in CSI:

- The description of the link by an outage rate and its outage probability, which aims at the transmission of delay sensitive data.
- The mean or ergodic information rate with the expectation of (4.18) w.r.t. the probability distribution of the unknown channel $\boldsymbol{h}$ and the observations $\boldsymbol{y}_\mathrm{T}$. Feasibility of this information rate requires a code long enough to include many statistically independent realizations of $\boldsymbol{h}$ and $\boldsymbol{y}_\mathrm{T}$.

Here, we aim at the mean information rate subject to an instantaneous power constraint on every realization of $\mathbf{P}(\boldsymbol{y}_\mathrm{T})$. We maximize the sum of the mean information rates to all receivers

$$\max_{\mathbf{P}} \mathrm{E}_{\boldsymbol{h},\boldsymbol{y}_\mathrm{T}}[\mathrm{R}_\mathrm{sum}(\mathbf{P}(\boldsymbol{y}_\mathrm{T});\boldsymbol{h})] \quad \text{s.t.} \quad \|\mathbf{P}(\boldsymbol{y}_\mathrm{T})\|_\mathrm{F}^2 \le P_\mathrm{tx}. \tag{4.27}$$

over all functions

$$\mathbf{P} : \mathbb{C}^{MNQ} \to \mathbb{C}^{M\times K}, \boldsymbol{P} = \mathbf{P}(\boldsymbol{y}_\mathrm{T}).$$

This is a variational problem. Because of the instantaneous power constraint, its argument is identical to an optimization w.r.t. $\boldsymbol{P} \in \mathbb{C}^{M\times K}$ given $\boldsymbol{y}_\mathrm{T}$

$$\max_{\boldsymbol{P}} \mathrm{E}_{\boldsymbol{h}|\boldsymbol{y}_\mathrm{T}}[\mathrm{R}_\mathrm{sum}(\boldsymbol{P};\boldsymbol{h})] \quad \text{s.t.} \quad \|\boldsymbol{P}\|_\mathrm{F}^2 \le P_\mathrm{tx}. \tag{4.28}$$

The first difficulty lies in finding an explicit expression for $\mathrm{E}_{\boldsymbol{h}|\boldsymbol{y}_\mathrm{T}}[\mathrm{R}_\mathrm{sum}(\boldsymbol{P};\boldsymbol{h})]$, i.e., $\mathrm{E}_{\boldsymbol{h}_k|\boldsymbol{y}_\mathrm{T}}[\mathrm{R}_k(\boldsymbol{P};\boldsymbol{h}_k)]$. With (4.22) and Jensen's inequality the mean rate to every receiver can be lower bounded by the mean MMSE and upper bounded by the mean SINR

$$\boxed{\begin{aligned}-\log_2\left[\mathrm{E}_{\boldsymbol{h}_k|\boldsymbol{y}_\mathrm{T}}[\mathrm{MMSE}_k(\boldsymbol{P};\boldsymbol{h}_k)]\right] &\leq \mathrm{E}_{\boldsymbol{h}_k|\boldsymbol{y}_\mathrm{T}}[\mathrm{R}_k(\boldsymbol{P};\boldsymbol{h}_k)] \\ &\leq \log_2\left[1+\mathrm{E}_{\boldsymbol{h}_k|\boldsymbol{y}_\mathrm{T}}[\mathrm{SINR}_k(\boldsymbol{P};\boldsymbol{h}_k)]\right].\end{aligned}} \tag{4.29}$$

Thus, $\mathrm{E}_{\boldsymbol{h}_k|\boldsymbol{y}_\mathrm{T}}[\mathrm{MMSE}_k(\boldsymbol{P};\boldsymbol{h}_k)]$ gives an inner bound on the mean rate region, which is certainly achievable, whereas the upper bound based on $\mathrm{E}_{\boldsymbol{h}_k|\boldsymbol{y}_\mathrm{T}}[\mathrm{SINR}_k(\boldsymbol{P};\boldsymbol{h}_k)]$ does not guarantee achievability. With these bounds, the mean sum rate is lower bounded by the mean sum of receivers' MMSEs and upper bounded by the sum rate corresponding to mean SINRs:

$$\begin{aligned}K\log_2[K] - K\log_2\left[\sum_{k=1}^{K}\mathrm{E}_{\boldsymbol{h}_k|\boldsymbol{y}_\mathrm{T}}[\mathrm{MMSE}_k(\boldsymbol{P};\boldsymbol{h}_k)]\right]& \\ \leq \mathrm{E}_{\boldsymbol{h}|\boldsymbol{y}_\mathrm{T}}[\mathrm{R}_\mathrm{sum}(\boldsymbol{P};\boldsymbol{h})]& \\ \leq \sum_{k=1}^{K}\log_2\left[1+\mathrm{E}_{\boldsymbol{h}_k|\boldsymbol{y}_\mathrm{T}}[\mathrm{SINR}_k(\boldsymbol{P};\boldsymbol{h}_k)]\right].&\end{aligned} \tag{4.30}$$

But the lower as well as the upper bound cannot be computed analytically either, because MMSE_k and SINR_k are ratios of quadratic forms in complex Gaussian random vectors $\boldsymbol{h}_k$ and no explicit expression for their first order moment has been found yet [161].[13]

The lower bound, although it may be loose, yields an information theoretic justification for minimizing the conditional mean estimate of the sum MMSE

$$\min_{\boldsymbol{P}}\sum_{k=1}^{K}\mathrm{E}_{\boldsymbol{h}_k|\boldsymbol{y}_\mathrm{T}}[\mathrm{MMSE}_k(\boldsymbol{P};\boldsymbol{h}_k)] \quad \text{s.t.} \quad \|\boldsymbol{P}\|_\mathrm{F}^2 \leq P_\mathrm{tx}. \tag{4.31}$$

Optimization based on a conditional mean estimate of the cost function can also be regarded as a robust optimization using the Bayesian paradigm (Appendix C).

The receivers' statistical signal model defining the MSE, which results in $\mathrm{MMSE}_k(\boldsymbol{P};\boldsymbol{h}_k)$, is

$$\mathrm{E}_{d_k,r_k}\left[\begin{bmatrix}d_k\\ r_k\end{bmatrix}\begin{bmatrix}d_k^* & r_k^*\end{bmatrix}\right] = \begin{bmatrix}1 & (\boldsymbol{h}_k^\mathrm{T}\boldsymbol{p}_k)^* \\ \boldsymbol{h}_k^\mathrm{T}\boldsymbol{p}_k & c_{r_k}\end{bmatrix}. \tag{4.32}$$

The LMMSE receiver corresponding to the lower bound assumes perfect knowledge of $\boldsymbol{h}_k^\mathrm{T}\boldsymbol{p}_k$ and c_{r_k}, which is available at the receiver for $\boldsymbol{Q}=\boldsymbol{P}$.

In summary, the mean (ergodic) information rate $\mathrm{E}_{\boldsymbol{h}_k,\boldsymbol{y}_\mathrm{T}}[\mathrm{R}_k(\mathbf{P}(\boldsymbol{y}_\mathrm{T});\boldsymbol{h}_k)]$ as performance measure is achievable under the following assumptions:

[13] The probability distribution of the SINR [139] suggests that an analytical expression would not be tractable.

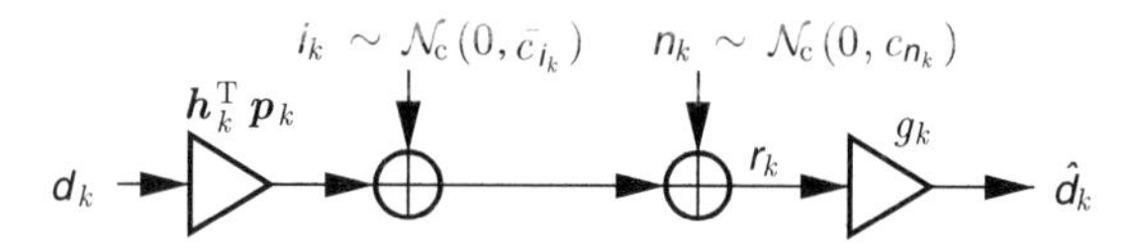

Fig. 4.10 Simplified AWGN fading BC model for the link to receiver k in Figure 4.8.

- Gaussian code books (coded data) of infinite length with rate equal to $\mathrm{E}_{\boldsymbol{h}_k,\boldsymbol{y}_\mathrm{T}}[\mathrm{R}_k(\mathbf{P}(\boldsymbol{y}_\mathrm{T});\boldsymbol{h}_k)]$,
- many statistically independent realizations of $\boldsymbol{h}$ and $\boldsymbol{y}_\mathrm{T}$ over the code length,
- knowledge of $\boldsymbol{h}_k^{\mathrm{T}}\boldsymbol{p}_k$ and $c_{i_k} + c_{n_k}$ at receiver k,
- receivers with ML decoding (or decoding based on typicality).

The conditional mean estimate $\mathrm{E}_{\boldsymbol{h}_k|\boldsymbol{y}_\mathrm{T}}[\mathrm{MMSE}_k(\boldsymbol{P};\boldsymbol{h}_k)]$ has a close relation to the mean information rate. Thinking in the framework of estimation theory (and not information theory), it assumes only LMMSE receivers.

4.3.2 Simplified Broadcast Channel Models: AWGN Fading BC, AWGN BC, and BC with Scaled Matched Filter Receivers

Now, we could pursue and search for approximations to $\mathrm{E}_{\boldsymbol{h}_k|\boldsymbol{y}_\mathrm{T}}[\mathrm{MMSE}_k(\boldsymbol{P};\boldsymbol{h}_k)]$ or $\mathrm{E}_{\boldsymbol{h}_k|\boldsymbol{y}_\mathrm{T}}[\mathrm{SINR}_k(\boldsymbol{P};\boldsymbol{h}_k)]$, e.g., as in [161] for the first order moment of a ratio of quadratic forms. We take a different approach and approximate the BC model in Figure 4.8 in order to obtain more tractable performance measures describing the approximated and simplified model. Thus, we expect to gain some more intuition regarding the value of the approximation.

The approximations are made w.r.t.

- the model of the interference ($\Rightarrow$ AWGN fading BC model),
- the model of the channel and the interference ($\Rightarrow$ AWGN BC model),
- and the transmitter's model of the receiver g_k (in the MSE framework).

AWGN Fading Broadcast Channel

In a first step, we choose a complex Gaussian model for the interference i_k (4.16) with same mean and variance, i.e., it has zero mean and variance

$$\bar{c}_{i_k} = \sum_{\substack{i=1\\i\neq k}}^{K} \mathrm{E}_{\boldsymbol{h}_k|\boldsymbol{y}_\mathrm{T}}\left[|\boldsymbol{h}_k^{\mathrm{T}}\boldsymbol{p}_i|^2\right] = \sum_{\substack{i=1\\i\neq k}}^{K} \boldsymbol{p}_i^{\mathrm{H}} \boldsymbol{R}_{\boldsymbol{h}_k|\boldsymbol{y}_\mathrm{T}}^{*} \boldsymbol{p}_i. \tag{4.33}$$

It is assumed to be statistically independent of the data signal $\boldsymbol{h}_k^{\mathrm{T}}\boldsymbol{p}_k d_k$. With P-CSI at the transmitter this yields a Rice fading channel with AWGN, where $\boldsymbol{h}_k^{\mathrm{T}}\boldsymbol{p}_k$ and $\bar{c}_{i_k} + c_{n_k}$ are perfectly known to the kth receiver (Figure 4.10). The mean information rate of this channel is bounded by (cf. (4.29))

$$\begin{aligned} -\log_2\left[\mathrm{E}_{\boldsymbol{h}_k|\boldsymbol{y}_{\mathrm{T}}}\left[\mathrm{MMSE}_k^{\mathrm{FAWGN}}(\boldsymbol{P};\boldsymbol{h}_k)\right]\right] &\leq \mathrm{E}_{\boldsymbol{h}_k|\boldsymbol{y}_{\mathrm{T}}}\left[\mathrm{R}_k^{\mathrm{FAWGN}}(\boldsymbol{P};\boldsymbol{h}_k)\right] \\ &\leq \overline{\mathrm{R}}_k^{\mathrm{AWGN}} = \log_2\left[1 + \overline{\mathrm{SINR}}_k(\boldsymbol{P})\right] \end{aligned} \quad (4.34)$$

according to Jensen's inequality and with definitions

$$\begin{aligned} \mathrm{R}_k^{\mathrm{FAWGN}}(\boldsymbol{P};\boldsymbol{h}_k) &= \log_2\left[1 + \frac{|\boldsymbol{h}_k^{\mathrm{T}}\boldsymbol{p}_k|^2}{\bar{c}_{i_k} + c_{n_k}}\right] \\ \mathrm{SINR}_k^{\mathrm{FAWGN}}(\boldsymbol{P};\boldsymbol{h}_k) &= \frac{|\boldsymbol{h}_k^{\mathrm{T}}\boldsymbol{p}_k|^2}{\bar{c}_{i_k} + c_{n_k}} \\ \mathrm{MMSE}_k^{\mathrm{FAWGN}}(\boldsymbol{P};\boldsymbol{h}_k) &= 1 - \frac{|\boldsymbol{h}_k^{\mathrm{T}}\boldsymbol{p}_k|^2}{|\boldsymbol{h}_k^{\mathrm{T}}\boldsymbol{p}_k|^2 + \bar{c}_{i_k} + c_{n_k}}. \end{aligned} \quad (4.35)$$

It is interesting that the upper bound depends on the widely employed SINR definition[14] (e.g. [19, 185])

$$\overline{\mathrm{SINR}}_k(\boldsymbol{P}) = \frac{\mathrm{E}_{\boldsymbol{h}_k|\boldsymbol{y}_{\mathrm{T}}}\left[|\boldsymbol{h}_k^{\mathrm{T}}\boldsymbol{p}_k|^2\right]}{\sum_{\substack{i=1\\i\neq k}}^{K}\mathrm{E}_{\boldsymbol{h}_k|\boldsymbol{y}_{\mathrm{T}}}\left[|\boldsymbol{h}_k^{\mathrm{T}}\boldsymbol{p}_i|^2\right] + c_{n_k}} = \frac{\boldsymbol{p}_k^{\mathrm{H}}\boldsymbol{R}_{\boldsymbol{h}_k|\boldsymbol{y}_{\mathrm{T}}}^{*}\boldsymbol{p}_k}{\bar{c}_{i_k} + c_{n_k}}, \quad (4.36)$$

where the expected value is taken in the numerator and denominator of (4.19) independently.[15]

For S-CSI with $\boldsymbol{\mu}_{\boldsymbol{h}} = \mathbf{0}$, we have Rayleigh fading channels to every receiver. Their mean information rate can be computed explicitly[16]

$$\mathrm{E}_{\boldsymbol{h}_k|\boldsymbol{y}_{\mathrm{T}}}\left[\mathrm{R}_k^{\mathrm{FAWGN}}(\boldsymbol{P};\boldsymbol{h}_k)\right] = \mathrm{E}_1\left(\overline{\mathrm{SINR}}_k(\boldsymbol{P})^{-1}\right)\exp\left(\overline{\mathrm{SINR}}_k(\boldsymbol{P})^{-1}\right)\log_2[\mathrm{e}].$$

Again the mean rate is determined by $\overline{\mathrm{SINR}}_k$.[17] The dependency on only $\overline{\mathrm{SINR}}_k$ has its origin in our Gaussian model for the interference, which is un-

[14] Typically, no reasoning is given in the literature about the relevance of this rather intuitive performance measure for describing the broadcast channel in Figure 4.8.

[15] This corresponds to the approximation of the first order moment for the SINR based on [161], which is accurate for P-CSI close to C-CSI (small error covariance matrix $\boldsymbol{C}_{\boldsymbol{h}|\boldsymbol{y}_{\mathrm{T}}}$).

[16] The exponential integral is defined as [2]

$$\mathrm{E}_1(x) = \int_x^{\infty}\frac{\exp(-t)}{t}\mathrm{d}t. \quad (4.37)$$

[17] Similarly, under the same assumption, the mean MMSE can be computed explicitly

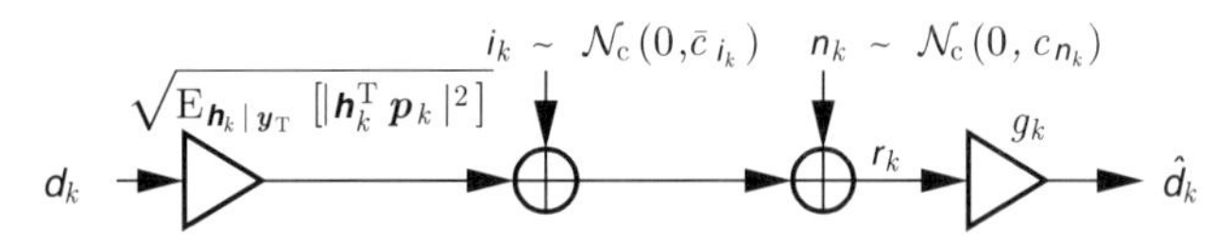

Fig. 4.11 Simplified AWGN BC model for the link to receiver k in Figure 4.8.

correlated from the received data signal. But due to the exponential integral the explicit expression of the mean rate for S-CSI is not very appealing for optimization.

The minimum $\mathrm{MMSE}_k^{\mathrm{FAWGN}}(\boldsymbol{P};\boldsymbol{h}_k)$ (4.35) relies on the statistical signal model

$$\mathrm{E}_{d_k,r_k}\left[\begin{bmatrix} d_k \\ r_k \end{bmatrix}\begin{bmatrix} d_k^* & r_k^* \end{bmatrix}\right] = \begin{bmatrix} 1 & (\boldsymbol{h}_k^\mathrm{T}\boldsymbol{p}_k)^* \\ \boldsymbol{h}_k^\mathrm{T}\boldsymbol{p}_k & |\boldsymbol{h}_k^\mathrm{T}\boldsymbol{p}_k|^2 + \bar{c}_{i_k} + c_{n_k} \end{bmatrix} \succeq 0 \tag{4.38}$$

of the receiver. This is an approximation to the original LMMSE receiver's model (4.32) with a mean variance of the interference.

AWGN Broadcast Channel

As the performance measures for the AWGN fading BC model (Figure 4.10) can only be evaluated numerically, we also make the Gaussian approximation for the data signal $\boldsymbol{h}_k^\mathrm{T}\boldsymbol{p}_k d_k$: It is zero-mean with variance $\mathrm{E}_{\boldsymbol{h}_k|\boldsymbol{y}_\mathrm{T}}\left[|\boldsymbol{h}_k^\mathrm{T}\boldsymbol{p}_k|^2\right]$ because of $\mathrm{E}[d_k] = 0$. This approximation yields the AWGN BC model in Figure 4.11 with channel gain given by the standard deviation $\mathrm{E}_{\boldsymbol{h}_k|\boldsymbol{y}_\mathrm{T}}\left[|\boldsymbol{h}_k^\mathrm{T}\boldsymbol{p}_k|^2\right]^{1/2}$. It is a system model based on equivalent Gaussian distributed signals having identical first and second order moments as the signals in Figure 4.8.[18]

The information rate for the AWGN BC model is

$$\boxed{\overline{\mathrm{R}}_k^{\mathrm{AWGN}}(\boldsymbol{P}) = \log_2\left[1 + \overline{\mathrm{SINR}}_k(\boldsymbol{P})\right]} \tag{4.39}$$

with $\overline{\mathrm{SINR}}_k(\boldsymbol{P})$ as defined in (4.36).

Switching again to the estimation theoretic perspective, the receiver g_k minimizing

$$\overline{\mathrm{MSE}}_k(\boldsymbol{P}, g_k) = \mathrm{E}_{r_k,d_k}\left[|g_k r_k - d_k|^2\right] \tag{4.40}$$

$$\mathrm{E}_{\boldsymbol{h}_k|\boldsymbol{y}_\mathrm{T}}\left[\mathrm{MMSE}_k^{\mathrm{FAWGN}}(\boldsymbol{P};\boldsymbol{h}_k)\right] = \overline{\mathrm{SINR}}_k(\boldsymbol{P})^{-1}\mathrm{E}_1\left(\overline{\mathrm{SINR}}_k(\boldsymbol{P})^{-1}\right)\exp\left(\overline{\mathrm{SINR}}_k(\boldsymbol{P})^{-1}\right)$$

as a function of $1/\overline{\mathrm{SINR}}_k(\boldsymbol{P})$.

[18] Similar approximations are made for system level considerations in a communication network. In [112, 243] a similar idea is pursued for optimization of linear precoding for rank-one covariance matrices $\boldsymbol{C}_{\boldsymbol{h}_k}$.

based on the receivers' statistical signal model

$$\mathrm{E}_{d_k,r_k}\left[\begin{bmatrix} d_k \\ r_k \end{bmatrix}\begin{bmatrix} d_k^* & r_k^* \end{bmatrix}\right] = \begin{bmatrix} 1 & \sqrt{\mathrm{E}_{\boldsymbol{h}_k|\boldsymbol{y}_\mathrm{T}}\left[|\boldsymbol{h}_k^\mathrm{T}\boldsymbol{p}_k|^2\right]} \\ \sqrt{\mathrm{E}_{\boldsymbol{h}_k|\boldsymbol{y}_\mathrm{T}}\left[|\boldsymbol{h}_k^\mathrm{T}\boldsymbol{p}_k|^2\right]} & \bar{c}_{r_k} \end{bmatrix}, \quad (4.41)$$

where $\bar{c}_{r_k} = \mathrm{E}_{\boldsymbol{h}_k|\boldsymbol{y}_\mathrm{T}}\left[|\boldsymbol{h}_k^\mathrm{T}\boldsymbol{p}_k|^2\right] + \bar{c}_{i_k} + c_{n_k}$, is

$$g_k = \frac{\sqrt{\mathrm{E}_{\boldsymbol{h}_k|\boldsymbol{y}_\mathrm{T}}\left[|\boldsymbol{h}_k^\mathrm{T}\boldsymbol{p}_k|^2\right]}}{\bar{c}_{r_k}}. \quad (4.42)$$

It yields the minimum

$$\begin{aligned}\overline{\mathrm{MMSE}}_k(\boldsymbol{P}) &= \min_{g_k} \overline{\mathrm{MSE}}_k(\boldsymbol{P}, g_k) = 1 - \frac{\mathrm{E}_{\boldsymbol{h}_k|\boldsymbol{y}_\mathrm{T}}\left[|\boldsymbol{h}_k^\mathrm{T}\boldsymbol{p}_k|^2\right]}{\bar{c}_{r_k}} \\ &= \frac{1}{1 + \overline{\mathrm{SINR}}_k(\boldsymbol{P})}\end{aligned} \quad (4.43)$$

with a relation between $\overline{\mathrm{MMSE}}_k$ and $\overline{\mathrm{SINR}}_k$ in analogy to (4.22) for C-CSI at the transmitter.

Back at the information theoretic point of view, maximization of the sum rate for the AWGN BC model

$$\max_{\boldsymbol{P}} \sum_{k=1}^{K} \overline{\mathrm{R}}_k^{\mathrm{AWGN}}(\boldsymbol{P}) \quad \text{s.t.} \quad \|\boldsymbol{P}\|_\mathrm{F}^2 \le P_\mathrm{tx} \quad (4.44)$$

can be lower bounded using (4.23)

$$\boxed{K \log_2[K] - K \log_2\left[\sum_{k=1}^{K} \overline{\mathrm{MMSE}}_k(\boldsymbol{P})\right] \le \sum_{k=1}^{K} \overline{\mathrm{R}}_k^{\mathrm{AWGN}}(\boldsymbol{P}).} \quad (4.45)$$

Therefore, maximization of the lower bound is equivalent to minimization of the sum MSE for this model:

$$\min_{\boldsymbol{P}} \sum_{k=1}^{K} \overline{\mathrm{MMSE}}_k(\boldsymbol{P}) \quad \text{s.t.} \quad \|\boldsymbol{P}\|_\mathrm{F}^2 \le P_\mathrm{tx}. \quad (4.46)$$

Irrespective of its relation to information rate, precoder design based on MSE is interesting on its own. The cost function is not convex,[19] but given in closed form. Moreover, it is closely related to the performance measure $\overline{\mathrm{SINR}}_k$, which is standard in the literature, via (4.43).

[19] For example, it is not convex in $\boldsymbol{p}_k$ keeping $\boldsymbol{p}_i, i \neq k$, constant.

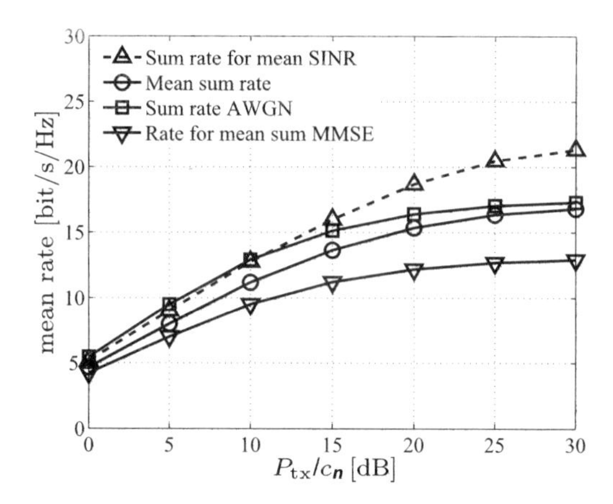

Fig. 4.12 Bounds on mean sum rate evaluated for the solution of (4.46) based on S-CSI with the algorithm from Section 4.4.1: Upper bound based on mean SINR (4.30), lower bound with mean sum MMSE (4.30), and the sum rate of the AWGN BC model (4.44) (Scenario 1, $M = K = 8$ and $\sigma = 5°$).

The relation of the information rate for the AWGN BC model to the rate for the original model in Figure 4.8 is unclear. Numerical evaluation of the sum rate shows that it is $\sum_{k=1}^{K} \overline{\mathrm{R}}_k^{\mathrm{AWGN}}(\boldsymbol{P})$ typically an upper bound to $\mathrm{E}_{\boldsymbol{h},\boldsymbol{y}_\mathrm{T}}[\mathrm{R}_\mathrm{sum}(\mathbf{P}(\boldsymbol{y}_\mathrm{T}); \boldsymbol{h})]$ (4.27).

Example 4.2. In Figure 4.12 we evaluate the different bounds on the mean sum rate (4.20) numerically for Scenario 1 defined in Section 4.4.4.[20] As precoder $\boldsymbol{P}$ we choose the solution given in Section 4.4.1 to problem (4.46) for S-CSI. The sum rate for mean SINR (4.30) (upper bound) and the rate for mean sum MMSE (4.30) (lower bound) are a good approximation for small SNR. The sum rate of the AWGN BC model (4.44) is also close to the true mean sum rate for small SNR, but the good approximation quality at high SNR depends on the selected $\boldsymbol{P}$. □

System Model with Scaled Matched Filter Receivers

The previous two models result from a Gaussian approximation of the interference and data signal. But there is also the possibility of assuming a more limited CSI at the receiver or receivers with limited signal processing capabilities. Following this concept, we construct an alternative approach to obtain a solution of (4.46) (cf. Section 4.4). It can be viewed as a mere mathematical trick, which is motivated by a relevant receiver. Let us start presenting some intuitive ideas.

[20] The bounds for the AWGN fading BC are not shown, because they are of minor importance in the sequel.

A common low-complexity receiver is a matched filter, which is matched to the data channel. The scalar matched filter, which is matched to $\boldsymbol{h}_k^{\mathrm{T}}\boldsymbol{p}_k$ and scaled by the mean variance of the received signal, is[21]

$$\boxed{g_k = \mathrm{g}_k^{\mathrm{cor}}(\boldsymbol{P}, \boldsymbol{h}_k) = \frac{(\boldsymbol{h}_k^{\mathrm{T}}\boldsymbol{p}_k)^*}{\bar{c}_{r_k}}.} \tag{4.47}$$

It is the solution of the optimization problem

$$\max_{g_k} \; 2\mathrm{Re}\big(g_k \boldsymbol{h}_k^{\mathrm{T}}\boldsymbol{p}_k\big) - |g_k|^2 \bar{c}_{r_k} \tag{4.48}$$

maximizing the real part of the crosscorrelation between $\hat{d}_k$ and d_k with a regularization of the norm of g_k. This is equivalent to minimizing

$$\boxed{\mathrm{COR}_k(\boldsymbol{P}, g_k; \boldsymbol{h}_k) = 1 + |g_k|^2 \bar{c}_{r_k} - 2\mathrm{Re}\big(g_k \boldsymbol{h}_k^{\mathrm{T}}\boldsymbol{p}_k\big),} \tag{4.49}$$

which yields

$$\boxed{\mathrm{MCOR}_k(\boldsymbol{P}; \boldsymbol{h}_k) = \min_{g_k} \mathrm{COR}_k(\boldsymbol{P}, g_k; \boldsymbol{h}_k) = 1 - \frac{|\boldsymbol{h}_k^{\mathrm{T}}\boldsymbol{p}_k|^2}{\bar{c}_{r_k}}.} \tag{4.50}$$

Note that the similarity (convexity and structure) of $\mathrm{COR}_k(\boldsymbol{P}, g_k; \boldsymbol{h}_k)$ with $\mathrm{MSE}_k(\boldsymbol{P}, g_k; \boldsymbol{h}_k)$ (4.24) is intended, but $\mathrm{MCOR}_k(\boldsymbol{P}; \boldsymbol{h}_k)$ is often negative.[22] But on average its minimum is positive, because of

$$\overline{\mathrm{MMSE}}_k(\boldsymbol{P}) = \mathrm{E}_{\boldsymbol{h}_k|\boldsymbol{y}_{\mathrm{T}}}[\mathrm{MCOR}_k(\boldsymbol{P}; \boldsymbol{h}_k)] = 1 - \frac{\mathrm{E}_{\boldsymbol{h}_k|\boldsymbol{y}_{\mathrm{T}}}\left[|\boldsymbol{h}_k^{\mathrm{T}}\boldsymbol{p}_k|^2\right]}{\bar{c}_{r_k}} \geq 0. \tag{4.52}$$

This relation and the similarity of $\mathrm{COR}_k(\boldsymbol{P}, g_k; \boldsymbol{h}_k)$ with $\mathrm{MSE}_k(\boldsymbol{P}, g_k; \boldsymbol{h}_k)$ can be exploited to find algorithms for solving (4.46) (Section 4.4) or to prove a duality (Section 4.5).

Moreover, $\mathrm{COR}_k(\boldsymbol{P}, g_k; \boldsymbol{h}_k)$ is a cost function, which yields a practically reasonable receiver (4.47) at its minimum. Assuming a receiver based on the scaled matched filter (MF), the transmitter can minimize the conditional mean estimate $\mathrm{E}_{\boldsymbol{h}_k|\boldsymbol{y}_{\mathrm{T}}}[\mathrm{MCOR}_k(\boldsymbol{P}; \boldsymbol{h}_k)]$ to improve the resulting re-

---

[21] It is realizable under the conditions of Section 4.2.2.

[22] It is negative whenever

$$\begin{bmatrix} 1 & (\boldsymbol{h}_k^{\mathrm{T}}\boldsymbol{p}_k)^* \\ \boldsymbol{h}_k^{\mathrm{T}}\boldsymbol{p}_k & \bar{c}_{r_k} \end{bmatrix} \tag{4.51}$$

is indefinite (if $\bar{c}_{r_k} < |\boldsymbol{h}_k^{\mathrm{T}}\boldsymbol{p}_k|^2$), i.e., is not a valid covariance matrix of the signals d_k and r_k. This is less likely for small $\overline{\mathrm{SINR}}_k$ at S-CSI or for P-CSI which is close to C-CSI.

ceiver cost.[23] But this argument is of limited interest, because the relevance of $\mathrm{COR}_k(\boldsymbol{P}, g_k; \boldsymbol{h}_k)$ for a communication link remains unclear: Only for large P_{tx} and $\boldsymbol{C}_{\boldsymbol{h}_k}$ with rank one, it yields the identical cost as the MMSE receiver.

4.3.3 Modeling of Receivers' Incomplete Channel State Information

Generally, the receivers' CSI can be regarded as complete relative to the uncertainty in CSI at the transmitter, because considerably more accurate information about $\boldsymbol{h}_k^{\mathrm{T}}\boldsymbol{p}_k$, $c_{n_k} + c_{i_k}$, and c_{r_k} is available to the kth receiver than to the transmitter about $\boldsymbol{H}$. But this is only true where the precoding for the training channel is $\boldsymbol{Q} = \boldsymbol{P}$ and a sufficient number of training symbols N_{t} is provided.

Knowing the number of training symbols N_{t}, the noise variances at the receiver c_{n_k}, and the precoder $\boldsymbol{Q}$ for the training channel, the transmitter has a rather accurate model of the receivers' CSI (cf. Section 4.2). This can be incorporated in the transmitter's cost function for optimizing $\boldsymbol{P}$.

Information theoretic results for dealing with incomplete CSI at the receiver provide lower bounds on the information rate [143, 90]. Instead of working with these lower bounds we discuss MSE-based cost functions to describe the receivers' signal processing capabilities.

Assume that the CSI at the kth receiver is given by the observation $\boldsymbol{y}_{\mathrm{f},k}$ (4.10) of the forward link training sequence and the conditional distribution $\mathrm{p}_{\boldsymbol{h}_k|\boldsymbol{y}_{\mathrm{f},k}}(\boldsymbol{h}_k|\boldsymbol{y}_{\mathrm{f},k})$. The receiver's signal processing g_k can be based on the conditional mean estimate of the MSE (4.24)

$$\mathrm{E}_{\boldsymbol{h}_k|\boldsymbol{y}_{\mathrm{f},k}}[\mathrm{MSE}_k(\boldsymbol{P}, g_k; \boldsymbol{h}_k)] = 1 + |g_k|^2 \sum_{i=1}^{K} \boldsymbol{p}_i^{\mathrm{H}} \boldsymbol{R}_{\boldsymbol{h}_k|\boldsymbol{y}_{\mathrm{f},k}}^{*} \boldsymbol{p}_i + |g_k|^2 c_{n_k} - 2\mathrm{Re}\left(g_k \hat{\boldsymbol{h}}_{\mathrm{f},k}^{\mathrm{T}} \boldsymbol{p}_k\right), \quad (4.53)$$

where $\hat{\boldsymbol{h}}_{\mathrm{f},k} = \mathrm{E}_{\boldsymbol{h}_k|\boldsymbol{y}_{\mathrm{f},k}}[\boldsymbol{h}_k]$ and $\boldsymbol{R}_{\boldsymbol{h}_k|\boldsymbol{y}_{\mathrm{f},k}} = \hat{\boldsymbol{h}}_{\mathrm{f},k}\hat{\boldsymbol{h}}_{\mathrm{f},k}^{\mathrm{H}} + \boldsymbol{C}_{\boldsymbol{h}_k|\boldsymbol{y}_{\mathrm{f},k}}$. The minimum cost which can be achieved by the receiver

$$g_k = \frac{(\hat{\boldsymbol{h}}_{\mathrm{f},k}^{\mathrm{T}} \boldsymbol{p}_k)^*}{\sum_{i=1}^{K} (|\hat{\boldsymbol{h}}_{\mathrm{f},k}^{\mathrm{T}} \boldsymbol{p}_i|^2 + \boldsymbol{p}_i^{\mathrm{H}} \boldsymbol{C}_{\boldsymbol{h}_k|\boldsymbol{y}_{\mathrm{f},k}}^{*} \boldsymbol{p}_i) + c_{n_k}} \quad (4.54)$$

is

---

[23] Compared to the implicit assumption of an LMMSE receiver (4.25) in (4.31), less information about r_k is exploited by (4.47)

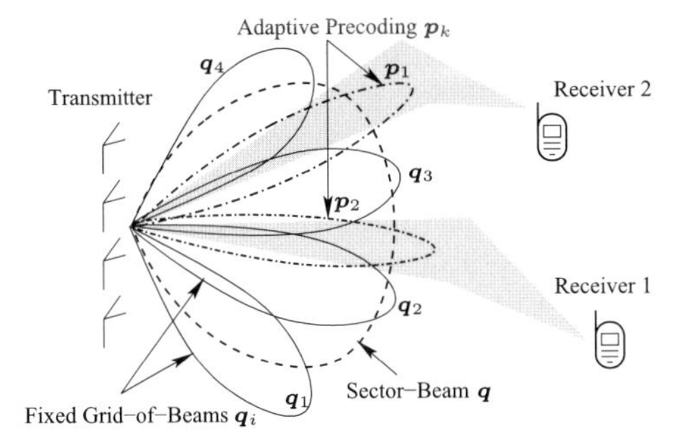

Fig. 4.13 Spatial characteristics of different fixed precoders $\boldsymbol{Q}$ ($\boldsymbol{Q} = \boldsymbol{q}$ for a sector-beam and $\boldsymbol{Q} = [\boldsymbol{q}_1, \boldsymbol{q}_2, \boldsymbol{q}_3, \boldsymbol{q}_4]$ for a fixed grid-of-beams) for the forward link training channel in comparison with (adaptive) linear precoding $\boldsymbol{P} = [\boldsymbol{p}_1, \boldsymbol{p}_2]$ in the data channel. (The spatial structure of the wireless channel is symbolized by the shaded area.)

$$\begin{aligned}\mathrm{MMSE}_k^{\mathrm{icsi}}(\boldsymbol{P}; \boldsymbol{y}_{\mathrm{f},k}) &= \min_{g_k} \mathrm{E}_{\boldsymbol{h}_k|\boldsymbol{y}_{\mathrm{f},k}}[\mathrm{MSE}_k(\boldsymbol{P}, g_k; \boldsymbol{h}_k)] \\ &= \frac{1}{1 + \mathrm{SINR}_k^{\mathrm{icsi}}(\boldsymbol{P}; \boldsymbol{y}_{\mathrm{f},k})}. \end{aligned} \tag{4.55}$$

The SINR definition

$$\mathrm{SINR}_k^{\mathrm{icsi}}(\boldsymbol{P}; \boldsymbol{y}_{\mathrm{f},k}) = \frac{|\hat{\boldsymbol{h}}_{\mathrm{f},k}^{\mathrm{T}} \boldsymbol{p}_k|^2}{\sum\limits_{\substack{i=1\\i\neq k}}^{K} \left(|\hat{\boldsymbol{h}}_{\mathrm{f},k}^{\mathrm{T}} \boldsymbol{p}_i|^2 + \boldsymbol{p}_i^{\mathrm{H}} \boldsymbol{C}_{\boldsymbol{h}_k|\boldsymbol{y}_{\mathrm{f},k}}^{*} \boldsymbol{p}_i \right) + \boldsymbol{p}_k^{\mathrm{H}} \boldsymbol{C}_{\boldsymbol{h}_k|\boldsymbol{y}_{\mathrm{f},k}}^{*} \boldsymbol{p}_k + c_{n_k}} \tag{4.56}$$

now includes the error covariance matrix $\boldsymbol{C}_{\boldsymbol{h}_k|\boldsymbol{y}_{\mathrm{f},k}}$ for the receiver's CSI in $\boldsymbol{h}_k$.[24]

Because the transmitter knows the distribution of $\boldsymbol{y}_{\mathrm{f},k}$ (or $\hat{\boldsymbol{h}}_{\mathrm{f},k}$), it can compute the mean of $\mathrm{MMSE}_k^{\mathrm{icsi}}(\boldsymbol{P}; \boldsymbol{y}_{\mathrm{f},k})$ w.r.t. this distribution. As before, an explicit expression for the conditional expectation does not exist and an approximation similar to the AWGN BC (Figure 4.11) or based on a scaled MF receiver can be found. The latter is equivalent to taking the conditional expectation of the numerator and denominator in $\mathrm{SINR}_k^{\mathrm{icsi}}$ separately.

Incomplete Channel State Information Due to Common Training Channel

An example for incomplete CSI at the receiver is a system with a common training channel (Section 4.2.2) if the precoding of the data and training signals is different, i.e., $\boldsymbol{P} \neq \boldsymbol{Q}$. Typically, also a dedicated training channel is available, but the transmit power in the common training channels is significantly larger than in the dedicated training channel. Therefore, channel estimates based on the former channels are more reliable. For this reason, the transmitter has to assume that a receiver mainly uses the CSI from the common training channel in practice (UMTS-FDD standard [1]).

At first glance, this observation requires transmission of the data over the same precoder $\boldsymbol{Q}$ as the training channel ($\boldsymbol{P} = \boldsymbol{Q}$) to provide the receivers with complete CSI. But this is not necessary, if the receivers' incomplete CSI is modeled appropriately and taken into account for optimization of an adaptive precoder $\boldsymbol{P}$. In [152] it is shown that such approaches still result in a significantly increased cell capacity.

We restrict the derivations to S-CSI, which is the relevant assumption for UMTS-FDD [1]. For this section, we assume Rice fading $\boldsymbol{\mu_h} \neq \boldsymbol{0}$ to elaborate on the differences between two approaches.

The following argumentation serves as an example for the construction of optimization criteria based on the concepts in the previous subsections.

As observation $\boldsymbol{y}_{\mathrm{f},k}$ we choose the noise free channel estimate from (4.11) based on the sth precoder $\boldsymbol{q}_s$

$$y_{\mathrm{f},k} = \boldsymbol{h}_k^{\mathrm{T}} \boldsymbol{q}_s. \tag{4.57}$$

Given $y_{\mathrm{f},k}$, the conditional mean estimate of $\boldsymbol{h}_k$ and its error covariance matrix required for the cost function (4.55) are

$$\hat{\boldsymbol{h}}_{\mathrm{f},k} = \boldsymbol{\mu}_{\boldsymbol{h}_k} + \boldsymbol{C}_{\boldsymbol{h}_k} \boldsymbol{q}_s^* (\boldsymbol{q}_s^{\mathrm{T}} \boldsymbol{C}_{\boldsymbol{h}_k} \boldsymbol{q}_s^*)^{-1} \left(\boldsymbol{h}_k^{\mathrm{T}} \boldsymbol{q}_s - \boldsymbol{\mu}_{\boldsymbol{h}_k}^{\mathrm{T}} \boldsymbol{q}_s \right) \tag{4.58}$$

$$\boldsymbol{C}_{\boldsymbol{h}_k | y_{\mathrm{f},k}} = \boldsymbol{C}_{\boldsymbol{h}_k} - \frac{\boldsymbol{C}_{\boldsymbol{h}_k} \boldsymbol{q}_s^* \boldsymbol{q}_s^{\mathrm{T}} \boldsymbol{C}_{\boldsymbol{h}_k}}{\boldsymbol{q}_s^{\mathrm{T}} \boldsymbol{C}_{\boldsymbol{h}_k} \boldsymbol{q}_s^*}. \tag{4.59}$$

Because the expected value of (4.55) cannot be obtained explicitly, we simplify the receiver model to a scaled MF (Section 4.3.2). It is the solution of[25]

---

[24] Alternatively, we could also consider an LMMSE receiver based on a ML channel estimate (4.11) with zero-mean estimation error and given error variance.

[25] Note that $\mathrm{E}_{y_{\mathrm{f},k}}[\boldsymbol{R}_{\boldsymbol{h}_k | y_{\mathrm{f},k}}] = \boldsymbol{R}_{\boldsymbol{h}_k}$.

$$\min_{g_k} \mathrm{COR}_k^{\mathrm{icsi}}(\boldsymbol{P}; y_{\mathrm{f},k}) = \min_{g_k} 1 + |g_k|^2 \sum_{i=1}^{K} \boldsymbol{p}_i^{\mathrm{H}} \boldsymbol{R}_{\boldsymbol{h}_k}^{*} \boldsymbol{p}_i + |g_k|^2 c_{n_k} - 2\mathrm{Re}\left(g_k \hat{\boldsymbol{h}}_{\mathrm{f},k}^{\mathrm{T}} \boldsymbol{p}_k \right). \quad (4.60)$$

The mean of its minimum reads

$$\mathrm{E}_{y_{\mathrm{f},k}}\left[\mathrm{MCOR}_k^{\mathrm{icsi}}(\boldsymbol{P}; y_{\mathrm{f},k})\right] = \mathrm{E}_{y_{\mathrm{f},k}}\left[1 - \frac{|\hat{\boldsymbol{h}}_{\mathrm{f},k}^{\mathrm{T}} \boldsymbol{p}_k|^2}{\bar{c}_{r_k}}\right] = \frac{1}{1 + \overline{\mathrm{SINR}}_k^{\mathrm{icsi}}(\boldsymbol{P})} \quad (4.61)$$

with

$$\overline{\mathrm{SINR}}_k^{\mathrm{icsi}}(\boldsymbol{P}) = \frac{\boldsymbol{p}_k^{\mathrm{H}} \boldsymbol{R}_{\hat{\boldsymbol{h}}_{\mathrm{f},k}}^{*} \boldsymbol{p}_k}{\sum_{i=1}^{K} \boldsymbol{p}_i^{\mathrm{H}} \boldsymbol{R}_{\boldsymbol{h}_k}^{*} \boldsymbol{p}_i - \boldsymbol{p}_k^{\mathrm{H}} \boldsymbol{R}_{\hat{\boldsymbol{h}}_{\mathrm{f},k}}^{*} \boldsymbol{p}_k + c_{n_k}}, \quad (4.62)$$

which is identical to taking the expectation in the numerator and denominator of (4.56) separately, and

$$\boldsymbol{R}_{\hat{\boldsymbol{h}}_{\mathrm{f},k}} = \boldsymbol{\mu}_{\boldsymbol{h}_k} \boldsymbol{\mu}_{\boldsymbol{h}_k}^{\mathrm{H}} + \frac{\boldsymbol{C}_{\boldsymbol{h}_k} \boldsymbol{q}_s^{*} \boldsymbol{q}_s^{\mathrm{T}} \boldsymbol{C}_{\boldsymbol{h}_k}}{\boldsymbol{q}_s^{\mathrm{T}} \boldsymbol{C}_{\boldsymbol{h}_k} \boldsymbol{q}_s^{*}}. \quad (4.63)$$

For $\boldsymbol{\mu}_{\boldsymbol{h}} = \boldsymbol{0}$, this SINR definition is equivalent to [36]. The general case yields the *rank-two* correlation matrix $\boldsymbol{R}_{\hat{\boldsymbol{h}}_{\mathrm{f},k}}$ for the kth receiver's channel.

Instead of a receiver model with a scaled MF and the conditional mean estimate (4.58), we suggest another approach: Choose a receiver model, whose first part depends only on the available CSI (4.57) and the second component is a scaling independent of the channel realization $\boldsymbol{h}_k$. We are free to choose a suitable model for the CSI-dependent component. Motivated by standardization (UMTS FDD mode [1]), the available CSI (4.57) from the common training channel is only utilized to compensate the phase. This results in the receiver model

$$\boxed{\mathrm{g}_k(\boldsymbol{h}_k) = \frac{(\boldsymbol{h}_k^{\mathrm{T}} \boldsymbol{q}_s)^{*}}{|\boldsymbol{h}_k^{\mathrm{T}} \boldsymbol{q}_s|} \beta_k.} \quad (4.64)$$

The scaling β_k models a slow automatic gain control at the receiver, which is optimized together with the precoder $\boldsymbol{P}$ based on S-CSI. The relevant cost function at the transmitter is the mean of the MSE (4.24)[26] resulting from this receiver model

[26] For rank-one channels and CDMA, a sum MSE criterion is developed in [244] based on an average signal model. An ad-hoc extension to general channels is presented in [243, 25].

$$
\begin{aligned}
\mathrm{E}_{\boldsymbol{h}_k}\left[\mathrm{MSE}_k\left(\boldsymbol{P}, \frac{(\boldsymbol{h}_k^{\mathrm{T}}\boldsymbol{q}_s)^*}{|\boldsymbol{h}_k^{\mathrm{T}}\boldsymbol{q}_s|}\beta_k; \boldsymbol{h}_k\right)\right] &= 1 + |\beta_k|^2\left(\sum_{i=1}^{K}\boldsymbol{p}_i^{\mathrm{H}}\boldsymbol{R}_{\boldsymbol{h}_k}^*\boldsymbol{p}_i + c_{n_k}\right) \\
&\quad - 2\mathrm{Re}\left(\beta_k \mathrm{E}_{\boldsymbol{h}_k}\left[\frac{(\boldsymbol{h}_k^{\mathrm{T}}\boldsymbol{q}_s)^*}{|\boldsymbol{h}_k^{\mathrm{T}}\boldsymbol{q}_s|}\boldsymbol{h}_k\right]^{\mathrm{T}}\boldsymbol{p}_k\right). \qquad (4.65)
\end{aligned}
$$

The expectation of the phase compensation and the channel $\boldsymbol{h}_k$ can be computed analytically as shown in Appendix D.2.

Minimizing (4.65) w.r.t. β_k yields

$$
\min_{\beta_k} \mathrm{E}_{\boldsymbol{h}_k}\left[\mathrm{MSE}_k\left(\boldsymbol{P}, \frac{(\boldsymbol{h}_k^{\mathrm{T}}\boldsymbol{q}_s)^*}{|\boldsymbol{h}_k^{\mathrm{T}}\boldsymbol{q}_s|}\beta_k; \boldsymbol{h}_k\right)\right] = \frac{1}{1 + \overline{\mathrm{SINR}}_k^{\mathrm{phase}}(\boldsymbol{P})} \qquad (4.66)
$$

with

$$
\overline{\mathrm{SINR}}_k^{\mathrm{phase}}(\boldsymbol{P}) = \frac{\left|\mathrm{E}_{\boldsymbol{h}_k}\left[\frac{(\boldsymbol{h}_k^{\mathrm{T}}\boldsymbol{q}_s)^*}{|\boldsymbol{h}_k^{\mathrm{T}}\boldsymbol{q}_s|}\boldsymbol{h}_k\right]^{\mathrm{T}}\boldsymbol{p}_k\right|^2}{\sum_{i=1}^{K}\boldsymbol{p}_i^{\mathrm{H}}\boldsymbol{R}_{\boldsymbol{h}_k}^*\boldsymbol{p}_i - \left|\mathrm{E}_{\boldsymbol{h}_k}\left[\frac{(\boldsymbol{h}_k^{\mathrm{T}}\boldsymbol{q}_s)^*}{|\boldsymbol{h}_k^{\mathrm{T}}\boldsymbol{q}_s|}\boldsymbol{h}_k\right]^{\mathrm{T}}\boldsymbol{p}_k\right|^2 + c_{n_k}}. \qquad (4.67)
$$

It differs from (4.62): The matrix substituting the correlation matrix of the kth data signal's channel in the numerator of (4.67) is of rank one (compare to (4.62) and (4.63)). For $\boldsymbol{\mu}_{\boldsymbol{h}} = \boldsymbol{0}$, we get

$$
\left|\mathrm{E}_{\boldsymbol{h}_k}\left[\frac{(\boldsymbol{h}_k^{\mathrm{T}}\boldsymbol{q}_s)^*}{|\boldsymbol{h}_k^{\mathrm{T}}\boldsymbol{q}_s|}\boldsymbol{h}_k\right]^{\mathrm{T}}\boldsymbol{p}_k\right|^2 = \frac{\pi}{4}(\boldsymbol{q}_s^{\mathrm{T}}\boldsymbol{C}_{\boldsymbol{h}_k}\boldsymbol{q}_s^*)^{-1}\left|\boldsymbol{q}_s^{\mathrm{H}}\boldsymbol{C}_{\boldsymbol{h}_k}^*\boldsymbol{p}_k\right|^2
$$

instead of $(\boldsymbol{q}_s^{\mathrm{T}}\boldsymbol{C}_{\boldsymbol{h}_k}\boldsymbol{q}_s^*)^{-1}|\boldsymbol{q}_s^{\mathrm{H}}\boldsymbol{C}_{\boldsymbol{h}_k}^*\boldsymbol{p}_k|^2$ in (4.62) (cf. Appendix D.2). The difference is the factor $\pi/4 \approx 0.79$ which results from a larger error in CSI at the receiver for only a phase correction.

This impact of the receiver model on the transmitter's SINR-definition is illustrated further in the following example.

Example 4.3. For a zero-mean channel with $\mathrm{rank}[\boldsymbol{C}_{\boldsymbol{h}_k}] = 1$, the error covariance matrix (4.59) for the conditional mean estimator is zero, because the channel can be estimated perfectly by the receiver.[27] Thus, (4.63) converges to $\boldsymbol{R}_{\boldsymbol{h}_k}$ and the transmitter's SINR definition (4.62) becomes identical to (4.36). On the other hand, for only a phase compensation at the receiver, the data signal's channel is always modeled as rank-one (cf. numerator of (4.67)) and $\overline{\mathrm{SINR}}_k^{\mathrm{phase}}(\boldsymbol{P})$ is not identical to (4.36) in general. □

The choice of an optimization criterion for precoding with a common training channel depends on the detailed requirements of the application, i.e., the

[27] Note that the observation in (4.57) is noise-free. Moreover, we assume that $\boldsymbol{q}_s$ is not orthogonal to the eigenvector of $\boldsymbol{C}_{\boldsymbol{h}_k}$ which corresponds to the non-zero eigenvalue.

most adequate model for the receiver. Moreover, rank-one matrices describing the desired receiver's channel as in the numerator of (4.67) reduce the computational complexity (Section 4.6.2 and [19]).

4.3.4 Summary of Sum Mean Square Error Performance Measures

Above, we show the relevance of MSE for the achievable data rate to every receiver. In particular, the sum of MSEs for all receivers gives a lower bound on the sum information rate (or throughput) to all receivers. Although it may be loose, this establishes the relevance of MSE for a communication system as an alternative performance measure to SINR. Moreover, defining simplified models for the BC (Section 4.3.2) we illustrated the relevance and intuition involved when using the suboptimum performance measure $\overline{\mathrm{SINR}}_k$ (4.36) for P-CSI and S-CSI: On the one hand, it can be seen as a heuristic with similar mathematical structure as SINR_k, on the other hand it describes the data rates for the fading AWGN and AWGN BC (Section 4.3.2). Numerical evaluation of the mean sum rate shows that the resulting suboptimum but analytically more tractable performance measures are good approximations for small SNR and small $K/M < 1$ whereas they are more loose in the interference-limited region (Figure 4.12).

The derived cost functions $\mathrm{E}_{\boldsymbol{h}_k|\boldsymbol{y}_\mathrm{T}}[\mathrm{MMSE}_k(\boldsymbol{P};\boldsymbol{h}_k)]$ and $\overline{\mathrm{MMSE}}_k(\boldsymbol{P})$ (equivalently $\overline{\mathrm{SINR}}_k(\boldsymbol{P})$) are related to the achievable data rate on the considered link. But an optimization of precoding based on a linear receiver model and the MSE is interesting on its own right. Optimization based on the MSE can be generalized to frequency selective channels with linear finite impulse response precoding [109, 113], multiple antennas at the receivers [146], and nonlinear precoding.

As an example, we now consider the sum MSEs (for these definitions). We compare the different possibilities for formulating optimization problems and, whenever existing, refer to the proposed algorithms in the literature.

Other important optimization problems are: Minimization of the transmit power under quality of service (QoS) requirements, which are based on the MSE expression [146]; Minimization of the maximum MSE (balancing of MSEs) among all receivers [101, 189, 146]. They are addressed in Section 4.6.

Complete Channel State Information at the Transmitter

Minimization of the sum of $\mathrm{MMSE}_k(\boldsymbol{P};\boldsymbol{h}_k)$ (4.26), i.e., maximization of a lower bound to the sum rate (4.20), reads

$$\min_{\boldsymbol{P}} \sum_{k=1}^{K} \mathrm{MMSE}_k(\boldsymbol{P}; \boldsymbol{h}_k) \quad \text{s.t.} \quad \|\boldsymbol{P}\|_\mathrm{F}^2 \leq P_\mathrm{tx}. \tag{4.68}$$

This is equivalent to a joint minimization of the sum of $\mathrm{MSE}_k(\boldsymbol{P}, g_k; \boldsymbol{h}_k)$ (4.24) with respect to the precoder $\boldsymbol{P} \in \mathbb{C}^{M \times K}$ and the linear receivers $g_k \in \mathbb{C}$

$$\min_{\boldsymbol{P}, \{g_k\}_{k=1}^K} \sum_{k=1}^{K} \mathrm{MSE}_k(\boldsymbol{P}, g_k; \boldsymbol{h}_k) \quad \text{s.t.} \quad \|\boldsymbol{P}\|_\mathrm{F}^2 \leq P_\mathrm{tx}. \tag{4.69}$$

When minimizing w.r.t. g_k first, we obtain (4.68).

This problem was first addressed by Joham [110, 111, 116] for equal receivers $g_k = \beta$, which yields an explicit solution derived in Appendix D.1. The generalization is presented in [102]: There, it is shown that an optimization following the principle of alternating variables (here: alternating between $\boldsymbol{P}$ and $\{g_k\}_{k=1}^K$) converges to the global optimum. A solution exploiting SINR duality between MAC and BC is proposed in [189]; a similar and computationally very efficient approach is followed in [7, 146] using a direct MSE duality.

The balancing of MSEs is addressed by [189, 101, 146], which is equivalent to a balancing of SINR_k [19, 185] because of (4.22).

Partial Channel State Information at the Transmitter

An ad hoc approach to linear precoding with P-CSI at the transmitter is based on (4.69) substituting $\boldsymbol{h}_k$ by its channel estimate or the predicted channel parameters $\hat{\boldsymbol{h}}_k$ (Chapter 2):

$$\min_{\boldsymbol{P}, \{g_k\}_{k=1}^K} \sum_{k=1}^{K} \mathrm{MSE}_k(\boldsymbol{P}, g_k; \hat{\boldsymbol{h}}_k) \quad \text{s.t.} \quad \|\boldsymbol{P}\|_\mathrm{F}^2 \leq P_\mathrm{tx}. \tag{4.70}$$

To solve it the same algorithms as for C-CSI can be applied.

Implicitly, two *assumptions* are made: The CSI at the transmitter and receivers is identical and the errors in $\hat{\boldsymbol{h}}_k$ are small (see Appendix C). Certainly, the first assumption is violated, because the receivers do not have to deal with outdated CSI.[28] Moreover, as the mobility and time-variance in the system increases the size of the prediction error increases and asymptotically only S-CSI is available at the transmitter.

The assumption of small errors is avoided performing a conditional mean estimate of the MSE given the observations $\boldsymbol{y}_\mathrm{T}$ [53]

[28] Assuming a block-wise constant channel.

$$\min_{\boldsymbol{P},\{g_k\}_{k=1}^K} \sum_{k=1}^K \mathrm{E}_{\boldsymbol{h}_k|\boldsymbol{y}_\mathrm{T}}[\mathrm{MSE}_k(\boldsymbol{P}, g_k; \boldsymbol{h}_k)] \quad \text{s.t.} \quad \|\boldsymbol{P}\|_\mathrm{F}^2 \leq P_\mathrm{tx}. \tag{4.71}$$

Still identical CSI at transmitter and receivers is assumed. Asymptotically, for S-CSI and $\mathrm{E}[\boldsymbol{h}_k] = \boldsymbol{0}$ it results in the trivial solution $\boldsymbol{P} = \boldsymbol{0}_{M\times K}$ [25, 58] (as for (4.70)), because the conditional mean estimate of $\boldsymbol{h}$ is $\hat{\boldsymbol{h}} = \boldsymbol{0}_P$.

The *asymmetry in CSI* between transmitter and receivers can be included modeling the receivers' CSI explicitly [58, 50]. As described in the previous sections, this is achieved in the most natural way within the MSE framework minimizing the MSE w.r.t. the receivers' processing, which is a function of $\boldsymbol{h}_k$ (cf. (4.25)): $\mathrm{MMSE}_k(\boldsymbol{P}; \boldsymbol{h}_k) = \min_{g_k} \mathrm{MSE}_k(\boldsymbol{P}, g_k; \boldsymbol{h}_k)$. Then, the transmitter is optimized based on the conditional mean estimate of $\mathrm{MMSE}_k(\boldsymbol{P}; \boldsymbol{h}_k)$ (4.73)

$$\min_{\boldsymbol{P}} \sum_{k=1}^K \mathrm{E}_{\boldsymbol{h}_k|\boldsymbol{y}_\mathrm{T}}[\mathrm{MMSE}_k(\boldsymbol{P}; \boldsymbol{h}_k)] \quad \text{s.t.} \quad \|\boldsymbol{P}\|_\mathrm{F}^2 \leq P_\mathrm{tx}. \tag{4.72}$$

The information theoretic implications are discussed in Section 4.3.1.

To reduce the complexity of this approach, we approximate the system model as in Figure 4.11 and maximize a lower bound to the corresponding mean sum rate (4.45)

$$\min_{\boldsymbol{P}} \sum_{k=1}^K \overline{\mathrm{MMSE}}_k(\boldsymbol{P}) \quad \text{s.t.} \quad \|\boldsymbol{P}\|_\mathrm{F}^2 \leq P_\mathrm{tx}. \tag{4.73}$$

The relation of $\overline{\mathrm{MMSE}}_k$ to COR_k and the conditional mean of MCOR_k (4.52) can be exploited to develop an algorithm for solving this problem, which we address together with (4.72) in Section 4.4.

Incomplete Channel State Information at Receivers

For incomplete CSI at the receivers due to a common training channel, we model the phase compensation at the receiver by $\mathrm{g}_k(\boldsymbol{h}_k) = \frac{(\boldsymbol{h}_k^\mathrm{T}\boldsymbol{q}_s)^*}{|\boldsymbol{h}_k^\mathrm{T}\boldsymbol{q}_s|}\beta_k$ (4.64). Minimization of the sum MSE with a common scalar $\beta_k = \beta$

$$\min_{\boldsymbol{P},\beta} \sum_{k=1}^K \mathrm{E}_{\boldsymbol{h}_k}\left[\mathrm{MSE}_k\left(\boldsymbol{P}, \frac{(\boldsymbol{h}_k^\mathrm{T}\boldsymbol{q}_s)^*}{|\boldsymbol{h}_k^\mathrm{T}\boldsymbol{q}_s|}\beta; \boldsymbol{h}_k\right)\right] \quad \text{s.t.} \quad \|\boldsymbol{P}\|_\mathrm{F}^2 \leq P_\mathrm{tx} \tag{4.74}$$

is proposed in [58, 7]. A solution can be given directly with Appendix D.1 (see also [116]).

For the SINR definition (4.62), power minimization is addressed in [36]. Balancing of the MSEs (4.65) and power minimization which is based on MSE duality and different receivers β_k is treated in [7, 8] for phase compensation at the receiver. We present a new proof for the MSE duality in Section 4.5.2 and briefly address the power minimization and the minimax MSE problems (balancing) in Section 4.6.

4.4 Optimization Based on the Sum Mean Square Error

In Section 4.3 we state that the mean MMSE gives a lower bound on the mean information rate (4.29) and the mean sum rate can be lower bounded by the sum of mean MMSEs (4.30). Maximization of this lower bound with a constraint on the total transmit power reads (compare (4.31) and (4.72))

$$\min_{\boldsymbol{P}} \sum_{k=1}^{K} \mathrm{E}_{\boldsymbol{h}_k|\boldsymbol{y}_\mathrm{T}}[\mathrm{MMSE}_k(\boldsymbol{P};\boldsymbol{h}_k)] \quad \text{s.t.} \quad \|\boldsymbol{P}\|_\mathrm{F}^2 \le P_\mathrm{tx}. \tag{4.75}$$

An explicit expression of the conditional expectation is not available. Therefore, we introduced the AWGN model for the broadcast channel (Section 4.3.2 and Figure 4.11). Maximization of the lower bound of its sum rate is equivalent to (compare (4.46) and (4.73))

$$\min_{\boldsymbol{P}} \sum_{k=1}^{K} \overline{\mathrm{MMSE}}_k(\boldsymbol{P}) \quad \text{s.t.} \quad \|\boldsymbol{P}\|_\mathrm{F}^2 \le P_\mathrm{tx}, \tag{4.76}$$

where $\overline{\mathrm{MMSE}}_k(\boldsymbol{P})$ is given in (4.43).

An analytical solution of both optimization problems, e.g., using KKT conditions, seems to be difficult. An alternative approach could be based on the BC-MAC dualities in Section 4.5, which we do not address here.[29] Instead we propose iterative solutions, which give insights into the optimum solution, but converge to stationarity points, i.e., not the global optimum in general. The underlying principle for these iterative solutions can be generalized to other systems, e.g., for nonlinear precoding (Section 5.3) and receivers with multiple antennas.

4.4.1 Alternating Optimization of Receiver Models and Transmitter

The algorithm exploits the observation that $\mathrm{MMSE}_k(\boldsymbol{P};\boldsymbol{h}_k)$ and $\overline{\mathrm{MMSE}}_k(\boldsymbol{P})$ result from models for the linear receivers g_k and the corresponding cost functions: The LMMSE receiver $g_k = \mathrm{g}_k^{\mathrm{mmse}}(\boldsymbol{P}^{(m)};\boldsymbol{h}_k)$ (4.25) min-

[29] Note that this relation has the potential for different and, maybe, computationally more efficient solutions.

imizes $\mathrm{MSE}_k(\boldsymbol{P}^{(m)}, g_k; \boldsymbol{h}_k)$ (4.24) and the scaled matched filter $g_k = \mathrm{g}_k^{\mathrm{cor}}(\boldsymbol{P}^{(m)}; \boldsymbol{h}_k)$ (4.47) minimizes $\mathrm{COR}_k(\boldsymbol{P}^{(m)}, g_k; \boldsymbol{h}_k)$ (4.49) for given $\boldsymbol{P}^{(m)}$.

We would like to alternate between optimization of the linear precoder $\boldsymbol{P}$ for fixed receivers $\{g_k^{(m-1)}\}_{k=1}^K$ from the previous iteration $m-1$ and vice versa given an initialization $\{g_k^{(0)}\}_{k=1}^K$. To obtain an explicit solution for $\boldsymbol{P}$ we introduce an extended receiver model $g_k = \beta \mathrm{g}_k(\boldsymbol{P}; \boldsymbol{h}_k)$ [116, 102]: The coherent part $\mathrm{g}_k(\boldsymbol{P}; \boldsymbol{h}_k)$ depends on the current channel realization $\boldsymbol{h}_k$; the new component is a scaling β which is common to all receivers and assumed to depend on the same CSI available at the transmitter. Formally, β adds an additional degree of freedom which enables an explicit solution for $\boldsymbol{P}$, if we optimize $\boldsymbol{P}$ jointly with β given $\mathrm{g}_k(\boldsymbol{h}_k)$. In the end, we show that $\beta \to 1$ after convergence of the iterative algorithm, i.e., the influence of β disappears in the solution.

For C-CSI, it is proved in [102, 99] that this alternating procedure reaches the global optimum (if not initialized at the boundary), although the convergence rate may be slow.

To ensure convergence of this alternating optimization it is necessary to optimize precoder and receivers consistently, i.e., based on the same type of cost function. Although only convergence to a stationary point is ensured for P-CSI, the advantages of an alternating optimization are as follows:

- Explicit solutions are obtained, which allow for further insights at the optimum.
- This paradigm can also be applied to nonlinear precoding (Chapter 5).
- An improvement over any given sub-optimum solution, which is used as initialization, is achieved.

Optimization Problem (4.75)

Assuming $\{g_k^{(m-1)}\}_{k=1}^K$ with $g_k^{(m-1)} = \mathrm{g}_k^{\mathrm{mmse}}(\boldsymbol{P}^{(m-1)}; \boldsymbol{h}_k)$ (4.25) for the receivers given by the previous iteration $m-1$, the optimum precoder together with the common scaling β at the receivers is obtained from

$$\min_{\boldsymbol{P},\beta} \sum_{k=1}^{K} \mathrm{E}_{\boldsymbol{h}_k|\boldsymbol{y}_{\mathrm{T}}}\left[\mathrm{MSE}_k(\boldsymbol{P}, \beta\, \mathrm{g}_k^{\mathrm{mmse}}(\boldsymbol{P}^{(m-1)}; \boldsymbol{h}_k); \boldsymbol{h}_k)\right] \quad \text{s.t.} \quad \|\boldsymbol{P}\|_{\mathrm{F}}^2 \le P_{\mathrm{tx}}. \tag{4.77}$$

This cost function can be written explicitly as

$$\sum_{k=1}^{K} \mathrm{E}_{\boldsymbol{h}_k|\boldsymbol{y}_\mathrm{T}}\left[\mathrm{MSE}_k(\boldsymbol{P}, \beta\, \mathrm{g}_k^{\mathrm{mmse}}(\boldsymbol{P}^{(m-1)}; \boldsymbol{h}_k); \boldsymbol{h}_k)\right]$$
$$= \sum_{k=1}^{K} \mathrm{E}_{\boldsymbol{h}_k|\boldsymbol{y}_\mathrm{T}}\left[\mathrm{E}_{r_k,d_k}\left[\left|\beta\, \mathrm{g}_k^{\mathrm{mmse}}(\boldsymbol{P}^{(m-1)}; \boldsymbol{h}_k) r_k - d_k\right|^2\right]\right] \quad (4.78)$$

which yields

$$K + \sum_{k=1}^{K} \mathrm{E}_{\boldsymbol{h}_k|\boldsymbol{y}_\mathrm{T}}\left[|\beta|^2 |\mathrm{g}_k^{\mathrm{mmse}}(\boldsymbol{P}^{(m-1)}; \boldsymbol{h}_k)|^2 \left(\sum_{i=1}^{K} |\boldsymbol{h}_k^\mathrm{T} \boldsymbol{p}_i|^2 + c_{n_k}\right)\right]$$
$$- \sum_{k=1}^{K} \mathrm{E}_{\boldsymbol{h}_k|\boldsymbol{y}_\mathrm{T}}\left[2\mathrm{Re}\left(\beta\, \mathrm{g}_k^{\mathrm{mmse}}(\boldsymbol{P}^{(m-1)}; \boldsymbol{h}_k) \boldsymbol{h}_k^\mathrm{T} \boldsymbol{p}_k\right)\right]$$
$$= K + |\beta|^2 \bar{G}^{(m-1)} + |\beta|^2 \mathrm{tr}\left[\boldsymbol{P}^\mathrm{H} \mathrm{E}_{\boldsymbol{h}|\boldsymbol{y}_\mathrm{T}}\left[\boldsymbol{H}^\mathrm{H} \mathbf{G}^{(m-1)}(\boldsymbol{h})^\mathrm{H} \mathbf{G}^{(m-1)}(\boldsymbol{h}) \boldsymbol{H}\right] \boldsymbol{P}\right]$$
$$- 2\mathrm{Re}\left(\beta\, \mathrm{tr}\left[\boldsymbol{H}_\mathbf{G}^{(m-1)} \boldsymbol{P}\right]\right). \quad (4.79)$$

It depends on the conditional mean estimate of the effective channel $\boldsymbol{H}_\mathbf{G}^{(m-1)} = \mathrm{E}_{\boldsymbol{h}|\boldsymbol{y}_\mathrm{T}}\left[\mathbf{G}^{(m-1)}(\boldsymbol{h}) \boldsymbol{H}\right]$ with receivers

$$\mathbf{G}^{(m-1)}(\boldsymbol{h}) = \mathbf{diag}\left[\mathrm{g}_1^{\mathrm{mmse}}(\boldsymbol{P}^{(m-1)}; \boldsymbol{h}_1), \ldots, \mathrm{g}_K^{\mathrm{mmse}}(\boldsymbol{P}^{(m-1)}; \boldsymbol{h}_K)\right]$$

and its Gram matrix

$$\boldsymbol{R}^{(m-1)} = \mathrm{E}_{\boldsymbol{h}|\boldsymbol{y}_\mathrm{T}}\left[\boldsymbol{H}^\mathrm{H} \mathbf{G}^{(m-1)}(\boldsymbol{h})^\mathrm{H} \mathbf{G}^{(m-1)}(\boldsymbol{h}) \boldsymbol{H}\right], \quad (4.80)$$

which cannot be computed in closed form. A numerical approximation of the multidimensional integrals can be performed in every step, which is another advantage of this alternating procedure.

The mean sum of noise variances after the receive filters with mean (power) gain $G_k^{(m-1)} = \mathrm{E}_{\boldsymbol{h}_k|\boldsymbol{y}_\mathrm{T}}[|\mathrm{g}_k^{\mathrm{mmse}}(\boldsymbol{P}^{(m-1)}; \boldsymbol{h}_k)|^2]$ is

$$\bar{G}^{(m-1)} = \sum_{k=1}^{K} c_{n_k} G_k^{(m-1)}. \quad (4.81)$$

In Appendix D.1 the solution of (4.77) is derived as (cf. (D.13))

$$\boxed{\boldsymbol{P}^{(m)} = \beta^{(m),-1} \left(\boldsymbol{R}^{(m-1)} + \frac{\bar{G}^{(m-1)}}{P_\mathrm{tx}} \boldsymbol{I}_M\right)^{-1} \mathrm{E}_{\boldsymbol{h}|\boldsymbol{y}_\mathrm{T}}\left[\mathbf{G}^{(m-1)}(\boldsymbol{h}) \boldsymbol{H}\right]^\mathrm{H}} \quad (4.82)$$

with the real-valued scaling $\beta^{(m)}$ in $\boldsymbol{P}^{(m)}$

$$\beta^{(m),2} = \frac{\mathrm{tr}\left[\boldsymbol{H}_{\mathrm{G}}^{(m-1),\mathrm{H}}\left(\boldsymbol{R}^{(m-1)} + \frac{\bar{G}^{(m-1)}}{P_{\mathrm{tx}}}\boldsymbol{I}_M\right)^{-2}\boldsymbol{H}_{\mathrm{G}}^{(m-1)}\right]}{P_{\mathrm{tx}}} \tag{4.83}$$

to ensure the transmit power constraint with equality.

Given the precoder $\boldsymbol{P}^{(m)}$ (4.82), the optimum receivers based on instantaneous CSI

$$\boxed{g_k^{(m)} = \mathrm{g}_k^{\mathrm{mmse}}(\boldsymbol{P}^{(m)};\boldsymbol{h}_k) = \beta^{(m),-1}\frac{(\boldsymbol{h}_k^{\mathrm{T}}\boldsymbol{p}_k^{(m)})^*}{\sum_{i=1}^K|\boldsymbol{h}_k^{\mathrm{T}}\boldsymbol{p}_i^{(m)}|^2 + c_{n_k}}} \tag{4.84}$$

achieve the minimum MSE

$$\mathrm{MMSE}_k(\boldsymbol{P}^{(m)};\boldsymbol{h}_k) = \min_{g_k}\mathrm{MSE}_k(\boldsymbol{P}^{(m)},\beta^{(m)}g_k;\boldsymbol{h}_k). \tag{4.85}$$

The scaling $\beta^{(m)}$ introduced artificially when optimizing $\boldsymbol{P}^{(m)}$ is canceled in this second step of every iteration.

In every iteration, the algorithm computes $\boldsymbol{P}^{(m)}$ (4.82) and $\{g_k^{(m)}\}_{k=1}^K$ (4.84), which are solutions of the optimization problems (4.77) and (4.85), respectively. Therefore, the mean sum of MMSEs $\sum_{k=1}^K \mathrm{E}_{\boldsymbol{h}_k|\boldsymbol{y}_{\mathrm{T}}}[\mathrm{MMSE}_k(\boldsymbol{P}^{(m)};\boldsymbol{h}_k)]$ is decreased or stays constant in every iteration

$$\sum_{k=1}^K \mathrm{E}_{\boldsymbol{h}_k|\boldsymbol{y}_{\mathrm{T}}}\left[\mathrm{MMSE}_k(\boldsymbol{P}^{(m)};\boldsymbol{h}_k)\right] \leq \sum_{k=1}^K \mathrm{E}_{\boldsymbol{h}_k|\boldsymbol{y}_{\mathrm{T}}}\left[\mathrm{MMSE}_k(\boldsymbol{P}^{(m-1)};\boldsymbol{h}_k)\right]. \tag{4.86}$$

Because it is bounded by zero from below it converges. But convergence to the global minimum is not ensured for P-CSI (for C-CSI it is proved in [99]). From convergence follows that $\beta^{(m)} \to 1$ for $m \to \infty$.[30]

The initialization is chosen as described in the next subsection for problem (4.76).

The *numerical complexity* for the solution of the system of linear equations (4.82) is of order $\mathrm{O}(M^3)$. The total computational load is dominated by the multidimensional numerical integration to compute the conditional mean with a complex Gaussian probability density: $K + KM + KM^2$ integrals in $2M$ real (M complex) dimensions (unlimited domain) with closely related integrands have to be evaluated. This involves a matrix product with order $\mathrm{O}(KM^2)$ and evaluations of $\mathbf{G}^{(m-1)}(\boldsymbol{h})$ per integration point. It can be performed by

[30] If $\beta^{(m)} \neq 1$, then $g_k^{(m-1)}$ necessarily changed in the previous iteration leading to a new $\boldsymbol{P}^{(m)} \neq \boldsymbol{P}^{(m-1)}$: $\beta^{(m)}$ ensures that the squared norm of the updated $\boldsymbol{P}^{(m)}$ equals P_{tx}. On the other hand, if the algorithm has converged and $\boldsymbol{P}^{(m)} = \boldsymbol{P}^{(m-1)}$, then $\beta^{(m)} = 1$.

- Monte-Carlo integration [62], which chooses the integration points pseudo-randomly, or
- monomial cubature rules (a multi-dimensional extension of Gauss-Hermite integration [172]) [38, 137].

Monomial cubature rules choose the integration points deterministically and are recommended for medium dimensions (8–15) of the integral [37]. They are designed to be exact for polynomials of a specified degree in the integration variable with Gaussian PDF; methods of high accuracy are given in [202].

Optimization Problem (4.76)

The second optimization problem (4.76) is solved choosing scaled matched filters as receiver models and the receivers' corresponding cost functions $\mathrm{COR}_k(\boldsymbol{P}, g_k; \boldsymbol{h}_k)$ ((4.49) in Section 4.3.2). From (4.52) we know that the conditional mean of the minimum is identical to $\overline{\mathrm{MMSE}}_k(\boldsymbol{P})$.

In the first step, we minimize the conditional mean estimate of the sum of receivers' cost functions COR_k given $\mathrm{g}_k^{\mathrm{cor}}(\boldsymbol{P}^{(m-1)}; \boldsymbol{h}_k)$ (4.47) from the previous iteration $m-1$:

$$\min_{\boldsymbol{P},\beta} \sum_{k=1}^{K} \mathrm{E}_{\boldsymbol{h}_k|\boldsymbol{y}_\mathrm{T}}\left[\mathrm{COR}_k\left(\boldsymbol{P}, \beta\, \mathrm{g}_k^{\mathrm{cor}}(\boldsymbol{P}^{(m-1)}; \boldsymbol{h}_k); \boldsymbol{h}_k\right)\right] \quad \text{s.t.} \quad \|\boldsymbol{P}\|_\mathrm{F}^2 \le P_\mathrm{tx}. \tag{4.87}$$

The scaling β at the receivers is introduced for the same reason as above. The cost function is

$$\begin{aligned}
&\sum_{k=1}^{K} \mathrm{E}_{\boldsymbol{h}_k|\boldsymbol{y}_\mathrm{T}}\left[\mathrm{COR}_k(\boldsymbol{P}, \beta\, \mathrm{g}_k^{\mathrm{cor}}(\boldsymbol{P}^{(m-1)}; \boldsymbol{h}_k); \boldsymbol{h}_k)\right] \\
&\quad = K + \sum_{k=1}^{K} \mathrm{E}_{\boldsymbol{h}_k|\boldsymbol{y}_\mathrm{T}}\left[|\beta|^2 \left|\mathrm{g}_k^{\mathrm{cor}}(\boldsymbol{P}^{(m-1)}; \boldsymbol{h}_k)\right|^2 \left(\sum_{i=1}^{K} \boldsymbol{p}_i^\mathrm{H} \boldsymbol{R}_{\boldsymbol{h}_k|\boldsymbol{y}_\mathrm{T}}^* \boldsymbol{p}_i + c_{n_k}\right)\right] \\
&\qquad\qquad - \sum_{k=1}^{K} \mathrm{E}_{\boldsymbol{h}_k|\boldsymbol{y}_\mathrm{T}}\left[2\mathrm{Re}\left(\beta\, \mathrm{g}_k^{\mathrm{cor}}(\boldsymbol{P}^{(m-1)}; \boldsymbol{h}_k) \boldsymbol{h}_k^\mathrm{T} \boldsymbol{p}_k\right)\right] \\
&\qquad\qquad = K + |\beta|^2 \bar{G}^{(m-1)} + |\beta|^2 \mathrm{tr}\left[\boldsymbol{P}^\mathrm{H} \left(\sum_{k=1}^{K} G_k^{(m-1)} \boldsymbol{R}_{\boldsymbol{h}_k|\boldsymbol{y}_\mathrm{T}}^*\right) \boldsymbol{P}\right] \\
&\qquad\qquad\qquad - 2\mathrm{Re}\left(\beta\, \mathrm{tr}\left[\mathrm{E}_{\boldsymbol{h}|\boldsymbol{y}_\mathrm{T}}\left[\mathbf{G}^{(m-1)}(\boldsymbol{h}) \boldsymbol{H}\right] \boldsymbol{P}\right]\right)
\end{aligned} \tag{4.88}$$

with $\bar{G}^{(m-1)} = \sum_{k=1}^{K} c_{n_k} G_k^{(m-1)}$, $[\mathbf{G}^{(m-1)}(\boldsymbol{h})]_{k,k} = \mathrm{g}_k^{\mathrm{cor}}(\boldsymbol{P}^{(m-1)}; \boldsymbol{h}_k)$, and

$$G_k^{(m-1)} = \mathrm{E}_{\boldsymbol{h}_k|\boldsymbol{y}_\mathrm{T}}\left[\left|[\mathbf{G}^{(m-1)}(\boldsymbol{h})]_{k,k}\right|^2\right] = \frac{\boldsymbol{p}_k^{(m-1),\mathrm{H}}\boldsymbol{R}_{\boldsymbol{h}_k|\boldsymbol{y}_\mathrm{T}}^*\boldsymbol{p}_k^{(m-1)}}{\bar{c}_{r_k}^{(m-1),2}} \tag{4.89}$$

$$\mathrm{E}_{\boldsymbol{h}_k|\boldsymbol{y}_\mathrm{T}}\left[[\mathbf{G}^{(m-1)}(\boldsymbol{h})]_{k,k}\boldsymbol{h}_k\right] = \frac{\boldsymbol{R}_{\boldsymbol{h}_k|\boldsymbol{y}_\mathrm{T}}\boldsymbol{p}_k^{(m-1),*}}{\bar{c}_{r_k}^{(m-1)}}. \tag{4.90}$$

All expectations can be computed analytically for this receiver model. Moreover, only the conditional first and second order moments are required and no assumption about the pdf $\mathrm{p}_{\boldsymbol{h}|\boldsymbol{y}_\mathrm{T}}(\boldsymbol{h}|\boldsymbol{y}_\mathrm{T})$ is made in contrast to problem (4.75).

The solution to (4.87) is derived in Appendix D.1 comparing (4.88) with (D.2). It reads

$$\boldsymbol{P}^{(m)} = \beta^{(m),-1}\left(\sum_{k=1}^{K} G_k^{(m-1)}\boldsymbol{R}_{\boldsymbol{h}_k|\boldsymbol{y}_\mathrm{T}}^* + \frac{\bar{G}^{(m-1)}}{P_\mathrm{tx}}\boldsymbol{I}_M\right)^{-1}\mathrm{E}_{\boldsymbol{h}|\boldsymbol{y}_\mathrm{T}}\left[\mathbf{G}^{(m-1)}(\boldsymbol{h})\boldsymbol{H}\right]^\mathrm{H} \tag{4.91}$$

with $\beta^{(m)}$ chosen to satisfy the power constraint with equality (cf. (D.13)).

In the second step of iteration m we determine the optimum receive filters by

$$\min_{g_k} \mathrm{COR}_k(\boldsymbol{P}^{(m)}, \beta^{(m)}g_k; \boldsymbol{h}_k) = \mathrm{MCOR}_k(\boldsymbol{P}^{(m)}; \boldsymbol{h}_k), \tag{4.92}$$

which are the scaled matched filters (4.47)

$$g_k^{(m)} = \mathrm{g}_k^{\mathrm{cor}}(\boldsymbol{P}^{(m)}; \boldsymbol{h}_k) = \beta^{(m),-1}\frac{(\boldsymbol{h}_k^\mathrm{T}\boldsymbol{p}_k^{(m)})^*}{\bar{c}_{r_k}^{(m)}} \tag{4.93}$$

based on the mean variance of the receive signal $\bar{c}_{r_k}^{(m)} = \sum_{i=1}^K \boldsymbol{p}_i^{(m),\mathrm{H}}\boldsymbol{R}_{\boldsymbol{h}_k|\boldsymbol{y}_\mathrm{T}}^*\boldsymbol{p}_i^{(m)} + c_{n_k}$. In fact, we determine the receivers as a function of the random vector $\boldsymbol{h}_k$, which we require for evaluating (4.88) in the next iteration.

In every iteration the cost function (4.76) decreases or stays constant:

$$\sum_{k=1}^{K}\overline{\mathrm{MMSE}}_k(\boldsymbol{P}^{(m)}) \le \sum_{k=1}^{K}\overline{\mathrm{MMSE}}_k(\boldsymbol{P}^{(m-1)}). \tag{4.94}$$

Asymptotically $\beta^{(m)} \to 1$ for $m \to \infty$.[31]

[31] Problem (4.76) with an equality constraint can be written as an unconstrained optimization problem in $\boldsymbol{P} = \bar{\boldsymbol{P}}\sqrt{P_\mathrm{tx}}/\|\bar{\boldsymbol{P}}\|_\mathrm{F}$. For this equivalent problem, the iteration above can be interpreted as a fixed point solution to the necessary condition on the gradient w.r.t. $\bar{\boldsymbol{P}}$ to be zero. The gradient for the kth column $\bar{\boldsymbol{p}}_k$ of $\bar{\boldsymbol{P}}$ reads (cf. (4.52), (4.43), and (4.36))

For P-CSI, the solution to problem (4.70) assuming equal[32] $g_k = \beta$ can serve as *initialization*, since we would like to obtain an improved performance w.r.t. this standard approach for P-CSI: This approach is based on the assumption of small errors (compare Section 4.3.4 and Appendix C) in the prediction $\hat{\boldsymbol{h}}$ of $\boldsymbol{h}$ from $\boldsymbol{y}_{\mathrm{T}}$.

The order of computational *complexity* is determined by the system of linear equations (4.91) and is $\mathrm{O}(M^3)$. It is significantly smaller than for optimizing (4.75), because all conditional mean expressions can be evaluated analytically.

4.4.2 From Complete to Statistical Channel State Information

Because we characterize the CSI at the transmitter by $\mathrm{p}_{\boldsymbol{h}|\boldsymbol{y}_{\mathrm{T}}}(\boldsymbol{h}|\boldsymbol{y}_{\mathrm{T}})$, a smooth transition from C-CSI to S-CSI is possible. Accurate estimation of the necessary parameters of $\mathrm{p}_{\boldsymbol{h}|\boldsymbol{y}_{\mathrm{T}}}(\boldsymbol{h}|\boldsymbol{y}_{\mathrm{T}})$ is addressed in Chapter 3.

The asymptotic case of *C-CSI* is reached for $\boldsymbol{C}_{\boldsymbol{v}} = \boldsymbol{0}_M$ (Section 4.1.3) and a time-invariant channel with constant autocovariance sequence $\mathrm{E}\left[\boldsymbol{h}[q]\boldsymbol{h}[q-i]^{\mathrm{H}}\right] = \boldsymbol{C}_{\boldsymbol{h}}$. Both optimization problems (4.75) and (4.76) coincide (cf. (4.69)) for this case. Their iterative solution is

$$\boxed{\boldsymbol{P}^{(m)} = \beta^{(m),-1}\left(\boldsymbol{H}_{\mathbf{G}}^{(m-1),\mathrm{H}}\boldsymbol{H}_{\mathbf{G}}^{(m-1)} + \frac{\bar{G}^{(m-1)}}{P_{\mathrm{tx}}}\boldsymbol{I}_M\right)^{-1}\boldsymbol{H}_{\mathbf{G}}^{(m-1),\mathrm{H}}} \tag{4.95}$$

with sum of noise variance after the receivers

$$\bar{G}^{(m-1)} = \sum_{k=1}^{K} c_{n_k}\left|\mathrm{g}_k^{\mathrm{mmse}}(\boldsymbol{P}^{(m-1)};\boldsymbol{h}_k)\right|^2, \tag{4.96}$$

effective channel $\boldsymbol{H}_{\mathbf{G}}^{(m-1)} = \mathbf{G}^{(m-1)}(\boldsymbol{h})\boldsymbol{H}$, and scaling of $\boldsymbol{P}^{(m)}$

$$\frac{\partial\sum_{i=1}^{K}\overline{\mathrm{MMSE}}_i\left(\bar{\boldsymbol{P}}\frac{\sqrt{P_{\mathrm{tx}}}}{\|\bar{\boldsymbol{P}}\|_{\mathrm{F}}}\right)}{\partial\bar{\boldsymbol{p}}_k^*} = \frac{\sum_{i=1}^{K}c_{n_i}G_i}{P_{\mathrm{tx}}}\bar{\boldsymbol{p}}_k + \sum_{i=1}^{K}G_i\boldsymbol{R}_{\boldsymbol{h}_i|\boldsymbol{y}_{\mathrm{T}}}^*\bar{\boldsymbol{p}}_k - \frac{1}{\bar{c}'_{r_k}}\boldsymbol{R}_{\boldsymbol{h}_k|\boldsymbol{y}_{\mathrm{T}}}^*\bar{\boldsymbol{p}}_k$$

with $G_i = \bar{\boldsymbol{p}}_i^{\mathrm{H}}\boldsymbol{R}_{\boldsymbol{h}_i|\boldsymbol{y}_{\mathrm{T}}}^*\bar{\boldsymbol{p}}_i/\bar{c}_{r_i}'^{,2}$ and $\bar{c}'_{r_i} = c_{n_i}\|\bar{\boldsymbol{P}}\|_{\mathrm{F}}^2/P_{\mathrm{tx}} + \sum_{j=1}^{K}\bar{\boldsymbol{p}}_j^{\mathrm{H}}\boldsymbol{R}_{\boldsymbol{h}_i|\boldsymbol{y}_{\mathrm{T}}}^*\bar{\boldsymbol{p}}_j$. The relation to (4.91) is evident. We proved convergence to a fixed point which is a local minimum of the cost function. In analogy, the same is true for problem (4.75) and its iterative solution (4.82) (assuming that the derivative and the integration can be interchanged, i.e., $c_{n_k} > 0$).

[32] The explicit solution is given in Appendix D.1.

$$\beta^{(m),2} = \frac{\operatorname{tr}\left[\boldsymbol{H}_{\mathrm{G}}^{(m-1)} \left(\boldsymbol{H}_{\mathrm{G}}^{(m-1),\mathrm{H}} \boldsymbol{H}_{\mathrm{G}}^{(m-1)} + \frac{\bar{G}^{(m-1)}}{P_{\mathrm{tx}}} \boldsymbol{I}_M\right)^{-2} \boldsymbol{H}_{\mathrm{G}}^{(m-1),\mathrm{H}}\right]}{P_{\mathrm{tx}}}. \tag{4.97}$$

In [102, 99] it is shown that the global optimum is reached if initialized with non-zero $\mathrm{g}_k^{\mathrm{mmse}}(\boldsymbol{P}^{(m-1)}; \boldsymbol{h}_k)$. An algorithm with faster convergence is derived in [146].

If $\boldsymbol{y}_{\mathrm{T}}$ is statistically independent of $\boldsymbol{h}$, the solutions from the previous section depend only on *S-CSI*. The iterative solution to problem (4.76) reads

$$\boldsymbol{P}^{(m)} = \beta^{(m),-1} \left(\sum_{k=1}^{K} G_k^{(m-1)} \boldsymbol{C}_{\boldsymbol{h}_k}^{*} + \frac{\bar{G}^{(m-1)}}{P_{\mathrm{tx}}} \boldsymbol{I}_M\right)^{-1} \mathrm{E}_{\boldsymbol{H}}\left[\mathbf{G}^{(m-1)}(\boldsymbol{h}) \boldsymbol{H}\right]^{\mathrm{H}} \tag{4.98}$$

with second order moment of the receive filters in iteration $m-1$

$$G_k^{(m-1)} = \frac{\boldsymbol{p}_k^{(m-1),\mathrm{H}} \boldsymbol{C}_{\boldsymbol{h}_k}^{*} \boldsymbol{p}_k^{(m-1)}}{\bar{c}_{r_k}^{(m-1),2}}, \tag{4.99}$$

$$\mathrm{E}_{\boldsymbol{h}_k|\boldsymbol{y}_{\mathrm{T}}}\left[[\mathbf{G}^{(m-1)}(\boldsymbol{h})]_{k,k} \boldsymbol{h}_k\right] = \frac{\boldsymbol{C}_{\boldsymbol{h}_k} \boldsymbol{p}_k^{(m-1),*}}{\bar{c}_{r_k}^{(m-1)}}, \tag{4.100}$$

and $\bar{G}^{(m-1)} = \sum_{k=1}^{K} c_{n_k} G_k^{(m-1)}$. The solution only depends on the second order moment $\boldsymbol{C}_{\boldsymbol{h}}$ of the channel.

For S-CSI, we choose the solution to problem (D.28) in Appendix D.3 as initialization. It is given in closed form under the assumption of zero-mean $\boldsymbol{h}$.

4.4.3 Examples

For insights in the iterative solution to minimizing the sum MSE (4.75) and (4.76), we consider two cases:

- P-CSI at the transmitter and channels with rank-one covariance matrices $\boldsymbol{C}_{\boldsymbol{h}_k}$,
- S-CSI and uncorrelated channels.

Rank-One Channels

If $\operatorname{rank}[\boldsymbol{C}_{\boldsymbol{h}_k}] = 1$, we can write (Karhunen-Loève expansion)

$$\boldsymbol{h}_k = x_k \boldsymbol{v}_k \tag{4.101}$$

defining $\mathrm{E}[|x_k|^2] = \mathrm{E}[\|\boldsymbol{h}_k\|_2^2]$ and $\|\boldsymbol{v}\|_2^2 = 1$. This results in

$$\boldsymbol{H} = \boldsymbol{X}\boldsymbol{V}^{\mathrm{T}} \tag{4.102}$$

where $\boldsymbol{X} = \mathbf{diag}\,[x_1, x_2, \ldots, x_K]$ and $\boldsymbol{v}_k$ is the kth column of $\boldsymbol{V} \in \mathbb{C}^{M\times K}$.

To apply this channel model to the solution based on the MMSE receiver model (4.82), we can simplify

$$\mathrm{E}_{\boldsymbol{h}_k|\boldsymbol{y}_\mathrm{T}}\left[\mathrm{g}_k^{\mathrm{mmse}}(\boldsymbol{P}^{(m-1)};\boldsymbol{h}_k)\boldsymbol{h}_k\right] = \underbrace{\mathrm{E}_{x_k|\boldsymbol{y}_\mathrm{T}}\left[\mathrm{g}_k^{\mathrm{mmse}}(\boldsymbol{P}^{(m-1)};x_k\boldsymbol{v}_k)x_k\right]}_{a_k}\boldsymbol{v}_k \tag{4.103}$$

$$\begin{aligned}\mathrm{E}_{\boldsymbol{h}|\boldsymbol{y}_\mathrm{T}}&\left[\boldsymbol{H}^\mathrm{H}\mathbf{G}^{(m-1)}(\boldsymbol{h})^\mathrm{H}\mathbf{G}^{(m-1)}(\boldsymbol{h})\boldsymbol{H}\right] = \\ &= \sum_{k=1}^K \boldsymbol{v}_k^*\boldsymbol{v}_k^\mathrm{T}\underbrace{\mathrm{E}_{x_k|\boldsymbol{y}_\mathrm{T}}\left[\left|\mathrm{g}_k^{\mathrm{mmse}}(\boldsymbol{P}^{(m-1)};x_k\boldsymbol{v}_k)\right|^2|x_k|^2\right]}_{b_k} = \boldsymbol{V}^*\boldsymbol{B}\boldsymbol{V}^\mathrm{T}.\end{aligned} \tag{4.104}$$

The relevant parameters for the solution based on the scaled MF receiver (4.91) are

$$\mathrm{E}_{\boldsymbol{h}_k|\boldsymbol{y}_\mathrm{T}}\left[\mathrm{g}_k^{\mathrm{cor}}(\boldsymbol{P}^{(m-1)};\boldsymbol{h}_k)\boldsymbol{h}_k\right] = \underbrace{\frac{\boldsymbol{v}_k^\mathrm{H}\boldsymbol{p}_k^{(m-1),*}\,\mathrm{E}_{x_k|\boldsymbol{y}_\mathrm{T}}\left[|x_k|^2\right]}{\bar{c}_{r_k}}}_{a_k}\boldsymbol{v}_k \tag{4.105}$$

and

$$\sum_{k=1}^K G_k^{(m-1)}\boldsymbol{R}^*_{\boldsymbol{h}_k|\boldsymbol{y}_\mathrm{T}} = \sum_{k=1}^K \underbrace{G_k^{(m-1)}\mathrm{E}_{x_k|\boldsymbol{y}_\mathrm{T}}\left[|x_k|^2\right]}_{b_k}\boldsymbol{v}_k^*\boldsymbol{v}_k^\mathrm{T} = \boldsymbol{V}^*\boldsymbol{B}\boldsymbol{V}^\mathrm{T} \tag{4.106}$$

with $G_k^{(m-1)} = \mathrm{E}_{x_k|\boldsymbol{y}_\mathrm{T}}[|x_k|^2]|\boldsymbol{v}_k^\mathrm{T}\boldsymbol{p}_k|^2/\bar{c}_{r_k}^2$, $\boldsymbol{A} = \mathbf{diag}\,[a_1,\ldots,a_K]$, and $\boldsymbol{B} = \mathbf{diag}\,[b_1,\ldots,b_K]$.

Therefore, both approaches result in

$$\boldsymbol{P}^{(m)} = \beta^{(m),-1}\left(\sum_{k=1}^K \boldsymbol{v}_k^*\boldsymbol{v}_k^\mathrm{T} b_k + \frac{\bar{G}^{(m-1)}}{P_\mathrm{tx}}\boldsymbol{I}_M\right)^{-1}\boldsymbol{V}^*\boldsymbol{A}^* \tag{4.107}$$

$$= \beta^{(m),-1}\boldsymbol{V}^*\left(\boldsymbol{V}^\mathrm{T}\boldsymbol{V}^* + \frac{\bar{G}^{(m-1)}}{P_\mathrm{tx}}\boldsymbol{B}^{-1}\right)^{-1}\boldsymbol{B}^{-1}\boldsymbol{A}^*, \tag{4.108}$$

where we applied the matrix inversion lemma (A.20) in the second equation.

For large P_{tx} and $K \leq M$, the overall channel is diagonal

$$\boldsymbol{G}^{(m)}\boldsymbol{H}\boldsymbol{P} = \boldsymbol{G}^{(m)}\boldsymbol{X}\boldsymbol{V}^{\mathrm{T}}\boldsymbol{P}^{(m)} = \beta^{(m),-1}\boldsymbol{G}^{(m)}\boldsymbol{X}\boldsymbol{V}^{\mathrm{T}}\boldsymbol{V}^{*}\left(\boldsymbol{V}^{\mathrm{T}}\boldsymbol{V}^{*}\right)^{-1}\boldsymbol{B}^{-1}\boldsymbol{A}^{*} \tag{4.109}$$

$$= \beta^{(m),-1}\boldsymbol{G}^{(m)}\boldsymbol{X}\boldsymbol{B}^{-1}\boldsymbol{A}^{*}, \tag{4.110}$$

because enough degrees of freedom are available to avoid any interference (zero-forcing).

We assume convergence of the iterative precoder optimization, i.e., $\beta^{(m)} \to 1$ for $m \to \infty$. For an MMSE receiver (and any zero-forcing precoder), $\boldsymbol{G}^{(m)}\boldsymbol{H}\boldsymbol{P} \to \boldsymbol{I}_K$ for high P_{tx}, because $\mathrm{g}_k^{\mathrm{mmse}}(\boldsymbol{P}^{(m)}; \boldsymbol{h}_k) = b_k/(x_k a_k^*)$.

For a scaled MF $\mathrm{g}_k^{\mathrm{cor}}(\boldsymbol{P}^{(m)}; \boldsymbol{h}_k) = x_k^*/(\mathrm{E}_{\mathsf{x}_k|\boldsymbol{y}_{\mathrm{T}}}[|\mathsf{x}_k|^2](\boldsymbol{v}_k^{\mathrm{T}}\boldsymbol{p}_k)$ at the receiver, the kth diagonal element of $\boldsymbol{G}^{(m)}\boldsymbol{H}\boldsymbol{P}$ converges to $|x_k|^2/\mathrm{E}_{\mathsf{x}_k|\boldsymbol{y}_{\mathrm{T}}}[|\mathsf{x}_k|^2]$.

The iterative solutions to both optimization problems (4.75) and (4.76) for rank-one channels differ only by the regularization term $\boldsymbol{B}^{-1}\bar{G}^{(m-1)}/P_{\mathrm{tx}}$ in the inverse and the allocation of the power, which is determined by a_k^*/b_k for large P_{tx} (cf. (4.108)). Note that performance criteria and solutions are identical for $P_{\mathrm{tx}} \to \infty$ and rank-one $\boldsymbol{C}_{\boldsymbol{h}_k}$.

Uncorrelated Channels

The other extreme scenario is given by S-CSI with $\boldsymbol{C}_{\boldsymbol{h}_k} = c_{\mathrm{h},k}\boldsymbol{I}_M$, i.e., spatially uncorrelated channels with different mean attenuation. Defining a normalized noise variance $c'_{n_k} = c_{n_k}/c_{\mathrm{h},k}$ and assuming $\|\boldsymbol{P}\|_{\mathrm{F}}^2 = P_{\mathrm{tx}}$, the MMSE for the AWGN BC model (Figure 4.11) reads

$$\overline{\mathrm{MMSE}}_k = 1 - \frac{\|\boldsymbol{p}_k\|_2^2}{\|\boldsymbol{P}\|_{\mathrm{F}}^2 + c'_{n_k}} = 1 - \frac{\|\boldsymbol{p}_k\|_2^2}{P_{\mathrm{tx}} + c'_{n_k}}. \tag{4.111}$$

The corresponding sum rate in (4.44) and (4.39) is

$$\sum_{k=1}^{K}\overline{\mathrm{R}}_k^{\mathrm{AWGN}} = \sum_{k=1}^{K}\left(\log_2\left[P_{\mathrm{tx}} + c'_{n_k}\right] - \log_2\left[P_{\mathrm{tx}} + c'_{n_k} - \|\boldsymbol{p}_k\|_2^2\right]\right). \tag{4.112}$$

Both performance measures depend only on the norm $\|\boldsymbol{p}_k\|_2^2$, i.e., the power allocation to every receiver.

It can be shown that the maximum sum rate is reached serving only the receiver with smallest c'_{n_k}, which yields

$$\sum_{k=1}^{K}\overline{\mathrm{R}}_k^{\mathrm{AWGN}} = \log_2\left[1 + \frac{P_{\mathrm{tx}}}{c'_{n_k}}\right]. \tag{4.113}$$

The choice of $\boldsymbol{p}_k$ is only constrained by $\|\boldsymbol{p}_k\|_2^2 = P_{\text{tx}}$ and any vector is equivalent on average, which is clear intuitively.

Minimizing the sum MSE

$$\sum_{k=1}^{K} \overline{\text{MMSE}}_k = K - \sum_{k=1}^{K} \frac{\|\boldsymbol{p}_k\|_2^2}{P_{\text{tx}} + c'_{n_k}} \tag{4.114}$$

yields the same solution as problem (4.112) for unequal c'_{n_k} with minimum

$$K - 1 \leq \sum_{k=1}^{K} \overline{\text{MMSE}}_k = K - \frac{P_{\text{tx}}}{P_{\text{tx}} + c'_n} \leq K. \tag{4.115}$$

Note that the sum MSE (4.114) depends only on $\|\boldsymbol{P}\|_{\text{F}}^2 = P_{\text{tx}}$ for identical $c'_{n_k} = c_n$. Thus, every initialization of the iterative algorithm in Section 4.4.1 is a stationary point. This is not true for the sum rate (4.112), which is maximized for selecting only one *arbitrary* receiver. This observation illustrates the deficiency of using this lower bound to the sum rate.[33]

4.4.4 Performance Evaluation

The proposed algorithms, which minimize the sum MSEs in (4.75) and (4.76), are evaluated for the *mean sum rate* $\text{E}_{\boldsymbol{h},\boldsymbol{y}_\text{T}}[\text{R}_{\text{sum}}(\mathbf{P}(\boldsymbol{y}_\text{T}); \boldsymbol{h})]$ (4.27) and different degrees of CSI (Table 4.1). In Section 5.5 we present a comparison with nonlinear precoding w.r.t. the mean uncoded bit error rate (BER).

The following linear precoders based on *sum MSE* are considered here and in Section 5.5: [34]

- For *C-CSI*, we solve (4.69) assuming MMSE receivers. The solution is given in [99, 102] and is a special case of Section 4.4.1 assuming C-CSI ("MMSE-Rx (C-CSI)"). As initialization we use the MMSE precoder from [116] (cf. Appendix D.1 with $\mathbf{G}^{(m-1)}(\boldsymbol{h}) = \boldsymbol{I}_K$) for identical receivers $g_k = \beta$ and C-CSI.
- The *heuristic* in (4.70), where a channel estimate is plugged into the cost function for C-CSI as if it was error-free (Appendix C), is applied assuming two different types of CSI: Outdated CSI $\hat{\boldsymbol{h}} = \text{E}_{\boldsymbol{h}[q']|\boldsymbol{y}[q']}[\boldsymbol{h}[q']]$ with $q' = q^{\text{tx}} - 1$ (O-CSI), i.e., the channel estimate from the last reverse link time slot is used without prediction, and P-CSI with the predicted channel $\hat{\boldsymbol{h}} = \text{E}_{\boldsymbol{h}|\boldsymbol{y}_\text{T}}[\boldsymbol{h}]$ ("Heur. MMSE-Rx (O-CSI)" and "Heur. MMSE-Rx (P-CSI)"). It is initialized in analogy to C-CSI but with O-CSI and P-CSI, respectively.

[33] A possible compensation would be to select the active receivers before designing the precoding $\boldsymbol{P}$.

[34] The names employed in the legends of the figures are given in quotation marks.

- For P-CSI, the transmitter's cost function is given by the *CM estimate* of the receivers' remaining cost (cf. (4.75) and (4.76)). The iterative algorithm in (4.82) and (4.84) for problem (4.75) with MMSE receivers ("MMSE-Rx (P-CSI)") is implemented using a straightforward Monte Carlo integration with 1000 integration points. The steps in the iterative algorithm in (4.91) and (4.93) assuming scaled matched filter (MF) receivers for problem (4.76) ("MF-Rx (P-CSI)") are already given in closed form. The same initialization is used for both, which is described in Section 4.4.1.
- For S-CSI, the iterative algorithm based on the CM estimate of the cost functions is a special case of P-CSI for both receiver models ("MMSE-Rx (S-CSI)" and "MF-Rx (S-CSI)"); see also (4.98) for the scaled matched filter receiver. It is initialized by the precoder from Appendix D.3.

For all MSE based algorithms, we choose a fixed number of 10 iterations to obtain a comparable complexity. We refer to the latter two solutions as *robust linear precoding* with P-CSI or S-CSI, respectively.

For comparison, we include algorithms which optimize *sum rate* directly:

- For C-CSI, we choose the precoder based on zero-forcing with a greedy receiver selection [59] ("Max. sum rate"). The comparison in [183] shows that it performs very close to the maximum sum rate with complex Gaussian data $\boldsymbol{d}[n]$.
- For P-CSI and S-CSI, the fixed point iteration in [31] finds a stationary point for maximization of the sum rate in the AWGN BC model (4.45) ("Max. sum rate (P-CSI)" and "Max. sum rate (S-CSI)").[35] Its complexity is of order $\mathrm{O}(M^3)$ per iteration. Contrary to C-CSI (Figure 4.9) we do not include receiver selection in order to allow for a direct comparison with the sum MSE approaches, which have the same order of complexity.[36] We choose the same initialization as for the sum MSE algorithms from Section 4.4.1 and limit the number of iterations to 40.

The alternative to linear precoding, which serves a maximum of K receivers in every time slot, is a *time division multiple access* (TDMA) strategy. In TDMA only a single receiver is served in every time slot. We choose TDMA with maximum throughput (TP) scheduling and a maximum SNR beamformer based on P-CSI ("TDMA max. TP (P-CSI)").

P-CSI is based on the observation of $Q = 5$ reverse link time slots with $N = 32$ training symbols per receiver (orthogonal training sequences with $\boldsymbol{S}^{\mathrm{H}}\boldsymbol{S} = N\boldsymbol{I}_{MK}$). The variances of the uncorrelated noise at the transmitter $\boldsymbol{C}_{\boldsymbol{v}} = c_n\boldsymbol{I}_M$ (Section 4.1.3) and receivers are identical $c_{n_k} = c_n$. The TDD system has the slot structure in Figure 4.2, i.e., "outdated CSI" (O-CSI) is outdated by one time slot. The receivers g_k are implemented as MMSE receivers (4.25) which have the necessary perfect CSI.

---

[35] No proof of convergence is given in [31].

[36] Including the receiver selection proposed in [31] the total algorithm has a numerical complexity of $\mathrm{O}(K^2M^3)$.

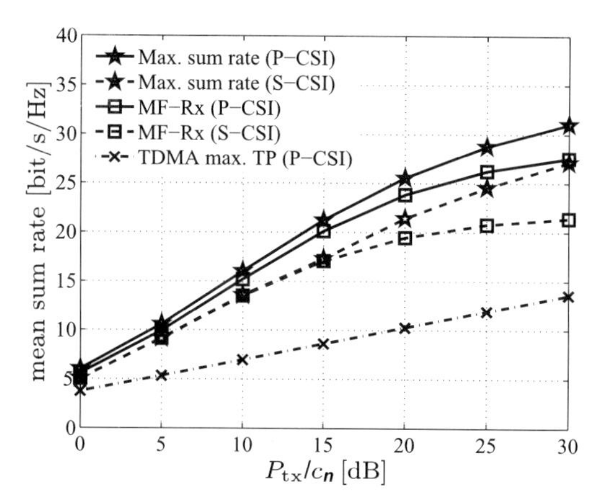

Fig. 4.14 Comparison of maximum sum rate and minimum sum MSE precoding based on the AWGN BC model with $f_{\max} = 0.2$ (Scenario 1).

The temporal channels correlations are described by Clarke's model (Appendix E and Figure 2.12) with a maximum Doppler frequency $f_{\max}$, which is normalized to the slot period T_{b}. For simplicity, the temporal correlations are identical for all elements in $\boldsymbol{h}$ (2.12). Two different stationary scenarios for the spatial covariance matrix with $M = 8$ transmit antennas are compared. For details and other implicit assumption see Appendix E. The SNR is defined as P_{tx}/c_n.

Scenario 1: We choose $K = 6$ receivers with the mean azimuth directions $\bar{\varphi} = [-45°, -27°, -9°, 9°, 27°, 45°]$ and Laplace angular power spectrum with spread $\sigma = 5°$.

Scenario 2: We have $K = 8$ receivers with mean azimuth directions $\bar{\varphi} = [-45°, -32.1°, -19.3°, -6.4°, 6.4°, 19.3°, 32.1°, 45°]$ and Laplace angular power spectrum with spread $\sigma = 0.5°$, i.e., $\mathrm{rank}[\boldsymbol{C}_{\boldsymbol{h}_k}] \approx 1$.

Discussion of Results

Because the lower bound to the sum rate, which depends on the sum MSE, is generally not tight, we expect a performance loss when minimizing the sum MSE compared to a direct maximization of the sum rate. For C-CSI, the loss is larger for increasing K/M and P_{tx}/c_n (Figure 4.12). This is also true for P-CSI and S-CSI (Figure 4.14). But the gain compared to TDMA is significant for all methods.

The algorithm for maximizing the sum rate of the AWGN BC model converges slowly and requires 30-40 iterations, i.e., the overall complexity is significantly larger than for the sum MSE approaches. But the latter are already outperformed in the first iterations, because the sum MSE is a lower bound to the sum rate (4.45) of the AWGN BC (compare with C-CSI in Figure 4.9).

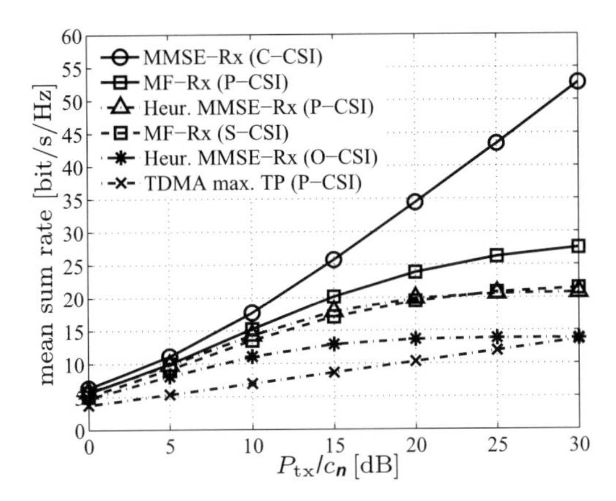

Fig. 4.15 Mean sum rate vs. P_{tx}/c_n for P-CSI with $f_{\mathrm{max}} = 0.2$ (Scenario 1).

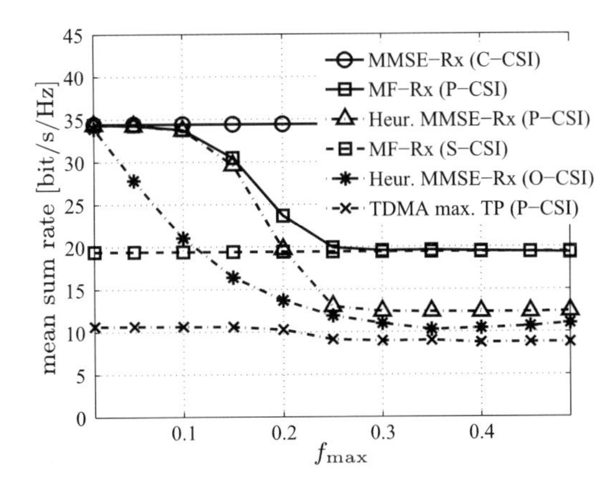

Fig. 4.16 Mean sum rate vs. f_{max} at $10\log_{10}(P_{\mathrm{tx}}/c_n) = 20\,\mathrm{dB}$ (Scenario 1).

Still the MSE is often considered as practically more relevant. For example, when transmitting data with a fixed constellation size, the MSE achieves a better performance.

With respect to the mean sum rate, the algorithms based on the scaled matched filter receiver model and the MMSE receiver show almost identical performance. Due to its simplicity, we only consider the solution based on the scaled matched filter.

The heuristic optimization of $\boldsymbol{P}$ based on sum MSE and P-CSI only applies a channel estimate to the cost function for C-CSI. For a maximum Doppler frequency below $f_{\mathrm{max}} = 0.15$, almost the whole gain in mean sum rate of robust linear precoding with P-CSI compared to the heuristic with O-CSI is due to channel prediction (Figures 4.15 and 4.16): Concerning the sum MSE as the cost function, the errors in CSI are still small enough and can be neglected (Appendix C); there is no performance advantage when we con-

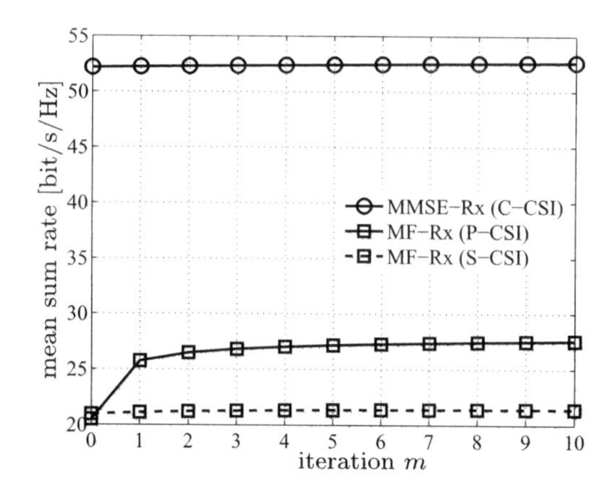

Fig. 4.17 Convergence in mean sum rate of alternating optimization with matched filter (MF) receiver model for C-CSI, P-CSI, and S-CSI (P-CSI with $f_{\max} = 0.2$, $10 \log_{10}(P_{\text{tx}}/c_n) = 30\,\text{dB}$, Scenario 1).

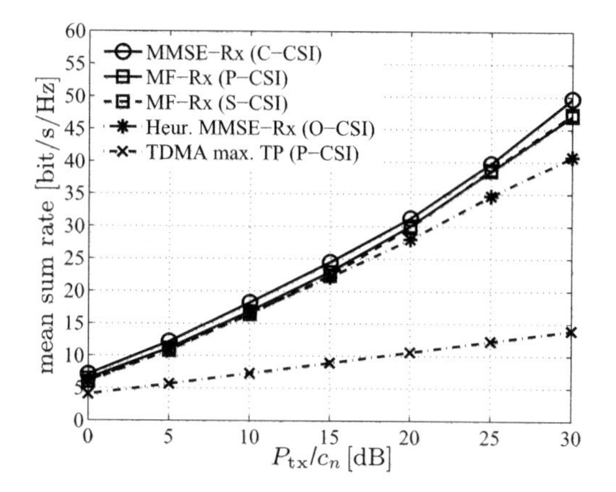

Fig. 4.18 Mean sum rate vs. P_{tx}/c_n for P-CSI with $f_{\max} = 0.2$ (Scenario 2).

sider the error covariance matrix in the optimization problem. Thus, accurate estimation of the channel predictor's parameters or its robust optimization is crucial (see Chapters 2 and 3). On the other hand, a systematic robust precoder design with P-CSI is important for $f_{\max} > 0.15$, where the heuristic optimization performs close to TDMA.

A meaningful precoder solution for S-CSI is obtained taking the expected cost (robust precoder). About twice the sum rate of TDMA can be achieved in this scenario. As $f_{\max}$ increases, a smooth transition from C-CSI to S-CSI is achieved when minimizing the CM estimate of the receivers' cost.

For this scenario, the robust precoder for P-CSI converges within three iterations (Figure 4.17). The initialization of the iterative algorithms for C-CSI and S-CSI is already close to the (local) optimum. For C-CSI, more iterations are required as K/M increases, but typically 10 iterations are sufficient.

If $\boldsymbol{C}_{\boldsymbol{h}_k}$ has full rank (as in scenario 1), the system is interference-limited and the sum rate saturates for high P_{tx}/c_n. Eventually, it is outperformed by TDMA, which is *not* interference-limited; therefore, its performance does not saturate. A proper selection of the receivers, which should be served in one time slot, would allow for convergence to TDMA at high P_{tx}/c_n. In [128] the conjecture is made that the multiplexing gain of the broadcast sum capacity for P-CSI is one, i.e., only one receiver is served at the same time. On the other hand, if $\sum_{k=1}^{K} \mathrm{rank}[\boldsymbol{C}_{\boldsymbol{h}_k}] \leq M$ and the eigenvectors corresponding to non-zero eigenvalues are linearly independent, the interference can be avoided by linear precoding and the sum rate does not saturate.

In scenario 2, the covariance matrices $\boldsymbol{C}_{\boldsymbol{h}_k}$ are close to rank one. Therefore, the performance for P-CSI and S-CSI is very close to C-CSI (Figure 4.18) for $M = K$. The heuristic degrades only at high P_{tx}/c_n because the channel estimates $\hat{\boldsymbol{h}}_k$ lie in the correct (almost) one-dimensional subspace.

In the next chapter (Section 5.5), we show the performance results for the mean uncoded bit error rate (BER), which is averaged over the K receivers, for comparison with nonlinear precoding. A fixed 16QAM (QAM: quadrature amplitude modulation) modulation is chosen. Minimizing the sum MSE corresponds to an optimization of the overall transmission quality without fairness constraints. Maximization of the sum rate is not a useful criterion to optimize precoding for a fixed modulation $\mathbb{D}$. Figures 5.10 with 5.16 give the uncoded BER for robust linear precoding with P-CSI and S-CSI in both scenarios.

Performance for Estimated Covariance Matrices

All previous performance results for linear precoding and P-CSI have assumed perfect knowledge of the channel covariance matrix $\boldsymbol{C}_{\boldsymbol{h}_{\mathrm{T}}}$ and the noise covariance matrix $\boldsymbol{C}_{\boldsymbol{v}} = c_n \boldsymbol{I}_M$. Now we apply the algorithms for estimating channel and noise covariance matrices which are derived in Chapter 3 to estimate the parameters of $\mathrm{p}_{\boldsymbol{h}|\boldsymbol{y}_{\mathrm{T}}}(\boldsymbol{h}|\boldsymbol{y}_{\mathrm{T}})$.

Only $B = 20$ correlated realizations $\boldsymbol{h}[q^{\mathrm{tx}} - \ell]$ are observed via $\boldsymbol{y}[q^{\mathrm{tx}} - \ell]$ for $\ell \in \{1, 3, \ldots, 2B - 1\}$, i.e., with a period of $N_{\mathrm{P}} = 2$. They are generated according to the model (2.12); the details are described above.

For estimation, $\boldsymbol{C}_{\boldsymbol{h}_{\mathrm{T}}}$ is assumed to have the structure in (3.98), i.e., the K receivers' channels are uncorrelated and have different temporal and spatial correlations.[37] We estimate the Toeplitz matrix $\boldsymbol{C}_{\mathrm{T},k}$, which describes the

[37] The real model (2.12) with block diagonal $\boldsymbol{C}_{\boldsymbol{h}}$ is a special case of (3.98).

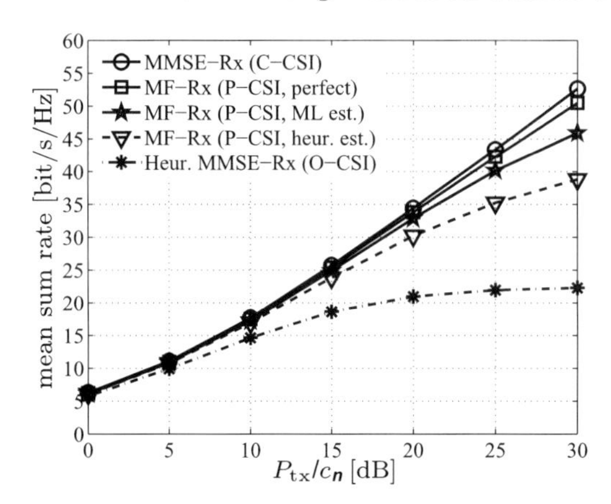

Fig. 4.19 Mean sum rate vs. $P_{\mathrm{tx}}/c_{\boldsymbol{n}}$ for P-CSI with $f_{\max} = 0.1$ for estimated covariance matrices (Scenario 1, $B = 20$).

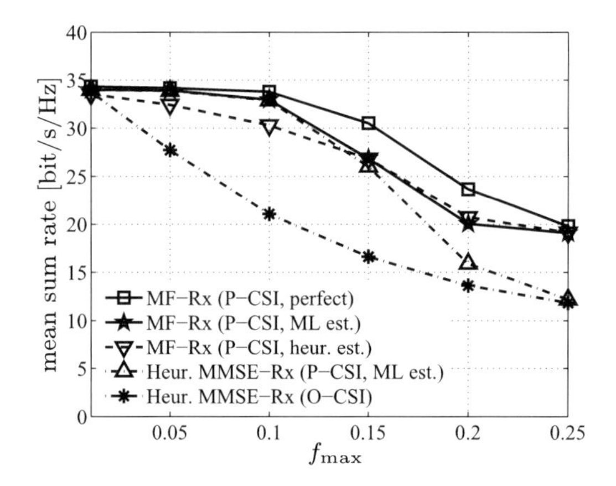

Fig. 4.20 Mean sum rate vs. $f_{\max}$ at $10\log_{10}(P_{\mathrm{tx}}/c_{\boldsymbol{n}}) = 20\,\mathrm{dB}$ for estimated covariance matrices (Scenario 1, $B = 20$).

temporal correlations, and $\boldsymbol{C}_{\boldsymbol{h}_k}$ separately as proposed in Section 3.3.6.[38] Two algorithms are compared:

- The *heuristic estimator* ("P-CSI, heur. est.") determines $\hat{\boldsymbol{C}}_{\boldsymbol{h}_k}$ according to (3.96) and $\hat{\boldsymbol{C}}_{\mathrm{T},k}$ with (3.86) for $I = 1$ and $P = 1$. $\hat{\boldsymbol{C}}_{\mathrm{T},k}$ is scaled to have ones on the diagonal as required by (3.98).
- The advanced estimator ("P-CSI, ML est.") exploits the conclusions from Chapter 3: $\hat{\boldsymbol{C}}_{\boldsymbol{h}_k}$ is obtained from (3.144) which is considerably less complex than the ML approach (Section 3.3); $\hat{\boldsymbol{C}}_{\mathrm{T},k}$ is the ML estimate from Section 3.3.5 with 10 iterations, which is also scaled to have ones on the diagonal.

[38] For example, when estimating $\boldsymbol{C}_{\boldsymbol{h}_k}$ we assume implicitly that $\boldsymbol{C}_{\mathrm{T},k} = \boldsymbol{I}_B$ and vice versa.

Finally, the samples of the autocovariance sequence[39] in $\hat{\boldsymbol{C}}_{\mathrm{T},k}$ are interpolated by solving the minimum norm completion problem (3.104) assuming $f_{\max} = 0.25$.

The noise variance c_n at the transmitter is estimated by (3.67) for $I = 1$. "P-CSI, perfect" denotes the performance for perfectly known covariance matrices.

At moderate to high SNR the advanced estimator for the channel correlations yields a significantly improved rate compared to the heuristic (Figure 4.19). For a maximum Doppler frequency $f_{\max}$ close to zero, the mean rate of all algorithms is comparable because this case is not very challenging (Figure 4.20). But for $f_{\max} \approx 0.1$ the advanced estimator yields a 10% gain in rate over the heuristic. For high $f_{\max}$, good channel prediction is impossible and the precoder relies only on the estimation quality of $\hat{\boldsymbol{C}}_{\boldsymbol{h}_k}$, which is sufficiently accurate also for the heuristic estimator (Figure 4.20).

With only $B = 20$ observations the channel covariance matrices can already be estimated with sufficient accuracy to ensure the performance gains expected by prediction and a robust precoder design. For $f_{\max} < 0.15$, prediction is the key component of the precoder to enhance its performance which requires a ML estimate of the temporal channel correlations. For high $f_{\max}$, the robust design based on an estimate of $\boldsymbol{C}_{\boldsymbol{h}_k}$ achieves a significantly larger mean sum rate than the heuristic precoder, which is also simulated based on the advanced estimator of the covariance matrices.

Comparing the maximum Doppler frequency $f_{\max}$ for a given sum rate requirement, the allowable mobility in the system is approximately doubled by the predictor and the robust design.

4.5 Mean Square Error Dualities of Broadcast and Multiple Access Channel

One general approach to solve constrained optimization problems defines a dual problem which has the same optimal value as the original problem. For example, the Lagrange dual problem can be used to find a solution to the original problem, if it is convex and feasible (strong duality and Slater's condition) [24].

To solve *optimization* problems for the *broadcast channel* (BC) it is advantageous to follow a similar idea. The construction of a dual problem is based on the intuition that the situation in the BC is related to the multiple access channel (MAC). Therefore, a MAC model is chosen which has the same achievable region for a certain performance measure as the BC model of interest. Examples for BC optimization problems whose solution can be obtained from the corresponding dual MAC are introduced in Section 4.6. For

[39] They are only given for a spacing of $N_{\mathrm{P}} = 2$.

Section 4.6, only a basic understanding of the BC-MAC duality is necessary and the details of the proofs for duality in this section can be skipped.

Consider the case of linear precoding (Figure 4.21) and C-CSI. For C-CSI, it is well known that the MAC model in Figure 4.22 has the same achievable region for the following performance measures under a sum power constraint $\|\boldsymbol{P}\|_{\mathrm{F}}^2 = \sum_{k=1}^{K} p_k^{\mathrm{mac},2} \leq P_{\mathrm{tx}}$: Achievable rate (4.18) [230], SINR (4.19) [185], and MSE (4.24) [7, 146, 221]. The rate for the data (complex Gaussian distributed) from the kth transmitter in the MAC is

$$\begin{aligned}\mathrm{R}_k^{\mathrm{mac}}(\boldsymbol{p}^{\mathrm{mac}}, \boldsymbol{u}_k^{\mathrm{mac}}; \boldsymbol{h}^{\mathrm{mac}}) &= \mathcal{I}(d_k^{\mathrm{mac}}; \hat{d}_k^{\mathrm{mac}}) \\ &= \log_2[1 + \mathrm{SINR}_k^{\mathrm{mac}}(\boldsymbol{p}^{\mathrm{mac}}, \boldsymbol{u}_k^{\mathrm{mac}}; \boldsymbol{h}^{\mathrm{mac}})], \qquad (4.116)\end{aligned}$$

which depends on

$$\mathrm{SINR}_k^{\mathrm{mac}}(\boldsymbol{p}^{\mathrm{mac}}, \boldsymbol{u}_k^{\mathrm{mac}}; \boldsymbol{h}^{\mathrm{mac}}) = \frac{p_k^{\mathrm{mac},2} |\boldsymbol{u}_k^{\mathrm{mac},\mathrm{T}} \boldsymbol{h}_k^{\mathrm{mac}}|^2}{\sum_{\substack{i=1 \\ i \neq k}}^{K} p_i^{\mathrm{mac},2} |\boldsymbol{u}_k^{\mathrm{mac},\mathrm{T}} \boldsymbol{h}_i^{\mathrm{mac}}|^2 + \|\boldsymbol{u}_k^{\mathrm{mac}}\|_2^2}. \qquad (4.117)$$

Similarly, $\mathrm{MSE}_k^{\mathrm{mac}}(\boldsymbol{p}^{\mathrm{mac}}, \boldsymbol{u}_k^{\mathrm{mac}}, g_k^{\mathrm{mac}}; \boldsymbol{h}^{\mathrm{mac}}) = \mathrm{E}_{n_k^{\mathrm{mac}}, \boldsymbol{d}^{\mathrm{mac}}}[|\hat{d}_k^{\mathrm{mac}} - d_k^{\mathrm{mac}}|^2]$ can be defined, which results in the relation

$$\min_{g_k^{\mathrm{mac}}} \mathrm{MSE}_k^{\mathrm{mac}}(\boldsymbol{p}^{\mathrm{mac}}, \boldsymbol{u}_k^{\mathrm{mac}}, g_k^{\mathrm{mac}}; \boldsymbol{h}^{\mathrm{mac}}) = 1 \Big/ (1 + \mathrm{SINR}_k^{\mathrm{mac}}(\boldsymbol{p}^{\mathrm{mac}}, \boldsymbol{u}_k^{\mathrm{mac}}; \boldsymbol{h}^{\mathrm{mac}})). \qquad (4.118)$$

The BC-MAC duality is valid under the following assumptions:

- The total transmit power is constrained by $\sum_{k=1}^{K} p_k^{\mathrm{mac},2} \leq P_{\mathrm{tx}}$.
- The dual MAC channel is defined as $\boldsymbol{h}_k^{\mathrm{mac}} = \boldsymbol{h}_k c_{n_k}^{-1/2}$ with uncorrelated noise $\boldsymbol{n}^{\mathrm{mac}}$ of variance one, where $\boldsymbol{h}_k^{\mathrm{mac}}$ is the kth column of $\boldsymbol{H}^{\mathrm{mac}} \in \mathbb{C}^{M \times K}$ and $\boldsymbol{h}^{\mathrm{mac}} = \mathbf{vec}[\boldsymbol{H}^{\mathrm{mac}}]$.
- Each data stream is decoded separately at the receiver.

This shows that the necessary MAC model is only a mathematical tool, i.e., a *virtual MAC*. A true MAC does not have a total power constraint, but individual power constraints, and the channel $\boldsymbol{H}^{\mathrm{mac}}$ would be $\boldsymbol{H}^{\mathrm{T}}$ if reciprocal.

To summarize we define MAC-BC duality as follows:

Definition 4.1. A BC and MAC model are dual regarding a set of BC and MAC performance measures if all values for the BC performance measures are achievable if and only if the same values are achievable for the MAC performance measures in the dual MAC model using the same total transmit power.

The important advantage in choosing a dual MAC model is that the K performance measures for the K data streams are only coupled in $\boldsymbol{p}^{\mathrm{mac}}$ and

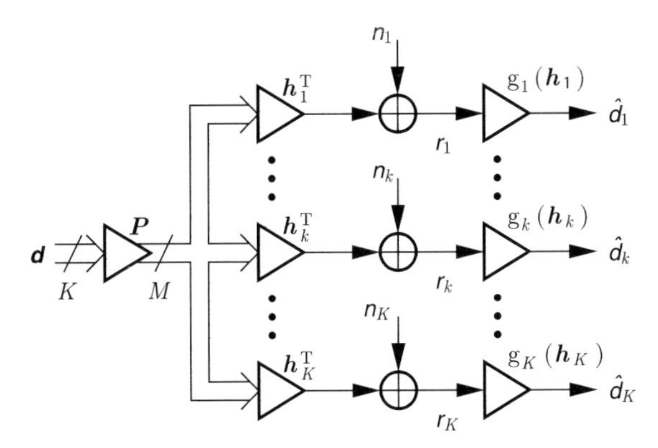

Fig. 4.21 Broadcast channel (BC) model with linear receivers (as in Figure 4.3).

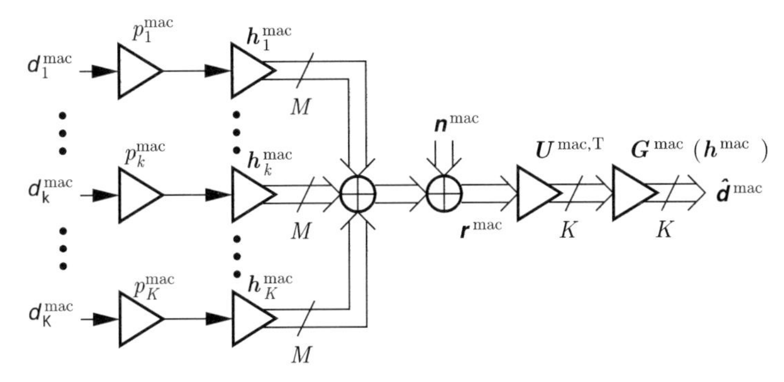

Fig. 4.22 Dual model for P-CSI with multiple access (MAC) structure and linear receivers corresponding to the BC model in Figure 4.21.

decoupled in $\boldsymbol{u}_k^{\mathrm{mac}}$ (cf. (4.117)). In the BC they are decoupled in g_k, but coupled in $\boldsymbol{p}_k$.

The following proofs of BC-MAC duality for P-CSI are carried out in analogy to Utschick and Joham [221], who show the MSE duality for the MIMO-BC and C-CSI in a more general setting.

4.5.1 Duality for AWGN Broadcast Channel Model

To prove duality for the BC (Figure 4.21) and the performance measure $\mathrm{E}_{\boldsymbol{h}_k|\boldsymbol{y}_{\mathrm{T}}}[\mathrm{COR}_k(\boldsymbol{P}, \mathrm{g}_k(\boldsymbol{h}_k); \boldsymbol{h}_k)]$ (4.49), we introduce the MAC model in Figure 4.22: Only the scalar filters $\mathbf{G}^{\mathrm{mac}}(\boldsymbol{h}^{\mathrm{mac}}) = \mathbf{diag}\left[\mathrm{g}_1^{\mathrm{mac}}(\boldsymbol{h}_1^{\mathrm{mac}}), \mathrm{g}_2^{\mathrm{mac}}(\boldsymbol{h}_2^{\mathrm{mac}}), \ldots, \mathrm{g}_K^{\mathrm{mac}}(\boldsymbol{h}_K^{\mathrm{mac}})\right]$ at the receiver have knowledge about the realization of $\boldsymbol{H}^{\mathrm{mac}}$, all other parameters are optimized based on P-CSI, i.e., the probability distribution of $\boldsymbol{H}^{\mathrm{mac}}$ given $\boldsymbol{y}_{\mathrm{T}}$. The dual MAC performance measure is $\mathrm{E}_{\boldsymbol{h}_k^{\mathrm{mac}}|\boldsymbol{y}_{\mathrm{T}}}[\mathrm{COR}_k^{\mathrm{mac}}(\boldsymbol{p}^{\mathrm{mac}}, \boldsymbol{u}_k^{\mathrm{mac}}, \mathrm{g}_k^{\mathrm{mac}}(\boldsymbol{h}_k^{\mathrm{mac}}); \boldsymbol{h}_k^{\mathrm{mac}})]$

where

$$\begin{aligned}\mathrm{COR}_k^{\mathrm{mac}}(\boldsymbol{p}^{\mathrm{mac}}, \boldsymbol{u}_k^{\mathrm{mac}}, \mathrm{g}_k^{\mathrm{mac}}(\boldsymbol{h}_k^{\mathrm{mac}}); \boldsymbol{h}_k^{\mathrm{mac}}) &= \\ = 1 + |\mathrm{g}_k^{\mathrm{mac}}(\boldsymbol{h}_k^{\mathrm{mac}})|^2 \boldsymbol{u}_k^{\mathrm{mac},\mathrm{T}} &\left(\boldsymbol{I}_M + \sum_{i=1}^{K} p_i^{\mathrm{mac},2} \boldsymbol{R}_{\boldsymbol{h}_i^{\mathrm{mac}}|\boldsymbol{y}_\mathrm{T}} \right) \boldsymbol{u}_k^{\mathrm{mac},*} \\ &- 2\mathrm{Re}\left(\mathrm{g}_k^{\mathrm{mac}}(\boldsymbol{h}_k^{\mathrm{mac}}) \boldsymbol{u}_k^{\mathrm{mac},\mathrm{T}} \boldsymbol{h}_k^{\mathrm{mac}} p_k^{\mathrm{mac}}\right). \quad (4.119)\end{aligned}$$

We define the linear receive filter $\boldsymbol{U}^{\mathrm{mac}} = [\boldsymbol{u}_1^{\mathrm{mac}}, \boldsymbol{u}_2^{\mathrm{mac}}, \ldots, \boldsymbol{u}_K^{\mathrm{mac}}] \in \mathbb{C}^{M \times K}$ based on P-CSI and the power allocation at the transmitters $\boldsymbol{p}^{\mathrm{mac}} = [p_1^{\mathrm{mac}}, p_2^{\mathrm{mac}}, \ldots, p_K^{\mathrm{mac}}]^\mathrm{T}$ with a constraint on the total transmit power $\sum_{k=1}^{K} p_k^{\mathrm{mac},2} \leq P_{\mathrm{tx}}$. The noise process $\boldsymbol{n}^{\mathrm{mac}}$ at the receivers is zero-mean with covariance matrix $\boldsymbol{C}_{\boldsymbol{n}^{\mathrm{mac}}} = \boldsymbol{I}_M$.

The minimum

$$\begin{aligned}\mathrm{MCOR}_k^{\mathrm{mac}}(\boldsymbol{p}^{\mathrm{mac}}, \boldsymbol{u}_k^{\mathrm{mac}}; \boldsymbol{h}_k^{\mathrm{mac}}) &= \\ &= 1 - \frac{p_k^{\mathrm{mac},2} |\boldsymbol{u}_k^{\mathrm{mac},\mathrm{T}} \boldsymbol{h}_k^{\mathrm{mac}}|^2}{\boldsymbol{u}_k^{\mathrm{mac},\mathrm{T}} \left(\boldsymbol{I}_M + \sum_{i=1}^{K} p_i^{\mathrm{mac},2} \boldsymbol{R}_{\boldsymbol{h}_i^{\mathrm{mac}}|\boldsymbol{y}_\mathrm{T}} \right) \boldsymbol{u}_k^{\mathrm{mac},*}} \quad (4.120)\end{aligned}$$

w.r.t. $g_k^{\mathrm{mac}} = \mathrm{g}_k^{\mathrm{mac}}(\boldsymbol{h}_k^{\mathrm{mac}})$ is achieved for

$$g_k^{\mathrm{mac}} = \mathrm{g}_k^{\mathrm{mac}}(\boldsymbol{h}_k^{\mathrm{mac}}) = \frac{\left(\boldsymbol{u}_k^{\mathrm{mac},\mathrm{T}} \boldsymbol{h}_k^{\mathrm{mac}} p_k^{\mathrm{mac}}\right)^*}{\boldsymbol{u}_k^{\mathrm{mac},\mathrm{T}} \left(\boldsymbol{I}_M + \sum_{i=1}^{K} p_i^{\mathrm{mac},2} \boldsymbol{R}_{\boldsymbol{h}_i^{\mathrm{mac}}|\boldsymbol{y}_\mathrm{T}} \right) \boldsymbol{u}_k^{\mathrm{mac},*}}. \quad (4.121)$$

The conditional mean of the minimum

$$\mathrm{E}_{\boldsymbol{h}_k^{\mathrm{mac}}|\boldsymbol{y}_\mathrm{T}}[\mathrm{MCOR}_k^{\mathrm{mac}}(\boldsymbol{p}^{\mathrm{mac}}, \boldsymbol{u}_k^{\mathrm{mac}}; \boldsymbol{h}_k^{\mathrm{mac}})] = \frac{1}{1 + \overline{\mathrm{SINR}}_k^{\mathrm{mac}}(\boldsymbol{u}_k^{\mathrm{mac}}, \boldsymbol{p}^{\mathrm{mac}})} \quad (4.122)$$

depends on the common SINR definition[40] for P-CSI

$$\overline{\mathrm{SINR}}_k^{\mathrm{mac}} = \frac{p_k^{\mathrm{mac},2} \boldsymbol{u}_k^{\mathrm{mac},\mathrm{T}} \boldsymbol{R}_{\boldsymbol{h}_k^{\mathrm{mac}}|\boldsymbol{y}_\mathrm{T}} \boldsymbol{u}_k^{\mathrm{mac},*}}{\boldsymbol{u}_k^{\mathrm{mac},\mathrm{T}} \left(\boldsymbol{I}_M + \sum_{\substack{i=1 \\ i \neq k}}^{K} p_i^{\mathrm{mac},2} \boldsymbol{R}_{\boldsymbol{h}_i^{\mathrm{mac}}|\boldsymbol{y}_\mathrm{T}} \right) \boldsymbol{u}_k^{\mathrm{mac},*}}. \quad (4.123)$$

Therefore, a duality for the cost functions $\mathrm{E}_{\boldsymbol{h}_k|\boldsymbol{y}_\mathrm{T}}[\mathrm{COR}_k(\boldsymbol{P}, \mathrm{g}_k(\boldsymbol{h}_k); \boldsymbol{h}_k)]$ and $\mathrm{E}_{\boldsymbol{h}_k^{\mathrm{mac}}|\boldsymbol{y}_\mathrm{T}}[\mathrm{COR}_k^{\mathrm{mac}}(\boldsymbol{p}^{\mathrm{mac}}, \boldsymbol{u}_k^{\mathrm{mac}}, \mathrm{g}_k^{\mathrm{mac}}(\boldsymbol{h}_k^{\mathrm{mac}}); \boldsymbol{h}_k^{\mathrm{mac}})]$ also proves a duality for the

[40] This SINR definition is also assumed in the duality theory by Schubert et al.[185]. Moreover it is often considered when optimizing beamforming at a receiver based on S-CSI, which is derived differently. Here we see that this criterion can be justified: It is the optimization criterion for a beamformer based on P-CSI which is followed by the scaled matched filter (4.121) with knowledge of $\boldsymbol{h}_k^{\mathrm{mac}}$.

AWGN BC model (Figure 4.11) w.r.t. its information rate $\overline{\mathrm{R}}_k^{\mathrm{AWGN}}$ (4.39) and $\overline{\mathrm{SINR}}_k$ (4.36).[41]

For the proof of duality, we have to constrain the possible filters to be consistent.

Definition 4.2. For the BC, a choice of $\mathrm{g}_k(\boldsymbol{h}_k)$ and $\boldsymbol{p}_k$ is *consistent*, if $\boldsymbol{p}_k = \mathbf{0}_M$ and $\mathrm{g}_k(\boldsymbol{h}_k) \equiv 0$ are only zero simultaneously. For the MAC, a choice of $\boldsymbol{u}_k^{\mathrm{mac}}$ and p_k^{mac} is consistent, if $p_k^{\mathrm{mac}} = 0$ and $\boldsymbol{u}_k^{\mathrm{mac}} = \mathbf{0}_M$ (or $\mathrm{g}_k^{\mathrm{mac}}(\boldsymbol{h}_k^{\mathrm{mac}}) \equiv 0$) only simultaneously.[42]

With this restriction we can summarize the BC-MAC duality in a theorem.

Theorem 4.1. *Assume a consistent choice of* $\mathrm{g}_k(\boldsymbol{h}_k)$ *and* $\boldsymbol{p}_k$ *for all* k *and, alternatively, of* $\boldsymbol{u}_k^{\mathrm{mac}}$ *and* p_k^{mac}. *Consider only the active receivers and transmitters with* $\boldsymbol{p}_k \neq \mathbf{0}_M$ *or* $p_k^{\mathrm{mac}} \neq 0$. *Then all values of* $\mathrm{E}_{\boldsymbol{h}_k|\boldsymbol{y}_\mathrm{T}}[\mathrm{COR}_k(\boldsymbol{P}, \mathrm{g}_k(\boldsymbol{h}_k); \boldsymbol{h}_k)], \forall k \in \{1, 2, \ldots, K\}$, *are achievable if and only if the same values are achievable for* $\mathrm{E}_{\boldsymbol{h}_k^{\mathrm{mac}}|\boldsymbol{y}_\mathrm{T}}[\mathrm{COR}_k^{\mathrm{mac}}(\mathrm{g}_k^{\mathrm{mac}}(\boldsymbol{h}_k^{\mathrm{mac}}), \boldsymbol{u}_k^{\mathrm{mac}}, \boldsymbol{p}^{\mathrm{mac}}; \boldsymbol{h}_k^{\mathrm{mac}})], \forall k \in \{1, 2, \ldots, K\}$, *with* $\|\boldsymbol{p}^{\mathrm{mac}}\|_2^2 = \|\boldsymbol{P}\|_\mathrm{F}^2$.

To prove this result we construct a transformation of the filters from MAC to BC [221], which yields identical performance measures

$$\boxed{\begin{aligned}&\mathrm{E}_{\boldsymbol{h}_k|\boldsymbol{y}_\mathrm{T}}[\mathrm{COR}_k(\boldsymbol{P}, \mathrm{g}_k(\boldsymbol{h}_k); \boldsymbol{h}_k)] = \\ &\quad = \mathrm{E}_{\boldsymbol{h}_k^{\mathrm{mac}}|\boldsymbol{y}_\mathrm{T}}[\mathrm{COR}_k^{\mathrm{mac}}(\boldsymbol{p}^{\mathrm{mac}}, \boldsymbol{u}_k^{\mathrm{mac}}, \mathrm{g}_k^{\mathrm{mac}}(\boldsymbol{h}_k^{\mathrm{mac}}); \boldsymbol{h}_k^{\mathrm{mac}})]\end{aligned}} \tag{4.127}$$

for the same total power $\|\boldsymbol{p}^{\mathrm{mac}}\|_2^2 = \|\boldsymbol{P}\|_\mathrm{F}^2$.

As the choice of a transformation is not unique [221], we introduce several constraints on the transformation. Firstly, the desired symbols d_k and d_k^{mac} experience the same total channel in the mean

---

[41] The dual AWGN MAC model is defined by

$$\hat{d}_k^{\mathrm{mac}} = \mathrm{g}_k^{\mathrm{mac}}(\boldsymbol{h}_k^{\mathrm{mac}}) \left(\sqrt{\mathrm{E}_{\boldsymbol{h}_k^{\mathrm{mac}}|\boldsymbol{y}_\mathrm{T}}[|\boldsymbol{u}_k^{\mathrm{mac},\mathrm{T}} \boldsymbol{h}_k^{\mathrm{mac}}|^2] p_k^{\mathrm{mac}}} d_k^{\mathrm{mac}} + i_k^{\mathrm{mac}} + n_k^{\mathrm{mac}} \right) \tag{4.124}$$

and

$$i_k^{\mathrm{mac}} = \sum_{\substack{i=1 \\ i \neq k}}^{K} \sqrt{\mathrm{E}_{\boldsymbol{h}_i^{\mathrm{mac}}|\boldsymbol{y}_\mathrm{T}}[|\boldsymbol{u}_k^{\mathrm{mac},\mathrm{T}} \boldsymbol{h}_i^{\mathrm{mac}}|^2] p_i^{\mathrm{mac}}} d_i^{\mathrm{mac}}. \tag{4.125}$$

Assuming complex Gaussian signaling this yields the dual MAC rate

$$\overline{\mathrm{R}}_k^{\mathrm{mac}}(\boldsymbol{p}^{\mathrm{mac}}, \boldsymbol{u}_k^{\mathrm{mac}}) = \log_2\left[1 + \overline{\mathrm{SINR}}_k^{\mathrm{mac}}(\boldsymbol{p}^{\mathrm{mac}}, \boldsymbol{u}_k^{\mathrm{mac}})\right] \tag{4.126}$$

which is a function of $\overline{\mathrm{SINR}}_k^{\mathrm{mac}}(\boldsymbol{p}^{\mathrm{mac}}, \boldsymbol{u}_k^{\mathrm{mac}})$. Implicitly, we assume a separate decoding of the K data streams in the dual MAC.

[42] If the transmit and receive filters are zero simultaneously, the receiver's performance measure is equal to one and duality is obvious. Therefore, we can exclude these data streams from the following duality and treat them separately.

$$\mathrm{E}_{\boldsymbol{h}_k^{\mathrm{mac}}|\boldsymbol{y}_\mathrm{T}}\left[\mathrm{Re}\left(\mathrm{g}_k^{\mathrm{mac}}(\boldsymbol{h}_k^{\mathrm{mac}})\boldsymbol{u}_k^{\mathrm{mac},\mathrm{T}}\boldsymbol{h}_k^{\mathrm{mac}}p_k^{\mathrm{mac}}\right)\right] = \mathrm{E}_{\boldsymbol{h}_k|\boldsymbol{y}_\mathrm{T}}\left[\mathrm{Re}\left(\mathrm{g}_k(\boldsymbol{h}_k)\boldsymbol{h}_k^\mathrm{T}\boldsymbol{p}_k\right)\right]. \tag{4.128}$$

Moreover, the power control in BC and MAC is different ($\|\boldsymbol{p}_k\|_2^2 \neq p_k^{\mathrm{mac},2}$) in general *and* the same transmit power is used. Therefore, we introduce additional degrees of freedom $\xi_k \in \mathbb{R}_{+,0}$ which satisfy $\xi_k^2 = p_k^{\mathrm{mac},2}/(\mathrm{E}_{\boldsymbol{h}_k|\boldsymbol{y}_\mathrm{T}}[|\mathrm{g}_k(\boldsymbol{h}_k)|^2]c_{n_k}) = \|\boldsymbol{p}_k\|_2^2/\|\boldsymbol{u}_k^{\mathrm{mac}}\|_2^2$. This results in the constraints

$$\xi_k^2\mathrm{E}_{\boldsymbol{h}_k|\boldsymbol{y}_\mathrm{T}}\left[|\mathrm{g}_k(\boldsymbol{h}_k)|^2\right]c_{n_k} = p_k^{\mathrm{mac},2} \tag{4.129}$$

$$\xi_k^2\|\boldsymbol{u}_k^{\mathrm{mac}}\|_2^2 = \|\boldsymbol{p}_k\|_2^2, \tag{4.130}$$

which yield a simple system of linear equations in $\{\xi_k^2\}_{k=1}^K$ below.

A straightforward choice of the MAC parameters satisfying (4.129) and (4.130) is

$$\boxed{\begin{aligned} p_k^{\mathrm{mac}} &= c_{n_k}^{1/2}\xi_k(\mathrm{E}_{\boldsymbol{h}_k|\boldsymbol{y}_\mathrm{T}}\left[|\mathrm{g}_k(\boldsymbol{h}_k)|^2\right])^{1/2} \\ \boldsymbol{u}_k^{\mathrm{mac}} &= \xi_k^{-1}\boldsymbol{p}_k. \end{aligned}} \tag{4.131}$$

To ensure (4.128) we set

$$\boldsymbol{h}_k^{\mathrm{mac}} = \boldsymbol{h}_k c_{n_k}^{-1/2} \tag{4.132}$$

$$\mathrm{g}_k^{\mathrm{mac}}(\boldsymbol{h}_k^{\mathrm{mac}}) = \mathrm{g}_k(\boldsymbol{h}_k)\mathrm{E}_{\boldsymbol{h}_k|\boldsymbol{y}_\mathrm{T}}\left[|\mathrm{g}_k(\boldsymbol{h}_k)|^2\right]^{-1/2}, \tag{4.133}$$

i.e., the dual MAC channel $\boldsymbol{H}^{\mathrm{mac}} = \boldsymbol{H}^\mathrm{T}\boldsymbol{C}_{\boldsymbol{n}}^{-1/2}$ is not reciprocal to the BC channel $\boldsymbol{H}$.

Now, we have to choose $\{\xi_k\}_{k=1}^K$ in order to guarantee (4.127). First, we rewrite

$$\begin{aligned} &\mathrm{E}_{\boldsymbol{h}_k^{\mathrm{mac}}|\boldsymbol{y}_\mathrm{T}}[\mathrm{COR}_k^{\mathrm{mac}}(\boldsymbol{p}^{\mathrm{mac}}, \boldsymbol{u}_k^{\mathrm{mac}}, \mathrm{g}_k^{\mathrm{mac}}(\boldsymbol{h}_k^{\mathrm{mac}}); \boldsymbol{h}_k^{\mathrm{mac}})] \\ &\quad = 1 + \xi_k^{-2}\boldsymbol{p}_k^\mathrm{T}\left(\boldsymbol{I}_M + \sum_{i=1}^K \xi_i^2\mathrm{E}_{\boldsymbol{h}_i|\boldsymbol{y}_\mathrm{T}}\left[|\mathrm{g}_i(\boldsymbol{h}_i)|^2\right]\boldsymbol{R}_{\boldsymbol{h}_i|\boldsymbol{y}_\mathrm{T}}\right)\boldsymbol{p}_k^* \\ &\qquad - 2\mathrm{E}_{\boldsymbol{h}_k|\boldsymbol{y}_\mathrm{T}}\left[\mathrm{Re}\left(\mathrm{g}_k(\boldsymbol{h}_k)\boldsymbol{h}_k^\mathrm{T}\boldsymbol{p}_k\right)\right] \end{aligned} \tag{4.134}$$

and apply it to (4.127) which yields (cf. (4.49))

$$\mathrm{E}_{\boldsymbol{h}_k|\boldsymbol{y}_\mathrm{T}}\left[|\mathrm{g}(\boldsymbol{h}_k)|^2\right]\bar{c}_{r_k} = \xi_k^{-2}\boldsymbol{p}_k^\mathrm{H}\left(\boldsymbol{I}_M + \sum_{i=1}^K \xi_i^2\mathrm{E}_{\boldsymbol{h}_i|\boldsymbol{y}_\mathrm{T}}\left[|\mathrm{g}_i(\boldsymbol{h}_i)|^2\right]\boldsymbol{R}_{\boldsymbol{h}_i|\boldsymbol{y}_\mathrm{T}}^*\right)\boldsymbol{p}_k \tag{4.135}$$

for $k = 1, 2, \ldots, K$. With $\bar{c}_{r_k} = c_{n_k} + \sum_{i=1}^K \boldsymbol{p}_i^\mathrm{H}\boldsymbol{R}_{\boldsymbol{h}_k|\boldsymbol{y}_\mathrm{T}}^*\boldsymbol{p}_i$ we obtain a system of linear equations in $\boldsymbol{\xi} = [\xi_1^2, \xi_2^2, \ldots, \xi_K^2]^\mathrm{T}$

$$\boldsymbol{W}\boldsymbol{\xi} = \boldsymbol{l}, \tag{4.136}$$

where $\boldsymbol{l} = \left[\|\boldsymbol{p}_1\|_2^2, \|\boldsymbol{p}_2\|_2^2, \ldots, \|\boldsymbol{p}_K\|_2^2\right]^{\mathrm{T}}$ and

$$[\boldsymbol{W}]_{u,v} = \begin{cases} -\mathrm{E}_{\boldsymbol{h}_v|\boldsymbol{y}_{\mathrm{T}}}\left[|\mathrm{g}_v(\boldsymbol{h}_v)|^2\right] \boldsymbol{p}_u^{\mathrm{H}} \boldsymbol{R}_{\boldsymbol{h}_v|\boldsymbol{y}_{\mathrm{T}}}^* \boldsymbol{p}_u, & u \neq v \\ \mathrm{E}_{\boldsymbol{h}_v|\boldsymbol{y}_{\mathrm{T}}}\left[|\mathrm{g}_v(\boldsymbol{h}_v)|^2\right] \bar{c}_{r_v} - \mathrm{E}_{\boldsymbol{h}_v|\boldsymbol{y}_{\mathrm{T}}}\left[|\mathrm{g}_v(\boldsymbol{h}_v)|^2\right] \boldsymbol{p}_v^{\mathrm{H}} \boldsymbol{R}_{\boldsymbol{h}_v|\boldsymbol{y}_{\mathrm{T}}}^* \boldsymbol{p}_v, & u = v \end{cases}. \tag{4.137}$$

Because of

$$[\boldsymbol{W}]_{v,v} > \sum_{u\neq v} \left|[\boldsymbol{W}]_{u,v}\right| = \sum_{u\neq v} \mathrm{E}_{\boldsymbol{h}_v|\boldsymbol{y}_{\mathrm{T}}}\left[|\mathrm{g}_v(\boldsymbol{h}_v)|^2\right] \boldsymbol{p}_u^{\mathrm{H}} \boldsymbol{R}_{\boldsymbol{h}_v|\boldsymbol{y}_{\mathrm{T}}}^* \boldsymbol{p}_u, \tag{4.138}$$

$\boldsymbol{W}$ is (column) diagonally dominant if $c_{n_k} > 0$, i.e., it is always invertible [95, p. 349].[43] Additionally, the diagonal of $\boldsymbol{W}$ has only positive elements, whereas all other elements are negative. This yields an inverse $\boldsymbol{W}^{-1}$ with non-negative entries [21, p. 137]. Because $\boldsymbol{l}$ is positive and $\boldsymbol{W}^{-1}$ is non-negative, $\boldsymbol{\xi}$ is always positive. We conclude that we can always find a dual MAC with parameters given in (4.131) to (4.133) and by (4.136) which achieves the same performance (4.127) as the given BC.

The transmit power in the BC is given by $\|\boldsymbol{l}\|_1$.[44] It is equal to taking the sum over all equations in (4.136). This yields $\|\boldsymbol{l}\|_1 = \sum_{k=1}^K \xi_k^2 \mathrm{E}_{\boldsymbol{h}_k|\boldsymbol{y}_{\mathrm{T}}}[|\mathrm{g}_k(\boldsymbol{h}_k)|^2] c_{n_k}$, which equals $\sum_{k=1}^K p_k^{\mathrm{mac},2}$ due to (4.131). Therefore, a MAC performance identical to the BC is reached with the same total transmit power.

In Appendix D.4 we prove the reverse direction: If some $\mathrm{E}_{\boldsymbol{h}_k^{\mathrm{mac}}|\boldsymbol{y}_{\mathrm{T}}}[\mathrm{COR}_k^{\mathrm{mac}}(\boldsymbol{p}^{\mathrm{mac}}, \boldsymbol{u}_k^{\mathrm{mac}}, \mathrm{g}_k^{\mathrm{mac}}(\boldsymbol{h}_k^{\mathrm{mac}}); \boldsymbol{h}_k^{\mathrm{mac}})]$ are achieved in the MAC, then the same values can be achieved for $\mathrm{E}_{\boldsymbol{h}_k|\boldsymbol{y}_{\mathrm{T}}}[\mathrm{COR}_k(\boldsymbol{P}, \mathrm{g}_k(\boldsymbol{h}_k); \boldsymbol{h}_k)]$ in the BC with the same total power.

Example 4.4. For given power allocation $\boldsymbol{p}^{\mathrm{mac}}$ and a scaled MF (4.121), the optimum beamformers $\boldsymbol{u}_k^{\mathrm{mac}}$ are the generalized eigenvectors maximizing $\overline{\mathrm{SINR}}_k^{\mathrm{mac}}$ (4.123). Due to (4.131), the corresponding precoding vectors $\boldsymbol{p}_k$ in the BC are the scaled generalized eigenvectors $\boldsymbol{u}_k^{\mathrm{mac}}$. □

4.5.2 Duality for Incomplete Channel State Information at Receivers

For incomplete CSI at the receivers due to a common training channel, we proposed a BC model in Section 4.3.3, which we summarize in Figure 4.23. It is based on a cascaded receiver: $\mathrm{g}_k(\boldsymbol{h}_k)$ is independent of $\boldsymbol{P}$ and relies on

[43] If $\mathrm{E}_{\boldsymbol{h}_v|\boldsymbol{y}_{\mathrm{T}}}[|\mathrm{g}_v(\boldsymbol{h}_v)|^2] = 0$ for some v, this receiver has to be treated separately.

[44] The 1-norm of a vector $\boldsymbol{l} = [l_1, l_2, \ldots, l_K]^{\mathrm{T}}$ is defined as $\|\boldsymbol{l}\|_1 = \sum_{k=1}^K |l_k|$.

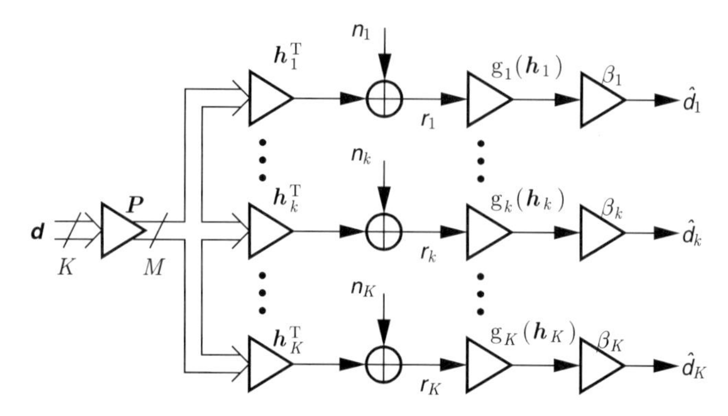

Fig. 4.23 BC model for incomplete CSI at receivers.

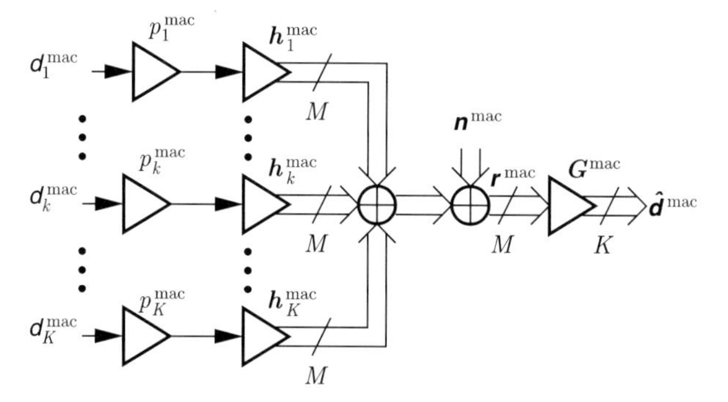

Fig. 4.24 Dual model with MAC structure corresponding to the BC model in Figure 4.23.

knowledge of $\boldsymbol{h}_k^{\mathrm{T}}\boldsymbol{q}_k$, whereas β_k depends on the transmitter's CSI and $\boldsymbol{P}$ modeling a slow automatic gain control. The MSE for this model reads (4.65)[45]

$$\begin{aligned}
\mathrm{E}_{\boldsymbol{h}_k|\boldsymbol{y}_{\mathrm{T}}}[\mathrm{MSE}_k(\boldsymbol{P},\beta_k \mathrm{g}_k(\boldsymbol{h}_k);\boldsymbol{h}_k)] &= 1 + |\beta_k|^2 \mathrm{E}_{\boldsymbol{h}_k|\boldsymbol{y}_{\mathrm{T}}}\left[|\mathrm{g}_k(\boldsymbol{h}_k)|^2\right] c_{n_k} \\
+ |\beta_k|^2 \sum_{i=1}^{K} \boldsymbol{p}_i^{\mathrm{H}} \mathrm{E}_{\boldsymbol{h}_k|\boldsymbol{y}_{\mathrm{T}}}\left[\boldsymbol{h}_k^* \boldsymbol{h}_k^{\mathrm{T}} |\mathrm{g}_k(\boldsymbol{h}_k)|^2\right] \boldsymbol{p}_i &- 2\mathrm{Re}\Big(\beta_k \mathrm{E}_{\boldsymbol{h}_k|\boldsymbol{y}_{\mathrm{T}}}[\mathrm{g}_k(\boldsymbol{h}_k)\boldsymbol{h}_k]^{\mathrm{T}} \boldsymbol{p}_k\Big).
\end{aligned} \tag{4.139}$$

As dual MAC we choose the structure in Figure 4.24 with P-CSI at the receiver $\boldsymbol{G}^{\mathrm{mac}} = [\boldsymbol{g}_1^{\mathrm{mac}}, \boldsymbol{g}_2^{\mathrm{mac}}, \ldots, \boldsymbol{g}_K^{\mathrm{mac}}]^{\mathrm{T}}$ and transmitter. The conditional mean estimate of $\mathrm{MSE}_k^{\mathrm{mac}}(\boldsymbol{p}^{\mathrm{mac}}, \boldsymbol{g}_k^{\mathrm{mac}}; \boldsymbol{h}^{\mathrm{mac}}) = \mathrm{E}_{\boldsymbol{n}^{\mathrm{mac}},\boldsymbol{d}^{\mathrm{mac}}}[\|\hat{\boldsymbol{d}}^{\mathrm{mac}} - \boldsymbol{d}^{\mathrm{mac}}\|_2^2]$

[45] Here we assume P-CSI at the transmitter instead of only S-CSI as in (4.65).

$$\begin{aligned}\mathrm{E}_{\boldsymbol{h}^{\mathrm{mac}}|\boldsymbol{y}_{\mathrm{T}}}[\mathrm{MSE}_k^{\mathrm{mac}}(\boldsymbol{p}^{\mathrm{mac}}, \boldsymbol{g}_k^{\mathrm{mac}}; \boldsymbol{H}^{\mathrm{mac}})] &= 1 + \|\boldsymbol{g}_k^{\mathrm{mac}}\|_2^2 \\ &+ \boldsymbol{g}_k^{\mathrm{mac},\mathrm{H}} \left(\sum_{i=1}^{K} p_i^{\mathrm{mac},2} \mathrm{E}_{\boldsymbol{h}_i^{\mathrm{mac}}|\boldsymbol{y}_{\mathrm{T}}} \left[\boldsymbol{h}_i^{\mathrm{mac},*} \boldsymbol{h}_i^{\mathrm{mac},\mathrm{T}} \right] \right) \boldsymbol{g}_k^{\mathrm{mac}} \\ &- 2\mathrm{Re}\Big(\boldsymbol{g}_k^{\mathrm{mac},\mathrm{T}} \mathrm{E}_{\boldsymbol{h}_k^{\mathrm{mac}}|\boldsymbol{y}_{\mathrm{T}}}[\boldsymbol{h}_k^{\mathrm{mac}}]\, p_k^{\mathrm{mac}} \Big) \end{aligned} \tag{4.140}$$

is chosen as dual performance measure ($\boldsymbol{C}_{\boldsymbol{n}^{\mathrm{mac}}} = \boldsymbol{I}_M$).

Again we have to restrict the transmit/receiver filters to the following class.

Definition 4.3. For the BC, a choice of β_k, $\mathrm{g}_k(\boldsymbol{h}_k) \not\equiv 0$, and $\boldsymbol{p}_k$ is consistent, if $\boldsymbol{p}_k = \boldsymbol{0}_M$ and $\beta_k = 0$ are only zero simultaneously. For the MAC, a choice of $\boldsymbol{g}_k^{\mathrm{mac}}$ and p_k^{mac} is consistent, if the same is true for $p_k^{\mathrm{mac}} = 0$ and $\boldsymbol{g}_k^{\mathrm{mac}} = \boldsymbol{0}_M$.

Duality is obvious for those performance measures corresponding to $\boldsymbol{p}_k = \boldsymbol{0}_M$ and $\beta_k = 0$ or $p_k^{\mathrm{mac}} = 0$ and $\boldsymbol{g}_k^{\mathrm{mac}} = \boldsymbol{0}_M$, respectively.

With this definition we can prove the following theorem:

Theorem 4.2. *Assume a consistent choice of β_k, $\mathrm{g}_k(\boldsymbol{h}_k)$, and $\boldsymbol{p}_k$ for all k and, alternatively, of $\boldsymbol{g}_k^{\mathrm{mac}}$ and p_k^{mac}. Consider only the active receivers and transmitters with $\boldsymbol{p}_k \neq \boldsymbol{0}_M$ or $p_k^{\mathrm{mac}} \neq 0$. Then all values of $\mathrm{E}_{\boldsymbol{h}_k|\boldsymbol{y}_{\mathrm{T}}}[\mathrm{MSE}_k(\boldsymbol{P}, \beta_k \mathrm{g}_k(\boldsymbol{h}_k); \boldsymbol{h}_k)], \forall k \in \{1, 2, \ldots, K\}$, are achievable if and only if the same values are achievable for $\mathrm{E}_{\boldsymbol{h}^{\mathrm{mac}}|\boldsymbol{y}_{\mathrm{T}}}[\mathrm{MSE}_k^{\mathrm{mac}}(\boldsymbol{p}^{\mathrm{mac}}, \boldsymbol{g}_k^{\mathrm{mac}}; \boldsymbol{H}^{\mathrm{mac}})], \forall k \in \{1, 2, \ldots, K\}$, with $\|\boldsymbol{p}^{\mathrm{mac}}\|_2^2 = \|\boldsymbol{P}\|_{\mathrm{F}}$.*

For given BC parameters, we construct a dual MAC with equal values of its performance measures

$$\boxed{\mathrm{E}_{\boldsymbol{h}_k|\boldsymbol{y}_{\mathrm{T}}}[\mathrm{MSE}_k(\boldsymbol{P}, \beta_k \mathrm{g}_k(\boldsymbol{h}_k); \boldsymbol{h}_k)] = \mathrm{E}_{\boldsymbol{h}^{\mathrm{mac}}|\boldsymbol{y}_{\mathrm{T}}}[\mathrm{MSE}_k^{\mathrm{mac}}(\boldsymbol{p}^{\mathrm{mac}}, \boldsymbol{g}_k^{\mathrm{mac}}; \boldsymbol{H}^{\mathrm{mac}})]} \tag{4.141}$$

in analogy to Section 4.5.1.

First we require the conditional mean of the total channel gain to be equal

$$\mathrm{Re}\big(\beta_k \mathrm{E}_{\boldsymbol{h}_k|\boldsymbol{y}_{\mathrm{T}}}\left[\mathrm{g}_k(\boldsymbol{h}_k)\boldsymbol{h}_k^{\mathrm{T}}\right] \boldsymbol{p}_k\big) = \mathrm{Re}\Big(\boldsymbol{g}_k^{\mathrm{mac},\mathrm{T}} \mathrm{E}_{\boldsymbol{h}_k^{\mathrm{mac}}|\boldsymbol{y}_{\mathrm{T}}}[\boldsymbol{h}_k^{\mathrm{mac}}]\, p_k^{\mathrm{mac}} \Big). \tag{4.142}$$

The power allocation is linked to the noise variance after the receive filters as

$$\xi_k^2 c_{n_k} |\beta_k|^2 \mathrm{E}_{\boldsymbol{h}_k|\boldsymbol{y}_{\mathrm{T}}}\left[|\mathrm{g}_k(\boldsymbol{h}_k)|^2\right] = |p_k^{\mathrm{mac}}|^2 \tag{4.143}$$

$$\xi_k^2 \|\boldsymbol{g}_k^{\mathrm{mac}}\|_2^2 = \|\boldsymbol{p}_k\|_2^2. \tag{4.144}$$

These three conditions are satisfied choosing

$$\boxed{\begin{aligned} p_k^{\mathrm{mac}} &= c_{n_k}^{1/2}\beta_k\xi_k \mathrm{E}_{\boldsymbol{h}_k|\boldsymbol{y}_\mathrm{T}}\left[|\mathrm{g}_k(\boldsymbol{h}_k)|^2\right]^{1/2} \\ \boldsymbol{g}_k^{\mathrm{mac}} &= \xi_k^{-1}\boldsymbol{p}_k \\ \boldsymbol{h}_k^{\mathrm{mac}} &= \mathrm{g}_k(\boldsymbol{h}_k)\boldsymbol{h}_k \mathrm{E}_{\boldsymbol{h}_k|\boldsymbol{y}_\mathrm{T}}\left[|\mathrm{g}_k(\boldsymbol{h}_k)|^2\right]^{-1/2} c_{n_k}^{-1/2}. \end{aligned}} \tag{4.145}$$

The detailed proof and the linear system of equations to compute $\{\xi_k^2\}_{k=1}^K \in \mathbb{R}_{+,0}$ are given in Appendix D.5.

4.6 Optimization with Quality of Service Constraints

Maximizing throughput (sum rate) or minimizing sum MSE does not provide any quality of service (QoS) guarantees to the receivers. For example, receivers with small SINR are not served for small P_{tx}. But very often a QoS requirement is associated with a data stream and it may be different for every receiver. We consider QoS measures which can be formulated or bounded in terms of the mean MSE (see Section 4.3) and give a concise overview of the related problems. The most common algorithms for their solution rely on the BC-MAC dualities from Section 4.5.

4.6.1 AWGN Broadcast Channel Model

In Section 4.3, the minimum MSE, i.e., $\overline{\mathrm{MMSE}}_k(\boldsymbol{P})$ (4.43), for the AWGN BC model turned out to be the most convenient measure among all presented alternatives.

One standard problem is to minimize the required resources, i.e., the total transmit power, while providing the desired QoS γ_k to every receiver. This *power minimization problem* reads

$$\min_{\boldsymbol{P}} \|\boldsymbol{P}\|_\mathrm{F}^2 \quad \text{s.t.} \quad \overline{\mathrm{MMSE}}_k(\boldsymbol{P}) \le \gamma_k,\ k = 1, 2, \ldots, K. \tag{4.146}$$

Because of the one-to-one relation (4.43) with $\overline{\mathrm{SINR}}_k(\boldsymbol{P})$ (4.36), we can transform the QoS requirements γ_k on $\overline{\mathrm{MMSE}}_k(\boldsymbol{P})$ to its equivalent SINR requirement $\gamma_k^{\mathrm{sinr}} = 1/\gamma_k - 1$. There are two standard approaches for solving (4.146):

- The problem is relaxed to a semidefinite program [19], whose solution is shown to coincide with the argument of (4.146).
- The dual MAC problem is solved iteratively alternating between power allocation and beamforming at the receiver $\boldsymbol{U}^{\mathrm{mac}}$. The algorithm by Schubert and Boche [185] converges to the global optimum.

In the sequel, we focus on the second solution based on the MAC-BC duality.

Because of the MAC-BC duality in Theorem 4.1, we can formulate the dual (virtual) MAC problem

$$\min_{\boldsymbol{p}^{\mathrm{mac}},\boldsymbol{U}^{\mathrm{mac}}} \|\boldsymbol{p}^{\mathrm{mac}}\|_2^2 \quad \text{s.t.} \quad \overline{\mathrm{SINR}}_k^{\mathrm{mac}}(\boldsymbol{u}_k^{\mathrm{mac}}, \boldsymbol{p}^{\mathrm{mac}}) \geq \gamma_k^{\mathrm{sinr}}, \ k = 1, 2, \dots, K, \tag{4.147}$$

which can now be solved with the algorithm in [185]. Note that (4.146) may be infeasible, which is detected by the algorithm.

Alternatively, we can maximize the margin to the QoS requirement γ_k for given total power constraint P_{tx}

$$\begin{aligned} \min_{\boldsymbol{P},\gamma_0} \gamma_0 \quad \text{s.t.} \quad & \|\boldsymbol{P}\|_{\mathrm{F}}^2 \leq P_{\mathrm{tx}}, \\ & \overline{\mathrm{MMSE}}_k(\boldsymbol{P}) \leq \gamma_0 \gamma_k, \ k = 1, 2, \dots, K. \end{aligned} \tag{4.148}$$

If this problem is feasible, the optimum variable γ_0 describing the margin lies in $0 \leq \gamma_0 \leq 1$. Otherwise, the problem is infeasible. The solution is equivalent to minimizing the normalized performance of the *worst* receiver

$$\min_{\boldsymbol{P}} \max_{k \in \{1,\dots,K\}} \frac{\overline{\mathrm{MMSE}}_k(\boldsymbol{P})}{\gamma_k} \quad \text{s.t.} \quad \|\boldsymbol{P}\|_{\mathrm{F}}^2 \leq P_{\mathrm{tx}}. \tag{4.149}$$

which can be shown [185] to yield a balancing among the relative MSEs $\overline{\mathrm{MMSE}}_k(\boldsymbol{P})/\gamma_k = \gamma_0$. For unequal γ_k, its optimum differs from balancing $\overline{\mathrm{SINR}}_k(\boldsymbol{P})$.

The corresponding problem in the dual MAC is

$$\begin{aligned} \min_{\boldsymbol{p}^{\mathrm{mac}},\boldsymbol{U}^{\mathrm{mac}},\gamma_0} \gamma_0 \quad \text{s.t.} \quad & \|\boldsymbol{p}^{\mathrm{mac}}\|_2^2 \leq P_{\mathrm{tx}}, \\ & \overline{\mathrm{MMSE}}_k^{\mathrm{mac}}(\boldsymbol{u}_k^{\mathrm{mac}}, \boldsymbol{p}^{\mathrm{mac}}) \leq \gamma_0 \gamma_k, \ k = 1, 2, \dots, K \end{aligned} \tag{4.150}$$

with $0 \leq \gamma_0 \leq 1$ if feasible and $\overline{\mathrm{MMSE}}_k^{\mathrm{mac}}(\boldsymbol{u}_k^{\mathrm{mac}}, \boldsymbol{p}^{\mathrm{mac}}) = 1/(1 + \overline{\mathrm{SINR}}_k^{\mathrm{mac}}(\boldsymbol{u}_k^{\mathrm{mac}}, \boldsymbol{p}^{\mathrm{mac}}))$ (4.122). If $\gamma_k = \gamma$, balancing of $\overline{\mathrm{MMSE}}_k$ is identical to balancing of $\overline{\mathrm{SINR}}_k$. Therefore, the algorithm by [185] gives a solution for equal $\gamma_k = \gamma$, which yields the optimum beamforming vectors for the BC based on the dual MAC problem and computes the optimum power allocation in the BC in a final step. Alternatively, we can use the BC-MAC transformation from Section 4.5.1.

4.6.2 Incomplete CSI at Receivers: Common Training Channel

For incomplete CSI at the receivers, which results from a common training channel in the forward link (Sections 4.2.2), we proposed to model the receivers' signal processing capabilities by $\beta_k g_k(\boldsymbol{h}_k)$ in Section 4.5.2.

For S-CSI, the corresponding power minimization problem is

$$\min_{\boldsymbol{P},\{\beta_k\}_{k=1}^K} \|\boldsymbol{P}\|_\mathrm{F}^2 \quad \text{s.t.} \quad \mathrm{E}_{\boldsymbol{h}_k}[\mathrm{MSE}_k(\boldsymbol{P},\beta_k g_k(\boldsymbol{h}_k);\boldsymbol{h}_k)] \le \gamma_k,\ k=1,\ldots,K. \tag{4.151}$$

The constraints in the BC are coupled in $\boldsymbol{P}$ which complicates its direct solution.

The formulation in the dual MAC results in constraints which are decoupled in $\boldsymbol{g}_k^\mathrm{mac}$ ($\boldsymbol{g}_k^\mathrm{mac}$ is related to the kth column $\boldsymbol{p}_k$ of $\boldsymbol{P}$ via (4.145)). The problem in the dual MAC with $\boldsymbol{p}^\mathrm{mac} = [p_1^\mathrm{mac},\ldots,p_K^\mathrm{mac}]^\mathrm{T}$ is

$$\min_{\boldsymbol{p}^\mathrm{mac},\{\boldsymbol{g}_k^\mathrm{mac}\}_{k=1}^K} \|\boldsymbol{p}^\mathrm{mac}\|_2^2 \quad \text{s.t.}$$
$$\mathrm{E}_{\boldsymbol{h}_k^\mathrm{mac}}[\mathrm{MSE}_k^\mathrm{mac}(\boldsymbol{p}^\mathrm{mac},\boldsymbol{g}_k^\mathrm{mac};\boldsymbol{h}_k^\mathrm{mac})] \le \gamma_k,\ k=1,\ldots,K. \tag{4.152}$$

Because of the decoupled constraints in $\boldsymbol{g}_k^\mathrm{mac}$, the optimum receiver $\boldsymbol{g}_k^\mathrm{mac}$ is necessary to minimize $\|\boldsymbol{p}^\mathrm{mac}\|_2^2$ and we can substitute the constraints by the minimum[46]

$$\min_{\boldsymbol{g}_k^\mathrm{mac}} \mathrm{E}_{\boldsymbol{h}_k^\mathrm{mac}}[\mathrm{MSE}_k^\mathrm{mac}(\boldsymbol{p}^\mathrm{mac},\boldsymbol{g}_k^\mathrm{mac};\boldsymbol{h}_k^\mathrm{mac})] =$$
$$= 1 - p_k^{\mathrm{mac},2}\mathrm{E}_{\boldsymbol{h}_k^\mathrm{mac}}[\boldsymbol{h}_k^\mathrm{mac}]^\mathrm{H}\left(\boldsymbol{I}_M + \sum_{i=1}^K p_i^{\mathrm{mac},2}\boldsymbol{R}_{\boldsymbol{h}_i^\mathrm{mac}}\right)^{-1}\mathrm{E}_{\boldsymbol{h}_k^\mathrm{mac}}[\boldsymbol{h}_k^\mathrm{mac}] \tag{4.153}$$

which is achieved for

$$\boldsymbol{g}_k^{\mathrm{mac},\mathrm{T}} = \mathrm{E}_{\boldsymbol{h}_k^\mathrm{mac}}[\boldsymbol{h}_k^\mathrm{mac}]^\mathrm{H}\left(\boldsymbol{I}_M + \sum_{i=1}^K p_i^{\mathrm{mac},2}\boldsymbol{R}_{\boldsymbol{h}_i^\mathrm{mac}}\right)^{-1} p_k^\mathrm{mac}.$$

With the argumentation from [236, App. II], we identify an interference function as defined by Yates [239]

[46] The dual MAC channel $\boldsymbol{h}_k^\mathrm{mac}$ for the common training channel defined in (4.145) is *not* zero-mean.

$$\mathrm{J}_k\left(\{p_i^{\mathrm{mac},2}\}_{i=1}^K;\gamma_k\right) = \frac{1-\gamma_k}{\mathrm{E}_{\boldsymbol{h}_k^{\mathrm{mac}}}[\boldsymbol{h}_k^{\mathrm{mac}}]^{\mathrm{H}}\left(\boldsymbol{I}_M + \sum_{i=1}^K p_i^{\mathrm{mac},2}\boldsymbol{R}_{\boldsymbol{h}_i^{\mathrm{mac}}}\right)^{-1}\mathrm{E}_{\boldsymbol{h}_k^{\mathrm{mac}}}[\boldsymbol{h}_k^{\mathrm{mac}}]}. \tag{4.154}$$

The MAC problem (4.152) after optimization w.r.t. $\{\boldsymbol{g}_k^{\mathrm{mac}}\}_{k=1}^K$ reduces to the power allocation problem

$$\min_{\boldsymbol{p}^{\mathrm{mac}}} \|\boldsymbol{p}^{\mathrm{mac}}\|_2^2 \quad \text{s.t.} \quad \mathrm{J}_k\left(\{p_i^{\mathrm{mac},2}\}_{i=1}^K;\gamma_k\right) \leq p_k^{\mathrm{mac},2},\ k=1,\ldots,K. \tag{4.155}$$

This is a classical problem in the literature.

Because all constraints are satisfied with equality in the optimum, the optimum power allocation is the solution of the K nonlinear equations

$$\mathrm{J}_k\left(\{p_i^{\mathrm{mac},2}\}_{i=1}^K;\gamma_k\right) = p_k^{\mathrm{mac},2}. \tag{4.156}$$

Yates [239] showed that the fixed point iteration on this equation converges to a unique fixed point and minimizes the total power $\|\boldsymbol{p}^{\mathrm{mac}}\|_2^2$ if the problem is feasible: Based on $\{p_i^{\mathrm{mac},(m-1)}\}_{i=1}^K$ we compute $p_k^{\mathrm{mac},(m),2} = \mathrm{J}_k(\{p_i^{\mathrm{mac}(m-1),2}\}_{i=1}^K;\gamma_k)$ in the mth iteration. But its convergence is slow and, for example, an approach similar to [185] as in [7] is recommended.

Similar to the AWGN BC model, we can also maximize the MSE margin in the BC channel under a total power constraint

$$\begin{aligned}\min_{\boldsymbol{P},\{\beta_k\}_{k=1}^K,\gamma_0} \gamma_0 \quad \text{s.t.} \quad & \|\boldsymbol{P}\|_{\mathrm{F}}^2 \leq P_{\mathrm{tx}}, \\ & \mathrm{E}_{\boldsymbol{h}_k}[\mathrm{MSE}_k(\boldsymbol{P},\beta_k \mathrm{g}_k(\boldsymbol{h}_k);\boldsymbol{h}_k)] \leq \gamma_0\gamma_k,\ k=1,\ldots,K.\end{aligned} \tag{4.157}$$

In the dual MAC the dual problem is

$$\begin{aligned}\min_{\boldsymbol{p}^{\mathrm{mac}},\boldsymbol{G}^{\mathrm{mac}},\gamma_0} \gamma_0 \quad \text{s.t.} \quad & \|\boldsymbol{p}^{\mathrm{mac}}\|_2^2 \leq P_{\mathrm{tx}}, \\ & \mathrm{E}_{\boldsymbol{h}_k^{\mathrm{mac}}}[\mathrm{MSE}_k^{\mathrm{mac}}(\boldsymbol{p}^{\mathrm{mac}},\boldsymbol{g}_k^{\mathrm{mac}};\boldsymbol{h}_k^{\mathrm{mac}})] \leq \gamma_0\gamma_k,\ k=1,\ldots,K\end{aligned}$$

with $0 \leq \gamma_0 \leq 1$ for feasibility. Again, the algorithm for its solution is in analogy to [185] (see [7] for details).

Alternatively to QoS constraints in terms of MSE expressions we can employ SINR expressions such as in (4.62) and (4.67). In Example 4.3, we illustrate the difference between both criteria. But the *optimum precoders* turn out to be identical in an interesting case despite the different values of the underlying SINR expressions: For $\boldsymbol{\mu_h} = \boldsymbol{0}$, the power minimization problem in analogy to (4.151) and the SINR-balancing in analogy to (4.157) for identical $\gamma_k = \gamma$ yield the *identical* precoders $\boldsymbol{P}$ no matter if the constraints are based on $\overline{\mathrm{SINR}}_k^{\mathrm{icsi}}(\boldsymbol{P})$ (4.62) or $\overline{\mathrm{SINR}}_k^{\mathrm{phase}}(\boldsymbol{P})$ (4.67). This can be proved investigating the eigenvectors of the matrix in the power balancing step in [185] and [36], which are identical for both SINR definitions.

Chapter 5
Nonlinear Precoding with Partial Channel State Information

Introduction to Chapter

Linear precoding with independent coding of the receivers' data does not achieve the capacity of the *vector broadcast channel* (BC) for C-CSI at the transmitter. It is achieved by successive encoding with *dirty paper coding* [232]. At high SNR, the achievable sum rate of linear precoding has the same slope (multiplexing gain) as dirty paper coding (e.g., [32, 106]). But there is an additional power offset, which increases with the ratio of the number of receivers to the number of transmit antennas K/M [106]. Perhaps surprisingly, the sum capacity of the broadcast channel with non-cooperative receivers is asymptotically (high transmit power) identical to the capacity of the same channel, but with cooperating receivers (point-to-point multiple-input multiple-output (MIMO) channel).

For dirty paper coding, the data streams for all receivers are *jointly* and sequentially coded: The interference from previously encoded data streams is known causally at the transmitter; thus, this knowledge can be used to code the next data stream such that its achievable data rate is the *same* as if the interference from previously coded data streams was not present (Writing on dirty paper [39]). Some practical aspects and an algorithm with performance close to capacity, which relies on dirty paper coding, is discussed in [210]. An implementation of writing on dirty paper for a single-input single-output (SISO) system with interference known at the transmitter is given by Erez and ten Brink [65].

The nonlinear precoding schemes considered in this chapter attempt to fill the performance gap between linear precoding and dirty paper coding, but with a smaller computational complexity than dirty paper coding. They are suboptimum, because they do not code the data sequence in time, i.e., they are "one-dimensional" [241].

The most prominent representative is *Tomlinson-Harashima precoding* (THP), which is proposed for the broadcast channel by [82, 241, 71]; originally, it was invented for mitigating intersymbol interference [213, 89]. In

[241] it is considered as the one-dimensional implementation of dirty paper coding. The first approach aims at a complete cancelation of the interference (zero-forcing) [71], which does not consider the variance of the receivers' noise. Joham et al.[109, 115, 108] propose a mean square error (MSE) based optimization, which includes optimization of the precoding order, a formulation for frequency selective channels based on finite impulse response filters, and optimization of the latency time. It is extended to a more general receiver model [100, 102] and to multiple antennas at the receivers [145]. An efficient implementation can be achieved using the symmetrically permuted Cholesky factorization, whose computational complexity is of the same (cubic) order as for linear MSE-based precoding [126, 127]. Optimization with quality service constraints is introduced in [187] for SINR and in [145] based on MSE. Duality between the multiple access channel (MAC) and the broadcast channel is proved in [230] for SINR and in [145] for MSE.

In [93] a more general structure of the precoder is introduced: A perturbation signal is optimized which is element of a multidimensional (in space) rectangular lattice. According to [184] we refer to this structure as *vector precoding*. As for THP, zero-forcing precoders were discussed first and the regularization which is typical for MSE-based schemes was introduced in an ad-hoc fashion. Typically, these zero-forcing precoders are outperformed in bit error rate (BER) by the less complex MMSE THP [184] for small to medium SNR. Finally, Schmidt et al.[184] introduce minimum MSE (MMSE) vector precoding. Contrary to THP, the general vector precoding achieves the full diversity order M of the system [209]. The computational complexity is determined by the algorithm for the search on the lattice; a suboptimum implementation requires a quartic order [237]. In [174], THP is shown to be a special case of MMSE vector precoding and a transition between both is proposed which allows for a further reduction of complexity.

For more information about important related issues such as lattices, shaping, lattice codes, and the capacity of a modulo channel see [69, 70] (they are not a prerequisite in the sequel).

The effect of small errors in CSI on zero-forcing THP is analyzed in [10]. But a zero-forcing design is known to be more sensitive to parameter errors than a regularized or MMSE optimization. Moreover, the uncertainty in CSI at the transmitter may be large because of the mobility in the wireless system which leads to a time-variant channel. It can be large enough that a robust optimization (Appendix C) is advantageous.

A robust optimization of zero-forcing THP is introduced by the author in [98], which includes MMSE prediction of the channel. It is generalized to MMSE THP in [57]. Both approaches are based on the Bayesian paradigm (Appendix C) and assume implicitly the same CSI at the transmitter as at the receiver. But the CSI is highly asymmetric, which has to be modeled appropriately. In [50, 49] the receivers' CSI is taken into account explicitly in the optimization which also yields a THP solution for S-CSI. For the definition of different categories of CSI see Table 4.1.

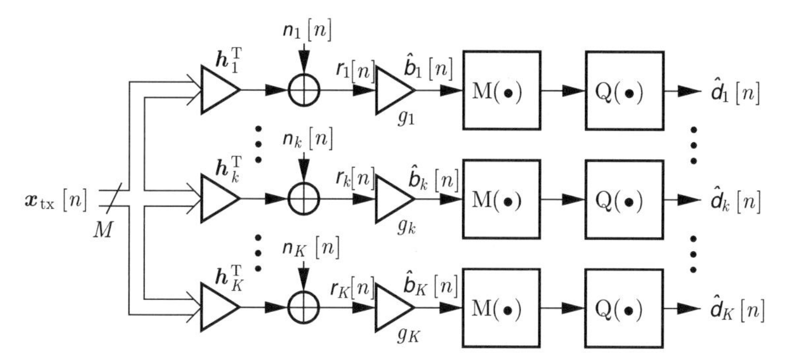

Fig. 5.1 Broadcast channel with nonlinear receivers, which include a modulo operator $\mathrm{M}(\hat{b}_k[n])$ (Figure 5.3).

Besides the development for the broadcast channel, THP based on P-CSI for *cooperative* receivers is treated in [135] for a SISO system. A heuristic solution is proposed for MIMO in [72]; for S-CSI another approach is given in [190].

Outline of Chapter

First, we derive the structure of the nonlinear vector precoder in Section 5.1. This includes the derivation of MMSE vector precoding and its relation to MMSE THP for C-CSI as an example. Optimization of nonlinear precoding with P-CSI requires new performance measures, which we introduce in Section 5.2 using the receiver models in analogy to linear precoding with P-CSI (Section 4.3). Conceptually, we also distinguish between vector precoding and THP. For these performance measures, we derive iterative algorithms minimizing the overall system efficiency, e.g., the sum MSE, and discuss the results for different examples (Section 5.3). For S-CSI, nonlinear precoding can be understood as nonlinear beamforming for communications. Precoding for the training channel to provide the receivers with the necessary CSI is addressed in Section 5.4. A detailed discussion of the performance for P-CSI and S-CSI is given in Section 5.5, which is based on the uncoded bit error rate (BER).

5.1 From Vector Precoding to Tomlinson-Harashima Precoding

To enable nonlinear precoding we have to introduce nonlinear receivers. The following modified receiver architecture (Figure 5.1) generates the necessary

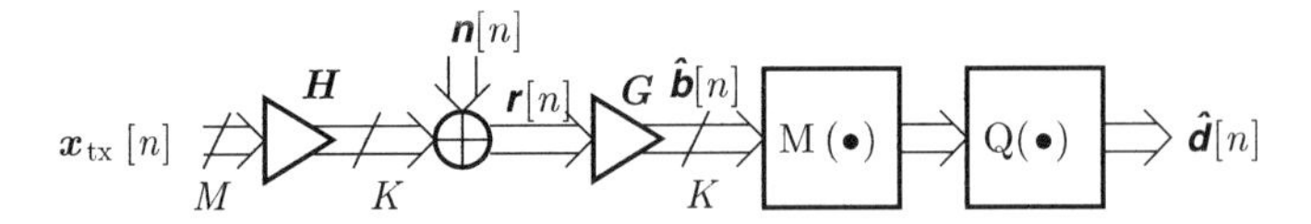

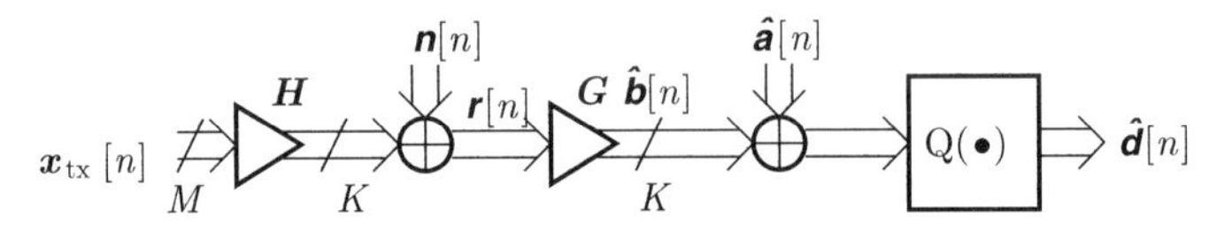

Fig. 5.2 Broadcast channel with nonlinear receivers in matrix-vector notation and representation of modulo operation by an auxiliary vector $\hat{\boldsymbol{a}}[n] \in \mathbb{L}^K$.

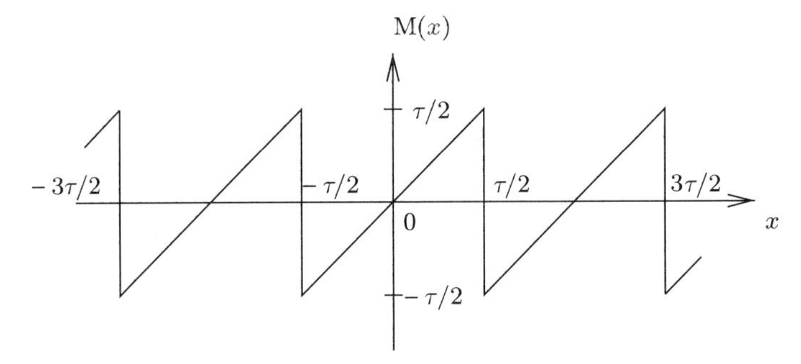

Fig. 5.3 Graph of the modulo-operator $\mathrm{M}(x)$ for real-valued arguments $x \in \mathbb{R}$.

additional degrees of freedom, which can be exploited for nonlinear precoding:[1]

- A modulo operator $\mathrm{M}(\hat{b}_k[n])$ with parameter τ (Figure 5.3 plots the function for real-valued input) is applied to the output of the linear processing (scaling) $g_k \in \mathbb{C}$ at the receivers. This nonlinearity of the receivers, which creates a modulo channel to every receiver, offers additional degrees of freedom for precoder design.[2]
- A nearest-neighbor quantizer $\mathrm{Q}(x) = \operatorname{argmin}_{y \in \mathbb{D}} |x - y|^2$ serves as a simple example[3] for the decision process on the transmitted data. It delivers hard decisions $\hat{d}_k[n] \in \mathbb{D}$ on the transmitted symbols $\boldsymbol{d}[n] = [d_1, [n], d_2[n], \ldots, d_K[n]]^{\mathrm{T}} \in \mathbb{D}^K$. It is the ML detector for uncoded data in $\mathbb{D}$ and a linear channel (without modulo operator) with $g_k = 1/(\boldsymbol{h}_k^{\mathrm{T}} \boldsymbol{p}_k)$ as in Figure 4.3, but it is not the optimum detector for the nonlinear modulo channel anymore due to the non-Gaussianity of the noise after the modulo operator. To simplify the receiver architecture we live with its suboptimal-

[1] The presentation in this section follows closely that of Joham et al.[114, 113].

[2] An information theoretic discussion of the general modulo channel compared to a linear AWGN channel is given in [70, 42].

[3] The decoding of coded data is not considered here.

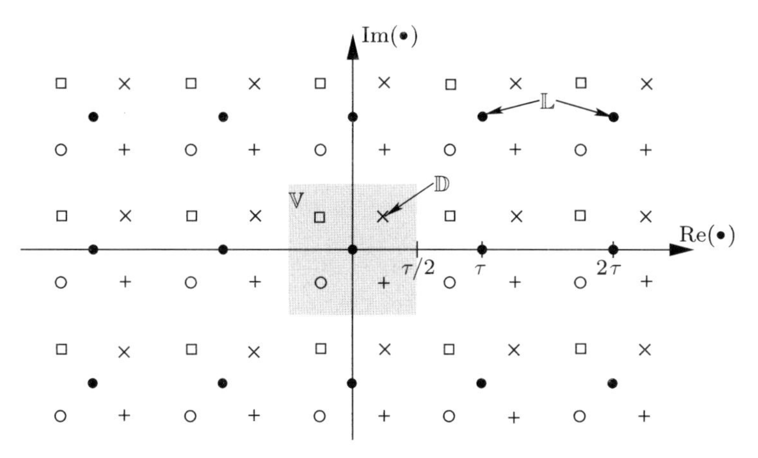

Fig. 5.4 Voronoi region $\mathbb{V}$ (shaded area), lattice $\mathbb{L}$ (filled circles), QPSK constellation $\mathbb{D}$ with variance one (4 different markers), and periodically repeated QPSK constellation for perturbations $a \in \mathbb{L}$.

ity. Contrary to the modulo operation, the quantizer will not influence our precoder design.[4]

The transmit signal $\boldsymbol{x}_{\mathrm{tx}}[n] \in \mathbb{C}^M$ is a nonlinear function of the modulated data $\boldsymbol{d}[n] \in \mathbb{D}^K$. To restrict the functional structure of the nonlinear precoder and to introduce a parameterization, we have to analyze the properties of the modulo operators at the receivers generating a nonlinear modulo channel.

The modulo operation (Figure 5.3) on $x \in \mathbb{C}$ can also be expressed as

$$\mathrm{M}(x) = x - \tau \left\lfloor \frac{\mathrm{Re}(x)}{\tau} + \frac{1}{2} \right\rfloor - \mathrm{j}\tau \left\lfloor \frac{\mathrm{Im}(x)}{\tau} + \frac{1}{2} \right\rfloor \in \mathbb{V}, \tag{5.1}$$

where the floor operation $\lfloor \bullet \rfloor$ gives the largest integer smaller than or equal to the argument and $\mathbb{V} = \{x + \mathrm{j}y | x, y \in [-\tau/2, \tau/2)\}$ is the fundamental Voronoi region[5] of the lattice $\mathbb{L} = \tau\mathbb{Z} + \mathrm{j}\tau\mathbb{Z}$ corresponding to the modulo operation (Figure 5.4).

In the matrix-vector notation of the BC (Figure 5.2), the modulo operation on a vector $\mathbf{M}(\hat{\boldsymbol{b}}[n])$ is defined element-wise, i.e., $\mathbf{M}(\hat{\boldsymbol{b}}[n]) = [\mathrm{M}(\hat{b}_1[n]), \ldots, \mathrm{M}(\hat{b}_K[n])]^{\mathrm{T}} \in \mathbb{V}^K$. As in Figure 5.2, the modulo operation can also be formulated with an auxiliary vector $\hat{\boldsymbol{a}}[n] \in \mathbb{L}^K$ on the $2K$-dimensional lattice $\mathbb{L}^K = \tau\mathbb{Z}^K + \mathrm{j}\tau\mathbb{Z}^K$.

The representation with an auxiliary variable $a \in \mathbb{L}$ shows an important property of the modulo operator: For any $x \in \mathbb{C}$, we have $\mathrm{M}(x + a) = \mathrm{M}(x)$

[4] In Figure 4.3 the symbol estimates $\hat{d}_k[n] \in \mathbb{C}$ are complex numbers; the quantizer is omitted, because the linear precoder based on MSE is optimized for the linear model.

[5] We introduce the common terminology for completeness, but do not require a deeper knowledge of lattice theory.

if $a \in \mathbb{L}$. The precoder can generate any signal $\hat{b}_k[n] + a_k[n]$ at the receiver with $a_k[n] \in \mathbb{L}$, which yields the same same signal at the modulo operator's output as $\hat{b}_k[n]$. For example, the estimates of the transmitted data $\hat{d}_k[n]$ are identical to $d_k[n]$, if the input of the modulo operator is $\hat{b}_k[n] = d_k[n] + a_k[n]$ for $a_k[n] \in \mathbb{L}$. Because of this property, it is sufficient to optimize the precoder such that $\hat{b}_k[n] \approx d_k[n] + a_k[n]$ (virtual desired signal [113]) is achieved. These *additional degrees of freedom* can be exploited by the precoder.

This observation allows for an optimization of the precoder based on the linear part of the models in Figures 5.1 and 5.2. This is not necessary, but simplifies the design process considerably, since the alternative is an optimization including the nonlinearity of the modulo operator or, additionally, the quantizer.

The corresponding structure of the transmitter is given in Figure 5.5, which is referred to as *vector precoder* [184]. Its degrees of freedom are a linear precoder $\boldsymbol{T} \in \mathbb{C}^{M \times K}$ and a perturbation vector $\boldsymbol{a}[n] \in \mathbb{L}^K$ on the lattice.

To define the receivers it remains to specify the parameter τ. It is chosen sufficiently large such that $\mathbb{D} \subset \mathbb{V}$. Because of this restriction on τ, any input from $\mathbb{D}$ is not changed by the modulo operation and the nearest-neighbor quantizer $\mathrm{Q}(\bullet)$ is applicable, which we have introduced already. We choose τ such that the symbol constellation seen by the receiver in $\hat{b}_k[n]$ without noise ($c_{n_k} = 0$) and distortions (e.g., $\boldsymbol{GHT} = \boldsymbol{I}_K$), is a periodic extension of $\mathbb{D}$ on $\mathbb{C}$ (Figure 5.4): For QPSK symbols ($\mathbb{D} = \{\exp(\mathrm{j}\mu\pi/4), \mu \in \{-3, -1, 1, 3\}\}$), it requires $\tau = 2\sqrt{2}$ and $\tau = 8/\sqrt{10}$ for rectangular 16QAM symbols with variance $c_{d_k} = 1$.

In the remainder of this section, we introduce the system models for nonlinear precoding (vector precoding and Tomlinson-Harashima precoding) of the data for the example of C-CSI at the transmitter and identical receivers $\boldsymbol{G} = \beta \boldsymbol{I}_K$, which allows for an explicit solution [184].

Vector Precoding

Optimization of the nonlinear precoder with structure depicted in Figure 5.5 differs from Chapter 4 in two aspects:

- If $\boldsymbol{d}[n]$ are known at the transmitter for the whole data block, i.e., for $n \in \{1, 2, \ldots, N_\mathrm{d}\}$, it is not necessary to use a stochastic model for $\boldsymbol{d}[n]$: In Chapter 4 we assume $\boldsymbol{d}[n]$ to be a zero-mean stationary random sequence $\boldsymbol{d}[n]$ with covariance matrix $\boldsymbol{C}_{\boldsymbol{d}} = \boldsymbol{I}_K$; but the optimization problem can also be formulated for given $\{\boldsymbol{d}[n]\}_{n=1}^{N_\mathrm{d}}$.
- The nonlinearity of the receivers induced by the modulo operations is not modeled explicitly. The additional degrees of freedom available are exploited considering only the linear part of the model in Figure 5.2 and choosing $d_k[n] + a_k[n]$ as the desired signal for $\hat{b}_k[n]$.

The MSE for receiver k is defined between the input of the modulo operators $\hat{b}_k[n]$ and the desired signal $b_k[n] = d_k[n] + a_k[n]$

$$\mathrm{MSE}_k^{\mathrm{VP}}\left(\boldsymbol{T}, \{\boldsymbol{a}[n]\}_{n=1}^{N_\mathrm{d}}, \beta; \boldsymbol{h}_k\right) = \frac{1}{N_\mathrm{d}} \sum_{n=1}^{N_\mathrm{d}} \mathrm{E}_{n_k}\left[\left|\hat{b}_k[n] - b_k[n]\right|^2\right], \tag{5.2}$$

where the expectation is performed w.r.t. the noise $n_k[n]$ and the arithmetic mean considers the given data $\boldsymbol{d}[n]$. The dependency on $\{\boldsymbol{d}[n]\}_{n=1}^{N_\mathrm{d}}$ is not denoted explicitly.

To optimize the performance of the total system we minimize the sum MSE (similar to Section 4.3.4). The optimization problem for the precoder with a constraint on the average transmit power for the given data block is

$$\min_{\boldsymbol{T}, \{\boldsymbol{a}[n]\}_{n=1}^{N_\mathrm{d}}, \beta} \sum_{k=1}^{K} \mathrm{MSE}_k^{\mathrm{VP}}\left(\boldsymbol{T}, \{\boldsymbol{a}[n]\}_{n=1}^{N_\mathrm{d}}, \beta; \boldsymbol{h}_k\right) \quad \text{s.t.} \quad \mathrm{tr}\left[\boldsymbol{T}\tilde{\boldsymbol{R}}_{\boldsymbol{b}}\boldsymbol{T}^\mathrm{H}\right] \leq P_\mathrm{tx}, \tag{5.3}$$

where $\tilde{\boldsymbol{R}}_{\boldsymbol{b}} = \sum_{n=1}^{N_\mathrm{d}} \boldsymbol{b}[n]\boldsymbol{b}[n]^\mathrm{H}/N_\mathrm{d}$ is the sample correlation matrix of $\boldsymbol{b}[n]$. With

$$\sum_{k=1}^{K} \mathrm{MSE}_k^{\mathrm{VP}}\left(\boldsymbol{T}, \{\boldsymbol{a}[n]\}_{n=1}^{N_\mathrm{d}}, \beta; \boldsymbol{h}_k\right) = \frac{1}{N_\mathrm{d}} \sum_{n=1}^{N_\mathrm{d}} \mathrm{E}_{\boldsymbol{n}}\left[\left\|\hat{\boldsymbol{b}}[n] - \boldsymbol{b}[n]\right\|^2\right] \tag{5.4}$$

and $\hat{\boldsymbol{b}}[n] = \beta\boldsymbol{H}\boldsymbol{T}\boldsymbol{b}[n] + \beta\boldsymbol{n}[n]$ we obtain the explicit expression for the cost function

$$\begin{aligned}\sum_{k=1}^{K} \mathrm{MSE}_k^{\mathrm{VP}}\left(\boldsymbol{T}, \{\boldsymbol{a}[n]\}_{n=1}^{N_\mathrm{d}}, \beta; \boldsymbol{h}_k\right) &= \mathrm{tr}\left[\tilde{\boldsymbol{R}}_{\boldsymbol{b}}\right] + |\beta|^2 \mathrm{tr}[\boldsymbol{C}_{\boldsymbol{n}}] \\ &\quad + |\beta|^2 \mathrm{tr}\left[\boldsymbol{H}^\mathrm{H}\boldsymbol{H}\boldsymbol{T}\tilde{\boldsymbol{R}}_{\boldsymbol{b}}\boldsymbol{T}^\mathrm{H}\right] - 2\mathrm{Re}\left(\beta \mathrm{tr}\left[\boldsymbol{H}\boldsymbol{T}\tilde{\boldsymbol{R}}_{\boldsymbol{b}}\right]\right).\end{aligned} \tag{5.5}$$

Solving (5.3) for $\boldsymbol{T}$ and β first, we get (Appendix D.1)

$$\boxed{\begin{aligned}\boldsymbol{T} &= \beta^{-1}\left(\boldsymbol{H}^\mathrm{H}\boldsymbol{H} + \frac{\mathrm{tr}[\boldsymbol{C}_{\boldsymbol{n}}]}{P_\mathrm{tx}}\boldsymbol{I}_M\right)^{-1}\boldsymbol{H}^\mathrm{H} \\ &= \beta^{-1}\boldsymbol{H}^\mathrm{H}\left(\boldsymbol{H}\boldsymbol{H}^\mathrm{H} + \frac{\mathrm{tr}[\boldsymbol{C}_{\boldsymbol{n}}]}{P_\mathrm{tx}}\boldsymbol{I}_K\right)^{-1}.\end{aligned}} \tag{5.6}$$

The scaling is required to be

$$\beta^2 = \mathrm{tr}\left[\tilde{\boldsymbol{R}}_{\boldsymbol{b}}\boldsymbol{H}\left(\boldsymbol{H}^\mathrm{H}\boldsymbol{H} + \frac{\mathrm{tr}[\boldsymbol{C}_{\boldsymbol{n}}]}{P_\mathrm{tx}}\boldsymbol{I}_M\right)^{-2}\boldsymbol{H}^\mathrm{H}\right] \Big/ P_\mathrm{tx} \tag{5.7}$$

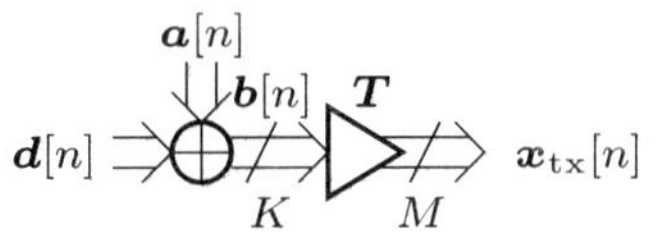

Fig. 5.5 Nonlinear precoding with a perturbation vector $\boldsymbol{a}[n] \in \tau\mathbb{L}^K + \mathrm{j}\tau\mathbb{L}^K$ on a $2K$ dimensional lattice for every $\boldsymbol{d}[n]$ before (linear) precoding with $\boldsymbol{T}$.

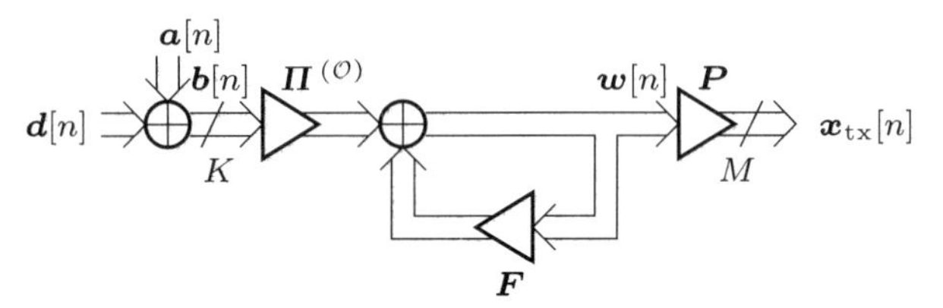

Fig. 5.6 Nonlinear precoding with equivalent representation of $\boldsymbol{T}$ as $\boldsymbol{T} = \boldsymbol{P}(\boldsymbol{I}_K - \boldsymbol{F})^{-1}\boldsymbol{\Pi}^{(\mathcal{O})}$.

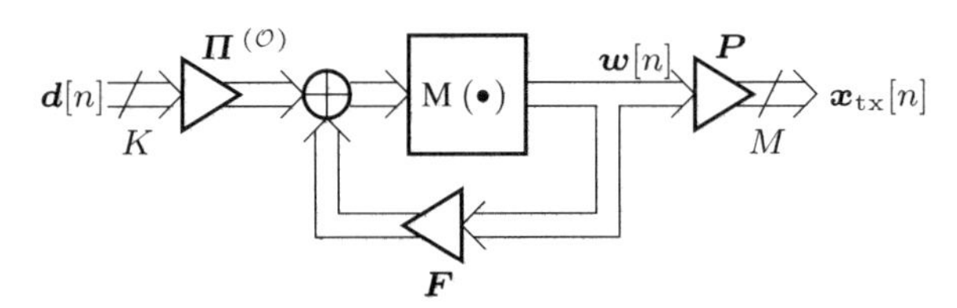

Fig. 5.7 Tomlinson-Harashima Precoding (THP) with ordering $\boldsymbol{\Pi}^{(\mathcal{O})}$: Equivalent representation of Figure 5.6 with modulo operator $\mathrm{M}(\bullet)$ instead of a perturbation vector $\boldsymbol{a}[n]$ for the restricted search on the lattice.

to satisfy the transmit power constraint. The minimum of (5.3) is a special case of (D.14) and reads (constraining $\boldsymbol{T}$ to meet the transmit power constraint)

$$\min_{\boldsymbol{T},\beta} \sum_{k=1}^{K} \mathrm{MSE}_k^{\mathrm{VP}}(\boldsymbol{T}, \{\boldsymbol{a}[n]\}_{n=1}^{N_\mathrm{d}}, \beta; \boldsymbol{h}_k) = \frac{\mathrm{tr}[\boldsymbol{C_n}]}{N_\mathrm{d} P_\mathrm{tx}} \sum_{n=1}^{N_\mathrm{d}} \|\boldsymbol{a}[n] + \boldsymbol{d}[n]\|_{\boldsymbol{\Phi}^{-1}}^2 \tag{5.8}$$

with weighting matrix $\boldsymbol{\Phi} = \boldsymbol{H}\boldsymbol{H}^\mathrm{H} + \frac{\mathrm{tr}[\boldsymbol{C_n}]}{P_\mathrm{tx}} \boldsymbol{I}_K$.

Optimization w.r.t. $\{\boldsymbol{a}[n]\}_{n=1}^{N_\mathrm{d}}$ is equivalent to a complete search on the $2K$-dimensional (real-valued) lattice generated by $\tau\boldsymbol{\Phi}^{-1/2}$ [184], i.e., the search for the closest point to $-\boldsymbol{\Phi}^{-1/2}\boldsymbol{d}[n]$. The computational complexity for this search is exponential in the number of receivers K, but it can be approximated resulting in a complexity of $\mathrm{O}(K^4)$ [174].

Restricted Search on Lattice

We define the Cholesky decomposition of the permuted matrix

$$\boldsymbol{\Pi}^{(\mathcal{O})}\boldsymbol{\Phi}^{-1}\boldsymbol{\Pi}^{(\mathcal{O}),\mathrm{T}} = \boldsymbol{L}^{\mathrm{H}}\boldsymbol{\Delta}\boldsymbol{L} \tag{5.9}$$

with permutation matrix $\boldsymbol{\Pi}^{(\mathcal{O})} = \sum_{i=1}^{K} \boldsymbol{e}_i\boldsymbol{e}_{p_i}^{\mathrm{T}}$, $\boldsymbol{\Pi}^{(\mathcal{O}),\mathrm{T}} = \boldsymbol{\Pi}^{(\mathcal{O}),-1}$, diagonal matrix $\boldsymbol{\Delta} = \mathbf{diag}[\delta_1, \ldots, \delta_K]$, and lower triangular matrix $\boldsymbol{L}$ with unit diagonal. It is referred to as symmetrically permuted Cholesky factorization [84, p. 148], which can be applied to find a good permutation $\mathcal{O}$ efficiently [126, 127]. The permutation is defined by the tuple $\mathcal{O} = [p_1, p_2, \ldots, p_K]$ with $p_i \in \{1, 2, \ldots, K\}$ (the data of the the p_ith receiver is coded in the ith step).

The transformation (5.6) can be factorized as[6]

$$\boldsymbol{T} = \boldsymbol{P}(\boldsymbol{I}_K - \boldsymbol{F})^{-1}\boldsymbol{\Pi}^{(\mathcal{O})} \tag{5.10}$$

with

$$\boldsymbol{P} = \beta^{-1}\boldsymbol{H}^{\mathrm{H}}\boldsymbol{\Pi}^{(\mathcal{O}),\mathrm{T}}\boldsymbol{L}^{\mathrm{H}}\boldsymbol{\Delta} \tag{5.11}$$

$$\boldsymbol{F} = \boldsymbol{I}_K - \boldsymbol{L}^{-1}. \tag{5.12}$$

This factorization yields an equivalent representation of the vector precoder depicted in Figure 5.6.

With (5.9) minimization of (5.8) reads

$$\boxed{\min_{\boldsymbol{a}[n]\in\mathbb{L}^K} \left\|\boldsymbol{\Delta}^{1/2}\boldsymbol{L}\boldsymbol{\Pi}^{(\mathcal{O})}\left(\boldsymbol{d}[n] + \boldsymbol{a}[n]\right)\right\|_2^2.} \tag{5.13}$$

An approach to reduce the computational complexity of solving (5.13) is a restriction of the search on the lattice. We use the heuristic presented in [175], which will lead us to THP. It exploits the lower triangular structure of $\boldsymbol{L}$: The kth element of $\boldsymbol{L}\boldsymbol{\Pi}^{(\mathcal{O})}\boldsymbol{a}[n]$ depends only on the first k elements of $\boldsymbol{a}^{(\mathcal{O})}[n] = \boldsymbol{\Pi}^{(\mathcal{O})}\boldsymbol{a}[n]$. We start with $i = 1$ and optimize for the ith element of $\boldsymbol{a}^{(\mathcal{O})}[n]$, which corresponds to the p_ith element of $\boldsymbol{a}[n]$ due to $\boldsymbol{e}_{p_i} = \boldsymbol{\Pi}^{(\mathcal{O}),\mathrm{T}}\boldsymbol{e}_i$. It is obtained minimizing the norm of the ith component of the Euclidean norm in (5.13)

$$\boxed{a_{p_i}[n] = \operatorname*{argmin}_{\bar{a}_{p_i}[n]\in\mathbb{L}} \left|\boldsymbol{e}_i^{\mathrm{T}}\boldsymbol{\Delta}^{1/2}\boldsymbol{L}\left(\boldsymbol{\Pi}^{(\mathcal{O})}\boldsymbol{d}[n] + \boldsymbol{a}^{(\mathcal{O})}[n]\right)\right|^2} \tag{5.14}$$

given the $i-1$ previous elements

$$\boldsymbol{a}^{(\mathcal{O})}[n] = [a_{p_1}[n], \ldots, a_{p_{i-1}}[n], \bar{a}_{p_i}[n], *, \ldots, *]^{\mathrm{T}}. \tag{5.15}$$

The last $K - i$ elements are arbitrary due to the structure of $\boldsymbol{L}$.

We would like to express (5.14) by a modulo operation. Using (5.12), we can rewrite

[6] This factorization is a heuristic, which is inspired by the structure of the solution for THP (D.68) [108, 115].

$$
\begin{aligned}
\boldsymbol{e}_i^{\mathrm{T}} \boldsymbol{\Delta}^{1/2} \boldsymbol{L} &= \delta_i^{1/2} \boldsymbol{e}_i^{\mathrm{T}} (\boldsymbol{I}_K - \boldsymbol{F})^{-1} = \delta_i^{1/2} \boldsymbol{e}_i^{\mathrm{T}} (\boldsymbol{I}_K - \boldsymbol{F} + \boldsymbol{F}) (\boldsymbol{I}_K - \boldsymbol{F})^{-1} \\
&= \delta_i^{1/2} \boldsymbol{e}_i^{\mathrm{T}} \left(\boldsymbol{I}_K + \boldsymbol{F} (\boldsymbol{I}_K - \boldsymbol{F})^{-1} \right)
\end{aligned}
\tag{5.16}
$$

Defining the signal $\boldsymbol{w}[n] = (\boldsymbol{I}_K - \boldsymbol{F})^{-1} \boldsymbol{\Pi}^{(\mathcal{O})} \boldsymbol{b}[n]$ problem (5.14) can be expressed in terms of a quantizer $\mathrm{Q}_{\mathbb{L}}(x)$ to the point on the 2-dimensional lattice $\mathbb{L}$ closest to $x \in \mathbb{C}$:

$$
\begin{aligned}
a_{p_i}[n] &= \underset{\bar{a}_{p_i}[n] \in \mathbb{L}}{\operatorname{argmin}}\, \delta_i \left| d_{p_i}[n] + \bar{a}_{p_i}[n] + \boldsymbol{e}_i^{\mathrm{T}} \boldsymbol{F} \boldsymbol{w}[n] \right|^2 \\
&= -\mathrm{Q}_{\mathbb{L}} \left(d_{p_i}[n] + \boldsymbol{e}_i^{\mathrm{T}} \boldsymbol{F} \boldsymbol{w}[n] \right).
\end{aligned}
\tag{5.17}
$$

Note that $w_i[n] = d_{p_i}[n] + \bar{a}_{p_i}[n] + \boldsymbol{e}_i^{\mathrm{T}} \boldsymbol{F} \boldsymbol{w}[n]$, which is the output of the modulo operator.

Since the first row of $\boldsymbol{F}$ is zero, we have $b_{p_1}[n] = d_{p_1}[n]$ for $i = 1$, i.e., $a_{p_1}[n] = 0$. Using the relation $\mathrm{Q}_{\mathbb{L}}(x) = x - \mathrm{M}(x)$ the perturbation signal is

$$
\boldsymbol{\Pi}^{(\mathcal{O})} \boldsymbol{a}[n] = \mathbf{M}\left(\boldsymbol{\Pi}^{(\mathcal{O})} \boldsymbol{d}[n] + \boldsymbol{F} \boldsymbol{w}[n] \right) - \boldsymbol{\Pi}^{(\mathcal{O})} \boldsymbol{d}[n] - \boldsymbol{F} \boldsymbol{w}[n] \tag{5.18}
$$

and the output of feedback loop in Figure 5.6 can be expressed by the modulo operation

$$
\boldsymbol{w}[n] = \boldsymbol{\Pi}^{(\mathcal{O})} \boldsymbol{b}[n] + \boldsymbol{F} \boldsymbol{w}[n] = \mathbf{M}\left(\boldsymbol{\Pi}^{(\mathcal{O})} \boldsymbol{d}[n] + \boldsymbol{F} \boldsymbol{w}[n] \right). \tag{5.19}
$$

This relation yields the representation of the vector precoder in Figure 5.7, which is the well known structure of THP for the BC. It is equivalent to Figures 5.5 and 5.6 if the lattice search for $\boldsymbol{a}[n]$ is restricted as in (5.14).

Note that the sample correlation matrix $\tilde{\boldsymbol{R}}_{\boldsymbol{w}} = \sum_{n=1}^{N_\mathrm{d}} \boldsymbol{w}[n] \boldsymbol{w}[n]^{\mathrm{H}} / N_\mathrm{d}$ of $\boldsymbol{w}[n]$ is not diagonal in general because of

$$
\tilde{\boldsymbol{R}}_{\boldsymbol{w}} = (\boldsymbol{I}_K - \boldsymbol{F})^{-1} \tilde{\boldsymbol{R}}_{\boldsymbol{b}} (\boldsymbol{I}_K - \boldsymbol{F})^{-\mathrm{H}}. \tag{5.20}
$$

Traditionally, a diagonal $\tilde{\boldsymbol{R}}_{\boldsymbol{w}}$ is assumed to optimize THP [69, 108]: It is obtained modeling the outputs a priori as statistically independent random variables which are uniformly distributed on $\mathbb{V}$ [69]. With this assumption, THP based on sum MSE and $\boldsymbol{G} = \beta \boldsymbol{I}_K$ yields a solution for the filters $\boldsymbol{P}$ and $\boldsymbol{F}$ [115, 108] (see also Appendix D.6) which is identical to (5.11) and (5.12) except for a different scaling of $\boldsymbol{P}$.[7] In the original derivation of THP the scaling β is obtained from the stochastic model for $\boldsymbol{w}[n]$. In the alternative derivation of THP above this assumption is not necessary and we have a solution for which the transmit power constraint for the given data holds exactly.

[7] This can be proved comparing (5.11) and (5.12) with the solutions given in [127].

Example 5.1. For large P_{tx}, the linear filter converges to $\boldsymbol{T} = \beta^{-1}\boldsymbol{H}^{\text{H}}(\boldsymbol{H}\boldsymbol{H}^{\text{H}})^{-1}$, i.e., it inverts the channel (zero-forcing). The lattice search is performed for the weight $\boldsymbol{\Phi} = \boldsymbol{H}\boldsymbol{H}^{\text{H}}$, i.e., the skewness of the resulting lattice is determined by the degree of orthogonality between the rows of $\boldsymbol{H}$.

This high-SNR solution can also be determined from (5.3) including the additional zero-forcing constraint $\beta\boldsymbol{H}\boldsymbol{T} = \boldsymbol{I}_K$. Applying the constraint to the cost function (5.5) directly, the problem is equivalent to minimizing β^2 under the same constraints (transmit power and zero-forcing): The necessary scaling β of $\boldsymbol{T}$ to satisfy the transmit power constraint is reduced; therefore the noise variance in $\hat{\boldsymbol{b}}[n]$ (Figure 5.1) at the receivers is minimized. The minimum scaling is $\beta^2 = \text{tr}[\tilde{\boldsymbol{R}}_{\boldsymbol{b}}(\boldsymbol{H}\boldsymbol{H}^{\text{H}})^{-1}]/P_{\text{tx}}$. Note that without a transmit power constraint, this solution is equivalent to minimizing the required transmit power (see also [93]).

For small P_{tx}, the diagonal matrix in $\boldsymbol{\Phi}$ dominates and we have approximately $\boldsymbol{L} = \boldsymbol{I}_K$ and $\boldsymbol{F} = \boldsymbol{0}$. Thus, for diagonal $\boldsymbol{\Phi}$ the lattice search yields $\boldsymbol{a}[n] = \boldsymbol{0}$, i.e., *linear* precoding of the data. □

Optimization of Precoding Order

Optimization of the precoding order would require to check all $K!$ possibilities for $\mathcal{O}$ to minimize (5.13) for the restricted search on the lattice (5.14), which is rather complex. The standard suboptimum solution is based on a greedy algorithm which can be formulated in terms of the symmetrically permuted Cholesky factorization (5.9). The first step of this factorization (5.9) (cf. [84, p. 148] and [127]) with an *upper* triangular $\boldsymbol{L}^{\text{H}}$ reads

$$\boldsymbol{\Pi}^{(\mathcal{O})}\boldsymbol{\Phi}_K^{-1}\boldsymbol{\Pi}^{(\mathcal{O}),\text{T}} = \begin{bmatrix} \boldsymbol{\Psi}_K & \boldsymbol{l}_K \\ \boldsymbol{l}_K^{\text{H}} & \delta_K \end{bmatrix} = \begin{bmatrix} \boldsymbol{I}_{K-1} & \boldsymbol{l}_K/\delta_K \\ \boldsymbol{0} & 1 \end{bmatrix} \begin{bmatrix} \boldsymbol{\Phi}_{K-1}^{-1} & \boldsymbol{0} \\ \boldsymbol{0} & \delta_K \end{bmatrix} \begin{bmatrix} \boldsymbol{I}_{K-1} & \boldsymbol{0} \\ \boldsymbol{l}_K^{\text{H}}/\delta_K & 1 \end{bmatrix} \tag{5.21}$$

with $\boldsymbol{\Phi}_K^{-1} = \boldsymbol{\Phi}^{-1}$ and $\boldsymbol{\Phi}_{K-1}^{-1} = \boldsymbol{\Psi}_K - \boldsymbol{l}_K\boldsymbol{l}_K^{\text{H}}/\delta_K$.

In the first iteration we are free to choose p_K in the permutation matrix $\boldsymbol{\Pi}^{(\mathcal{O})}$, i.e., δ_K corresponding to the maximum or minimum element on the diagonal of $\boldsymbol{\Phi}^{-1}$. Thus, in the first step we decide for the data stream to be precoded last. We decide to choose p_K as the index of the smallest diagonal element of $\boldsymbol{\Phi}^{-1}$, which has the smallest contribution (5.14) in (5.13): The best data stream is precoded last. Similarly, we proceed with $\boldsymbol{\Phi}_{K-1}^{-1}$.

This choice of the best data stream becomes clear in the following example:

Example 5.2. The first column of $\boldsymbol{P}$ (5.11) is $\boldsymbol{p}_1 = \beta^{-1}\boldsymbol{h}_{p_1}^*\delta_1$, which is a matched filter w.r.t. the channel of the p_1th receiver precoded first. On the other hand, the last column of $\boldsymbol{P}$ is $\boldsymbol{p}_K = \beta^{-1}\boldsymbol{H}^{\text{H}}\boldsymbol{\Pi}^{(\mathcal{O}),\text{T}}\boldsymbol{L}^{\text{H}}\boldsymbol{e}_K\delta_K$. At high P_{tx} we have $\boldsymbol{H}\boldsymbol{p}_K = \beta^{-1}\boldsymbol{\Pi}^{(\mathcal{O}),\text{T}}\boldsymbol{L}^{-1}\boldsymbol{e}_K = \beta^{-1}\boldsymbol{e}_{p_K}$, because

$$\begin{aligned} \boldsymbol{p}_K &= \beta^{-1}\boldsymbol{H}^{\mathrm{H}}\boldsymbol{\Pi}^{(\mathcal{O}),\mathrm{T}}\boldsymbol{L}^{\mathrm{H}}\boldsymbol{\Delta}\boldsymbol{e}_K \\ &= \beta^{-1}\boldsymbol{H}^{\mathrm{H}}\boldsymbol{\Phi}^{-1}\boldsymbol{\Pi}^{(\mathcal{O}),\mathrm{T}}\boldsymbol{L}^{-1}e_K \\ &= \beta^{-1}\boldsymbol{H}^{\mathrm{H}}(\boldsymbol{H}\boldsymbol{H}^{\mathrm{H}})^{-1}\boldsymbol{\Pi}^{(\mathcal{O}),\mathrm{T}}\boldsymbol{L}^{-1}e_K \end{aligned}$$

from (5.9) for $K \leq M$: Interference from the p_Kth data stream to all other receivers is avoided; the performance gain in MSE is largest if we choose the receiver with the best channel to be precoded last, i.e., it creates the smallest contribution to the MSE. □

Because the precoding order is determined while computing the symmetrically permuted Cholesky factorization (5.9), the computational complexity, which results from computing the inverse $\boldsymbol{\Phi}^{-1}$ and the Cholesky factorization (5.9), is of order $\mathrm{O}(K^3)$. Note that this is the same order of complexity as for linear precoding in Section 4.4 (see [127] for details).

5.2 Performance Measures for Partial Channel State Information

In the previous section we introduce the system models for nonlinear precoding (vector precoding in Figure 5.5 and THP in Figure 5.7) assuming C-CSI at the transmitter and identical receivers $\boldsymbol{G} = \beta\boldsymbol{I}_K$. In the sequel we consider the general case:

- We assume P-CSI at the transmitter (Table 4.1) based on the model of the reverse link training channel in a TDD system (Section 4.1) and
- different scaling at the receivers $\boldsymbol{G} = \mathbf{diag}[g_1, \ldots, g_K]$.

The MSE criterion for optimizing the scaling g_k prior to the modulo operation at the receivers is shown to be *necessary* to reach the Shannon capacity for the AWGN channel on a single link with a modulo receiver (modulo channel) as in Figure 5.1 [42]. Of course, although the MSE criterion is necessary for optimizing the receivers, a capacity achieving precoding (coding) is not based on MSE.

Because only few information theoretic results about the broadcast channel with modulo receivers or P-CSI at the transmitter are known, the following argumentation is pursued from the point of view of estimation theory based on MSE. The steps are motivated by linear precoding in Section 4.3.

We first discuss the cost functions for optimization of nonlinear precoding with P-CSI assuming the optimum MMSE scaling at the receivers (Section 5.2.1). To obtain a performance criterion which can be given explicitly, we assume a scaled matched filter at the receivers in Section 5.2.2 for optimization of precoding. Optimization of vector precoding (Figure 5.5) and THP (Figure 5.7) are treated separately.

Note that a relation between vector precoding and THP, which we present in Section 5.1, has not been found yet for the general cost functions below.

5.2.1 MMSE Receivers

Vector Precoding

Given $\boldsymbol{h}_k$, the MSE between the signal at the input of the modulo operator $\hat{b}_k[n] = g_k \boldsymbol{h}_k^{\mathrm{T}} \boldsymbol{T} \boldsymbol{b}[n] + g_k n_k[n]$ (Figure 5.1) and the (virtual) desired signal $b_k[n] = d_k[n] + a_k[n]$ (Figure 5.5) is

$$\begin{aligned}\mathrm{MSE}_k^{\mathrm{VP}}\Big(\boldsymbol{T}, \{\boldsymbol{a}[n]\}_{n=1}^{N_\mathrm{d}}, g_k; \boldsymbol{h}_k\Big) &= \frac{1}{N_\mathrm{d}} \sum_{n=1}^{N_\mathrm{d}} \mathrm{E}_{n_k}\left[\left|\hat{b}_k[n] - b_k[n]\right|^2\right] \\ &= \tilde{r}_{b_k} + |g_k|^2 c_{n_k} + |g_k|^2 \boldsymbol{h}_k^{\mathrm{T}} \boldsymbol{T} \tilde{\boldsymbol{R}}_{\boldsymbol{b}} \boldsymbol{T}^{\mathrm{H}} \boldsymbol{h}_k^* - \\ &\quad - 2\mathrm{Re}\big(g_k \boldsymbol{h}_k^{\mathrm{T}} \boldsymbol{T} \tilde{\boldsymbol{r}}_{\boldsymbol{b}b_k}\big)\end{aligned} \tag{5.22}$$

with sample crosscorrelation $\tilde{\boldsymbol{r}}_{\boldsymbol{b}b_k} = \tilde{\boldsymbol{R}}_{\boldsymbol{b}} \boldsymbol{e}_k$ and second order moment $\tilde{r}_{b_k} = [\tilde{\boldsymbol{R}}_{\boldsymbol{b}}]_{k,k}$ given the data $\{\boldsymbol{d}[n]\}_{n=1}^{N_\mathrm{d}}$ to be transmitted ($\tilde{\boldsymbol{R}}_{\boldsymbol{b}} = \sum_{n=1}^{N_\mathrm{d}} \boldsymbol{b}[n]\boldsymbol{b}[n]^{\mathrm{H}}/N_\mathrm{d}$).

Because the receivers are assumed to have perfect knowledge of the necessary system parameters, given this knowledge the minimum of (5.22) w.r.t. g_k is

$$\begin{aligned}\mathrm{MMSE}_k^{\mathrm{VP}}\Big(\boldsymbol{T}, \{\boldsymbol{a}[n]\}_{n=1}^{N_\mathrm{d}}; \boldsymbol{h}_k\Big) &= \min_{g_k} \mathrm{MSE}_k^{\mathrm{VP}}\Big(\boldsymbol{T}, \{\boldsymbol{a}[n]\}_{n=1}^{N_\mathrm{d}}, g_k; \boldsymbol{h}_k\Big) \\ &= \tilde{r}_{b_k} - \frac{|\boldsymbol{h}_k^{\mathrm{T}} \boldsymbol{T} \tilde{\boldsymbol{r}}_{\boldsymbol{b}b_k}|^2}{\boldsymbol{h}_k^{\mathrm{T}} \boldsymbol{T} \tilde{\boldsymbol{R}}_{\boldsymbol{b}} \boldsymbol{T}^{\mathrm{H}} \boldsymbol{h}_k^* + c_{n_k}}.\end{aligned} \tag{5.23}$$

The MMSE receiver

$$g_k = \mathrm{g}_k^{\mathrm{mmse}}(\boldsymbol{h}_k) = \frac{(\boldsymbol{h}_k^{\mathrm{T}} \boldsymbol{T} \tilde{\boldsymbol{r}}_{\boldsymbol{b}b_k})^*}{\boldsymbol{h}_k^{\mathrm{T}} \boldsymbol{T} \tilde{\boldsymbol{R}}_{\boldsymbol{b}} \boldsymbol{T}^{\mathrm{H}} \boldsymbol{h}_k^* + c_{n_k}} \tag{5.24}$$

is realizable if $\boldsymbol{h}_k^{\mathrm{T}} \boldsymbol{T} \tilde{\boldsymbol{r}}_{\boldsymbol{b}b_k}$ can be estimated at the receivers which is discussed in Section 5.4. The denominator in (5.24) is the variance of the receive signal c_{r_k}.

Optimization of vector precoding is based on the conditional mean estimate of (5.23) (compare Section 4.3.1 and Appendix C)

$$\boxed{\mathrm{E}_{\boldsymbol{h}_k|\boldsymbol{y}_\mathrm{T}}\left[\mathrm{MMSE}_k^{\mathrm{VP}}\Big(\boldsymbol{T}, \{\boldsymbol{a}[n]\}_{n=1}^{N_\mathrm{d}}; \boldsymbol{h}_k\Big)\right].} \tag{5.25}$$

An explicit expression is not known due to the ratio of quadratic forms with correlated non-zero mean random vectors $\boldsymbol{h}_k$ in (5.23).[8]

Tomlinson-Harashima Precoding

The MSE for THP (Figure 5.7) is derived from its equivalent representation in Figure 5.6. The nonlinearity of the precoder, i.e., the modulo operators or, equivalently, the restricted lattice search, is described by a stochastic model for the output $\boldsymbol{w}[n]$ of the modulo operators at the transmitter. In [69] it is shown that the outputs can be approximated as statistically independent, zero-mean, and uniformly distributed on $\mathbb{V}$, which results in the covariance matrix

$$\boldsymbol{C}_{\boldsymbol{w}} = \mathbf{diag}\left[c_{w_1}, \ldots, c_{w_K}\right] \tag{5.26}$$

with $c_{w_1} = c_{d_1} = 1$ and $c_{w_k} = \tau^2/6, k > 1$ ($w_k[n]$ is the kth element of $\boldsymbol{w}[n]$). It is a good model for C-CSI and zero-forcing THP (MMSE THP at high P_{tx}) [108, 69], where the approximation becomes more exact as the constellation size of the modulation alphabet increases. We make this assumption also for the MSE criterion and P-CSI, where it is ad-hoc. Its validity is discussed in Section 5.5.

The feedback filter $\boldsymbol{F}$ is constrained to be lower triangular with zero diagonal to be realizable; this requirement is often referred to as *spatial causality*.

For the kth receiver, the MSE is defined as

$$\mathrm{MSE}_k^{\mathrm{THP}}(\boldsymbol{P}, \tilde{\boldsymbol{f}}_{o_k}, g_k; \boldsymbol{h}_k) = \mathrm{E}_{\boldsymbol{w}, n_k}\left[\left|\hat{b}_k[n] - b_k[n]\right|^2\right], \tag{5.27}$$

where the expectation is now also taken w.r.t. $\boldsymbol{w}[n]$. We define the inverse function of $k = p_i$ as $i = o_k$, i.e., the kth receiver's data stream is the o_kth data stream to be precoded, and the response of the permutation matrix $\boldsymbol{e}_{o_k} = \boldsymbol{\Pi}^{(\mathcal{O})}\boldsymbol{e}_k$ and $\boldsymbol{e}_{p_i} = \boldsymbol{\Pi}^{(\mathcal{O}),\mathrm{T}}\boldsymbol{e}_i$. Moreover, the ith row of $\boldsymbol{F}$ is denoted $\tilde{\boldsymbol{f}}_i^{\mathrm{T}}$.

With the expressions for the signals related to the kth receiver's data stream

$$\boldsymbol{b}[n] = \boldsymbol{\Pi}^{(\mathcal{O}),\mathrm{T}}(\boldsymbol{I}_K - \boldsymbol{F})\boldsymbol{w}[n] \tag{5.28}$$

$$b_k[n] = (\boldsymbol{e}_{o_k}^{\mathrm{T}} - \tilde{\boldsymbol{f}}_{o_k}^{\mathrm{T}})\boldsymbol{w}[n] \tag{5.29}$$

$$\hat{b}_k[n] = g_k \boldsymbol{h}_k^{\mathrm{T}} \boldsymbol{P} \boldsymbol{w}[n] + g_k n_k[n] \tag{5.30}$$

the MSE (5.27) reads

[8] This is also the case for linear precoding (cf. (4.31)).

$$
\begin{aligned}
\mathrm{MSE}_k^{\mathrm{THP}}(\boldsymbol{P}, \tilde{\boldsymbol{f}}_{o_k}, g_k; \boldsymbol{h}_k) &= |g_k|^2 c_{n_k} + \mathrm{E}_{\boldsymbol{w}}\left[\left|\left(g_k \boldsymbol{h}_k^{\mathrm{T}} \boldsymbol{P} - \boldsymbol{e}_{o_k}^{\mathrm{T}} + \tilde{\boldsymbol{f}}_{o_k}^{\mathrm{T}}\right) \boldsymbol{w}[n]\right|^2\right] \\
&= |g_k|^2 c_{n_k} + \sum_{i=1}^{K} c_{w_i} \left|g_k \boldsymbol{h}_k^{\mathrm{T}} \boldsymbol{p}_i - \boldsymbol{e}_{o_k}^{\mathrm{T}} \boldsymbol{e}_i + f_{o_k,i}\right|^2
\end{aligned}
\tag{5.31}
$$

with $f_{o_k,i} = [\boldsymbol{F}]_{o_k,i}$ ($\boldsymbol{n}[n]$ and $\boldsymbol{w}[n]$ are uncorrelated). Before treating the case of P-CSI, we discuss the MSE for C-CSI as an example.

Example 5.3. Because transmitter and receiver have the same CSI (C-CSI), we can start optimizing for $\tilde{\boldsymbol{f}}_{o_k}$ keeping $\mathcal{O}$, $\boldsymbol{P}$, and g_k fixed. Minimizing (5.31) w.r.t. $\tilde{\boldsymbol{f}}_{o_k}$ under the constraint on $\boldsymbol{F}$ to be lower triangular with zero diagonal, we obtain $f_{o_k,i} = -g_k \boldsymbol{h}_k^{\mathrm{T}} \boldsymbol{p}_i, i < o_k$, and $f_{o_k,i} = 0, i \geq o_k$. The feedback filter for the kth data stream removes all interference from previously encoded data streams in the cost function, because

$$
\tilde{\boldsymbol{f}}_{o_k}^{\mathrm{T}} = \left[-g_k \boldsymbol{h}_k^{\mathrm{T}} \left[\boldsymbol{p}_1, \ldots, \boldsymbol{p}_{o_k-1}\right], \boldsymbol{0}_{1\times(K-o_k+1)}\right]. \tag{5.32}
$$

The minimum is

$$
\begin{aligned}
\min_{\tilde{\boldsymbol{f}}_{o_k}} \mathrm{MSE}_k^{\mathrm{THP}}(\boldsymbol{P}, \tilde{\boldsymbol{f}}_{o_k}, g_k; \boldsymbol{h}_k) = c_{w_{o_k}} + |g_k|^2 c_{n_k} + \sum_{i=o_k}^{K} c_{w_i} |g_k|^2 |\boldsymbol{h}_k^{\mathrm{T}} \boldsymbol{p}_i|^2 \\
- 2 c_{w_{o_k}} \mathrm{Re}\left(g_k \boldsymbol{h}_k^{\mathrm{T}} \boldsymbol{p}_{o_k}\right).
\end{aligned}
\tag{5.33}
$$

Minimization w.r.t. the receiver g_k yields

$$
g_k = c_{w_{o_k}} \left(\boldsymbol{h}_k^{\mathrm{T}} \boldsymbol{p}_{o_k}\right)^* \Big/ \left(\sum_{i=o_k}^{K} c_{w_i} |\boldsymbol{h}_k^{\mathrm{T}} \boldsymbol{p}_i|^2 + c_{n_k}\right). \tag{5.34}
$$

The minimum can be expressed in terms of

$$
\mathrm{SINR}_k^{\mathrm{THP}}(\boldsymbol{P}; \boldsymbol{h}_k) = \frac{c_{w_{o_k}} |\boldsymbol{h}_k^{\mathrm{T}} \boldsymbol{p}_{o_k}|^2}{\sum_{i=o_k+1}^{K} c_{w_i} |\boldsymbol{h}_k^{\mathrm{T}} \boldsymbol{p}_i|^2 + c_{n_k}} \tag{5.35}
$$

as

$$
\min_{\tilde{\boldsymbol{f}}_{o_k}, g_k} \mathrm{MSE}_k^{\mathrm{THP}}(\boldsymbol{P}, \tilde{\boldsymbol{f}}_{o_k}, g_k; \boldsymbol{h}_k) = c_{w_{o_k}} \frac{1}{1 + \mathrm{SINR}_k^{\mathrm{THP}}(\boldsymbol{P}; \boldsymbol{h}_k)}. \tag{5.36}
$$

Therefore, the interference from other data streams, which has been removed by $\tilde{\boldsymbol{f}}_{o_k}^{\mathrm{T}}$ already, is not considered anymore for optimization of the forward filter $\boldsymbol{P}$.

Because of the structure of $\mathrm{SINR}_k^{\mathrm{THP}}(\boldsymbol{P}; \boldsymbol{h}_k)$, any value of $\mathrm{SINR}_k^{\mathrm{THP}}$ is achievable if P_{tx} is arbitrary, i.e., $P_{\mathrm{tx}} \to \infty$ results in arbitrarily large

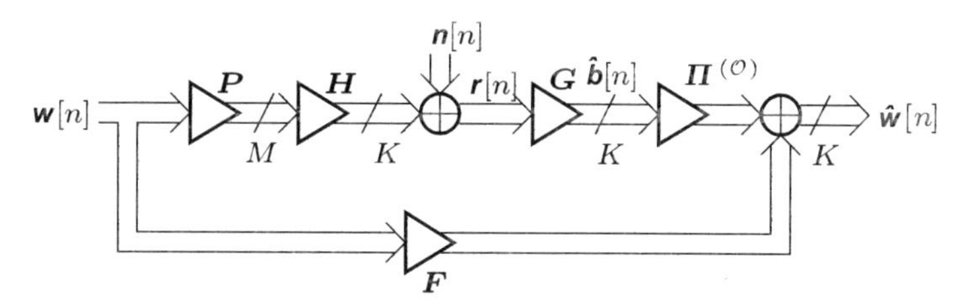

Fig. 5.8 Representation of THP with identical estimation error $\hat{\boldsymbol{w}}[n]-\boldsymbol{w}[n]=\hat{\boldsymbol{b}}[n]-\boldsymbol{b}[n]$ as in Figure 5.6.

$\mathrm{SINR}_k^{\mathrm{THP}}$ — even for $K > M$ (e.g., [186]). This is impossible for linear precoding. To verify this, consider $\mathrm{SINR}_k^{\mathrm{THP}}(\boldsymbol{P};\boldsymbol{h}_k)$ for $o_k = K$, i.e., the receiver to be precoded last: Any $\mathrm{SINR}_k^{\mathrm{THP}}(\boldsymbol{P};\boldsymbol{h}_k) = c_{w_K}|\boldsymbol{h}_k^{\mathrm{T}}\boldsymbol{p}_K|^2/c_{n_k}$ can be ensured only by choosing the necessary power allocation $\|\boldsymbol{p}_K\|_2$; For receiver $o_i = K-1$, we have $\mathrm{SINR}_i^{\mathrm{THP}}(\boldsymbol{P};\boldsymbol{h}_i) = c_{w_{K-1}}|\boldsymbol{h}_i^{\mathrm{T}}\boldsymbol{p}_{K-1}|^2/(c_{w_K}|\boldsymbol{h}_i^{\mathrm{T}}\boldsymbol{p}_K|^2+c_{n_i})$ and any value can be achieved choosing $\|\boldsymbol{p}_{K-1}\|_2$.

The SINR definition (5.35) also parameterizes the sum-capacity of the BC with C-CSI, which can be achieved by dirty paper precoding [230] (see also [186]).[9] We will see that it is generally not possible to derive a relation similar to (5.36) for P-CSI from MSE.

The representation of THP given in Figure 5.8 is equivalent to Figure 5.6 regarding the error signal used for the MSE definition: the error $\hat{w}_k[n]-w_k[n]$ in Figure 5.8 is identical to $\hat{b}_k[n]-b_k[n]$ based on Figure 5.6. Figure 5.8 gives the intuition for interpreting the solution for THP with C-CSI: The precoder is optimized as if $\boldsymbol{w}[n]$ was partially known to the receiver via the lower triangular filter $\boldsymbol{F}$. □

For P-CSI at the transmitter, we have to optimize g_k first because of the asymmetry of CSI in the system. Intuitively, we expect that a complete cancelation of the interference from already precoded data streams is impossible for P-CSI: From Figure 5.8 we note that $\boldsymbol{F}$ cannot cancel the interference completely, if it does not know the channel $\boldsymbol{H}$ perfectly.

The MMSE receiver is given by

$$g_k = \mathrm{g}_k^{\mathrm{mmse}}(\boldsymbol{h}_k) = \frac{\left(\boldsymbol{h}_k^{\mathrm{T}}\boldsymbol{P}\boldsymbol{C}_{\boldsymbol{w}}\left(\boldsymbol{e}_{o_k}-\tilde{\boldsymbol{f}}_{o_k}^*\right)\right)^*}{c_{r_k}} \tag{5.37}$$

with variance of the receive signal $c_{r_k} = \sum_{i=1}^K c_{w_i}|\boldsymbol{h}_k^{\mathrm{T}}\boldsymbol{p}_i|^2 + c_{n_k}$. The THP parameters are optimized based on the conditional mean estimate

$$\mathrm{E}_{\boldsymbol{h}_k|\boldsymbol{y}_{\mathrm{T}}}\left[\mathrm{MMSE}_k^{\mathrm{THP}}(\boldsymbol{P},\tilde{\boldsymbol{f}}_{o_k};\boldsymbol{h}_k)\right] \tag{5.38}$$

[9] $\{c_{w_i}\|\boldsymbol{p}_i\|_2^2\}_{i=1}^K$ are the variances of independent data streams, which have to be optimized to achieve sum capacity.

of the minimum MSE achieved by the kth receiver's processing (5.37)

$$\begin{aligned}\mathrm{MMSE}_k^{\mathrm{THP}}(\boldsymbol{P}, \tilde{\boldsymbol{f}}_{o_k}; \boldsymbol{h}_k) &= \min_{g_k} \mathrm{MSE}_k^{\mathrm{THP}}(\boldsymbol{P}, \tilde{\boldsymbol{f}}_{o_k}, g_k; \boldsymbol{h}_k) \\ &= \left(\boldsymbol{e}_{o_k}^{\mathrm{T}} - \tilde{\boldsymbol{f}}_{o_k}^{\mathrm{T}}\right)\left(\boldsymbol{C}_{\boldsymbol{w}} - \frac{\boldsymbol{C}_{\boldsymbol{w}}\boldsymbol{P}\boldsymbol{h}_k^*\boldsymbol{h}_k^{\mathrm{T}}\boldsymbol{P}\boldsymbol{C}_{\boldsymbol{w}}}{c_{r_k}}\right)\left(\boldsymbol{e}_{o_k} - \tilde{\boldsymbol{f}}_{o_k}^*\right). \end{aligned} \tag{5.39}$$

Although the conditional expectation cannot be computed in closed form, an explicit solution for $\tilde{\boldsymbol{f}}_{o_k}$ is possible (constraining its last $K - o_k + 1$ elements to be zero), because the matrix in the quadratic form (5.39) is independent of $\tilde{\boldsymbol{f}}_{o_k}$. But optimization w.r.t. $\boldsymbol{P}$ under the total power constraint $\mathrm{tr}[\boldsymbol{P}\boldsymbol{C}_{\boldsymbol{w}}\boldsymbol{P}^{\mathrm{H}}] \leq P_{\mathrm{tx}}$ remains difficult.

Example 5.4. For $\mathrm{rank}[\boldsymbol{C}_{\boldsymbol{h}_k}] = 1$, define $\boldsymbol{h}_k = x_k \boldsymbol{v}_k$ (compare Section 4.4.3). If $c_{n_k} = 0$, $c_{r_k} = |x_k|^2 \sum_{i=1}^{K} c_{w_i} |\boldsymbol{v}_k^{\mathrm{T}}\boldsymbol{p}_i|^2$ and the minimum MSE (5.39) becomes independent of x_k: The MMSE receiver (5.37) is inversely proportional to x_k, i.e., $\mathrm{g}_k^{\mathrm{mmse}}(\boldsymbol{h}_k) \propto 1/x_k$, and removes the uncertainty in CSI of the effective channel $\mathrm{g}_k^{\mathrm{mmse}}(\boldsymbol{h}_k)\boldsymbol{h}_k$, which is now completely known to the transmitter.

The minimum of the cost function (5.38) has the same form as for C-CSI

$$\min_{\tilde{\boldsymbol{f}}_{o_k}} \mathrm{MMSE}_k^{\mathrm{THP}}(\boldsymbol{P}, \tilde{\boldsymbol{f}}_{o_k}; \boldsymbol{h}_k) = c_{w_{o_k}} \frac{1}{1 + \overline{\mathrm{SIR}}_k^{\mathrm{THP}}(\boldsymbol{P})} \tag{5.40}$$

for $c_{n_k} = 0$ and SIR (signal-to-interference ratio) defined by

$$\overline{\mathrm{SIR}}_k^{\mathrm{THP}}(\boldsymbol{P}) = \frac{c_{w_{o_k}} |\boldsymbol{v}_k^{\mathrm{T}}\boldsymbol{p}_{o_k}|^2}{\sum_{i=o_k+1}^{K} c_{w_i} |\boldsymbol{v}_k^{\mathrm{T}}\boldsymbol{p}_i|^2}. \tag{5.41}$$

As for C-CSI, any $\overline{\mathrm{SIR}}_k^{\mathrm{THP}}(\boldsymbol{P})$ is feasible and the resulting MSE can be made arbitrarily small at the expense of an increasing P_{tx}. This is true even for $K > M$. □

5.2.2 Scaled Matched Filter Receivers

Because of the difficulties to obtain a tractable cost function based on the assumption of MMSE receivers, which are also motivated by results in information theory, we follow a similar strategy as for linear precoding in Section 4.3.2: The receivers g_k are modeled by scaled matched filters. This approximation is made to reduce the computational complexity. In case the *true* g_k applied at the receivers minimize the MSE (5.22) or (5.27), the scaled matched filters are only an approximation; this model mismatch results in a performance loss (Section 5.5).

Vector Precoding

For vector precoding (Figure 5.5), the scaled matched filter (compare to (4.47))

$$\mathrm{g}_k^{\mathrm{cor}}(\boldsymbol{h}_k) = \frac{(\boldsymbol{h}_k^{\mathrm{T}} \boldsymbol{T} \tilde{\boldsymbol{r}}_{\boldsymbol{b}b_k})^*}{\bar{c}_{r_k}} \tag{5.42}$$

with mean variance of the receive signal $\bar{c}_{r_k} = \mathrm{tr}[\boldsymbol{T}\tilde{\boldsymbol{R}}_{\boldsymbol{b}}\boldsymbol{T}^{\mathrm{H}}\boldsymbol{R}_{\boldsymbol{h}_k|\boldsymbol{y}_{\mathrm{T}}}^*] + c_{n_k}$ minimizes

$$\mathrm{COR}_k^{\mathrm{VP}}\Big(\boldsymbol{T}, \{\boldsymbol{a}[n]\}_{n=1}^{N_{\mathrm{d}}}, g_k; \boldsymbol{h}_k\Big) = \tilde{r}_{b_k} + |g_k|^2 \bar{c}_{r_k} - 2\mathrm{Re}\big(g_k \boldsymbol{h}_k^{\mathrm{T}} \boldsymbol{T} \tilde{\boldsymbol{r}}_{\boldsymbol{b}b_k}\big). \tag{5.43}$$

Maximizing $-\mathrm{COR}_k^{\mathrm{VP}}$ can be interpreted as maximization of the real part of the (sample) crosscorrelation $g_k \boldsymbol{h}_k^{\mathrm{T}} \boldsymbol{T} \tilde{\boldsymbol{r}}_{\boldsymbol{b}b_k}$ between $\hat{b}_k[n]$ and $b_k[n]$ with regularization of the norm $|g_k|^2$ and regularization parameter $\bar{c}_{r_k}$.

The minimum of (5.43) is

$$\mathrm{MCOR}_k^{\mathrm{VP}}\Big(\boldsymbol{T}, \{\boldsymbol{a}[n]\}_{n=1}^{N_{\mathrm{d}}}; \boldsymbol{h}_k\Big) = \tilde{r}_{b_k} - \frac{|\boldsymbol{h}_k^{\mathrm{T}} \boldsymbol{T} \tilde{\boldsymbol{r}}_{\boldsymbol{b}b_k}|^2}{\bar{c}_{r_k}}, \tag{5.44}$$

which may be negative. But its conditional mean estimate

$$\boxed{\mathrm{E}_{\boldsymbol{h}_k|\boldsymbol{y}_{\mathrm{T}}}\Big[\mathrm{MCOR}_k^{\mathrm{VP}}\Big(\boldsymbol{T}, \{\boldsymbol{a}[n]\}_{n=1}^{N_{\mathrm{d}}}; \boldsymbol{h}_k\Big)\Big] = \tilde{r}_{b_k} - \frac{\tilde{\boldsymbol{r}}_{\boldsymbol{b}b_k}^{\mathrm{H}} \boldsymbol{T}^{\mathrm{H}} \boldsymbol{R}_{\boldsymbol{h}_k|\boldsymbol{y}_{\mathrm{T}}}^* \boldsymbol{T} \tilde{\boldsymbol{r}}_{\boldsymbol{b}b_k}}{\bar{c}_{r_k}} \geq 0} \tag{5.45}$$

is always positive. It is the transmitter's estimate of the receiver's cost resulting from the scalar matched filter (5.42).

Tomlinson-Harashima Precoding

With the same model for $\boldsymbol{w}[n]$ as in Section 5.2.1, we define the cost function (Figure 5.6)

$$\begin{aligned}\mathrm{COR}_k^{\mathrm{THP}}(\boldsymbol{P}, \tilde{\boldsymbol{f}}_{o_k}, g_k; \boldsymbol{h}_k) = |g_k|^2 \bar{c}_{r_k} + \left(\boldsymbol{e}_{o_k}^{\mathrm{T}} - \tilde{\boldsymbol{f}}_{o_k}^{\mathrm{T}}\right) \boldsymbol{C}_{\boldsymbol{w}} \left(\boldsymbol{e}_{o_k} - \tilde{\boldsymbol{f}}_{o_k}^*\right) \\ - 2\mathrm{Re}\Big(g_k \boldsymbol{h}_k^{\mathrm{T}} \boldsymbol{P} \boldsymbol{C}_{\boldsymbol{w}} \left(\boldsymbol{e}_{o_k} - \tilde{\boldsymbol{f}}_{o_k}^*\right)\Big),\end{aligned} \tag{5.46}$$

which yields the matched filter receiver (compare to the MMSE receiver in (5.37))

$$\mathrm{g}_k^{\mathrm{cor}}(\boldsymbol{h}_k) = \frac{\left(\boldsymbol{h}_k^{\mathrm{T}} \boldsymbol{P} \boldsymbol{C}_{\boldsymbol{w}} \left(\boldsymbol{e}_{o_k} - \tilde{\boldsymbol{f}}_{o_k}^*\right)\right)^*}{\bar{c}_{r_k}} \tag{5.47}$$

scaled by the mean variance of the receive signal.

$$\bar{c}_{r_k} = \sum_{i=1}^{K} c_{w_i} \boldsymbol{p}_i^{\mathrm{H}} \boldsymbol{R}_{\boldsymbol{h}_k|\boldsymbol{y}_{\mathrm{T}}}^* \boldsymbol{p}_i + c_{n_k} = \mathrm{tr}\left[\boldsymbol{P}\boldsymbol{C}_{\boldsymbol{w}}\boldsymbol{P}^{\mathrm{H}}\boldsymbol{R}_{\boldsymbol{h}_k|\boldsymbol{y}_{\mathrm{T}}}^*\right] + c_{n_k}. \tag{5.48}$$

The conditional mean estimate of the minimum $\mathrm{MCOR}_k^{\mathrm{THP}}(\boldsymbol{P}, \tilde{\boldsymbol{f}}_{o_k}; \boldsymbol{h}_k) = \min_{g_k} \mathrm{COR}_k^{\mathrm{THP}}(\boldsymbol{P}, \tilde{\boldsymbol{f}}_{o_k}, g_k; \boldsymbol{h}_k)$ is

$$\begin{aligned}
&\mathrm{E}_{\boldsymbol{h}_k|\boldsymbol{y}_{\mathrm{T}}}\left[\mathrm{MCOR}_k^{\mathrm{THP}}(\boldsymbol{P}, \tilde{\boldsymbol{f}}_{o_k}; \boldsymbol{h}_k)\right] = \\
&= \left(\boldsymbol{e}_{o_k}^{\mathrm{T}} - \tilde{\boldsymbol{f}}_{o_k}^{\mathrm{T}}\right)\left(\boldsymbol{C}_{\boldsymbol{w}} - \frac{\boldsymbol{C}_{\boldsymbol{w}}\boldsymbol{P}\boldsymbol{R}_{\boldsymbol{h}_k|\boldsymbol{y}_{\mathrm{T}}}^*\boldsymbol{P}\boldsymbol{C}_{\boldsymbol{w}}}{\bar{c}_{r_k}}\right)\left(\boldsymbol{e}_{o_k} - \tilde{\boldsymbol{f}}_{o_k}^*\right).
\end{aligned} \tag{5.49}$$

Minimization w.r.t. the feedback filter parameters $\tilde{\boldsymbol{f}}_{o_k}$ can be performed explicitly, which results in an explicit cost function for optimizing $\boldsymbol{P}$ and $\mathcal{O}$.

To obtain $\tilde{\boldsymbol{f}}_{o_k}$ we minimize $\mathrm{E}_{\boldsymbol{h}_k|\boldsymbol{y}_{\mathrm{T}}}[\mathrm{COR}_k^{\mathrm{THP}}(\boldsymbol{P}, \tilde{\boldsymbol{f}}_{o_k}, \mathrm{g}_k(\boldsymbol{h}_k); \boldsymbol{h}_k)]$ for an arbitrary $\mathrm{g}_k(\boldsymbol{h}_k)$, which yields

$$\tilde{\boldsymbol{f}}_{o_k}^{\mathrm{T}} = -\mathrm{E}_{\boldsymbol{h}_k|\boldsymbol{y}_{\mathrm{T}}}\left[\mathrm{g}_k(\boldsymbol{h}_k)\boldsymbol{h}_k^{\mathrm{T}}\right]\boldsymbol{P}_{o_k} \tag{5.50}$$

with $\boldsymbol{P}_{o_k} = [\boldsymbol{p}_1, \ldots, \boldsymbol{p}_{o_k-1}, \boldsymbol{0}_{M\times(K-o_k+1)}]$. We apply the optimum receiver (5.47) parameterized in terms of $\tilde{\boldsymbol{f}}_{o_k}$ to (5.50) and solve for the feedback filter

$$\tilde{\boldsymbol{f}}_{o_k}^{\mathrm{T}} = -c_{w_{o_k}} \boldsymbol{p}_{o_k}^{\mathrm{H}} \boldsymbol{R}_{\boldsymbol{h}_k|\boldsymbol{y}_{\mathrm{T}}}^* \boldsymbol{P}_{o_k} \left(\bar{c}_{r_k}\boldsymbol{I}_K - \boldsymbol{C}_{\boldsymbol{w}}\boldsymbol{P}^{\mathrm{H}}\boldsymbol{R}_{\boldsymbol{h}_k|\boldsymbol{y}_{\mathrm{T}}}^*\boldsymbol{P}_{o_k}\right)^{-1}.$$

By means of the matrix inversion lemma (Appendix A.2.2)[10], it can be written as[11]

$$\tilde{\boldsymbol{f}}_{o_k}^{\mathrm{T}} = -c_{w_{o_k}} \boldsymbol{p}_{o_k}^{\mathrm{H}} \left(\bar{c}_{r_k}\boldsymbol{I}_M - \boldsymbol{R}_{\boldsymbol{h}_k|\boldsymbol{y}_{\mathrm{T}}}^* \sum_{i=1}^{o_k-1} c_{w_i}\boldsymbol{p}_i\boldsymbol{p}_i^{\mathrm{H}}\right)^{-1} \boldsymbol{R}_{\boldsymbol{h}_k|\boldsymbol{y}_{\mathrm{T}}}^* \boldsymbol{P}_{o_k}. \tag{5.51}$$

Contrary to linear precoding with P-CSI (4.43) and THP with C-CSI (5.36), a parameterization of (5.49) in terms of an SINR-like parameter, i.e., a ratio of quadratic forms, is not possible without further assumptions.

---

[10] It yields $\boldsymbol{B}(\alpha\boldsymbol{I} + \boldsymbol{A}\boldsymbol{B})^{-1} = (\alpha\boldsymbol{I} + \boldsymbol{B}\boldsymbol{A})^{-1}\boldsymbol{B}$ for some matrices $\boldsymbol{A}, \boldsymbol{B}$. Here, we set $\boldsymbol{A} = \boldsymbol{C}_{\boldsymbol{w}}\boldsymbol{P}^{\mathrm{H}}$ and $\boldsymbol{B} = \boldsymbol{R}_{\boldsymbol{h}_k|\boldsymbol{y}_{\mathrm{T}}}^*\boldsymbol{P}_{o_k}$.

[11] Note that $\boldsymbol{P}_{o_k}\boldsymbol{C}_{\boldsymbol{w}}\boldsymbol{P}^{\mathrm{H}} = \sum_{i=1}^{o_k-1} c_{w_i}\boldsymbol{p}_i\boldsymbol{p}_i^{\mathrm{H}}$ because the last columns of $\boldsymbol{P}_{o_k}$ are zero.

Example 5.5. If $\mathrm{rank}[\boldsymbol{C}_{\boldsymbol{h}_k}] = 1$, we have $\boldsymbol{R}_{\boldsymbol{h}_k|\boldsymbol{y}_\mathrm{T}} = \bar{\boldsymbol{v}}_k \bar{\boldsymbol{v}}_k^\mathrm{H}$ where $\bar{\boldsymbol{v}}_k$ is a scaled version of the eigenvector $\boldsymbol{v}_k$ in $\boldsymbol{C}_{\boldsymbol{h}_k} = \mathrm{E}[|x_k|^2]\boldsymbol{v}_k\boldsymbol{v}_k^\mathrm{H}$. Thus, (5.51) simplifies to

$$\tilde{\boldsymbol{f}}_{o_k}^\mathrm{T} = -g_k' \bar{\boldsymbol{v}}_k^\mathrm{T} \boldsymbol{P}_{o_k} \tag{5.52}$$

where we identify $g_k' = (\bar{\boldsymbol{v}}^\mathrm{T} \boldsymbol{p}_{o_k})^* c_{w_{o_k}} / (\sum_{i=o_k}^K c_{w_i} |\bar{\boldsymbol{v}}_k^\mathrm{H} \boldsymbol{p}_i|^2 + c_{n_k})$ similar to C-CSI (5.34).[12] The minimum

$$\min_{\tilde{\boldsymbol{f}}_{o_k}} \mathrm{E}_{\boldsymbol{h}_k|\boldsymbol{y}_\mathrm{T}} \left[\mathrm{MCOR}_k^\mathrm{THP}(\boldsymbol{P}, \tilde{\boldsymbol{f}}_{o_k}; \boldsymbol{h}_k)\right] = c_{w_{o_k}} \frac{1}{1 + \overline{\mathrm{SINR}}_k^\mathrm{THP}(\boldsymbol{P})} \tag{5.53}$$

can be formulated in terms of

$$\overline{\mathrm{SINR}}_k^\mathrm{THP}(\boldsymbol{P}) = \frac{c_{w_{o_k}} |\bar{\boldsymbol{v}}_k^\mathrm{T} \boldsymbol{p}_{o_k}|^2}{\sum_{i=o_k+1}^K c_{w_i} |\bar{\boldsymbol{v}}_k^\mathrm{T} \boldsymbol{p}_i|^2 + c_{n_k}}. \tag{5.54}$$

The solution of the feedback filter as well as the minimum of the cost function show a structure identical to C-CSI (cf. (5.32) and (5.36)). Any $\overline{\mathrm{SINR}}_k^\mathrm{THP}(\boldsymbol{P})$ is achievable for an arbitrary P_tx. This is not the case for $\mathrm{rank}[\boldsymbol{C}_{\boldsymbol{h}_k}] > 1$.

For an optimum MMSE receiver, this behavior is only achieved for $c_{n_k} = 0$. Therefore, for large P_tx and rank-one $\boldsymbol{C}_{\boldsymbol{h}_k}$, the achievable cost (5.53) with a scaled matched filter receiver is identical to (5.40) for the optimum MMSE receiver. But if P_tx is finite and the true receiver is an MMSE receiver, the performance criterion (5.38) is more accurate than (5.53) and gives an appropriate model for the uncertainty in CSI: This example shows that the uncertainty in CSI is not described completely by the cost function (5.49).

□

5.3 Optimization Based on Sum Performance Measures

The proposed performance measures for P-CSI are either based on the optimum MMSE receiver or a suboptimum scaled matched filter receiver. To optimize the performance of the overall system (similar to maximizing the mean sum rate for linear precoding in Section 4.4), we minimize the conditional mean estimate of the sum MSE resulting from an MMSE receiver or of the sum performance measure for a scaled matched filter receiver under a total transmit power constraint.

For vector precoding, the optimization problem is

[12] We use the relation $(\alpha \boldsymbol{I} + \boldsymbol{a}\boldsymbol{b}^\mathrm{T})^{-1}\boldsymbol{a} = \boldsymbol{a}(\alpha + \boldsymbol{b}^\mathrm{T}\boldsymbol{a})^{-1}$, for some vectors $\boldsymbol{a}, \boldsymbol{b}$, and choose $\boldsymbol{a} = \bar{\boldsymbol{v}}_k^*$ and $\boldsymbol{b} = \bar{\boldsymbol{v}}_k^\mathrm{T} \sum_{i=1}^{o_k - 1} \boldsymbol{p}_i \boldsymbol{p}_i^\mathrm{H}$ and apply the definition of $\bar{c}_{r_k}$.

$$\min_{\boldsymbol{T},\{\boldsymbol{a}[n]\}_{n=1}^{N_\mathrm{d}}} \sum_{k=1}^{K} \mathrm{E}_{\boldsymbol{h}_k|\boldsymbol{y}_\mathrm{T}}\left[\mathrm{MMSE}_k^\mathrm{VP}\left(\boldsymbol{T},\{\boldsymbol{a}[n]\}_{n=1}^{N_\mathrm{d}};\boldsymbol{h}_k\right)\right] \text{ s.t. } \mathrm{tr}\left[\boldsymbol{T}\tilde{\boldsymbol{R}}_{\boldsymbol{b}}\boldsymbol{T}^\mathrm{H}\right] \le P_\mathrm{tx} \tag{5.55}$$

based on (5.25) and

$$\min_{\boldsymbol{T},\{\boldsymbol{a}[n]\}_{n=1}^{N_\mathrm{d}}} \sum_{k=1}^{K} \mathrm{E}_{\boldsymbol{h}_k|\boldsymbol{y}_\mathrm{T}}\left[\mathrm{MCOR}_k^\mathrm{VP}\left(\boldsymbol{T},\{\boldsymbol{a}[n]\}_{n=1}^{N_\mathrm{d}};\boldsymbol{h}_k\right)\right] \text{ s.t. } \mathrm{tr}\left[\boldsymbol{T}\tilde{\boldsymbol{R}}_{\boldsymbol{b}}\boldsymbol{T}^\mathrm{H}\right] \le P_\mathrm{tx} \tag{5.56}$$

for (5.45). For THP, we have (cf. (5.38))

$$\min_{\boldsymbol{P},\boldsymbol{F},\mathcal{O}} \sum_{k=1}^{K} \mathrm{E}_{\boldsymbol{h}_k|\boldsymbol{y}_\mathrm{T}}\left[\mathrm{MMSE}_k^\mathrm{THP}(\boldsymbol{P},\tilde{\boldsymbol{f}}_{o_k};\boldsymbol{h}_k)\right] \quad \text{s.t.} \quad \mathrm{tr}\left[\boldsymbol{P}\boldsymbol{C}_{\boldsymbol{w}}\boldsymbol{P}^\mathrm{H}\right] \le P_\mathrm{tx} \tag{5.57}$$

and

$$\min_{\boldsymbol{P},\boldsymbol{F},\mathcal{O}} \sum_{k=1}^{K} \mathrm{E}_{\boldsymbol{h}_k|\boldsymbol{y}_\mathrm{T}}\left[\mathrm{MCOR}_k^\mathrm{THP}(\boldsymbol{P},\tilde{\boldsymbol{f}}_{o_k};\boldsymbol{h}_k)\right] \quad \text{s.t.} \quad \mathrm{tr}\left[\boldsymbol{P}\boldsymbol{C}_{\boldsymbol{w}}\boldsymbol{P}^\mathrm{H}\right] \le P_\mathrm{tx} \tag{5.58}$$

from (5.49).

Both performance measures proposed for THP in Section 5.2 yield an explicit solution for the feedback filter $\boldsymbol{F}$ in terms of $\boldsymbol{P}$ and $\mathcal{O}$. But the resulting problem (5.57) cannot be given in explicit form and an explicit solution of (5.58) seems impossible. A standard numerical solution would not allow for further insights. For vector precoding, the situation is more severe, since the lattice search is very complex and it is not known whether a heuristic as in Section 5.1 can be found.

Therefore, we resort to the same principle as for linear precoding (Section 4.4): Given an initialization for the precoder parameters, we alternate between optimizing the nonlinear precoder and the receiver models. Thus, we improve performance over the initialization, obtain explicit expressions for every iteration, and reach a stationary point of the corresponding cost function. In contrast to linear precoding the global minimum is not reached for C-CSI and an arbitrary initialization. This procedure also allows for a solution of (5.55) and (5.57) by means of numerical integration.

5.3.1 Alternating Optimization of Receiver Models and Transmitter

Vector Precoding

For solving (5.55) as well as (5.56) iteratively, it is sufficient to consider one cost function

$$\boxed{\begin{aligned}\mathrm{F}^{\mathrm{VP},(m)}\Big(\boldsymbol{T},\{\boldsymbol{a}[n]\}_{n=1}^{N_\mathrm{d}},\beta\Big) &= \mathrm{tr}\left[\tilde{\boldsymbol{R}}_{\boldsymbol{b}}\right] + |\beta|^2\bar{G}^{(m-1)} \\ &+|\beta|^2\mathrm{tr}\left[\boldsymbol{R}^{(m-1)}\boldsymbol{T}\tilde{\boldsymbol{R}}_{\boldsymbol{b}}\boldsymbol{T}^{\mathrm{H}}\right] - 2\mathrm{Re}\Big(\beta\boldsymbol{H}_{\mathbf{G}}^{(m-1)}\boldsymbol{T}\tilde{\boldsymbol{R}}_{\boldsymbol{b}}\Big).\end{aligned}} \tag{5.59}$$

It is equivalent to

$$\sum_{k=1}^{K}\mathrm{E}_{\boldsymbol{h}_k|\boldsymbol{y}_\mathrm{T}}\left[\mathrm{MSE}_k^{\mathrm{VP}}\Big(\boldsymbol{T},\{\boldsymbol{a}[n]\}_{n=1}^{N_\mathrm{d}},\beta \mathrm{g}_k^{\mathrm{mmse},(m-1)}(\boldsymbol{h}_k);\boldsymbol{h}_k\Big)\right] \tag{5.60}$$

for

$$\begin{aligned}\boldsymbol{H}_{\mathbf{G}}^{(m-1)} &= \mathrm{E}_{\boldsymbol{h}|\boldsymbol{y}_\mathrm{T}}\left[\mathbf{G}^{(m-1)}(\boldsymbol{h})\boldsymbol{H}\right] \\ \boldsymbol{R}^{(m-1)} &= \mathrm{E}_{\boldsymbol{h}|\boldsymbol{y}_\mathrm{T}}\left[\boldsymbol{H}^{\mathrm{H}}\mathbf{G}^{(m-1)}(\boldsymbol{h})^{\mathrm{H}}\mathbf{G}^{(m-1)}(\boldsymbol{h})\boldsymbol{H}\right] \\ \bar{G}^{(m-1)} &= \sum_{k=1}^{K}\mathrm{E}_{\boldsymbol{h}_k|\boldsymbol{y}_\mathrm{T}}\left[\left|\mathrm{g}_k^{\mathrm{mmse},(m-1)}(\boldsymbol{h}_k)\right|^2\right]c_{n_k} \\ \mathbf{G}^{(m-1)}(\boldsymbol{h}) &= \mathbf{diag}\left[\mathrm{g}_1^{\mathrm{mmse},(m-1)}(\boldsymbol{h}_1),\ldots,\mathrm{g}_K^{\mathrm{mmse},(m-1)}(\boldsymbol{h}_K)\right].\end{aligned} \tag{5.61}$$

Minimization of (5.60) w.r.t. the function $\mathrm{g}_k^{\mathrm{mmse},(m-1)}(\boldsymbol{h}_k)$, e.g., before taking the conditional expectation, yields (5.55).

The cost function for scaled matched filter receivers

$$\sum_{k=1}^{K}\mathrm{E}_{\boldsymbol{h}_k|\boldsymbol{y}_\mathrm{T}}\left[\mathrm{COR}_k^{\mathrm{VP}}\Big(\boldsymbol{T},\{\boldsymbol{a}[n]\}_{n=1}^{N_\mathrm{d}},\beta \mathrm{g}_k^{\mathrm{cor},(m-1)}(\boldsymbol{h}_k);\boldsymbol{h}_k\Big)\right] \tag{5.62}$$

can also be written as (5.59) defining

$$\begin{aligned}
\boldsymbol{H}_{\mathrm{G}}^{(m-1)} &= \mathrm{E}_{\boldsymbol{h}|\boldsymbol{y}_{\mathrm{T}}}\left[\mathbf{G}^{(m-1)}(\boldsymbol{h})\boldsymbol{H}\right] \\
\boldsymbol{R}^{(m-1)} &= \sum_{k=1}^{K} \mathrm{E}_{\boldsymbol{h}_k|\boldsymbol{y}_{\mathrm{T}}}\left[\left|\mathrm{g}_k^{\mathrm{cor},(m-1)}(\boldsymbol{h}_k)\right|^2\right] \boldsymbol{R}_{\boldsymbol{h}_k|\boldsymbol{y}_{\mathrm{T}}}^{*} \\
\bar{G}^{(m-1)} &= \sum_{k=1}^{K} \mathrm{E}_{\boldsymbol{h}_k|\boldsymbol{y}_{\mathrm{T}}}\left[\left|\mathrm{g}_k^{\mathrm{cor},(m-1)}(\boldsymbol{h}_k)\right|^2\right] c_{n_k} \\
\mathbf{G}^{(m-1)}(\boldsymbol{h}) &= \mathbf{diag}\left[\mathrm{g}_1^{\mathrm{cor},(m-1)}(\boldsymbol{h}_1), \ldots, \mathrm{g}_K^{\mathrm{cor},(m-1)}(\boldsymbol{h}_K)\right].
\end{aligned} \tag{5.63}$$

For this performance measure, all parameters can be given in closed form. For the optimum function $\mathrm{g}_k^{\mathrm{cor},(m-1)}(\boldsymbol{h}_k)$ from below (cf. (5.71)), the cost function (5.62) is identical to (5.56).

In (5.60) and (5.62) we introduced the additional degree of freedom β to obtain an explicit solution for $\boldsymbol{T}$ and enable an optimization alternating between transmitter and receivers.

First we set the receiver model $\mathbf{G}^{(m-1)}(\boldsymbol{h})$ equal to the optimum receiver from the previous iteration. Minimization of (5.59) w.r.t. $\boldsymbol{T}$ and β subject to the total power constraint $\mathrm{tr}[\boldsymbol{T}\tilde{\boldsymbol{R}}_{\boldsymbol{b}}\boldsymbol{T}^{\mathrm{H}}] \leq P_{\mathrm{tx}}$ yields

$$\boxed{\begin{aligned}
\boldsymbol{T}^{(m)} &= \beta^{(m),-1}\left(\boldsymbol{R}^{(m-1)} + \frac{\bar{G}^{(m-1)}}{P_{\mathrm{tx}}}\boldsymbol{I}_M\right)^{-1} \boldsymbol{H}_{\mathrm{G}}^{(m-1),\mathrm{H}} \\
&= \beta^{(m),-1}\left(\boldsymbol{\mathcal{X}}^{(m-1)} + \frac{\bar{G}^{(m-1)}}{P_{\mathrm{tx}}}\boldsymbol{I}_M\right)^{-1} \boldsymbol{H}_{\mathrm{G}}^{(m-1),\mathrm{H}}\boldsymbol{\Phi}^{(m-1),-1} \\
\beta^{(m),2} &= \mathrm{tr}\left[\tilde{\boldsymbol{R}}_{\boldsymbol{b}}\boldsymbol{H}_{\mathrm{G}}^{(m-1)}\left(\boldsymbol{R}^{(m-1)} + \frac{\bar{G}^{(m-1)}}{P_{\mathrm{tx}}}\boldsymbol{I}_M\right)^{-2} \boldsymbol{H}_{\mathrm{G}}^{(m-1),\mathrm{H}}\right] / P_{\mathrm{tx}}
\end{aligned}} \tag{5.64}$$

with

$$\boldsymbol{\Phi}^{(m-1)} = \boldsymbol{H}_{\mathrm{G}}^{(m-1)}\left(\boldsymbol{\mathcal{X}}^{(m-1)} + \frac{\bar{G}^{(m-1)}}{P_{\mathrm{tx}}}\boldsymbol{I}_M\right)^{-1} \boldsymbol{H}_{\mathrm{G}}^{(m-1),\mathrm{H}} + \boldsymbol{I}_K. \tag{5.65}$$

and[13]

$$\boldsymbol{\mathcal{X}}^{(m-1)} = \boldsymbol{R}^{(m-1)} - \boldsymbol{H}_{\mathrm{G}}^{(m-1),\mathrm{H}}\boldsymbol{H}_{\mathrm{G}}^{(m-1)}. \tag{5.66}$$

This result is obtained directly from Appendix D.1 comparing (5.59) with (D.2).

---

[13] $\boldsymbol{\mathcal{X}}^{(m-1)}$ can be considered as the covariance matrix of the additional correlated noise due to errors in CSI for P-CSI.

The minimum of (5.59)

$$\mathrm{F}^{\mathrm{VP},(m)}\Big(\{\boldsymbol{a}[n]\}_{n=1}^{N_\mathrm{d}}\Big) = \min_{\boldsymbol{T},\beta} \mathrm{F}^{\mathrm{VP},(m)}\Big(\boldsymbol{T},\{\boldsymbol{a}[n]\}_{n=1}^{N_\mathrm{d}},\beta\Big) \quad \text{s.t. } \operatorname{tr}\Big[\boldsymbol{T}\tilde{\boldsymbol{R}}_{\boldsymbol{b}}\boldsymbol{T}^\mathrm{H}\Big] \leq P_\mathrm{tx} \tag{5.67}$$

is given by (D.14)

$$\boxed{\begin{aligned}\mathrm{F}^{\mathrm{VP},(m)}\Big(\{\boldsymbol{a}[n]\}_{n=1}^{N_\mathrm{d}}\Big) &= \frac{1}{N_\mathrm{d}}\sum_{n=1}^{N_\mathrm{d}} \|\boldsymbol{a}[n]+\boldsymbol{d}[n]\|^2_{\boldsymbol{\Phi}^{(m-1),-1}} \\ &= \frac{1}{N_\mathrm{d}}\sum_{n=1}^{N_\mathrm{d}} \Big\|\boldsymbol{\Delta}^{(m-1),\frac{1}{2}}\boldsymbol{L}^{(m-1)}\boldsymbol{\Pi}^{(\mathcal{O})}(\boldsymbol{a}[n]+\boldsymbol{d}[n])\Big\|_2^2.\end{aligned}} \tag{5.68}$$

Thus, the optimization w.r.t. the perturbation vectors $\{\boldsymbol{a}[n]\}_{n=1}^{N_\mathrm{d}}$ on the lattice $\mathbb{L}^K$ is equivalent to the case of C-CSI in (5.13) based on the symmetrically permuted Cholesky factorization

$$\boldsymbol{\Pi}^{(\mathcal{O}),\mathrm{T}}\boldsymbol{\Phi}^{(m-1),-1}\boldsymbol{\Pi}^{(\mathcal{O})} = \boldsymbol{L}^{(m-1),\mathrm{H}}\boldsymbol{\Delta}^{(m-1)}\boldsymbol{L}^{(m-1)}, \tag{5.69}$$

where $\boldsymbol{L}^{(m-1)}$ is lower triangular with unit diagonal.

For optimization of $\{\boldsymbol{a}[n]\}_{n=1}^{N_\mathrm{d}}$, we propose two algorithms:

- If the lattice search is too complex to be executed in every iteration, we can perform it only in the last iteration.
- The same heuristic as proposed in Equations (5.11) to (5.14) is applied in every iteration, which yields THP with structure in Figure 5.7. The argumentation is identical to Equations (5.9) up to (5.21).

We follow the second approach in the sequel.

Given $\boldsymbol{T}^{(m)}$ from (5.64) and, possibly, an update of the perturbation vectors $\{\boldsymbol{a}^{(m)}[n]\}_{n=1}^{N_\mathrm{d}}$, which is included in $\tilde{\boldsymbol{R}}_{\boldsymbol{b}}^{(m)} = \sum_{n=1}^{N_\mathrm{d}} \boldsymbol{b}^{(m)}[n]\boldsymbol{b}^{(m)}[n]^\mathrm{H}/N_\mathrm{d}$ ($\boldsymbol{b}^{(m)}[n] = \boldsymbol{d}[n]+\boldsymbol{a}^{(m)}[n]$), we optimize the receivers in the next step. Note that we are actually looking for the optimum receiver as a function of the unknown channel $\boldsymbol{h}_k$, which is a random vector for P-CSI at the transmitter. For cost function (5.60), it is given by (5.24)

$$g_k^{\mathrm{mmse},(m)}(\boldsymbol{h}_k) = \beta^{(m),-1}\frac{\Big(\boldsymbol{h}_k^\mathrm{T}\boldsymbol{T}^{(m)}\tilde{\boldsymbol{r}}_{\boldsymbol{b}b_k}^{(m)}\Big)^*}{\boldsymbol{h}_k^\mathrm{T}\boldsymbol{T}^{(m)}\tilde{\boldsymbol{R}}_{\boldsymbol{b}}^{(m)}\boldsymbol{T}^{(m),\mathrm{H}}\boldsymbol{h}_k^* + c_{n_k}} \tag{5.70}$$

and for (5.62) by (5.42)

$$g_k^{\mathrm{cor},(m)}(\boldsymbol{h}_k) = \beta^{(m),-1} \frac{\left(\boldsymbol{h}_k^{\mathrm{T}} \boldsymbol{T}^{(m)} \tilde{\boldsymbol{r}}_{\boldsymbol{b}b_k}^{(m)}\right)^*}{\mathrm{tr}\left[\boldsymbol{T}^{(m)} \tilde{\boldsymbol{R}}_{\boldsymbol{b}}^{(m)} \boldsymbol{T}^{(m),\mathrm{H}} \boldsymbol{R}_{\boldsymbol{h}_k|\boldsymbol{y}_\mathrm{T}}^*\right] + c_{n_k}}. \tag{5.71}$$

As for linear precoding in Section 4.4, the scaling $\beta^{(m)}$, which was introduced artificially in the first step of every iteration (5.64), is removed by the optimum receivers. Based on the improved receivers we compute the parameters of the general cost function (5.59) and continue with (5.64) followed by the optimization of $\boldsymbol{a}[n]$.

The computational *complexity* per iteration is similar to the situation of C-CSI (Section 5.1): With a lattice search based on the permuted Cholesky factorization (5.69) according to Equations (5.11) up to (5.14) (in analogy to [127]), the total number of operations is in the order of $\mathrm{O}(K^3 + M^3)$. The cubic order in M results from the inverse in $\boldsymbol{\Phi}^{(m-1)}$ (5.65).

As for linear precoding (Section 4.4.1), the expressions for MMSE receivers require a multidimensional numerical integration, which increases the complexity considerably. This is not necessary for (5.62), which can be given in closed form as a function of the second order moments of $\mathrm{p}_{\boldsymbol{h}|\boldsymbol{y}_\mathrm{T}}(\boldsymbol{h}|\boldsymbol{y}_\mathrm{T})$.

In *summary*, our iterative algorithm proceeds as follows:

- We require an initialization for the receivers ((5.70) or (5.71)) or equivalently for $\boldsymbol{T}^{(0)}$ and $\{\boldsymbol{a}^{(0)}[n]\}_{n=1}^{N_\mathrm{d}}$.
- In every iteration, we compute the precoder parameters $\boldsymbol{T}^{(m)}$ (5.64) and $\{\boldsymbol{a}^{(m)}[n]\}_{n=1}^{N_\mathrm{d}}$ (5.68) first. The suboptimum (THP-like) computation of $\{\boldsymbol{a}^{(m)}[n]\}_{n=1}^{N_\mathrm{d}}$ involves an optimization of the ordering.
- It follows an update of the receivers (5.70) or (5.71), which yields the new cost function (5.59) for the next iteration.
- Convergence can be controlled by the distance of $\beta^{(m)}$ to 1. Alternatively, we stop after a fixed number of iterations, i.e., for a fixed amount of computational complexity.

Because the cost functions are bounded by zero, the algorithm converges to a stationary point. An improved or equal performance is guaranteed in every step, i.e.,

$$\begin{aligned}\sum_{k=1}^{K} &\mathrm{E}_{\boldsymbol{h}_k|\boldsymbol{y}_\mathrm{T}}\left[\mathrm{MMSE}_k^{\mathrm{VP}}\left(\boldsymbol{T}^{(m)}, \{\boldsymbol{a}^{(m)}[n]\}_{n=1}^{N_\mathrm{d}}; \boldsymbol{h}_k\right)\right] \\ &\leq \sum_{k=1}^{K} \mathrm{E}_{\boldsymbol{h}_k|\boldsymbol{y}_\mathrm{T}}\left[\mathrm{MMSE}_k^{\mathrm{VP}}\left(\boldsymbol{T}^{(m-1)}, \{\boldsymbol{a}^{(m-1)}[n]\}_{n=1}^{N_\mathrm{d}}; \boldsymbol{h}_k\right)\right]\end{aligned} \tag{5.72}$$

for (5.60) and for the suboptimum matched filter receiver (5.62)

$$\sum_{k=1}^{K} \mathrm{E}_{\boldsymbol{h}_k|\boldsymbol{y}_\mathrm{T}}\left[\mathrm{MCOR}_k^{\mathrm{VP}}\left(\boldsymbol{T}^{(m)}, \{\boldsymbol{a}^{(m)}[n]\}_{n=1}^{N_\mathrm{d}}; \boldsymbol{h}_k\right)\right]$$
$$\leq \sum_{k=1}^{K} \mathrm{E}_{\boldsymbol{h}_k|\boldsymbol{y}_\mathrm{T}}\left[\mathrm{MCOR}_k^{\mathrm{VP}}\left(\boldsymbol{T}^{(m-1)}, \{\boldsymbol{a}^{(m-1)}[n]\}_{n=1}^{N_\mathrm{d}}; \boldsymbol{h}_k\right)\right]. \quad (5.73)$$

In both cases we have $\beta^{(m)} \rightarrow 1$ as $m \rightarrow \infty$.

For the proposed algorithm, we have to choose an *initialization* of the receivers (5.70) and (5.71) for P-CSI. We choose them based on the vector precoder derived in Section 5.1 ($\boldsymbol{G} = \beta \boldsymbol{I}_K$), where we simply plug in the conditional mean estimate $\hat{\boldsymbol{H}}$ from $\boldsymbol{y}_\mathrm{T}$ for $\boldsymbol{H}$ as if it was perfectly known (Appendix C). For S-CSI and zero mean $\boldsymbol{h}$, we take the linear precoder in Appendix D.3 for $\boldsymbol{T}^{(0)}$ and set $\tilde{\boldsymbol{R}}_{\boldsymbol{b}}^{(0)} = \boldsymbol{I}_K$.

Tomlinson-Harashima Precoding

For THP, the iterative algorithm is very similar. The general cost function

$$\mathrm{F}^{\mathrm{THP},(m)}(\boldsymbol{P}, \boldsymbol{F}, \beta, \mathcal{O}) = |\beta|^2 \bar{G}^{(m-1)} + |\beta|^2 \mathrm{tr}\left[\boldsymbol{P}\boldsymbol{C}_{\boldsymbol{w}}\boldsymbol{P}^\mathrm{H}\boldsymbol{R}^{(m-1)}\right]$$
$$-2\mathrm{Re}\left(\beta\, \mathrm{tr}\left[\boldsymbol{H}_\mathbf{G}^{(m-1)}\boldsymbol{P}\boldsymbol{C}_{\boldsymbol{w}}(\boldsymbol{I}_K - \boldsymbol{F})^\mathrm{H}\boldsymbol{\Pi}^{(\mathcal{O})}\right]\right) + \mathrm{tr}\left[(\boldsymbol{I}_K - \boldsymbol{F})^\mathrm{H}\boldsymbol{C}_{\boldsymbol{w}}(\boldsymbol{I}_K - \boldsymbol{F})\right] \quad (5.74)$$

can be written as

$$\boxed{\begin{aligned}\mathrm{F}^{\mathrm{THP},(m)}(\boldsymbol{P}, \boldsymbol{F}, \beta, \mathcal{O}) &= |\beta|^2 \bar{G}^{(m-1)} + |\beta|^2 \mathrm{tr}\left[\boldsymbol{P}^\mathrm{H}\boldsymbol{\mathcal{X}}^{(m-1)}\boldsymbol{P}\boldsymbol{C}_{\boldsymbol{w}}\right] \\ &+ \mathrm{E}_{\boldsymbol{w}}\left[\left\|\left(\beta\boldsymbol{\Pi}^{(\mathcal{O})}\boldsymbol{H}_\mathbf{G}^{(m-1)}\boldsymbol{P} - (\boldsymbol{I}_K - \boldsymbol{F})\right)\boldsymbol{w}[n]\right\|_2^2\right]\end{aligned}} \quad (5.75)$$

with definitions

$$\boldsymbol{H}_\mathbf{G}^{(m-1)} = \mathrm{E}_{\boldsymbol{h}|\boldsymbol{y}_\mathrm{T}}\left[\mathbf{G}^{(m-1)}(\boldsymbol{h})\boldsymbol{H}\right] \quad (5.76)$$
$$\boldsymbol{\mathcal{X}}^{(m-1)} = \boldsymbol{R}^{(m-1)} - \boldsymbol{H}_\mathbf{G}^{(m-1),\mathrm{H}}\boldsymbol{H}_\mathbf{G}^{(m-1)}, \quad (5.77)$$

which are valid for both cost functions below. With appropriate definitions of $\boldsymbol{R}^{(m-1)}$, $\bar{G}^{(m-1)}$, and $\mathbf{G}^{(m-1)}(\boldsymbol{h})$, it includes the cost functions

$$\sum_{k=1}^{K} \mathrm{E}_{\boldsymbol{h}|\boldsymbol{y}_\mathrm{T}}\left[\mathrm{MSE}_k^{\mathrm{THP}}\left(\boldsymbol{P}, \tilde{\boldsymbol{f}}_{o_k}, \beta \mathrm{g}_k^{\mathrm{mmse},(m-1)}(\boldsymbol{h}_k); \boldsymbol{h}_k\right)\right] \quad \text{and} \tag{5.78}$$

$$\sum_{k=1}^{K} \mathrm{E}_{\boldsymbol{h}|\boldsymbol{y}_\mathrm{T}}\left[\mathrm{COR}_k^{\mathrm{THP}}\left(\boldsymbol{P}, \tilde{\boldsymbol{f}}_{o_k}, \beta \mathrm{g}_k^{\mathrm{cor},(m-1)}(\boldsymbol{h}_k); \boldsymbol{h}_k\right)\right], \tag{5.79}$$

which result in (5.57) and (5.58), respectively, when minimizing w.r.t. the receiver functions $\mathrm{g}_k(\boldsymbol{h}_k)$ (before taking the conditional expectation).

For (5.78), the parameters in (5.75) are defined as

$$\boldsymbol{R}^{(m-1)} = \mathrm{E}_{\boldsymbol{h}|\boldsymbol{y}_\mathrm{T}}\left[\boldsymbol{H}^\mathrm{H}\mathbf{G}^{(m-1)}(\boldsymbol{h})^\mathrm{H}\mathbf{G}^{(m-1)}(\boldsymbol{h})\boldsymbol{H}\right] \tag{5.80}$$

$$\bar{G}^{(m-1)} = \sum_{k=1}^{K} c_{n_k} \mathrm{E}_{\boldsymbol{h}_k|\boldsymbol{y}_\mathrm{T}}\left[\left|\mathrm{g}_k^{\mathrm{mmse},(m-1)}(\boldsymbol{h}_k)\right|^2\right] \tag{5.81}$$

$$\mathbf{G}^{(m-1)}(\boldsymbol{h}) = \mathbf{diag}\left[\mathrm{g}_1^{\mathrm{mmse},(m-1)}(\boldsymbol{h}_1), \ldots, \mathrm{g}_K^{\mathrm{mmse},(m-1)}(\boldsymbol{h}_K)\right]. \tag{5.82}$$

In this case $\boldsymbol{\mathcal{X}}^{(m-1)}$ (5.77) is the sum of (complex conjugate) error covariance matrices for the conditional mean estimate of the effective channel $\mathrm{g}_k^{\mathrm{mmse},(m-1)}(\boldsymbol{h}_k)\boldsymbol{h}_k$.

For (5.79), we obtain the explicit expressions for the parameters in (5.74)

$$\boldsymbol{R}^{(m-1)} = \sum_{k=1}^{K} \mathrm{E}_{\boldsymbol{h}_k|\boldsymbol{y}_\mathrm{T}}\left[\left|\mathrm{g}_k^{\mathrm{cor},(m-1)}(\boldsymbol{h}_k)\right|^2\right] \boldsymbol{R}_{\boldsymbol{h}_k|\boldsymbol{y}_\mathrm{T}}^* \tag{5.83}$$

$$\bar{G}^{(m-1)} = \sum_{k=1}^{K} c_{n_k} \mathrm{E}_{\boldsymbol{h}_k|\boldsymbol{y}_\mathrm{T}}\left[\left|\mathrm{g}_k^{\mathrm{cor},(m-1)}(\boldsymbol{h}_k)\right|^2\right] \tag{5.84}$$

$$\mathbf{G}^{(m-1)}(\boldsymbol{h}) = \mathbf{diag}\left[\mathrm{g}_1^{\mathrm{cor},(m-1)}(\boldsymbol{h}_1), \ldots, \mathrm{g}_K^{\mathrm{cor},(m-1)}(\boldsymbol{h}_K)\right]. \tag{5.85}$$

Comparing the new cost function (5.75) with the cost function derived from C-CSI, where we plug in $\mathrm{E}_{\boldsymbol{h}|\boldsymbol{y}_\mathrm{T}}\left[\mathbf{G}^{(m-1)}(\boldsymbol{h})\boldsymbol{H}\right]$ for $\boldsymbol{H}$, we have an additional regularization term for $\boldsymbol{P}$. It regularizes the weighted norm of $\boldsymbol{P}$ with weight given by $\boldsymbol{\mathcal{X}}^{(m-1)}$.

The solution for $\boldsymbol{F}^{(m)}$, $\boldsymbol{P}^{(m)}$, and $\beta^{(m)}$ for a fixed ordering $\mathcal{O}$ and for given receivers $\mathbf{G}^{(m-1)}(\boldsymbol{h})$ is derived in Appendix D.6 with assumption (5.26). The kth columns of $\boldsymbol{F}^{(m)}$ and $\boldsymbol{P}^{(m)}$ are

$$\boldsymbol{f}_k^{(m)} = -\beta^{(m)} \begin{bmatrix} \mathbf{0}_{k\times M} \\ \boldsymbol{B}_k^{(\mathcal{O}),(m-1)} \end{bmatrix} \boldsymbol{p}_k^{(m)}, \tag{5.86}$$

and

$$\boldsymbol{p}_k^{(m)} = \beta^{(m),-1} \left(\boldsymbol{\mathcal{R}}_k^{(m-1)} + \frac{\bar{G}^{(m-1)}}{P_{\mathrm{tx}}} \boldsymbol{I}_M \right)^{-1} \boldsymbol{A}_k^{(\mathcal{O}),(m-1),\mathrm{H}} \boldsymbol{e}_k \tag{5.87}$$

with $\boldsymbol{\mathcal{R}}_k^{(m-1)} = \boldsymbol{\mathcal{X}}^{(m-1)} + \boldsymbol{A}_k^{(\mathcal{O}),(m-1),\mathrm{H}} \boldsymbol{A}_k^{(\mathcal{O}),(m-1)}$. For every receiver, the permuted channel is partitioned as

$$\boldsymbol{\Pi}^{(\mathcal{O})} \boldsymbol{H}_{\mathbf{G}}^{(m-1)} = \begin{bmatrix} \boldsymbol{A}_k^{(\mathcal{O}),(m-1)} \\ \boldsymbol{B}_k^{(\mathcal{O}),(m-1)} \end{bmatrix}, \tag{5.88}$$

where $\boldsymbol{A}_k^{(\mathcal{O}),(m-1)} \in \mathbb{C}^{k \times M}$ contains the conditional mean estimate of the effective channels to already precoded receivers and $\boldsymbol{B}_k^{(\mathcal{O}),(m-1)} \in \mathbb{C}^{(K-k)\times M}$ to receivers which are precoded in the $k+1$th to Kth precoding step. The regularization term with $\boldsymbol{\mathcal{X}}^{(m-1)}$ in (5.75) yields a non-diagonal loading in the inverse of $\boldsymbol{P}^{(m)}$ (5.87).

The precoding order $\mathcal{O}^{(m)} = [p_1^{(m)}, \ldots, p_K^{(m)}]$ for this iteration is determined from the minimum $\mathrm{F}^{\mathrm{THP},(m)}(\mathcal{O})$ of (5.75), which is given by (D.72). We observe that the Kth term in the sum $(i = K)$ depends only on $\boldsymbol{A}_K^{(\mathcal{O}),(m-1)} = \boldsymbol{H}_{\mathbf{G}}^{(m-1)}$ and p_K and is independent of $p_i, i < K$. Therefore, we start with $i = K$ and choose the best receiver to be precoded last: $p_K^{(m)}$ is the argument of

$$\max_k \boldsymbol{e}_k^{\mathrm{T}} \boldsymbol{H}_{\mathbf{G}}^{(m-1)} \left(\boldsymbol{\mathcal{X}}^{(m-1)} + \boldsymbol{H}_{\mathbf{G}}^{(m-1),\mathrm{H}} \boldsymbol{H}_{\mathbf{G}}^{(m-1)} + \frac{\bar{G}^{(m-1)}}{P_{\mathrm{tx}}} \boldsymbol{I}_M \right)^{-1} \boldsymbol{H}_{\mathbf{G}}^{(m-1),\mathrm{H}} \boldsymbol{e}_k, \tag{5.89}$$

which minimizes the contribution of the $i = K$th term to the total remaining MSE (D.72). Similarly, we continue with the $(i-1)$th term, which depends only on $p_i^{(m)}, i \in \{K-1, K\}$, but not on p_i for $i < K-2$.

In summary, $\mathcal{O}^{(m)}$ is found by the arguments of

$$\max_k \boldsymbol{e}_k^{\mathrm{T}} \boldsymbol{A}_i^{(\mathcal{O}),(m-1)} \left(\boldsymbol{\mathcal{R}}_i^{(m-1)} + \frac{\bar{G}^{(m-1)}}{P_{\mathrm{tx}}} \boldsymbol{I}_M \right)^{-1} \boldsymbol{A}_i^{(\mathcal{O}),(m-1),\mathrm{H}} \boldsymbol{e}_k \tag{5.90}$$

for $i = K, K-1, \ldots, 1$, where the maximization is over $k \in \{1, \ldots, K\} \setminus \{p_{i+1}^{(m)}, \ldots, p_K^{(m)}\}$ and yields $p_i^{(m)}$. This requires computing an inverse in every step, which yields a computational *complexity* of order $\mathrm{O}(K^4)$.

But the minimum $\mathrm{F}^{\mathrm{THP},(m)}(\mathcal{O})$ can also be expressed in terms of the symmetrically permuted Cholesky factorization (5.69):

$$\mathrm{F}^{\mathrm{THP},(m)}(\mathcal{O}) = \mathrm{tr}\left[\boldsymbol{\Delta}^{(m-1)} \boldsymbol{C}_{\boldsymbol{w}}\right] = \sum_{k=1}^{K} \delta_k^{(m-1)} c_{w_k}. \tag{5.91}$$

The derivation of this expression requires a different notation of the optimization problem and is omitted here because it can be carried out in complete analogy to [127], where it is given for C-CSI. The resulting efficient algorithm yields the same ordering as (5.90). Its computational complexity for computing $\boldsymbol{F}^{(m)}$, $\boldsymbol{P}^{(m)}$, and $\mathcal{O}^{(m)}$ is $\mathrm{O}\big(K^3 + M^3\big)$. With $p_k^{(m)}$ we also obtain its inverse function $o_k^{(m)}$ implicitly.

In the last step of iteration m we compute the optimum receivers as a function of $\boldsymbol{h}_k$. They are given by (5.37)

$$\mathrm{g}_k^{\mathrm{mmse},(m)}(\boldsymbol{h}_k) = \beta^{(m),-1} \frac{\left(\boldsymbol{h}_k^{\mathrm{T}} \boldsymbol{P}^{(m)} \boldsymbol{C}_{\boldsymbol{w}} \left(\boldsymbol{e}_{o_k^{(m)}} - \tilde{\boldsymbol{f}}_{o_k^{(m)}}^{(m),*}\right)\right)^*}{\sum_{i=1}^{K} c_{w_i} \left|\boldsymbol{h}_k^{\mathrm{T}} \boldsymbol{p}_i^{(m)}\right|^2 + c_{n_k}} \tag{5.92}$$

and (5.47)

$$\mathrm{g}_k^{\mathrm{cor},(m)}(\boldsymbol{h}_k) = \beta^{(m),-1} \frac{\left(\boldsymbol{h}_k^{\mathrm{T}} \boldsymbol{P}^{(m)} \boldsymbol{C}_{\boldsymbol{w}} \left(\boldsymbol{e}_{o_k^{(m)}} - \tilde{\boldsymbol{f}}_{o_k^{(m)}}^{(m),*}\right)\right)^*}{\operatorname{tr}\left[\boldsymbol{P}^{(m)} \boldsymbol{C}_{\boldsymbol{w}} \boldsymbol{P}^{(m),\mathrm{H}} \boldsymbol{R}_{\boldsymbol{h}_k|\boldsymbol{y}_{\mathrm{T}}}^*\right] + c_{n_k}}. \tag{5.93}$$

Again, *convergence* to a stationary point is guaranteed and the total cost decreases or stays constant in every iteration:

$$\sum_{k=1}^{K} \mathrm{E}_{\boldsymbol{h}_k|\boldsymbol{y}_{\mathrm{T}}}\left[\mathrm{MMSE}_k^{\mathrm{THP}}\left(\boldsymbol{P}^{(m)}, \tilde{\boldsymbol{f}}_{o_k^{(m)}}^{(m)}; \boldsymbol{h}_k\right)\right] \leq \sum_{k=1}^{K} \mathrm{E}_{\boldsymbol{h}_k|\boldsymbol{y}_{\mathrm{T}}}\left[\mathrm{MMSE}_k^{\mathrm{THP}}\left(\boldsymbol{P}^{(m-1)}, \tilde{\boldsymbol{f}}_{o_k^{(m-1)}}^{(m-1)}; \boldsymbol{h}_k\right)\right] \tag{5.94}$$

$$\sum_{k=1}^{K} \mathrm{E}_{\boldsymbol{h}_k|\boldsymbol{y}_{\mathrm{T}}}\left[\mathrm{MCOR}_k^{\mathrm{THP}}\left(\boldsymbol{P}^{(m)}, \tilde{\boldsymbol{f}}_{o_k^{(m)}}^{(m)}; \boldsymbol{h}_k\right)\right] \leq \sum_{k=1}^{K} \mathrm{E}_{\boldsymbol{h}_k|\boldsymbol{y}_{\mathrm{T}}}\left[\mathrm{MCOR}_k^{\mathrm{THP}}\left(\boldsymbol{P}^{(m-1)}, \tilde{\boldsymbol{f}}_{o_k^{(m-1)}}^{(m-1)}; \boldsymbol{h}_k\right)\right]. \tag{5.95}$$

Asymptotically, $\beta^{(m)} \to 1$ as $m \to \infty$.

The *initialization* is chosen similar to vector precoding above: For P-CSI, we first compute $\boldsymbol{F}^{(0)}$, $\boldsymbol{P}^{(0)}$ and $\mathcal{O}^{(0)}$ applying the conditional mean estimate $\hat{\boldsymbol{H}}$ from $\boldsymbol{y}_{\mathrm{T}}$ to the THP for C-CSI with $\boldsymbol{G} = \beta \boldsymbol{I}_K$, which is a special case of Appendix D.6; for S-CSI we set $\boldsymbol{F}^{(0)} = \boldsymbol{0}_K$, $\boldsymbol{\Pi}^{(\mathcal{O})} = \boldsymbol{I}_K$, and take $\boldsymbol{P}^{(0)}$ equal to the linear precoder from Appendix D.3.

5.3.2 From Complete to Statistical Channel State Information

The algorithms for optimization of vector precoding with the restricted lattice search and THP which we derive in the previous section are closely related. We summarize their differences and similarities:

- The optimized parameters of vector precoding depend on $\{\boldsymbol{d}[n]\}_{n=1}^{N_\mathrm{d}}$, whereas THP is independent of the data and relies on the stochastic model (5.26) for $\boldsymbol{w}[n]$.
- The cost functions (5.59) and (5.75) are only equivalent for non-diagonal $\boldsymbol{C}_{\boldsymbol{w}} = (\boldsymbol{I}_K - \boldsymbol{F})^{-1}\tilde{\boldsymbol{R}}_{\boldsymbol{b}}(\boldsymbol{I}_K - \boldsymbol{F})^{-\mathrm{H}}$ as in (5.20) and identical receiver functions $\mathbf{G}^{(m-1)}(\boldsymbol{h})$. For the assumption of diagonal $\boldsymbol{C}_{\boldsymbol{w}}$ and identical $\mathbf{G}^{(m-1)}(\boldsymbol{h})$, (5.75) is an approximation of (5.59) for large P_tx and large constellation size $\mathbb{D}$ if the suboptimum THP-like lattice search is performed for vector precoding.
- The formulation of vector precoding is more general: The lattice search can be improved as described in [174].
- Comparing the solutions for vector precoding with the THP-like lattice search and THP for identical $\mathbf{G}^{(m-1)}(\boldsymbol{h})$, i.e., same initialization or solution from the previous iteration, we conclude:
  - For the given data, vector precoding satisfies the transmit power constraint *exactly*.
  - If the solution for $\boldsymbol{P}^{(m)}$ in THP is scaled to satisfy the transmit power constraint, THP is identical to vector precoding, i.e., yields the same transmit signal $\boldsymbol{x}_\mathrm{tx}[n]$. Note that also the same precoding order is chosen for both precoders.

Although they are closely related and the vector precoding formulation is more general, we presented the THP formulation for the following reasons: On the one hand it is more widely used and on the other hand its solution allows for additional insights in the effect of the modulo operation, i.e., the suboptimum THP-like lattice search. In the sequel, we focus on the interpretation of both nonlinear precoders.

Complete Channel State Information

For C-CSI, the MMSE receiver and scaled matched filter are identical. Therefore, the corresponding optimization problems for vector precoding or THP are also identical. The necessary parameters depend on the channel realization $\boldsymbol{H}$

$$\boldsymbol{H}_{\mathbf{G}}^{(m-1)} = \mathbf{G}^{(m-1)}(\boldsymbol{h})\boldsymbol{H} \tag{5.96}$$

$$\boldsymbol{R}^{(m-1)} = \boldsymbol{H}^{\mathrm{H}}\mathbf{G}^{(m-1)}(\boldsymbol{h})^{\mathrm{H}}\mathbf{G}^{(m-1)}(\boldsymbol{h})\boldsymbol{H} = \boldsymbol{H}_{\mathbf{G}}^{(m-1),\mathrm{H}}\boldsymbol{H}_{\mathbf{G}}^{(m-1)} \tag{5.97}$$

$$\bar{G}^{(m-1)} = \sum_{k=1}^{K} \left| \mathrm{g}_k^{\mathrm{mmse},(m-1)}(\boldsymbol{h}_k) \right|^2 c_{n_k} \tag{5.98}$$

$$\boldsymbol{\Phi}^{(m-1)} = \frac{P_{\mathrm{tx}}}{\bar{G}^{(m-1)}} \left(\boldsymbol{H}_{\mathbf{G}}^{(m-1)}\boldsymbol{H}_{\mathbf{G}}^{(m-1),\mathrm{H}} + \frac{\bar{G}^{(m-1)}}{P_{\mathrm{tx}}}\boldsymbol{I}_K \right). \tag{5.99}$$

The matrix $\boldsymbol{\mathcal{X}}^{(m-1)} = \mathbf{0}$ determining the regularization of the solution vanishes. For MMSE receivers, $\boldsymbol{\mathcal{X}}^{(m-1)}$ is the sum of the error covariance matrices for the effective channels, which is zero for C-CSI by definition.

For vector precoding, the linear filter reads

$$\boldsymbol{T}^{(m)} = \beta^{(m),-1} \left(\boldsymbol{H}_{\mathbf{G}}^{(m-1),\mathrm{H}}\boldsymbol{H}_{\mathbf{G}}^{(m-1)} + \frac{\bar{G}^{(m-1)}}{P_{\mathrm{tx}}}\boldsymbol{I}_M \right)^{-1} \boldsymbol{H}_{\mathbf{G}}^{(m-1),\mathrm{H}} \tag{5.100}$$

$$= \beta^{(m),-1}\boldsymbol{H}_{\mathbf{G}}^{(m-1),\mathrm{H}} \left(\boldsymbol{H}_{\mathbf{G}}^{(m-1)}\boldsymbol{H}_{\mathbf{G}}^{(m-1),\mathrm{H}} + \frac{\bar{G}^{(m-1)}}{P_{\mathrm{tx}}}\boldsymbol{I}_K \right)^{-1} \tag{5.101}$$

with $\beta^{(m)}$ chosen to satisfy $\mathrm{tr}[\boldsymbol{T}^{(m)}\tilde{\boldsymbol{R}}_{\boldsymbol{b}}\boldsymbol{T}^{(m),\mathrm{H}}] = P_{\mathrm{tx}}$.[14] The nonlinearity of the precoder results from the optimization of $\boldsymbol{a}[n]$.

For THP, the effect of the modulo operators is described by $\boldsymbol{C}_{\boldsymbol{w}}$. If assumption (5.26) holds, e.g., for large P_{tx} and size of $\mathbb{D}$, THP allows for an interpretation of the solutions and the effect of the perturbation vector.

For C-CSI, the THP solution (5.86) and (5.87) for an optimized ordering $\mathcal{O} = \mathcal{O}^{(m)}$ simplifies to

$$\boldsymbol{f}_k^{(m)} = -\beta^{(m)} \begin{bmatrix} \mathbf{0}_{k \times M} \\ \boldsymbol{B}_k^{(\mathcal{O}),(m-1)} \end{bmatrix} \boldsymbol{p}_k^{(m)} \tag{5.102}$$

$$\boldsymbol{p}_k^{(m)} = \beta^{(m),-1} \left(\boldsymbol{A}_k^{(\mathcal{O}),(m-1),\mathrm{H}}\boldsymbol{A}_k^{(\mathcal{O}),(m-1)} + \frac{\bar{G}^{(m-1)}}{P_{\mathrm{tx}}}\boldsymbol{I}_M \right)^{-1} \boldsymbol{A}_k^{(\mathcal{O}),(m-1),\mathrm{H}}\boldsymbol{e}_k, \tag{5.103}$$

where $\beta^{(m)}$ is determined by $\mathrm{tr}[\boldsymbol{P}^{(m)}\boldsymbol{C}_{\boldsymbol{w}}\boldsymbol{P}^{(m),\mathrm{H}}] = P_{\mathrm{tx}}$ and $\boldsymbol{A}_k^{(\mathcal{O}),(m-1)} \in \mathbb{C}^{k \times M}$ contains the channels to already precoded receivers and $\boldsymbol{B}_k^{(\mathcal{O}),(m-1)} \in \mathbb{C}^{(K-k) \times M}$ to receivers which are precoded in the $(k+1)$th to Kth step. Therefore, our choice of $\boldsymbol{a}[n]$, i.e., the modulo operations in Figure 5.7, can be interpreted as follows:

- $\mathbf{f}_k^{(m)}$ cancels the interference from the kth precoded receiver to all subsequent receivers with channel $\boldsymbol{B}_k^{(\mathcal{O}),(m-1)}$.

[14] For $\mathbf{G}^{(m-1)}(\boldsymbol{h}) = \boldsymbol{I}_K$ we obtain our result (5.6) in the introduction.

- Therefore, the forward filter $\boldsymbol{p}_k^{(m)}$ for the kth precoded data stream has to avoid interference only to the already precoded receivers with channel $\boldsymbol{A}_k^{(\mathcal{O}),(m-1)}$, because the interference from the kth receiver's data stream has not been known in precoding steps 1 to $k-1$ due to causality. Numerically, this reduces the condition number of the matrix to be "inverted" in $\boldsymbol{p}_k^{(m)}$.[15] For $k=1$, the full diversity and antenna gain is achieved because $\boldsymbol{p}_1^{(m)} \propto \boldsymbol{h}_{p_1}^*$.

For small P_{tx}, this interpretation is not valid, because $\boldsymbol{a}[n] = \boldsymbol{0}$, i.e., vector precoding and THP are equivalent to linear precoding. If the modulo operators are not removed from the receivers, which would require additional signaling to the receivers, performance is worse than for linear precoding due to the modulo loss [70, 108].

Partial and Statistical Channel State Information

For P-CSI and in the limit for S-CSI, the channel $\mathbf{G}^{(m-1)}(\boldsymbol{h})\boldsymbol{H}$ is not known perfectly.[16] Therefore, the solutions are given in terms of the conditional mean estimate of the total channel $\boldsymbol{H}_{\mathbf{G}}^{(m-1)} = \mathrm{E}_{\boldsymbol{h}|\boldsymbol{y}_{\mathrm{T}}}[\mathbf{G}^{(m-1)}(\boldsymbol{h})\boldsymbol{H}]$ and the regularization matrix $\boldsymbol{\mathcal{X}}^{(m-1)}$. For MMSE receivers, $\boldsymbol{\mathcal{X}}^{(m-1)}$ is the sum of error covariance matrices for the estimates $\mathrm{E}_{\boldsymbol{h}_k|\boldsymbol{y}_{\mathrm{T}}}[\mathrm{g}_k^{\mathrm{mmse},(m-1)}(\boldsymbol{h}_k)\boldsymbol{h}_k]$.

To judge the effect on nonlinear precoding we discuss the THP solution (5.86) and (5.87). Contrary to C-CSI, the feedback filter $\boldsymbol{F}^{(m)}$ cannot cancel the interference and the forward filter also has to compensate the interference to subsequently precoded data streams, which is described by $\boldsymbol{\mathcal{X}}^{(m-1)}$ for $\mathcal{O} = \mathcal{O}^{(m)}$. As long as (5.26) is valid the effect of $\boldsymbol{a}[n]$ can be understood as a partial cancelation of the interference described by $\boldsymbol{B}_k^{(\mathcal{O}),(m-1)}$. It results in (cf. (5.87))

$$\boldsymbol{\mathcal{X}}^{(m-1)} + \boldsymbol{A}_k^{(\mathcal{O}),(m-1),\mathrm{H}}\boldsymbol{A}_k^{(\mathcal{O}),(m-1)} = \boldsymbol{R}^{(m-1)} - \boldsymbol{B}_k^{(\mathcal{O}),(m-1),\mathrm{H}}\boldsymbol{B}_k^{(\mathcal{O}),(m-1)}, \tag{5.104}$$

which differs in its last term from the corresponding term in (4.82) for linear precoding.

For $K \leq M$ and good channels, i.e., spatially well separated receivers, it is likely that the precoder chooses $\boldsymbol{a}[n] = \boldsymbol{0}$ for S-CSI — even for large P_{tx}. If this is the case, our interpretation of THP is not valid and nonlinear precoding is identical to linear precoding.

---

[15] Note that c_{w_k} is close to one for increasing size of the constellation $\mathbb{D}$, i.e., the input of $\boldsymbol{P}^{(m)}$ has the same variance as to a linear precoder $\boldsymbol{P}$ (Section 4.1).

[16] Note that the precoder always operates with a receiver model $\mathbf{G}^{(m-1)}(\boldsymbol{h})$, which can deviate from the implemented MMSE receiver, for example, if the iterative procedure has not converged yet or the scaled matched filter model is assumed.

Example 5.6. Let us consider the example of channels with rank-one covariance matrices $\boldsymbol{C}_{\boldsymbol{h}_k}$. As in Section 4.4.3, the corresponding channel is $\boldsymbol{H} = \boldsymbol{X}\boldsymbol{V}^{\mathrm{T}}$ with $\boldsymbol{X} = \mathbf{diag}[\mathsf{x}_1, \ldots, \mathsf{x}_K]$ and $\boldsymbol{V} \in \mathbb{C}^{M\times K}$, whose columns are the array steering vectors (Appendix E).

For this channel, $\boldsymbol{\mathcal{X}}^{(m-1)}$ (5.77) simplifies to

$$\boldsymbol{\mathcal{X}}^{(m-1)} = \boldsymbol{V}^{*}\boldsymbol{\Xi}^{(m-1)}\boldsymbol{V}^{\mathrm{T}}$$

with $\boldsymbol{\Xi}^{(m-1)} = \mathbf{diag}\left[\xi_1^{(m-1)}, \ldots, \xi_K^{(m-1)}\right]$.

For MMSE receivers and vector precoding, we have

$$\xi_k^{(m-1)} = \mathrm{E}_{\mathsf{x}_k, \boldsymbol{h}_k|\boldsymbol{y}_{\mathrm{T}}}\left[|\mathsf{x}_k|^2 \left|\mathrm{g}_k^{\mathrm{mmse},(m-1)}(\boldsymbol{h}_k)\right|^2\right] - \left|\mathrm{E}_{\mathsf{x}_k, \boldsymbol{h}_k|\boldsymbol{y}_{\mathrm{T}}}\left[\mathsf{x}_k \mathrm{g}_k^{\mathrm{mmse},(m-1)}(\boldsymbol{h}_k)\right]\right|^2.$$

From (5.24)

$$\mathrm{g}_k^{\mathrm{mmse},(m-1)}(\boldsymbol{h}_k) = \beta^{(m-1),-1}\frac{\mathsf{x}_k^{*}\left(\boldsymbol{v}_k^{\mathrm{T}}\boldsymbol{T}^{(m-1)}\tilde{\boldsymbol{r}}_{\boldsymbol{b}b_k}^{(m-1)}\right)^{*}}{|\mathsf{x}_k|^2\boldsymbol{v}_k^{\mathrm{T}}\boldsymbol{T}^{(m-1)}\tilde{\boldsymbol{R}}_{\boldsymbol{b}}^{(m-1)}\boldsymbol{T}^{(m-1),\mathrm{H}}\boldsymbol{v}_k^{*} + c_{n_k}},$$

we observe that the *effective* channel $\mathbf{G}^{(m-1)}(\boldsymbol{h})\boldsymbol{H}$ is known to the transmitter for large P_{tx} (small c_{n_k}), because the receivers remove x_k and the rank-1 channel becomes deterministic (cf. Example 5.4). From $\xi_k^{(m-1)} = 0$ follows that the error covariance matrix $\boldsymbol{\mathcal{X}}^{(m-1)} = \mathbf{0}$ is zero.

For scaled matched filter receivers (5.42), $\boldsymbol{\mathcal{X}}^{(m-1)}$ is *not* equal to the sum of error covariance matrices of the estimated channel $\mathrm{E}_{\boldsymbol{h}_k|\boldsymbol{y}_{\mathrm{T}}}[\mathrm{g}_k^{\mathrm{cor},(m-1)}(\boldsymbol{h}_k)\boldsymbol{h}_k]$, because $\boldsymbol{R}^{(m-1)}$ in (5.63) is not the sum of correlation matrices of $\mathrm{g}_k^{\mathrm{cor},(m-1)}(\boldsymbol{h}_k)\boldsymbol{h}_k$.[17] In this case, we have $\xi_k^{(m-1)} = 0$ for *all* P_{tx}, which results in $\boldsymbol{\mathcal{X}}^{(m-1)} = \mathbf{0}$ and $\boldsymbol{R}^{(m-1)} = \boldsymbol{H}_{\mathbf{G}}^{(m-1),\mathrm{H}}\boldsymbol{H}_{\mathbf{G}}^{(m-1)}$. But note that the total channel is not known perfectly, since the error covariance matrix of its estimate is not zero.

If $\boldsymbol{\mathcal{X}}^{(m-1)} = \mathbf{0}_{M\times M}$, the search on the lattice (5.68) is determined by

$$\boldsymbol{\Phi}^{(m-1)} = \frac{P_{\mathrm{tx}}}{\bar{G}^{(m-1)}}\left(\boldsymbol{X}_{\mathbf{G}}^{(m-1)}\boldsymbol{V}^{\mathrm{T}}\boldsymbol{V}^{*}\boldsymbol{X}_{\mathbf{G}}^{(m-1),*} + \frac{\bar{G}^{(m-1)}}{P_{\mathrm{tx}}}\boldsymbol{I}_K\right)$$

[17] With [148], $\boldsymbol{\mathcal{X}}^{(m-1)}$ can be shown to be equal to one of the terms in the expression for the sum of error covariance matrices. On the other hand, $\boldsymbol{R}^{(m-1)}$ in (5.61) for MMSE receivers is the sum of correlation matrices of $\mathrm{g}_k^{\mathrm{mmse},(m-1)}(\boldsymbol{h}_k)\boldsymbol{h}_k$; with its definition in (5.66), $\boldsymbol{\mathcal{X}}^{(m-1)}$ is identical to the sum of error covariance matrices.

with $\boldsymbol{X}_{\mathbf{G}}^{(m-1)} = \mathrm{E}_{\boldsymbol{h}|\boldsymbol{y}_\mathrm{T}}[\mathbf{G}^{(m-1)}(\boldsymbol{h})\boldsymbol{X}]$ and mainly influenced by the spatial separation of the receivers.

For rank-1 channels, we conclude:

- The channel estimation errors are modeled correctly when considering MMSE receiver models. This yields an improved interference cancelation at medium to high P_tx compared to the optimization based on a scaled MF receiver.
- For $P_\mathrm{tx} \to \infty$, the solutions for MF receivers and MMSE receivers converge.
- As for C-CSI (5.103), the advantage of nonlinear precoding for scaled matched filter receivers compared to linear precoding becomes more evident considering THP: $\boldsymbol{p}_k^{(m)}$ only considers interference to already precoded data streams, which results in an improved condition number of the matrix to be "inverted". But in contrast to C-CSI, the interference cannot be canceled completely by $\boldsymbol{F}^{(m)}$ for finite P_tx. Still this leads to the full antenna gain for the p_1th receiver because $\boldsymbol{p}_1 \propto \boldsymbol{v}_{p_1}^*$. Note that this interpretation is not valid anymore, if $\boldsymbol{a}[n] = \mathbf{0}$ due to the "good" properties of $\boldsymbol{H}$, i.e., $\boldsymbol{V}$. □

5.4 Precoding for the Training Channel

Optimization of vector precoding is based on the assumption of the receivers (5.24) and (5.42). Their realization requires the knowledge of $\boldsymbol{h}_k^\mathrm{T}\boldsymbol{T}\tilde{\boldsymbol{r}}_{\boldsymbol{b}b_k}$ as well as c_{r_k} and $\bar{c}_{r_k}$, respectively. The variance of the receive signal can be easily estimated. The crosscorrelation $\boldsymbol{h}_k^\mathrm{T}\boldsymbol{T}\tilde{\boldsymbol{r}}_{\boldsymbol{b}b_k}$ can be signaled to the receivers by precoding of the training sequence (Section 4.1.2) with

$$\boldsymbol{Q} = \boldsymbol{T}\tilde{\boldsymbol{R}}_{\boldsymbol{b}}. \tag{5.105}$$

Now the necessary crosscorrelation terms can be estimated at the receivers (cf. (4.11)). Contrary to linear precoding, where $\boldsymbol{Q} = \boldsymbol{P}$, the required transmit power is not equal to the transmit power in the data channel, i.e., $\|\boldsymbol{Q}\|_\mathrm{F}^2 \neq P_\mathrm{tx} = \mathrm{tr}[\boldsymbol{T}\tilde{\boldsymbol{R}}_{\boldsymbol{b}}\boldsymbol{T}^\mathrm{H}]$. Moreover, $\|\boldsymbol{Q}\|_\mathrm{F}^2$ is not constant, but depends on the channel and the data. A cumulative distribution of $\|\boldsymbol{Q}\|_\mathrm{F}^2$ is shown in Figure 5.9.

The same is true for THP, where precoding of the training sequence with

$$\boldsymbol{Q} = \boldsymbol{P}\boldsymbol{C}_{\boldsymbol{w}}(\boldsymbol{I}_K - \boldsymbol{F}^\mathrm{H})\boldsymbol{\Pi}^{(\mathcal{O})} \tag{5.106}$$

ensures the realizability of the assumed receivers (5.37) and (5.47).

We can scale $\boldsymbol{Q}$ by a constant to satisfy the transmit power constraint P_tx^t in the "worst case". If this scaling is known to the receivers, it can be compensated. But the channel estimation quality at the receivers degrades.

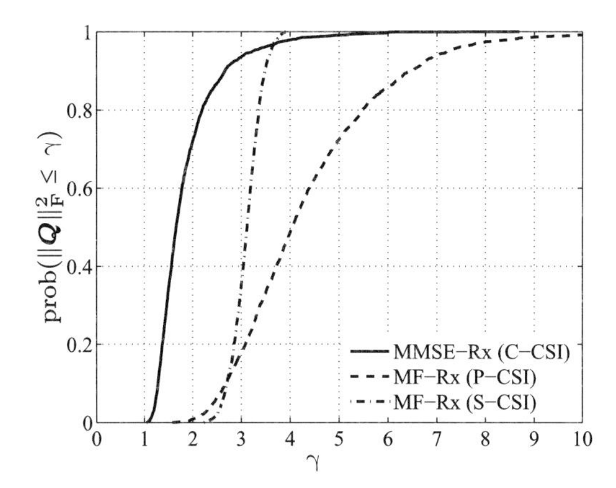

Fig. 5.9 Cumulative distribution function of $\|\boldsymbol{Q}\|_F^2$ (5.105) at $10\log_{10}(P_{tx}/c_n) =$ 20 dB and $f_{max} = 0.2$ for P-CSI (Scenario 1).

As in Figure 5.9 the worst case $\|\boldsymbol{Q}\|_F^2$ can exceed $P_{tx}^t = P_{tx} = 1$ (cf. (4.7)) significantly. Note also that the distribution of $\|\boldsymbol{Q}\|_F^2$ depends critically on the scenario and SNR: For small and large SNR, the distribution is concentrated around $P_{tx}^t = 1$; it is more widespread for larger K/M or increased angular spread.

5.5 Performance Evaluation

For a complexity comparable to linear precoding, we implement vector precoding with the THP-like lattice search, which is equivalent to THP with the appropriate scaling of $\boldsymbol{P}$. Therefore, we refer to the *nonlinear precoders* as THP and the solution from Section 5.3 as robust THP. The following variations are considered:

- For C-CSI, the iterative solution in Section 5.3.2 with MMSE receivers is chosen ("THP: MMSE-Rx (C-CSI)"). It is initialized with the THP solution for identical receivers $g_k = \beta$ from [115], which is obtained for $\mathbf{G}^{(m-1)}(\boldsymbol{h}) = \boldsymbol{I}_K$. Contrary to linear precoding, it does not converge to the global minimum in general.
- Heuristic nonlinear precoders are obtained from the solution for C-CSI with outdated CSI (O-CSI) $\hat{\boldsymbol{h}} = \mathrm{E}_{\boldsymbol{h}[q']|\boldsymbol{y}[q']}[\boldsymbol{h}[q']]$ with $q' = q^{tx} - 1$ ("THP: Heur. MMSE-Rx (O-CSI)") or the predicted CSI $\hat{\boldsymbol{h}} = \mathrm{E}_{\boldsymbol{h}|\boldsymbol{y}_T}[\boldsymbol{h}]$ ("THP: Heur. MMSE-Rx (P-CSI)") instead of $\boldsymbol{h}$, i.e., the CSI is treated as if it was error-free. The initialization is also chosen in analogy to C-CSI but with O-CSI and P-CSI, respectively.
- As an example for the systematic approach to robust nonlinear precoding with P-CSI, we choose the iterative solution to (5.55) assuming MMSE

receivers ("THP: MMSE-Rx (P-CSI)") and to (5.56) assuming scaled matched filter (MF) receivers ("THP: MF-Rx (P-CSI)"). More details and the initialization are described in Section 5.3.1.
- For both receiver models, the algorithms for S-CSI are a special case of P-CSI ("THP: MMSE-Rx (S-CSI)" and "THP: MF-Rx (S-CSI)") with initialization from Appendix D.3.

A fixed number of 10 iterations is chosen for all precoders, i.e., the computational complexity of all nonlinear and linear MSE-related precoders is comparable. The necessary numerical integration for precoding with P-CSI or S-CSI assuming MMSE receivers is implemented as a Monte-Carlo integration with 1000 integration points.

The nonlinear precoders are compared with the *linear MSE precoders* from Section 4.4. The system parameters and the first two simulation scenarios are identical to Section 4.4.4. We repeat the definitions for convenience.

The temporal channels correlations are described by Clarke's model (Appendix E and Figure 2.12) with a maximum Doppler frequency $f_{\max}$, which is normalized to the slot period T_b. For simplicity, the temporal correlations are identical for all elements in $\boldsymbol{h}$ (2.12). Four different stationary scenarios for the spatial covariance matrix are compared. For details and other implicit assumptions see Appendix E. The SNR is defined as P_{tx}/c_n.

Scenario 1: We choose $M = 8$ transmit antennas and $K = 6$ receivers with the mean azimuth directions $\bar{\varphi} = [-45°, -27°, -9°, 9°, 27°, 45°]$ and Laplace angular power spectrum with spread $\sigma = 5°$.

Scenario 2: We have $M = 8$ transmit antennas and $K = 8$ receivers with mean azimuth directions $\bar{\varphi} = [-45°, -32.1°, -19.3°, -6.4°, 6.4°, 19.3°, 32.1°, 45°]$ and Laplace angular power spectrum with spread $\sigma = 0.5°$, i.e., $\mathrm{rank}[\boldsymbol{C}_{\boldsymbol{h}_k}] \approx 1$.

Scenario 3: We restrict $\mathrm{rank}[\boldsymbol{C}_{\boldsymbol{h}_k}] = 2$ with $M = 8$, $K = 4$, i.e., $\sum_{k=1}^{K} \mathrm{rank}[\boldsymbol{C}_{\boldsymbol{h}_k}] = M$. The two azimuth directions $(\varphi_{k,1}, \varphi_{k,2})$ per receiver from (E.3) with $W = 2$ are

$$\{(-10.7°, -21.4°), (-3.6°, -7.2°), (3.6°, 7.2°), (10.7°, 21.4°)\}.$$

Scenario 4 simulates an overloaded system with $M = 2$ and $K = 3$ for $\bar{\varphi} = [-19.3°, -6.4°, 6.4°]$ and different angular spread σ (Appendix E).[18]

The optimum MMSE receivers (5.24) are implemented with perfect channel knowledge. As modulation alphabet $\mathbb{D}$ for the data signals d_k we choose rectangular 16QAM for all receivers.

[18] The mean angles correspond to the central three receivers of Scenario 2.

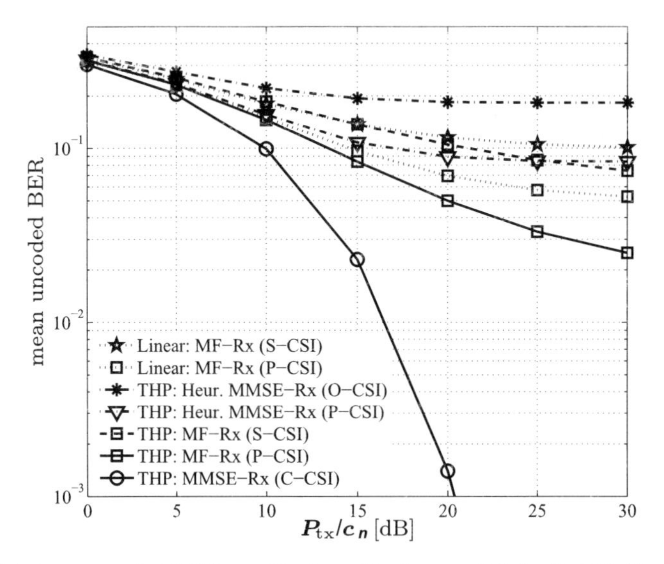

Fig. 5.10 Mean uncoded BER vs. P_{tx}/c_n with $f_{\max} = 0.2$ for P-CSI (Scenario 1).

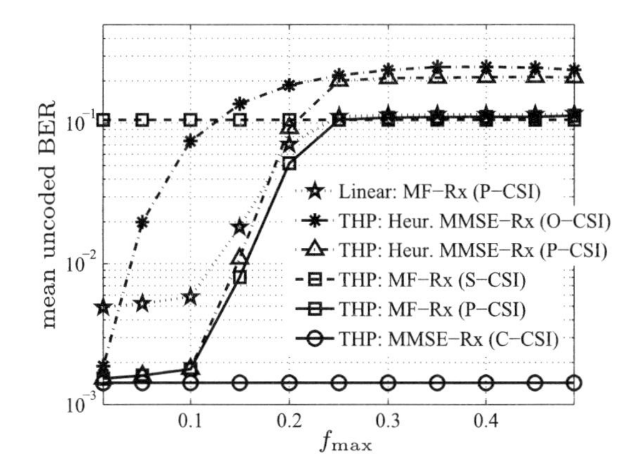

Fig. 5.11 Mean uncoded BER vs. $f_{\max}$ for $10\log_{10}(P_{\mathrm{tx}}/c_n) = 20\,\mathrm{dB}$ (Scenario 1).

Discussion of Results

The mean uncoded bit error rate (BER), which is an average w.r.t. the K receivers, serves as a simple performance measure.

In *scenario 1*, all $\boldsymbol{C}_{\boldsymbol{h}_k}$ have full rank. In a typical example the largest eigenvalue is about 5 times larger than the second and 30 times larger than the third largest eigenvalue. Therefore, the transmitter does not know the effective channel (including the receivers processing g_k) perfectly. Because $\sum_{k=1}^{K} \mathrm{rank}[\boldsymbol{C}_{\boldsymbol{h}_k}] = KM > M$, the channels do not have enough structure to allow for a full cancelation of the interference as for C-CSI. The BER

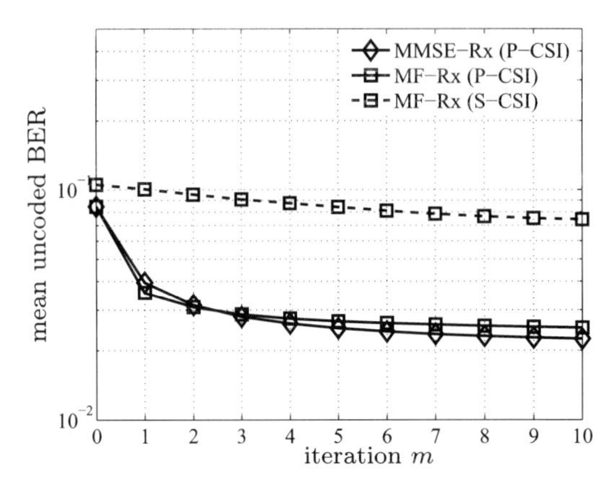

Fig. 5.12 Convergence in mean BER of alternating optimization at $10\log_{10}(P_{\mathrm{tx}}/c_n) = 30\,\mathrm{dB}$ (Scenario 1, 16QAM).

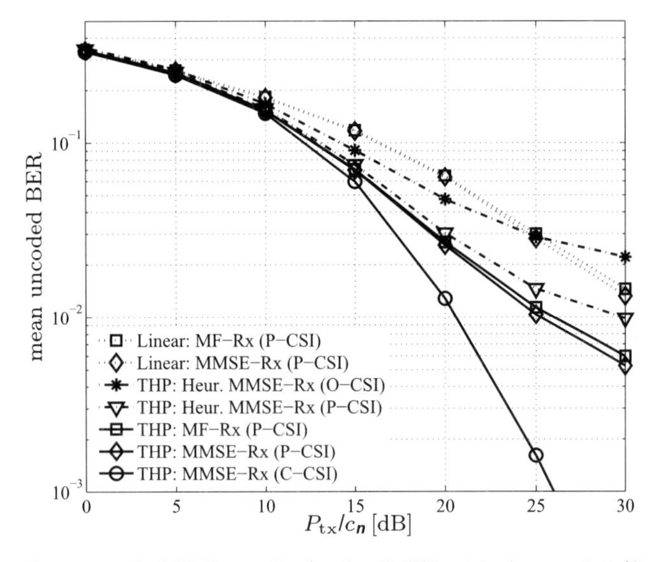

Fig. 5.13 Mean uncoded BER vs. P_{tx}/c_n for P-CSI with $f_{\mathrm{max}} = 0.2$ (Scenario 2).

saturates for high P_{tx}/c_n (Figure 5.10). The heuristic optimization for P-CSI does not consider the interference sufficiently and the performance can be improved by the robust optimization of THP based on the CM estimate of the receivers' cost, which is based on scaled matched filter receivers. Due to the insufficient channel structure, robust THP can only achieve a small additional reduction of the interference and the BER saturates at a lower BER.

For increasing Doppler frequency, the heuristic design performs worse than linear precoding (Figure 5.11). The robust solution for P-CSI ensures a performance which is always better than linear precoding at high P_{tx}/c_n. As for

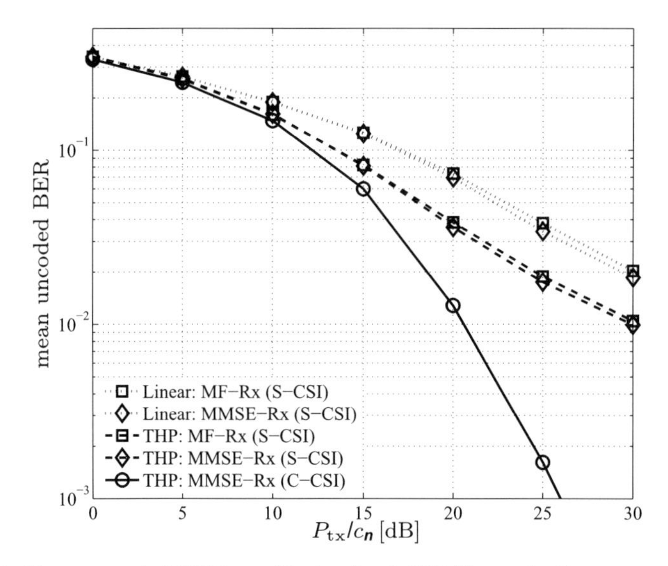

Fig. 5.14 Mean uncoded BER vs. P_{tx}/c_n for S-CSI (Scenario 2).

linear precoding (Section 4.4.4), the prediction errors are small enough as long as $f_{\max} \leq 0.15$ to yield a performance of the heuristic, which is very close to our systematic approach.

For this scenario, approximately 10 iterations are necessary for convergence (Figure 5.12). For S-CSI and scenario 2, the algorithms converge in the first iteration.

The interference can already be canceled sufficiently by linear precoding in *scenario 2*, for which the ratio of the first to the second eigenvalue of $\boldsymbol{C}_{\boldsymbol{h}_k}$ is about 300. Because the receivers' channels are close to rank one, the probability is very high that the channel realization for a receiver and its predicted channel lie in the same one-dimensional subspace. Thus, the performance degradation by the heuristic THP optimization is rather small (Figure 5.13). But for S-CSI, the THP optimization based on the mean performance measures gains 5 dB at high P_{tx}/c_n compared to linear precoding from Section 4.4 (Figure 5.14). Figure 5.15 shows the fast transition between a performance close to C-CSI and S-CSI which is controlled by the knowledge of the autocovariance sequence.

Optimization based on an MMSE receiver model results in a small performance improvement at high P_{tx}/c_n: For example, the discussion in Section 5.3.2 shows that for S-CSI and $\mathrm{rank}[\boldsymbol{C}_{\boldsymbol{h}_k}] = 1$ the estimation error of the effective channel is not zero, but the regularization matrix $\boldsymbol{\mathcal{X}}^{(m-1)}$ is zero for optimization based on the scaled matched filter receiver model. Therefore, the small degradation results from the mismatch to the implemented MMSE receiver and the incomplete description of the error in CSI by $\boldsymbol{\mathcal{X}}^{(m-1)}$.

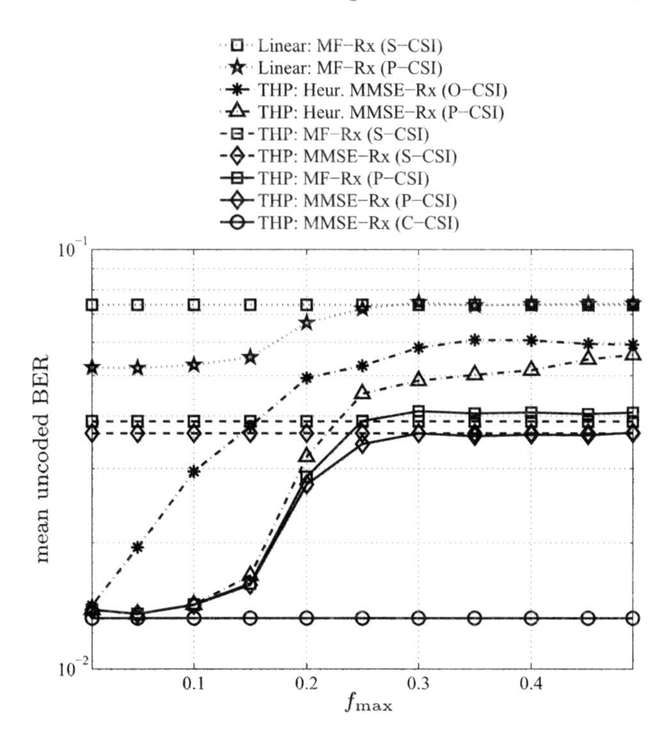

Fig. 5.15 Mean uncoded BER vs. $f_{\max}$ for $10\log_{10}(P_{\text{tx}}/c_n) = 20\,\text{dB}$ (Scenario 2).

The performance in *scenario 3* (Figure 5.16) confirms that the gains of THP over linear precoding for P-CSI and S-CSI are largest, when the channel has significant structure. Robust THP is not limited by interference, which is the case for its heuristic design. Note the cross-over point between heuristic THP and linear precoding for P-CSI at 10 dB.

For $\text{rank}[\boldsymbol{C}_{\boldsymbol{h}_k}] = 1$ $(\sigma = 0°)$, we show in Section 5.2.1 that any MSE is realizable even for S-CSI at the expense of an increased transmit power. Therefore, the system is not interference-limited. This is confirmed by Figure 5.17 for *scenario 4*, i.e., $K = 3 > M = 2$: Robust THP with scaled MF receivers does not saturate whereas linear precoding is interference-limited with a BER larger than 10^{-1}. With increasing angular spread σ the performance of robust THP degrades and the BER saturates at a significantly smaller BER than linear precoding. Robust THP requires about 30 iterations for an overloaded system; this considerably slower convergence rate is caused by our alternating optimization method. In principle, it can be improved optimizing the explicit cost function in Section 5.2.2 for $\boldsymbol{P}$ directly.

Whether the optimization of nonlinear precoding chooses an element $a_k[n]$ of the perturbation vector $\boldsymbol{a}[n]$ to be non-zero depends on the system parameters and channel properties. The gain of THP over linear precoding is influenced by the probability of $a_k[n]$ to be zero, which is shown in Figure 5.18 for

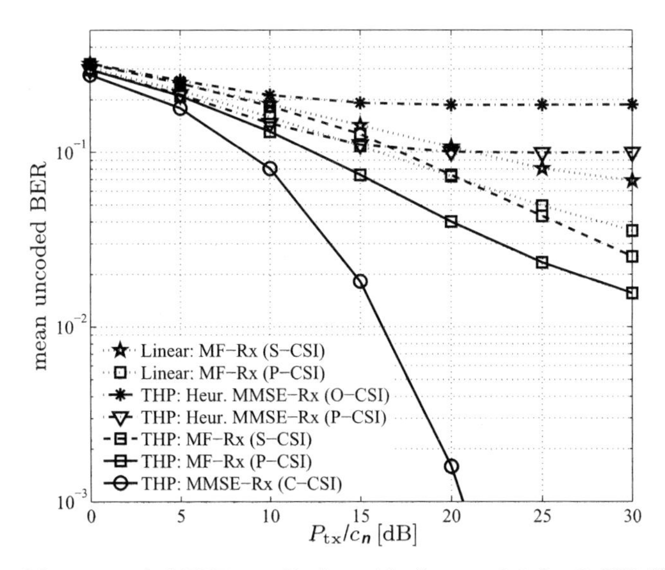

Fig. 5.16 Mean uncoded BER vs. P_{tx}/c_n with $f_{max} = 0.2$ for P-CSI (Scenario 3, $M = 8$, $K = 4$).

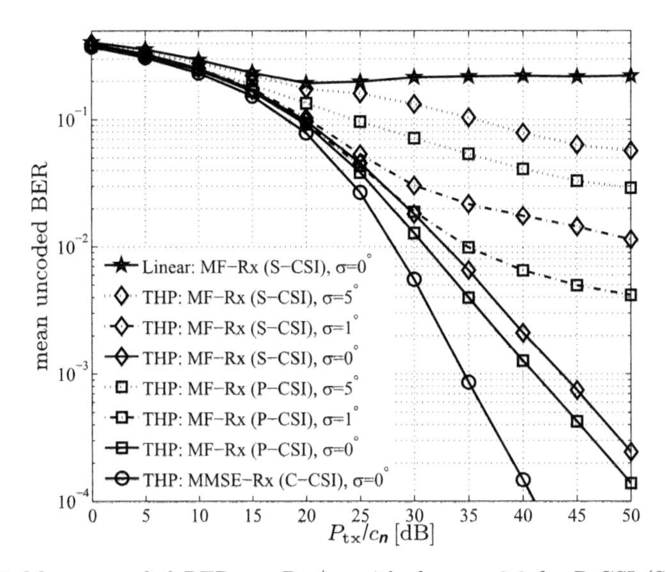

Fig. 5.17 Mean uncoded BER vs. P_{tx}/c_n with $f_{max} = 0.2$ for P-CSI (Scenario 4, $M = 2$, $K = 3$).

different parameters (Scenario 2). Inherently, the THP-like lattice search always chooses the first element of $\boldsymbol{\Pi}^{(\mathcal{O})}\boldsymbol{a}[n]$ to be zero, i.e., the probability to be zero is always larger than $1/K$. For small P_{tx}/c_n, the probability converges to one and the nonlinear precoder is linear: In BER we see a performance degradation compared to linear precoding due to the modulo operators at the

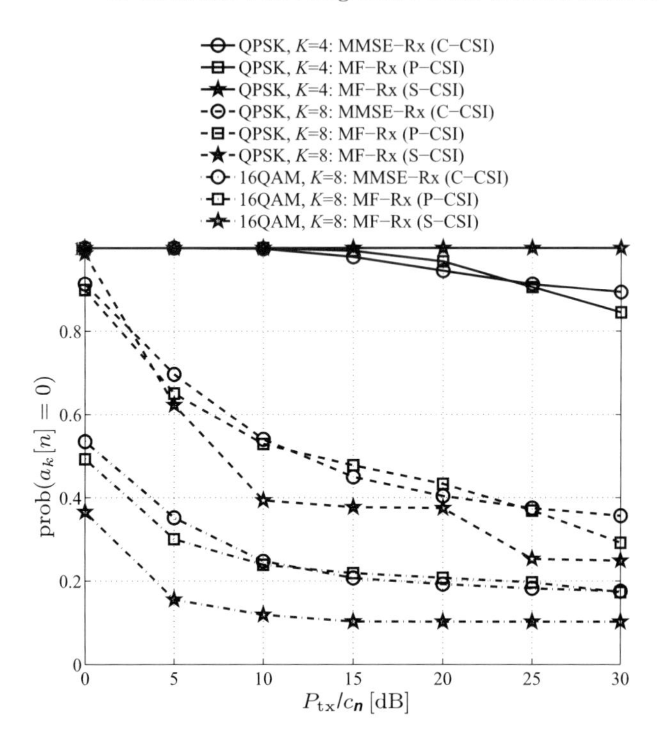

Fig. 5.18 Probability that a modulo operation at the transmitter is inactive (linear), i.e., $a_k[n] = 0$ (Scenario 2, $M = 8$).

receivers. Due to the larger amplitudes of the symbols d_k, the probability of $a_k[n]$ being zero is smaller for 16QAM than for QPSK. If we also reduce the number of receivers to $K = 4$, which are equally spaced in azimuth between $\pm 45°$, the optimum precoder becomes linear with a high probability.

To summarize for $K/M \leq 1$, gains are only large if

- it is difficult to separate the receivers, e.g., for small differences in the receivers' azimuth directions and K close to M,
- the channels show sufficient structure in the sense that $\mathrm{rank}[\boldsymbol{C}_{\boldsymbol{h}_k}] \leq M$, and
- the size of the QAM modulation alphabet $\mathbb{D}$ is large, e.g., $|\mathbb{D}| \geq 16$.

This is in accordance to the model of $\boldsymbol{C}_{\boldsymbol{w}}$ (5.26), which is more accurate for larger modulation alphabet and smaller probability for zero elements in $\boldsymbol{a}[n]$. Moreover, information theoretic results confirm that the gap between linear precoding and dirty paper precoding increases for larger K/M (cf. [106] for $K \leq M$).

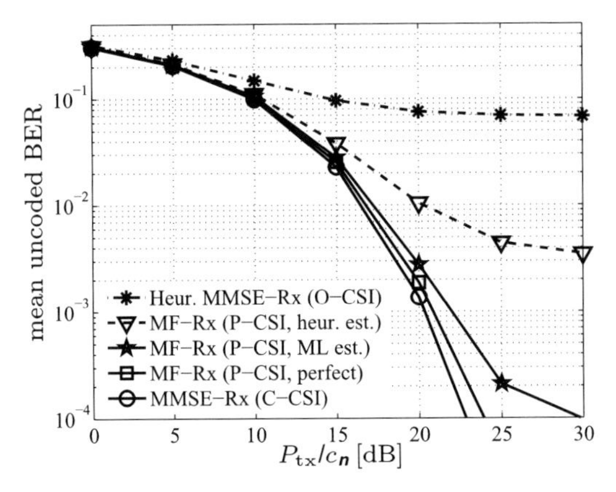

Fig. 5.19 Mean uncoded BER vs. $P_{\mathrm{tx}}/c_{\boldsymbol{n}}$ of THP with $f_{\max} = 0.1$ for P-CSI with estimated covariance matrices (Scenario 1, $B = 20$).

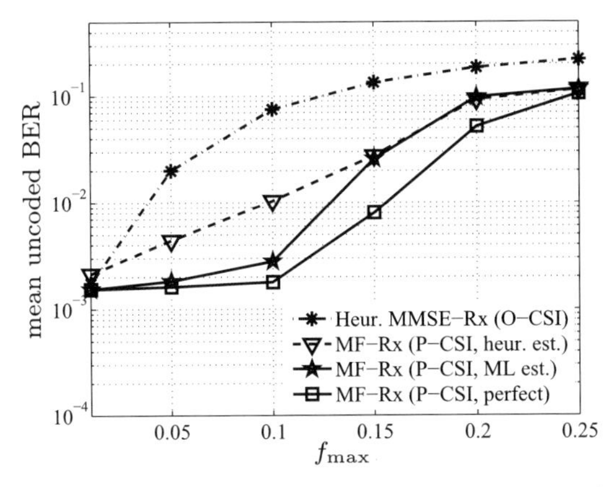

Fig. 5.20 Mean uncoded BER vs. $f_{\max}$ of THP at $10\log_{10}(P_{\mathrm{tx}}/c_{n}) = 20\,\mathrm{dB}$ with estimated covariance matrices (Scenario 1, $B = 20$).

Performance for Estimated Covariance Matrices

All previous performance results for nonlinear precoding and P-CSI have assumed perfect knowledge of the channel covariance matrix $\boldsymbol{C}_{\boldsymbol{h}_{\mathrm{T}}}$ and the noise covariance matrix $\boldsymbol{C}_{\boldsymbol{v}} = c_n \boldsymbol{I}_M$. As for linear precoding, we apply the algorithms for estimating channel and noise covariance matrices which are derived in Chapter 3 to estimate the parameters of $\mathrm{p}_{\boldsymbol{h}|\boldsymbol{y}_{\mathrm{T}}}(\boldsymbol{h}|\boldsymbol{y}_{\mathrm{T}})$.

The model and the estimators of the covariance matrices are described at the end of Section 4.4.4 (Page 165): Only $B = 20$ correlated realizations $\boldsymbol{h}[q^{\mathrm{tx}} - \ell]$ are observed via $\boldsymbol{y}[q^{\mathrm{tx}} - \ell]$ for $\ell \in \{1, 3, \ldots, 2B - 1\}$, i.e., with a period of $N_{\mathrm{P}} = 2$.

The heuristic and an advanced estimator ("ML est.") for the channel covariance matrix are compared; the details can be found on page 165. Only the mean uncoded BER for THP is shown in Figures 5.19 and 5.20.

The loss in BER of THP based on the heuristic covariance estimator compared to the advanced method from Chapter 3 is significant for $f_{\max} = 0.1$ (Figure 5.19): The saturation level is at about $2 \cdot 10^{-3}$ compared to (below) 10^{-4}. THP based on the advanced estimator performs close to a perfect knowledge of the channel correlations.

With the advanced estimator a performance close to C-CSI is achieved up to $f_{\max} \approx 0.1$ (Figure 5.20). For $f_{\max} \geq 0.15$, both estimators yield a similar BER of THP, because only the knowledge of the spatial channel covariance matrix is relevant whose estimation accuracy in terms of BER is comparable for both approaches.

Appendix A
Mathematical Background

A.1 Complex Gaussian Random Vectors

A *complex Gaussian random vector* $\mathbf{x} \in \mathbb{C}^N$ is defined as a random vector with statistically independent Gaussian distributed real and imaginary part, whose probability distributions are identical [164, p. 198], [124]. We denote it as $\mathbf{x} \sim \mathcal{N}_c(\boldsymbol{\mu}_x, \boldsymbol{C}_x)$ with mean $\boldsymbol{\mu}_x = \mathrm{E}[\mathbf{x}]$ and covariance matrix $\boldsymbol{C}_x = \boldsymbol{C}_{xx} = \mathrm{E}\left[(\mathbf{x} - \boldsymbol{\mu}_x)(\mathbf{x} - \boldsymbol{\mu}_x)^{\mathrm{H}}\right]$. Its probability density reads

$$\mathrm{p}_x(\boldsymbol{x}) = \frac{1}{\pi^N \det[\boldsymbol{C}_x]} \exp\left[-(\boldsymbol{x} - \boldsymbol{\mu}_x)^{\mathrm{H}} \boldsymbol{C}_x^{-1} (\boldsymbol{x} - \boldsymbol{\mu}_x)\right]. \tag{A.1}$$

Consider the jointly complex Gaussian random vector $\mathbf{z} = [\mathbf{x}^{\mathrm{T}}, \mathbf{y}^{\mathrm{T}}]^{\mathrm{T}}$. Additionally to the parameters of $\mathrm{p}_x(\boldsymbol{x})$, its probability density is characterized by $\boldsymbol{C}_y$, $\boldsymbol{\mu}_y$, and the crosscovariance matrix $\boldsymbol{C}_{xy} = \mathrm{E}\left[(\mathbf{x} - \boldsymbol{\mu}_x)(\mathbf{y} - \boldsymbol{\mu}_y)^{\mathrm{H}}\right]$ with $\boldsymbol{C}_{xy} = \boldsymbol{C}_{yx}^{\mathrm{H}}$. The parameters of the corresponding *conditional probability density* $\mathrm{p}_{x|y}(\boldsymbol{x}|\boldsymbol{y})$ are (e.g., [124])

$$\boldsymbol{\mu}_{x|y} = \mathrm{E}_{x|y}[\mathbf{x}] = \boldsymbol{\mu}_x + \boldsymbol{C}_{xy} \boldsymbol{C}_y^{-1} (\boldsymbol{y} - \boldsymbol{\mu}_y) \tag{A.2}$$

$$\boldsymbol{C}_{x|y} = \boldsymbol{C}_x - \boldsymbol{C}_{xy} \boldsymbol{C}_y^{-1} \boldsymbol{C}_{yx} = \mathbf{K}_{\boldsymbol{C}_y}(\boldsymbol{C}_z), \tag{A.3}$$

where $\mathbf{K}_\bullet(\bullet)$ denotes the Schur complement (Section A.2.2).

Useful expressions for higher order moments and the moments of some nonlinear functions of complex Gaussian random vectors are derived in [148] and [208].

A.2 Matrix Calculus

A.2.1 Properties of Trace and Kronecker Product

The *trace* of the square matrix $\boldsymbol{A} \in \mathbb{C}^{N\times N}$ with ijth element $[\boldsymbol{A}]_{i,j} = a_{i,j}$ is

$$\mathrm{tr}[\boldsymbol{A}] \triangleq \sum_{i=1}^{N} a_{i,i}. \tag{A.4}$$

It is invariant w.r.t. a circulant permutation of its argument

$$\mathrm{tr}[\boldsymbol{BC}] = \mathrm{tr}[\boldsymbol{CB}], \tag{A.5}$$

where $\boldsymbol{B} \in \mathbb{C}^{M\times N}$ and $\boldsymbol{C} \in \mathbb{C}^{N\times M}$. Thus, we have $\mathrm{tr}[\boldsymbol{A}] = \sum_{i=1}^{N} \lambda_i$ for the eigenvalues λ_i of $\boldsymbol{A}$.

The *Kronecker product* of two matrices $\boldsymbol{A} \in \mathbb{C}^{M\times N}$ and $\boldsymbol{B} \in \mathbb{C}^{P\times Q}$ is defined as

$$\boldsymbol{A} \otimes \boldsymbol{B} \triangleq \begin{bmatrix} a_{1,1}\boldsymbol{B} & \dots & a_{1,n}\boldsymbol{B} \\ \vdots & \ddots & \vdots \\ a_{M,1}\boldsymbol{B} & \dots & a_{M,N}\boldsymbol{B} \end{bmatrix} \in \mathbb{C}^{MP\times NQ}. \tag{A.6}$$

We summarize some of its properties from [26] ($\boldsymbol{M} \in \mathbb{C}^{M\times M}$, $\boldsymbol{N} \in \mathbb{C}^{N\times N}$, $\boldsymbol{C} \in \mathbb{C}^{N\times R}$, and $\boldsymbol{D} \in \mathbb{C}^{Q\times S}$):

$$(\boldsymbol{A} \otimes \boldsymbol{B})^{\mathrm{T}} = \boldsymbol{A}^{\mathrm{T}} \otimes \boldsymbol{A}^{\mathrm{T}}, \tag{A.7}$$

$$\boldsymbol{A} \otimes \alpha = \alpha \otimes \boldsymbol{A} = \alpha \boldsymbol{A}, \ \alpha \in \mathbb{C}, \tag{A.8}$$

$$\boldsymbol{a}^{\mathrm{T}} \otimes \boldsymbol{b} = \boldsymbol{b} \otimes \boldsymbol{a}^{\mathrm{T}} \in \mathbb{C}^{N\times M}, \ \boldsymbol{a} \in \mathbb{C}^{M}, \ \boldsymbol{b} \in \mathbb{C}^{N}, \tag{A.9}$$

$$(\boldsymbol{A} \otimes \boldsymbol{B})(\boldsymbol{C} \otimes \boldsymbol{D}) = (\boldsymbol{AC}) \otimes (\boldsymbol{BD}), \tag{A.10}$$

$$(\boldsymbol{M} \otimes \boldsymbol{N})^{-1} = \boldsymbol{M}^{-1} \otimes \boldsymbol{N}^{-1}, \ \boldsymbol{M}, \ \boldsymbol{N} \text{ regular}, \tag{A.11}$$

$$\mathrm{tr}[\boldsymbol{M} \otimes \boldsymbol{N}] = \mathrm{tr}[\boldsymbol{M}]\,\mathrm{tr}[\boldsymbol{N}], \tag{A.12}$$

$$\det[\boldsymbol{M} \otimes \boldsymbol{N}] = \det[\boldsymbol{M}]^{N} \det[\boldsymbol{N}]^{M}. \tag{A.13}$$

The operation

$$\mathbf{vec}[\boldsymbol{A}] \triangleq \begin{bmatrix} \boldsymbol{a}_1 \\ \vdots \\ \boldsymbol{a}_N \end{bmatrix} \in \mathbb{C}^{MN} \tag{A.14}$$

stacks the columns of the matrix $\boldsymbol{A} = [\boldsymbol{a}_1, \dots, \boldsymbol{a}_N] \in \mathbb{C}^{M\times N}$ in one vector. It is related to the Kronecker product by

$$\mathbf{vec}[\boldsymbol{ABC}] = (\boldsymbol{C}^{\mathrm{T}} \otimes \boldsymbol{A})\mathbf{vec}[\boldsymbol{B}] \in \mathbb{C}^{MQ} \tag{A.15}$$

with $\boldsymbol{B} \in \mathbb{C}^{N\times P}$ and $\boldsymbol{C} \in \mathbb{C}^{P\times Q}$.

A.2.2 Schur Complement and Matrix Inversion Lemma

Define the portioning of

$$\boldsymbol{A} = \begin{bmatrix} \boldsymbol{A}_{1,1} & \boldsymbol{A}_{1,2} \\ \boldsymbol{A}_{2,1} & \boldsymbol{A}_{2,2} \end{bmatrix} \in \mathbb{C}^{M+N\times M+N} \tag{A.16}$$

with submatrices $\boldsymbol{A}_{1,1} \in \mathbb{C}^{M\times M}$, $\boldsymbol{A}_{1,2} \in \mathbb{C}^{M\times N}$, $\boldsymbol{A}_{2,1} \in \mathbb{C}^{N\times M}$, and $\boldsymbol{A}_{2,2} \in \mathbb{C}^{N\times N}$.

The *Schur complement* of $\boldsymbol{A}_{1,1}$ in $\boldsymbol{A}$ (e.g., [121]) is defined as

$$\mathbf{K}_{\boldsymbol{A}_{1,1}}(\boldsymbol{A}) \triangleq \boldsymbol{A}_{2,2} - \boldsymbol{A}_{2,1}\boldsymbol{A}_{1,1}^{-1}\boldsymbol{A}_{1,2} \in \mathbb{C}^{N\times N}, \tag{A.17}$$

similarly, the Schur complement of $\boldsymbol{A}_{2,2}$ in $\boldsymbol{A}$

$$\mathbf{K}_{\boldsymbol{A}_{2,2}}(\boldsymbol{A}) \triangleq \boldsymbol{A}_{1,1} - \boldsymbol{A}_{1,2}\boldsymbol{A}_{2,2}^{-1}\boldsymbol{A}_{2,1} \in \mathbb{C}^{M\times M}. \tag{A.18}$$

Two important properties regarding the positive (semi-) definiteness of the Schur complement are stated by the following theorems where $\boldsymbol{B}$ is partitioned as $\boldsymbol{A}$.

Theorem A.1 (Theorem 3.1 in [132]). *If* $\boldsymbol{A} \succeq \boldsymbol{B}$, *then* $\mathbf{K}_{\boldsymbol{A}_{1,1}}(\boldsymbol{A}) \succeq \mathbf{K}_{\boldsymbol{B}_{1,1}}(\boldsymbol{B})$.

In [24, p.651] equivalent characterizations of a positive (semi-) definite Schur complement in terms of the positive (semi-) definiteness of the block matrix $\boldsymbol{A}$ is given:

Theorem A.2.

1. $\boldsymbol{A} \succ 0$ *if and only if* $\mathbf{K}_{\boldsymbol{A}_{1,1}}(\boldsymbol{A}) \succ 0$ *and* $\boldsymbol{A}_{1,1} \succ 0$.
2. *If* $\boldsymbol{A}_{1,1} \succ 0$, *then* $\boldsymbol{A} \succeq 0$ *if and only if* $\mathbf{K}_{\boldsymbol{A}_{1,1}}(\boldsymbol{A}) \succeq 0$.

Moreover, throughout the book we require the *matrix inversion lemma* for the matrices $\boldsymbol{A} \in \mathbb{C}^{N\times N}$, $\boldsymbol{B} \in \mathbb{C}^{N\times M}$, $\boldsymbol{C} \in \mathbb{C}^{M\times M}$, and $\boldsymbol{D} \in \mathbb{C}^{M\times N}$:

$$(\boldsymbol{A} + \boldsymbol{BCD})^{-1} = \boldsymbol{A}^{-1} - \boldsymbol{A}^{-1}\boldsymbol{B}(\boldsymbol{D}\boldsymbol{A}^{-1}\boldsymbol{B} + \boldsymbol{C}^{-1})^{-1}\boldsymbol{D}\boldsymbol{A}^{-1}. \tag{A.19}$$

It can be used to prove [124]

$$(\boldsymbol{C}^{-1} + \boldsymbol{B}^{\mathrm{H}}\boldsymbol{R}^{-1}\boldsymbol{B})^{-1}\boldsymbol{B}^{\mathrm{H}}\boldsymbol{R}^{-1} = \boldsymbol{C}\boldsymbol{B}^{\mathrm{H}}(\boldsymbol{B}\boldsymbol{C}\boldsymbol{B}^{\mathrm{H}} + \boldsymbol{R})^{-1} \tag{A.20}$$

with $\boldsymbol{B} \in \mathbb{C}^{M \times N}$ and invertible $\boldsymbol{C} \in \mathbb{C}^{N \times N}$ and $\boldsymbol{R} \in \mathbb{C}^{M \times M}$.

A.2.3 Wirtinger Calculus and Matrix Gradients

To determine the gradient of complex-valued functions $\mathrm{f}(\boldsymbol{z}) : \mathbb{C}^M \rightarrow \mathbb{C}$ with a complex argument $\boldsymbol{z} = \boldsymbol{x} + \mathrm{j}\boldsymbol{y} \in \mathbb{C}^M$ and $\boldsymbol{x}, \boldsymbol{y} \in \mathbb{R}^M$, we could proceed considering it is as a function from $\mathbb{R}^{2M} \rightarrow \mathbb{R}^2$. But very often it is more efficient to define a calculus which works with $\boldsymbol{z}$ directly. In the *Wirtinger calculus* [177] the partial derivative w.r.t. $\boldsymbol{z}$ and its complex conjugate $\boldsymbol{z}^*$ are defined in terms of the partial derivative w.r.t. to the real and imaginary part $\boldsymbol{x}$ and $\boldsymbol{y}$, respectively (see also [108] and references):

$$\frac{\partial}{\partial \boldsymbol{z}} \triangleq \frac{1}{2}\left(\frac{\partial}{\partial \boldsymbol{x}} - \mathrm{j}\frac{\partial}{\partial \boldsymbol{y}}\right) \tag{A.21}$$

$$\frac{\partial}{\partial \boldsymbol{z}^*} \triangleq \frac{1}{2}\left(\frac{\partial}{\partial \boldsymbol{x}} + \mathrm{j}\frac{\partial}{\partial \boldsymbol{y}}\right). \tag{A.22}$$

Formally, we treat the function $\mathrm{f}(\boldsymbol{z})$ as a function $\mathrm{f}(\boldsymbol{z}, \boldsymbol{z}^*)$ in two variables $\boldsymbol{z}$ and $\boldsymbol{z}^*$. With this definition most of the basic properties of the (real) partial derivative still hold: linearity, product rule, and the rule for the derivative of a ratio of two functions. Additionally we have $\partial \mathrm{f}(\boldsymbol{z})^*/\partial \boldsymbol{z} = (\partial \mathrm{f}(\boldsymbol{z})/\partial \boldsymbol{z}^*)^*$ and $\partial \mathrm{f}(\boldsymbol{z})^*/\partial \boldsymbol{z}^* = (\partial \mathrm{f}(\boldsymbol{z})/\partial \boldsymbol{z})^*$. Note that the chain rule is slightly different but still consistent with the interpretation of $\mathrm{f}(\boldsymbol{z})$ as a function in $\boldsymbol{z}$ and $\boldsymbol{z}^*$ [177]. If $\mathrm{f}(\boldsymbol{z})$ does not depend on $\boldsymbol{z}^*$, then $\partial \mathrm{f}(\boldsymbol{z})/\partial \boldsymbol{z}^* = \boldsymbol{0}$.

For functions $\mathrm{f} : \mathbb{C}^{M \times N} \rightarrow \mathbb{C}$ whose argument is a matrix $\boldsymbol{X} \in \mathbb{C}^{M \times N}$ with elements $x_{m,n} = [\boldsymbol{X}]_{m,n}$, it is convenient to work directly with the following *matrix calculus*, which follows from the definition of the Wirtinger derivative above:

$$\frac{\partial \mathrm{f}(\boldsymbol{X})}{\partial \boldsymbol{X}} \triangleq \begin{bmatrix} \frac{\partial \mathrm{f}(\boldsymbol{X})}{\partial x_{1,1}} & \cdots & \frac{\partial \mathrm{f}(\boldsymbol{X})}{\partial x_{1,N}} \\ \vdots & & \vdots \\ \frac{\partial \mathrm{f}(\boldsymbol{X})}{\partial x_{M,1}} & \cdots & \frac{\partial \mathrm{f}(\boldsymbol{X})}{\partial x_{M,N}} \end{bmatrix} \in \mathbb{C}^{M \times N}. \tag{A.23}$$

Defining $\boldsymbol{A} \in \mathbb{C}^{M \times N}$, $\boldsymbol{B} \in \mathbb{C}^{N \times M}$, $\boldsymbol{C} \in \mathbb{C}^{M \times M}$, and $\boldsymbol{R} \in \mathbb{C}^{N \times N}$ the partial derivatives for a number of useful functions are [124, 182]:

$$\frac{\partial \mathrm{tr}[\boldsymbol{A}\boldsymbol{B}]}{\partial \boldsymbol{A}} = \boldsymbol{B}^{\mathrm{T}} \tag{A.24}$$

$$\frac{\partial \mathrm{tr}\left[\boldsymbol{A}\boldsymbol{R}\boldsymbol{A}^{\mathrm{H}}\right]}{\partial \boldsymbol{A}^*} = \boldsymbol{A}\boldsymbol{R} \tag{A.25}$$

$$\frac{\partial \mathrm{tr}[\boldsymbol{A}\boldsymbol{B}]}{\partial \boldsymbol{A}^*} = \boldsymbol{0}_{M\times N} \tag{A.26}$$

$$\frac{\partial \mathrm{tr}[\boldsymbol{A}\boldsymbol{B}\boldsymbol{A}\boldsymbol{C}]}{\partial \boldsymbol{A}} = \boldsymbol{C}^{\mathrm{T}}\boldsymbol{A}^{\mathrm{T}}\boldsymbol{B}^{\mathrm{T}} + \boldsymbol{B}^{\mathrm{T}}\boldsymbol{A}^{\mathrm{T}}\boldsymbol{C}^{\mathrm{T}}. \tag{A.27}$$

For $M = N$, we get

$$\frac{\partial \mathrm{tr}\left[\boldsymbol{A}^{-1}\boldsymbol{B}\right]}{\partial \boldsymbol{A}} = -\left(\boldsymbol{A}^{-1}\boldsymbol{B}\boldsymbol{A}^{-1}\right)^{\mathrm{T}}. \tag{A.28}$$

If $\boldsymbol{A}$ is invertible and $\boldsymbol{A} = \boldsymbol{A}^{\mathrm{H}}$,

$$\frac{\partial \det[\boldsymbol{A}]}{\partial \boldsymbol{A}} = \det[\boldsymbol{A}]\,\boldsymbol{A}^{-\mathrm{T}} \tag{A.29}$$

$$\frac{\partial \ln[\det[\boldsymbol{A}]]}{\partial \boldsymbol{A}} = \boldsymbol{A}^{-\mathrm{T}}. \tag{A.30}$$

For Hermitian and regular $\boldsymbol{C}(\theta)$ with $\theta \in \mathbb{C}$, it can be shown that [124]

$$\frac{\partial \ln[\det[\boldsymbol{C}(\theta)]]}{\partial \theta} = \mathrm{tr}\left[\boldsymbol{C}^{-1}(\theta)\frac{\partial \boldsymbol{C}(\theta)}{\partial \theta}\right] \tag{A.31}$$

$$\frac{\partial \boldsymbol{C}^{-1}(\theta)}{\partial \theta} = -\boldsymbol{C}^{-1}(\theta)\frac{\partial \boldsymbol{C}(\theta)}{\partial \theta}\boldsymbol{C}^{-1}(\theta). \tag{A.32}$$

A.3 Optimization and Karush-Kuhn-Tucker Conditions

Consider the general nonlinear optimization problem

$$\min_{\boldsymbol{x}} \mathrm{F}(\boldsymbol{x}) \quad \text{s.t.} \quad \mathbf{g}(\boldsymbol{x}) \leq \boldsymbol{0}_S,\; \mathbf{h}(\boldsymbol{x}) = \boldsymbol{0}_P \tag{A.33}$$

with optimization variables $\boldsymbol{x} \in \mathbb{C}^M$, cost function $\mathrm{F}(\boldsymbol{x}) : \mathbb{C}^M \to \mathbb{R}$, inequality constraints $\mathbf{g}(\boldsymbol{x}) : \mathbb{C}^M \to \mathbb{R}^S$, and equality constraints $\mathbf{h}(\boldsymbol{x}) : \mathbb{C}^M \to \mathbb{C}^P$. The inequality is defined elementwise for a vector, i.e., $\mathrm{g}_i(\boldsymbol{x}) \leq 0$ for $\mathbf{g}(\boldsymbol{x}) = [\mathrm{g}_1(\boldsymbol{x}), \ldots, \mathrm{g}_S(\boldsymbol{x})]^{\mathrm{T}}$.

The corresponding Lagrange function $\mathrm{L} : \mathbb{C}^M \times \mathbb{R}^S_{+,0} \times \mathbb{C}^P \to \mathbb{R}$ reads [24]

$$\mathrm{L}(\boldsymbol{x}, \boldsymbol{\lambda}, \boldsymbol{\nu}) = \mathrm{F}(\boldsymbol{x}) + \boldsymbol{\lambda}^{\mathrm{T}}\mathbf{g}(\boldsymbol{x}) + 2\mathrm{Re}\big(\boldsymbol{\nu}^{\mathrm{T}}\mathbf{h}(\boldsymbol{x})\big) \tag{A.34}$$

with Lagrange variables $\boldsymbol{\lambda} \in \mathbb{R}^S_{+,0}$ and $\boldsymbol{\nu} \in \mathbb{C}^P$. If L is differentiable, the *Karush-Kuhn-Tucker (KKT)* conditions

$$\begin{aligned}
\mathbf{g}(\boldsymbol{x}) &\leq \mathbf{0}_S \\
\mathbf{h}(\boldsymbol{x}) &= \mathbf{0}_P \\
\boldsymbol{\lambda} &\geq \mathbf{0}_S \\
\lambda_i \mathrm{g}_i(\boldsymbol{x}) &= 0, \ i = 1, 2, \ldots, S \\
\frac{\partial}{\partial \boldsymbol{x}} \mathrm{L}(\boldsymbol{x}, \boldsymbol{\lambda}, \boldsymbol{\nu}) &= \mathbf{0}_M
\end{aligned} \tag{A.35}$$

are necessary for a minimal point of problem (A.33). Alternatively, the partial derivative of the Lagrange function can be evaluated as $\frac{\partial}{\partial \boldsymbol{x}^*} \mathrm{L}(\boldsymbol{x}, \boldsymbol{\lambda}, \boldsymbol{\nu}) = \mathbf{0}_M$ according to the Wirtinger calculus (Appendix A.2.3) for complex variables $\boldsymbol{x}$.[1]

If the optimization problem is given in terms of a matrix $\boldsymbol{X} \in \mathbb{C}^{M \times N}$ with equality constraint $\mathbf{H}(\boldsymbol{X}) : \mathbb{C}^{M \times N} \rightarrow \mathbb{C}^{P \times Q}$, it is often more convenient to work with the equivalent Lagrange function

$$\mathrm{L}(\boldsymbol{X}, \boldsymbol{\lambda}, \boldsymbol{N}) = \mathrm{F}(\boldsymbol{X}) + \boldsymbol{\lambda}^{\mathrm{T}} \mathbf{g}(\boldsymbol{X}) + 2\mathrm{Re}(\mathrm{tr}[\boldsymbol{N}\mathbf{H}(\boldsymbol{X})]) \tag{A.36}$$

with Lagrange variables $\boldsymbol{N} \in \mathbb{C}^{Q \times P}$ and the corresponding KKT conditions.

[1] $(\frac{\partial}{\partial \boldsymbol{x}} \mathrm{L}(\boldsymbol{x}, \boldsymbol{\lambda}, \boldsymbol{\nu}))^* = \frac{\partial}{\partial \boldsymbol{x}^*} \mathrm{L}(\boldsymbol{x}, \boldsymbol{\lambda}, \boldsymbol{\nu})$ because L is real-valued.

Appendix B
Completion of Covariance Matrices and Extension of Sequences

B.1 Completion of Toeplitz Covariance Matrices

Matrix completion problems deal with *partial matrices*, whose entries are only specified for a subset of the elements. The unknown elements of these matrices have to be completed subject to additional constraints [130]. An important issue is the existence of a completion. If it exists and is not unique, the unknown elements can be chosen according to a suitable criterion, e.g., maximum entropy. Here, we are interested in partial covariance matrices and their completions which have to satisfy the positive semidefinite constraint.

A fully specified matrix is positive semidefinite only if all its principal submatrices are positive semidefinite. A *partial positive semidefinite matrix* is Hermitian regarding the specified entries and all its fully specified principal submatrices[1] are positive semidefinite [117]. From this follows that a partial matrix has a positive semidefinite completion only if it is partially positive semidefinite.

The covariance matrix

$$
\boldsymbol{C}_{\mathrm{T}} = \begin{bmatrix} c[0] & c[1] \dots & c[P(B-1)] \\ c[1]^* & c[0] \ddots & \vdots \\ \vdots & \ddots \ \ddots & c[1] \\ c[P(B-1)]^* \dots & c[1]^* & c[0] \end{bmatrix} \tag{B.1}
$$

of interest in Section 3.4 is Hermitian Toeplitz with first row $[c[0], c[1], \dots, c[P(B-1)]]$. Only every Pth sub-diagonal including the main diagonal is given and the corresponding partial matrix is denoted $\boldsymbol{C}_{\mathrm{T}}^{(P)}$. For $P = 2$ and $B = 3$, the partial matrix is (question marks symbolize the un-

[1] A principal submatrix is obtained deleting a subset of the columns and the corresponding rows of a matrix. The main diagonal of a principal submatrix is on the diagonal of the original matrix.

known elements)

$$
\boldsymbol{C}_{\mathrm{T}}^{(2)} = \begin{bmatrix} c[0] & ? & c[2] & ? & c[4] \\ ? & c[0] & ? & c[2] & ? \\ c[2]^* & ? & c[0] & ? & c[2] \\ ? & c[2]^* & ? & c[0] & ? \\ c[4]^* & ? & c[2]^* & ? & c[0] \end{bmatrix}. \tag{B.2}
$$

All fully specified principal submatrices of $\boldsymbol{C}_{\mathrm{T}}^{(P)}$ are the principal submatrices of

$$
\bar{\boldsymbol{C}}_{\mathrm{T}} = \begin{bmatrix} c[0] & c[P] & \dots & c[P(B-1)] \\ c[P]^* & c[0] & \ddots & \vdots \\ \vdots & \ddots & \ddots & c[P] \\ c[P(B-1)]^* & \dots & c[P]^* & c[0] \end{bmatrix}, \tag{B.3}
$$

which itself is a principal submatrix of $\boldsymbol{C}_{\mathrm{T}}^{(P)}$. Thus, the known elements of $\boldsymbol{C}_{\mathrm{T}}$ have to form a positive semidefinite Toeplitz matrix $\bar{\boldsymbol{C}}_{\mathrm{T}}$. This necessary condition makes sense intuitively, because its elements $\{c[Pi]\}_{i=0}^{B-1}$ form the decimated sequence w.r.t. $c[\ell]$.

A set of specified positions in a partial matrix is called a *pattern*. For the Toeplitz matrix $\boldsymbol{C}_{\mathrm{T}}$, it is defined by

$$
\mathcal{P} = \{\ell | c[\ell] \text{ is specified for } \ell > 0\} \cup \{0\} \tag{B.4}
$$

and includes the main diagonal to avoid the trivial case[2]. A pattern $\mathcal{P}$ is said to be *positive semidefinite completable* if for *every* partial positive semidefinite Toeplitz matrix with pattern $\mathcal{P}$ a completion to a positive semidefinite Toeplitz matrix exists.

Considering all possible patterns $\mathcal{P}$ for $\boldsymbol{C}_{\mathrm{T}}$ the following result is given by Johnson et al.[117]:

Theorem B.1 (Johnson et al.[117]). *A pattern* $\mathcal{P} \subseteq \{0, 1, \dots, P(B-1)\}$ *is positive semidefinite completable if and only if* $\mathcal{P} = \{0, P, P2, \dots, P(B-1)\}$.

Alternatively to the proof given in [117], graph-theoretic arguments (e.g. [131]) lead to the same result. In general the completion is not unique. For example, the completion with zeros, i.e., $c[Pi - \ell] = 0$ for $\ell \in \{1, 2, \dots, P-1\}$ and $i \in \{1, 2, \dots, B-1\}$, yields a positive definite Toeplitz covariance matrix [117].

---

[2] If the main diagonal was not specified, a positive definite completion could always be found.

B.2 Band-Limited Positive Semidefinite Extension of Sequences

Given a *partial sequence* $c[\ell]$ on the interval $\mathcal{I} = \{0, 1, \ldots, N\}$, its *extension* to $\mathbb{Z}$ is a sequence $\hat{c}[\ell]$ with $\hat{c}[\ell] = c[\ell]$ on $\mathcal{I}$. This is also called an *extrapolation.*

We are interested in extensions to positive semidefinite sequences which have a positive real Fourier transform, i.e., a (power) spectral density

$$\mathrm{S}(\omega) \geq 0, \tag{B.5}$$

or in general a spectral distribution $\mathrm{s}(\omega)$ with $\mathrm{S}(\omega) = \mathrm{ds}(\omega)/\mathrm{d}\omega$. Necessarily, we assume that the given partial sequence $c[\ell] = c[-\ell]^*$ is conjugate symmetric as is its positive semidefinite extension.

Conditions for the existence and uniqueness of a positive semidefinite extension are summarized in the following theorem (e.g. [6]).

Theorem B.2. *A positive semidefinite extension of* $\{c[\ell]\}_{\ell=0}^{N}$ *exists if and only if the Toeplitz matrix* $\boldsymbol{C}_\mathrm{T}$ *with first row* $[c[0], c[1], \ldots, c[N]]$ *is positive semidefinite. The extension is unique, if* $\boldsymbol{C}_\mathrm{T}$ *is singular. For singular* $\boldsymbol{C}_\mathrm{T}$ *with rank* $R \leq N$*, it is composed of* R *zero-phase complex sinusoids.*

A *band-limited extension* with maximum frequency $\omega_{\max} = 2\pi f_{\max}$ has a Fourier transform, which is only non-zero on $\Omega = [-\omega_{\max}, \omega_{\max}] \subseteq [-\pi, \pi]$. It is easy to construct an extension which is band-limited and *not* positive semidefinite: Given any band-limited sequence a linear FIR filter can be constructed, which creates a (necessarily) band-limited sequence that is identical to $c[\ell]$ on $\mathcal{I}$ [181]. This sequence is not necessarily positive semidefinite even if the filter input is a band-limited positive semidefinite sequence.

Necessary and sufficient conditions for the existence and uniqueness of a band-limited positive semidefinite extension are presented by Arun and Potter [6]. They propose a linear filter to test a sequence $\{c[\ell]\}_{\ell=-\infty}^{\infty}$ for being positive semidefinite and band-limited:

$$\boxed{c'[\ell] = c[\ell-1] - 2\cos(\omega_{\max})c[\ell] + c[\ell+1].} \tag{B.6}$$

The frequency response of this filter is only positive on $[-\omega_{\max}, \omega_{\max}]$ (Figure B.1), i.e., if the input sequence is not band-limited to $\omega_{\max}$ or indefinite on Ω, the output is an indefinite sequence. A test for band-limited positive semidefinite extendibility can be derived based on the in- and output of this filter. It results in the following characterization.

Theorem B.3 (Arun and Potter [6]).

1. *The sequence* $\{c[\ell]\}_{\ell=0}^{N}$ *has a positive semidefinite extension band-limited to* $\omega_{\max}$ *if and only if the* $(N+1) \times (N+1)$ *Toeplitz matrix* $\boldsymbol{C}_\mathrm{T}$ *with first row* $[c[0], c[1], \ldots, c[N]]$ *and the* $N \times N$ *Toeplitz matrix* $\boldsymbol{C}'_\mathrm{T}$ *with first row*

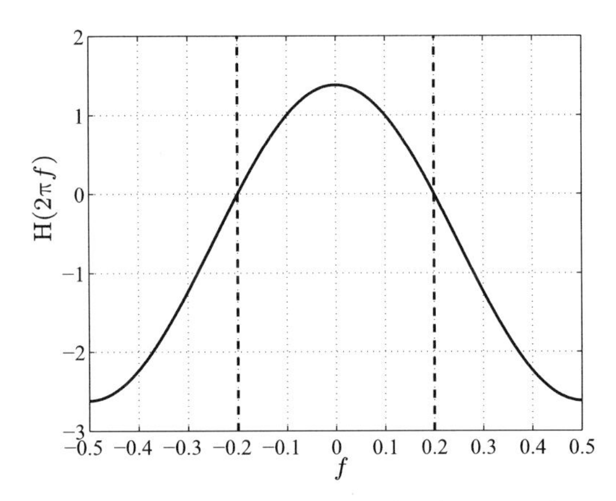

Fig. B.1 Frequency response $\mathrm{H}(2\pi f) = 2\cos(2\pi f) - 2\cos(2\pi f_{\mathrm{max}})$ of (B.6) for $f_{\mathrm{max}} = 0.2$ ($\omega_{\mathrm{max}} = 0.4\pi$).

$[c'[0], c'[1], \ldots, c'[N-1]]$ *are both positive semidefinite, where* $c'[\ell]$ *is given by* (B.6).

2. *If this partial sequence is extendible in this sense, the extrapolation is unique if and only if* $\boldsymbol{C}_{\mathrm{T}}$ *or* $\boldsymbol{C}'_{\mathrm{T}}$, *or both, is singular.*

We illustrate this theorem with two examples: The first is required for our argumentation in Section 2.4.3.2, the second example for Section 3.4.

Example B.1. Given the sequence $c[0] = 1$ and $c[\ell] = 0, \ell \in \{1, 2, \ldots, B\}$, for which maximum frequencies f_{max} does a positive semidefinite band-limited extension exist?

The output in (B.6) is

$$c'[0] = -2\cos(\omega_{\mathrm{max}}) \tag{B.7}$$

$$c'[1] = 1 \tag{B.8}$$

$$c'[\ell] = 0, \ell \geq 2, \tag{B.9}$$

which yields a tridiagonal matrix $\boldsymbol{C}'_{\mathrm{T}}$. $\boldsymbol{C}_{\mathrm{T}}$ from Theorem B.3 is always positive semidefinite. Necessary conditions for a positive semidefinite $\boldsymbol{C}'_{\mathrm{T}}$ are

- $c'[0] = -2\cos(\omega_{\mathrm{max}}) \geq 0$, which requires $\omega_{\mathrm{max}} \geq \pi/2$ ($f_{\mathrm{max}} \geq 0.25$) and
- $c'[0] = -2\cos(\omega_{\mathrm{max}}) \geq |c'[1]| = 1$, which is given for $\omega_{\mathrm{max}} \geq 2\pi/3$ ($f_{\mathrm{max}} \geq 1/3$).

The exact value of f_{max} for sufficiency depends on B: For $B = 1$, we have

$$\boldsymbol{C}'_{\mathrm{T}} = -2\cos(\omega_{\mathrm{max}}) \geq 0 \tag{B.10}$$

and $f_{\mathrm{max}} \geq 0.25$ is also sufficient. For $B = 2$, it yields

$$\boldsymbol{C}'_{\mathrm{T}} = \begin{bmatrix} -2\cos(\omega_{\max}) & 1 \\ 1 & -2\cos(\omega_{\max}) \end{bmatrix}, \tag{B.11}$$

which is also positive semidefinite for $f_{\max} \geq 1/3$. The necessary and sufficient $f_{\max}$ increases further for $B = 3$: $f_{\max} \geq 0.375$. For $B \rightarrow \infty$, we have $f_{\max} \rightarrow 0.5$, i.e., no band-limited positive semidefinite extension exists in the limit. □

Example B.2. The partial decimated sequence $c[0], c[P], \ldots, c[P(B-1)]$ is given. Does the sequence $c[0], 0, \ldots, 0, c[P], \ldots, 0, \ldots, 0, c[P(B-1)]$, which is the given sequence interpolated with $P-1$ zeros, have a positive semidefinite and band-limited extension?

For $P = 2$, the output in (B.6) is

$$c'[0] = -2\cos(\omega_{\max})c[0] \tag{B.12}$$

$$c'[1] = c[0] + c[2] \tag{B.13}$$

$$c'[2] = -2\cos(\omega_{\max})c[2], \quad \text{etc.} \tag{B.14}$$

Due to $c'[0] \geq 0$ a necessary condition is $f_{\max} \geq 0.25$, which coincides with the condition of the sampling theorem for the reconstruction of $c[\ell]$ from the sequence decimated by $P = 2$.

For $P = 3$, we obtain a second necessary condition $c'[1] = c[0] \leq c'[0]$, which is always true for $f_{\max} \geq 1/3$. □

B.3 Generalized Band-Limited Trigonometric Moment Problem

Combining the results on *completion* of partial positive semidefinite Toeplitz matrices (Section B.1) and the band-limited *extension* of conjugate symmetric sequences (Section B.2), we can solve the following generalized trigonometric moment problem.

Problem Statement: What are the necessary and sufficient conditions for the existence of a (band-limited) spectral distribution $\mathrm{s}(\omega)$, which satisfies the interpolation conditions

$$c[\ell] = \frac{1}{2\pi} \int_{\Omega} \exp(\mathrm{j}\omega\ell) \mathrm{ds}(\omega), \quad \ell \in \mathcal{P} \subseteq \{0, 1, \ldots, N\},\ \{0, N\} \subseteq \mathcal{P} \tag{B.15}$$

for $\Omega = [-\omega_{\max}, \omega_{\max}]$ with $\omega_{\max} = 2\pi f_{\max}$? And is it unique?

The constraint $N \in \mathcal{P}$ is imposed without loss of generality to simplify the discussion: Else the existence has to be proved for an effectively shorter partial sequence. Moreover, we define $\boldsymbol{C}_{\mathrm{T}}^{(\mathcal{P})}$ as the partial and Hermitian Toeplitz matrix for $c[\ell], \ell \in \mathcal{P}$.

The band-limited spectral distribution $\mathrm{s}(\omega)$ is constant outside Ω with $\mathrm{s}(-\omega_{\max}) = 0$ and $\mathrm{s}(\omega_{\max}) = 2\pi c[0]$. If $\mathrm{s}(\omega)$ is differentiable, this corresponds to the existence of a band-limited positive real (power) spectral density $\mathrm{S}(\omega) = \frac{\mathrm{ds}(\omega)}{\mathrm{d}\omega} \geq 0$, which is non-zero only on Ω, with

$$c[\ell] = \frac{1}{2\pi} \int_{-\omega_{\max}}^{\omega_{\max}} \exp(\mathrm{j}\omega\ell)\mathrm{S}(\omega)\mathrm{d}\omega, \quad \ell \in \mathcal{P} \subseteq \{0, 1, \ldots, N\}, \ \{0, N\} \subseteq \mathcal{P}. \tag{B.16}$$

Given a suitable spectral distribution the extension of the sequence $c[\ell], \ell \in \mathbb{Z}$ can be computed and vice versa.

We summarize the conditions in a theorem.

Theorem B.4. *Given* $c[\ell], \ell \in \mathcal{P}$, *as in* (B.15) *and a maximum frequency* $f_{\max}$. *For every partially positive semidefinite* $\boldsymbol{C}_{\mathrm{T}}^{(\mathcal{P})}$, *a spectral distribution* $\mathrm{s}(\omega)$ *satisfying the interpolation conditions* (B.15) *for* $\Omega = [-\omega_{\max}, \omega_{\max}]$ *with* $\omega_{\max} = 2\pi f_{\max} \leq \pi$ *exists if and only if*

1. $\mathcal{P}$ *is a periodic pattern*

$$\mathcal{P} = \{\ell | \ell = Pi, i \in \{0, 1, \ldots, B-1\}, N = P(B-1)\}, \tag{B.17}$$

2. and the following additional condition holds for the case $f_{\max} < \frac{1}{2P}$: $\bar{\boldsymbol{C}}_{\mathrm{T}}'$ *with first row* $[c'[0], c'[P], \ldots, c'[P(B-2)]]$ *is positive semidefinite, where* $c'[Pi]$ *is defined as* $c'[Pi] = c[P(i-1)] - 2\cos(2\pi f'_{\max})c[Pi] + c[P(i+1)]$ *for* $f'_{\max} = f_{\max}P$.

Under these conditions, the spectral distribution is unique if and only if the Toeplitz matrix $\bar{\boldsymbol{C}}_{\mathrm{T}}$ *or* $\bar{\boldsymbol{C}}_{\mathrm{T}}'$, *or both, is singular and* $f_{\max} \leq \frac{1}{2P}$.

Proof. The proof follows directly from Theorems B.1, B.3, and a reasoning similar to the sampling theorem. We distinguish four cases of $f_{\max}$:

1) $f_{\max} = 1/2$: This corresponds to no band-limitation and is related to Theorem B.1, which shows that a positive definite matrix completion for *all* partially positive semidefinite $\boldsymbol{C}_{\mathrm{T}}^{(\mathcal{P})}$ exists if and only if the pattern $\mathcal{P}$ is periodic. Then partial positive semidefiniteness is equivalent to $\bar{\boldsymbol{C}}_{\mathrm{T}} \succeq 0$ (B.3). Given a positive semidefinite completion, a positive semidefinite extension of this sequence can be found [6]. The *extension* is unique given a singular *completion* $\boldsymbol{C}_{\mathrm{T}}$, but the matrix completion itself is not unique.

A periodic pattern is also necessary for $f_{\max} < 1/2$, because a positive semidefinite $\boldsymbol{C}_{\mathrm{T}}$ is necessary for existence of a (band-limited) positive semidefinite sequence extension. Therefore, we restrict the following argumentation to $\mathcal{P} = \{\ell | \ell = Pi, i \in \{0, 1, \ldots, B-1\}, N = P(B-1)\}$. (Note that Theorem B.4 requires existence for *every* partially positive semidefinite $\boldsymbol{C}_{\mathrm{T}}^{(\mathcal{P})}$.) For the remaining cases with $f_{\max} < 1/2$, we first proof (band-limited) extendibility of the decimated sequence followed by arguments based on interpolation because of the periodicity of $\mathcal{P}$.

2) $1/(2P) < f_{\max} < 1/2$: If and only if $\bar{\boldsymbol{C}}_{\mathrm{T}}$ is positive semidefinite, a positive semidefinite extension of $c[Pi], i \in \{0, 1, \ldots, B-1\}$, exists for $i \in \mathbb{Z}$. We can generate a band-limited interpolation $c[\ell], \ell \in \mathbb{Z}$, which is not unique: For example, the interpolated sequence $c[\ell]$ with maximum frequency $1/(2P)$, which is trivially band-limited to $f_{\max} > 1/(2P)$.

3) $f_{\max} = 1/(2P)$: The condition of the sampling theorem is satisfied. A positive semidefinite extension $c[Pi], i \in \mathbb{Z}$, exists for positive semidefinite $\bar{\boldsymbol{C}}_{\mathrm{T}}$. It is unique if and only if $\bar{\boldsymbol{C}}_{\mathrm{T}}$ is singular. The interpolated sequence $c[\ell], \ell \in \mathbb{Z}$, can be reconstructed from the band-limited spectral distribution of the extended decimated sequence $c[Pi], i \in \mathbb{Z}$, and is unique.

4) $f_{\max} < 1/(2P)$: If and only if $\bar{\boldsymbol{C}}_{\mathrm{T}}$ and $\bar{\boldsymbol{C}}'_{\mathrm{T}}$ based on $c[Pi], i \in \{0, 1, \ldots, B-1\}$, and $c'[Pi]$ for maximum frequency $f'_{\max} = f_{\max}P < 1/2$ (as defined in the Theorem)[3] are positive semidefinite, a band-limited extension $c[Pi], i \in \mathbb{Z}$, exists. It is unique if and only if one of the matrices, or both, is singular. From its spectral distribution follows a unique interpolation $c[\ell], \ell \in \mathbb{Z}$. □

[3] To test the *decimated* sequence for positive semidefinite extendibility, we have to consider the maximum frequency $f'_{\max}$ of the P-fold decimated sequence $c[Pi]$.

Appendix C
Robust Optimization from the Perspective of Estimation Theory

Consider the optimization problem

$$\mathbf{x}_{\text{opt}}(\boldsymbol{h}) = \underset{\boldsymbol{x}}{\operatorname{argmin}}\, \mathrm{F}(\boldsymbol{x};\boldsymbol{h}) \quad \text{s.t.} \quad \boldsymbol{x} \in \mathbb{X}, \tag{C.1}$$

where the cost function depends on an unknown parameter $\boldsymbol{h}$ and the constraint set $\mathbb{X}$ is independent of $\boldsymbol{h}$. Furthermore, an observation $\boldsymbol{y}$ with the probability density $\mathrm{p}_{\boldsymbol{y}}(\boldsymbol{y};\boldsymbol{h})$ is available.

The question arises, whether we should estimate the unknown parameter $\boldsymbol{h}$ and plug this estimate into the cost function *or* estimate the cost function itself. In the first case an additional question arises: What is the optimum estimator for $\boldsymbol{h}$ in the context of this optimization problem?

The Maximum Likelihood Invariance Principle

Based on a stochastic model of $\boldsymbol{y}$ we can obtain the maximum likelihood (ML) estimate for $\boldsymbol{h}$:

$$\hat{\boldsymbol{h}}_{\text{ML}} = \underset{\boldsymbol{h}}{\operatorname{argmax}}\, \mathrm{p}_{\boldsymbol{y}}(\boldsymbol{y};\boldsymbol{h})\,. \tag{C.2}$$

But we are only interested in $\mathrm{F}(\boldsymbol{x};\boldsymbol{h})$ and not explicitly in $\boldsymbol{h}$. According to the ML invariance principle, the ML estimate of a function of $\boldsymbol{h}$ is simply obtained by plugging $\hat{\boldsymbol{h}}_{\text{ML}}$ into this function [124], because the probability distribution of the observation is invariant w.r.t. our actual parameter of interest, i.e., whether we want to estimate $\boldsymbol{h}$ or $\mathrm{F}(\boldsymbol{x};\boldsymbol{h})$. Therefore, the ML estimate of the cost function is

$$\hat{\mathrm{F}}_{\text{ML}}(\boldsymbol{x};\boldsymbol{y}) = \mathrm{F}(\boldsymbol{x};\hat{\boldsymbol{h}}_{\text{ML}}) \tag{C.3}$$

and yields the optimization problem

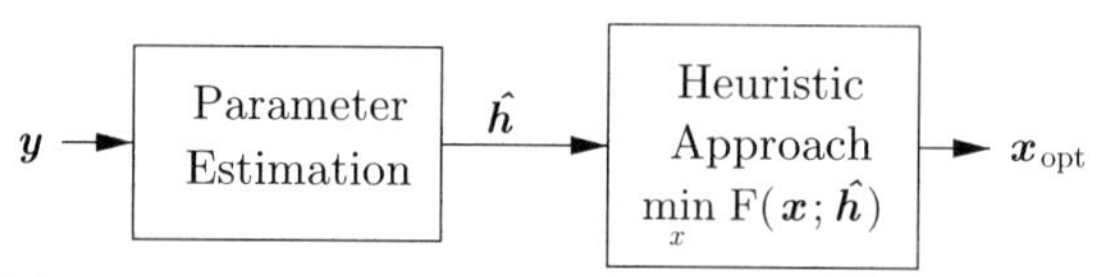

(a) Heuristic Approach: Plug in estimate $\hat{\boldsymbol{h}}$ as if it was error-free.

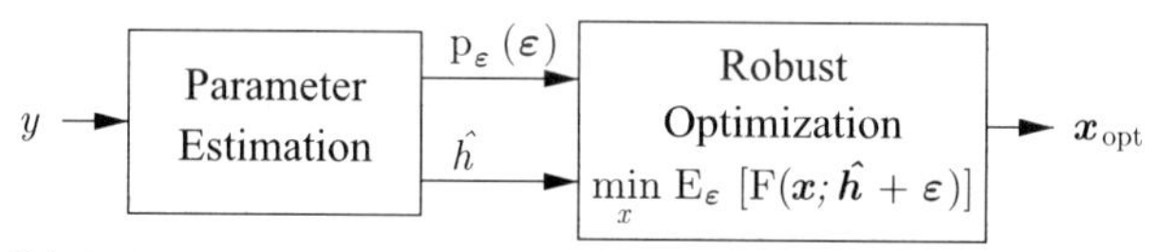

(b) Robust optimization (C.6): The interface between parameter estimation and the application is enhanced and provides also a probability distribution $\mathrm{p}_{\boldsymbol{\varepsilon}}(\boldsymbol{\varepsilon})$ or covariance matrix $\boldsymbol{C}_{\boldsymbol{\varepsilon}}$ of the estimation error $\boldsymbol{\varepsilon}$.

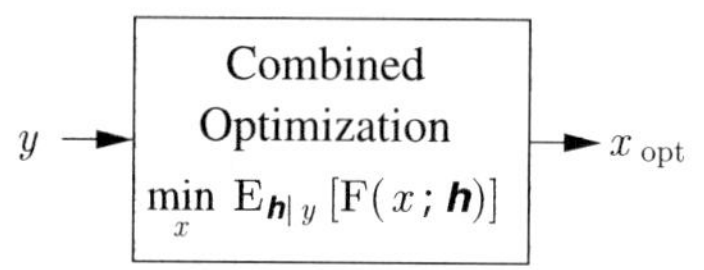

(c) Combined estimation: Parameters $\boldsymbol{h}$ are not estimated explicitly and the observation $\boldsymbol{y}$ is directly applied to cost function following the Bayesian principle (C.5).

Fig. C.1 Optimization for an application with cost function $\mathrm{F}(\boldsymbol{x}; \boldsymbol{h})$ and unknown parameters $\boldsymbol{h}$ which can be estimated from the observations $\boldsymbol{y}$.

$$\mathbf{x}_{\mathrm{opt}}^{\mathrm{ML}}(\hat{\boldsymbol{h}}_{\mathrm{ML}}) = \operatorname*{argmin}_{\boldsymbol{x} \in \mathbb{X}} \mathrm{F}(\boldsymbol{x}; \hat{\boldsymbol{h}}_{\mathrm{ML}}). \tag{C.4}$$

Many results in the literature follow this approach — but some follow this principle, although their estimate $\hat{\boldsymbol{h}}$ does not maximize the likelihood (see first heuristic approach below). What are alternative methods?

The Bayesian Principle

Given the a posteriori density $\mathrm{p}_{\boldsymbol{h}|\boldsymbol{y}}(\boldsymbol{h}|\boldsymbol{y})$, the estimate minimizing the MSE w.r.t. $\boldsymbol{h}$ is the conditional mean (CM) $\hat{\boldsymbol{h}}_{\mathrm{CM}} = \mathrm{E}_{\boldsymbol{h}|\boldsymbol{y}}[\boldsymbol{h}]$ (Section 2.2.1) [124]. Because we are not interested in $\boldsymbol{h}$ directly, we optimize for the cost function with minimum MSE. This yields the CM estimate of the cost function

$$\hat{\mathrm{F}}_{\mathrm{CM}}(\boldsymbol{x}; \boldsymbol{y}) = \mathrm{E}_{\boldsymbol{h}|\boldsymbol{y}}[\mathrm{F}(\boldsymbol{x}; \boldsymbol{h})]\,. \tag{C.5}$$

It is only identical to plugging the CM estimate of $\boldsymbol{h}$ into the cost function (see first heuristic below), i.e., $\mathrm{F}(\boldsymbol{x}; \hat{\boldsymbol{h}}_{\mathrm{CM}})$, if $\mathrm{F}(\boldsymbol{x}; \boldsymbol{h})$ is an affine Function in $\boldsymbol{h}$.

Obviously the CM estimator of $\boldsymbol{h}$ is a necessary step in this approach, but it is not sufficient: For example, also the error covariance matrix $\boldsymbol{C}_{\boldsymbol{h}|\boldsymbol{y}}$ is taken into account in the generally nonlinear estimator of F (C.5).

We can view (C.5) as a combined optimization of parameter estimation and the corresponding application described by $\mathrm{F}(\boldsymbol{x}; \boldsymbol{h})$ and $\mathbb{X}$ (Figure C.1(c)). If $\mathrm{p}_{\boldsymbol{h}|\boldsymbol{y}}(\boldsymbol{h}|\boldsymbol{y})$ is complex Gaussian, additionally to the estimate $\hat{\boldsymbol{h}}_{\mathrm{CM}}$ its error covariance matrix has to be passed from the parameter estimator to the application which is represented by $\mathrm{F}(\boldsymbol{x}; \boldsymbol{h})$ (Figure C.1(b)).

There are numerous applications for this principle in wireless communications: Examples are a receiver design under channel uncertainties [217, 43, 51, 55] and the optimization of the transmitter (Chapters 4 and 5; see also [179, 119, 245])

Robust Optimization

Another approach is known from static stochastic programming [171, 24]: Given an estimate $\hat{\boldsymbol{h}}$ and a stochastic model for its estimation error $\boldsymbol{\varepsilon} = \boldsymbol{h} - \hat{\boldsymbol{h}}$ the expected cost is minimized

$$\min_{\boldsymbol{x} \in \mathbb{X}} \mathrm{E}_{\boldsymbol{\varepsilon}}\left[\mathrm{F}(\boldsymbol{x}; \hat{\boldsymbol{h}} + \boldsymbol{\varepsilon})\right]. \tag{C.6}$$

This is identical to (C.5), if $\boldsymbol{\varepsilon}$ is statistically independent of $\boldsymbol{y}$

$$\mathrm{E}_{\boldsymbol{\varepsilon}|\boldsymbol{y}}\left[\mathrm{F}(\boldsymbol{x}; \hat{\boldsymbol{h}} + \boldsymbol{\varepsilon})\right] = \mathrm{E}_{\boldsymbol{\varepsilon}}\left[\mathrm{F}(\boldsymbol{x}; \hat{\boldsymbol{h}} + \boldsymbol{\varepsilon})\right], \tag{C.7}$$

i.e., all a priori information included in $\mathrm{p}_{\boldsymbol{\varepsilon}|\boldsymbol{y}}(\boldsymbol{\varepsilon}|\boldsymbol{y})$ has already been exploited in $\hat{\boldsymbol{h}}$.

Heuristic Approach: Plug Estimate into Cost Function

As mentioned above, the standard method for dealing with an optimization problem such as (C.1) applies the estimated parameters $\hat{\boldsymbol{h}}$ as if they were the true parameters (Figure C.1(a)). Thus, if knowledge about the size and structure of the error in $\hat{\boldsymbol{h}}$ is available, it is neglected.

Consider the Taylor expansion of $\mathrm{F}(\boldsymbol{x}; \hat{\boldsymbol{h}} + \boldsymbol{\varepsilon})$ at $\boldsymbol{\varepsilon} = \mathrm{E}[\boldsymbol{\varepsilon}]$. Assuming an error $\boldsymbol{\varepsilon}$ small enough such that the expansion up to the linear terms

$$\mathrm{F}(\boldsymbol{x};\hat{\boldsymbol{h}}+\boldsymbol{\varepsilon}) \approx \mathrm{F}(\boldsymbol{x};\hat{\boldsymbol{h}}+\mathrm{E}[\boldsymbol{\varepsilon}]) + \left.\frac{\partial \mathrm{F}(\boldsymbol{x};\hat{\boldsymbol{h}}+\boldsymbol{\varepsilon})}{\partial \boldsymbol{\varepsilon}^{\mathrm{T}}}\right|_{\boldsymbol{\varepsilon}=\mathrm{E}[\boldsymbol{\varepsilon}]}(\boldsymbol{\varepsilon}-\mathrm{E}[\boldsymbol{\varepsilon}]) + \left.\frac{\partial \mathrm{F}(\boldsymbol{x};\hat{\boldsymbol{h}}+\boldsymbol{\varepsilon})}{\partial \boldsymbol{\varepsilon}^{\mathrm{H}}}\right|_{\boldsymbol{\varepsilon}=\mathrm{E}[\boldsymbol{\varepsilon}]}(\boldsymbol{\varepsilon}^{*}-\mathrm{E}[\boldsymbol{\varepsilon}]^{*}) \quad (\mathrm{C.8})$$

is sufficient, the following approximation is valid:

$$\mathrm{E}_{\boldsymbol{\varepsilon}}\left[\mathrm{F}(\boldsymbol{x};\hat{\boldsymbol{h}}+\boldsymbol{\varepsilon})\right] \approx \mathrm{F}(\boldsymbol{x};\hat{\boldsymbol{h}}+\mathrm{E}_{\boldsymbol{\varepsilon}}[\boldsymbol{\varepsilon}]). \quad (\mathrm{C.9})$$

Similarly, for an error covariance matrix $\boldsymbol{C}_{\boldsymbol{h}|\boldsymbol{y}}$ of sufficiently small norm, a Taylor expansion of $\mathrm{F}(\boldsymbol{x};\boldsymbol{h})$ at $\boldsymbol{h}=\hat{\boldsymbol{h}}_{\mathrm{CM}}$ can be terminated after the linear term and yields

$$\mathrm{E}_{\boldsymbol{h}|\boldsymbol{y}}[\mathrm{F}(\boldsymbol{x};\boldsymbol{h})] \approx \mathrm{F}(\boldsymbol{x};\hat{\boldsymbol{h}}_{\mathrm{CM}}). \quad (\mathrm{C.10})$$

We conclude that the naive heuristic which uses a parameter estimate as if it was error-free remains valid for small errors. What is considered "small", is determined by the properties of $\mathrm{F}(\boldsymbol{x};\boldsymbol{h})$ in $\boldsymbol{h}$, i.e., the accuracy of the linear approximation (C.8).

Heuristic Approach: Estimate Functions of Parameters in Solution

Finding a closed form expression for the expectation of the cost function as in (C.5) and (C.6) can be difficult. Alternatively we could also estimate the argument of the original optimization problem (C.1)

$$\mathrm{E}_{\boldsymbol{h}|\boldsymbol{y}}[\mathbf{x}_{\mathrm{opt}}(\boldsymbol{h})], \quad (\mathrm{C.11})$$

which is usually not less complex (e.g., [104]). If a numerical approximation is not an option for practical reasons and the cost function or the solution $\mathbf{x}_{\mathrm{opt}}(\boldsymbol{h})$ are cascaded functions in $\boldsymbol{h}$, we can also estimate a partial function of the cascade: For example, if $\mathrm{F}(\boldsymbol{x};\boldsymbol{h})$ or $\mathbf{x}_{\mathrm{opt}}(\boldsymbol{h})$ depends on $\boldsymbol{h}$ via a quadratic form, we may estimate the quadratic form directly and plug the solution into the cost function or solution, respectively [52]. This will also reduce the estimation error of the cost function.

Appendix D
Detailed Derivations for Precoding with Partial CSI

D.1 Linear Precoding Based on Sum Mean Square Error

For the solution of the optimization problem

$$\min_{\boldsymbol{P},\beta} \mathrm{F}(\boldsymbol{P},\beta) \quad \text{s.t.} \quad \mathrm{tr}\left[\boldsymbol{P}\boldsymbol{C}_{\boldsymbol{d}}\boldsymbol{P}^{\mathrm{H}}\right] \leq P_{\mathrm{tx}} \tag{D.1}$$

with cost function[1] (non-singular $\boldsymbol{C}_{\boldsymbol{d}}$)

$$\mathrm{F}(\boldsymbol{P},\beta) = \mathrm{tr}[\boldsymbol{C}_{\boldsymbol{d}}] + |\beta|^2\bar{G} + |\beta|^2\mathrm{tr}\left[\boldsymbol{C}_{\boldsymbol{d}}\boldsymbol{P}^{\mathrm{H}}\boldsymbol{R}\boldsymbol{P}\right] - 2\mathrm{Re}(\mathrm{tr}[\beta\boldsymbol{H}_{\mathrm{G}}\boldsymbol{P}\boldsymbol{C}_{\boldsymbol{d}}]) \tag{D.2}$$

and effective channel $\boldsymbol{H}_{\mathrm{G}} = \mathrm{E}_{\boldsymbol{h}|\boldsymbol{y}_{\mathrm{T}}}[\mathbf{G}(\boldsymbol{h})\boldsymbol{H}]$, we form the Lagrange function

$$\mathrm{L}(\boldsymbol{P},\beta,\mu) = \mathrm{F}(\boldsymbol{P},\beta) + \mu\left(\mathrm{tr}\left[\boldsymbol{P}^{\mathrm{H}}\boldsymbol{P}\boldsymbol{C}_{\boldsymbol{d}}\right] - P_{\mathrm{tx}}\right). \tag{D.3}$$

The definitions of the parameters $\bar{G}$ and $\boldsymbol{R}$ vary depending on the specific problem and are defined in Chapter 4. The necessary KKT conditions (Appendix A.3) for this non-convex problem are

$$\frac{\partial \mathrm{L}}{\partial \boldsymbol{P}^*} = |\beta|^2\boldsymbol{R}\boldsymbol{P}\boldsymbol{C}_{\boldsymbol{d}} - \beta^*\boldsymbol{H}_{\mathrm{G}}^{\mathrm{H}}\boldsymbol{C}_{\boldsymbol{d}} + \mu\boldsymbol{P}\boldsymbol{C}_{\boldsymbol{d}} = \mathbf{0}_{M\times K} \tag{D.4}$$

$$\frac{\partial \mathrm{L}}{\partial \beta^*} = \beta\bar{G} + \beta\mathrm{tr}\left[\boldsymbol{P}^{\mathrm{H}}\boldsymbol{R}\boldsymbol{P}\boldsymbol{C}_{\boldsymbol{d}}\right] - \mathrm{tr}\left[\boldsymbol{P}^{\mathrm{H}}\boldsymbol{H}_{\mathrm{G}}^{\mathrm{H}}\boldsymbol{C}_{\boldsymbol{d}}\right] = 0 \tag{D.5}$$

$$\mathrm{tr}\left[\boldsymbol{P}^{\mathrm{H}}\boldsymbol{P}\boldsymbol{C}_{\boldsymbol{d}}\right] \leq P_{\mathrm{tx}} \tag{D.6}$$

$$\mu(\mathrm{tr}\left[\boldsymbol{P}^{\mathrm{H}}\boldsymbol{P}\boldsymbol{C}_{\boldsymbol{d}}\right] - P_{\mathrm{tx}}) = 0. \tag{D.7}$$

As in [116, 108] we first solve (D.5) for β

[1] For example, it is equivalent to the sum MSE $\mathrm{E}_{\boldsymbol{d},\boldsymbol{n},\boldsymbol{h}|\boldsymbol{y}_{\mathrm{T}}}[\|\hat{\boldsymbol{d}} - \boldsymbol{d}\|_2^2]$ with $\hat{\boldsymbol{d}} = \beta\mathbf{G}(\boldsymbol{h})\boldsymbol{H}\boldsymbol{P}\boldsymbol{d} + \beta\mathbf{G}(\boldsymbol{h})\boldsymbol{n}$ based on the model in Figure 4.4.

$$\beta = \mathrm{tr}\left[\boldsymbol{P}^{\mathrm{H}}\boldsymbol{H}_{\mathrm{G}}^{\mathrm{H}}\boldsymbol{C}_{\boldsymbol{d}}\right] / \left(\bar{G} + \mathrm{tr}\left[\boldsymbol{P}^{\mathrm{H}}\boldsymbol{R}\boldsymbol{P}\boldsymbol{C}_{\boldsymbol{d}}\right]\right) \tag{D.8}$$

and substitute it in (D.4). The resulting expression is multiplied by $\boldsymbol{P}^{\mathrm{H}}$ from the left. Applying the trace operation we obtain

$$\frac{\left|\mathrm{tr}\left[\boldsymbol{P}^{\mathrm{H}}\boldsymbol{H}_{\mathrm{G}}^{\mathrm{H}}\boldsymbol{C}_{\boldsymbol{d}}\right]\right|^2}{\left(\bar{G} + \mathrm{tr}[\boldsymbol{P}^{\mathrm{H}}\boldsymbol{R}\boldsymbol{P}\boldsymbol{C}_{\boldsymbol{d}}]\right)^2}\mathrm{tr}\left[\boldsymbol{P}^{\mathrm{H}}\boldsymbol{R}\boldsymbol{P}\boldsymbol{C}_{\boldsymbol{d}}\right] - \frac{\left|\mathrm{tr}\left[\boldsymbol{P}^{\mathrm{H}}\boldsymbol{H}_{\mathrm{G}}^{\mathrm{H}}\boldsymbol{C}_{\boldsymbol{d}}\right]\right|^2}{\bar{G} + \mathrm{tr}[\boldsymbol{P}^{\mathrm{H}}\boldsymbol{R}\boldsymbol{P}\boldsymbol{C}_{\boldsymbol{d}}]} + \mu\mathrm{tr}\left[\boldsymbol{P}^{\mathrm{H}}\boldsymbol{P}\boldsymbol{C}_{\boldsymbol{d}}\right] = 0. \tag{D.9}$$

Identifying expression (D.8) for β it simplifies to

$$|\beta|^2\bar{G} = \mu\mathrm{tr}\left[\boldsymbol{P}^{\mathrm{H}}\boldsymbol{P}\boldsymbol{C}_{\boldsymbol{d}}\right]. \tag{D.10}$$

Excluding the meaningless solution $\boldsymbol{P} = \boldsymbol{0}_{M\times K}$ and $\beta = 0$ for the KKT conditions we have $\mu > 0$. With $\mathrm{tr}[\boldsymbol{P}\boldsymbol{C}_{\boldsymbol{d}}\boldsymbol{P}^{\mathrm{H}}] = P_{\mathrm{tx}}$ the Lagrange multiplier is given by

$$\mu = \frac{|\beta|^2\bar{G}}{P_{\mathrm{tx}}}. \tag{D.11}$$

Application to the first condition (D.4) gives

$$\boldsymbol{P} = \frac{\beta^*}{|\beta|^2}\left(\boldsymbol{R} + \frac{\bar{G}}{P_{\mathrm{tx}}}\boldsymbol{I}_M\right)^{-1}\boldsymbol{H}_{\mathrm{G}}^{\mathrm{H}}. \tag{D.12}$$

Assuming $\beta \in \mathbb{R}$ the solution to (D.1) reads

$$\boxed{\begin{aligned} \boldsymbol{P} &= \beta^{-1}\left(\boldsymbol{R} + \frac{\bar{G}}{P_{\mathrm{tx}}}\boldsymbol{I}_M\right)^{-1}\boldsymbol{H}_{\mathrm{G}}^{\mathrm{H}} \\ \beta &= \left(\mathrm{tr}\left[\boldsymbol{H}_{\mathrm{G}}\left(\boldsymbol{R} + \frac{\bar{G}}{P_{\mathrm{tx}}}\boldsymbol{I}_M\right)^{-2}\boldsymbol{H}_{\mathrm{G}}^{\mathrm{H}}\boldsymbol{C}_{\boldsymbol{d}}\right]\right)^{1/2} \Big/ P_{\mathrm{tx}}^{1/2}. \end{aligned}} \tag{D.13}$$

The expression for the minimum of (D.1) can be simplified factoring out $\boldsymbol{H}_{\mathrm{G}}(\boldsymbol{R}+\frac{\bar{G}}{P_{\mathrm{tx}}}\boldsymbol{I}_M)^{-1}$ to the left and $(\boldsymbol{R}+\frac{\bar{G}}{P_{\mathrm{tx}}}\boldsymbol{I}_M)^{-1}\boldsymbol{H}_{\mathrm{G}}^{\mathrm{H}}\boldsymbol{C}_{\boldsymbol{d}}$ to the right in the trace of the second, third, and fourth term in (D.2). Finally, $\min_{\boldsymbol{P},\beta}\mathrm{F}(\boldsymbol{P},\beta)$ can be written as

$$\operatorname{tr}[\boldsymbol{C}_{\boldsymbol{d}}] - \operatorname{tr}\left[\boldsymbol{H}_{\mathrm{G}}\left(\boldsymbol{R}+\frac{\bar{G}}{P_{\mathrm{tx}}}\boldsymbol{I}_M\right)^{-1}\boldsymbol{H}_{\mathrm{G}}^{\mathrm{H}}\boldsymbol{C}_{\boldsymbol{d}}\right]$$
$$= \operatorname{tr}\left[\boldsymbol{C}_{\boldsymbol{d}}\left(\boldsymbol{H}_{\mathrm{G}}\left(\boldsymbol{\mathcal{X}}+\frac{\bar{G}}{P_{\mathrm{tx}}}\boldsymbol{I}_M\right)^{-1}\boldsymbol{H}_{\mathrm{G}}^{\mathrm{H}}+\boldsymbol{I}_K\right)^{-1}\right] \quad \text{(D.14)}$$

with $\boldsymbol{\mathcal{X}} = \boldsymbol{R} - \boldsymbol{H}_{\mathrm{G}}^{\mathrm{H}}\boldsymbol{H}_{\mathrm{G}}$, where we applied the matrix inversion lemma (A.20) with $\boldsymbol{C}^{-1} = \boldsymbol{\mathcal{X}} + \frac{\bar{G}}{P_{\mathrm{tx}}}\boldsymbol{I}_M$ and $\boldsymbol{B} = \boldsymbol{H}_{\mathrm{G}}$ in the second step.

D.2 Conditional Mean for Phase Compensation at the Receiver

For the receiver model from Section 4.3.3

$$\mathrm{g}_k(\boldsymbol{h}_k) = \frac{(\boldsymbol{h}_k^{\mathrm{T}}\boldsymbol{q}_s)^*}{|\boldsymbol{h}_k^{\mathrm{T}}\boldsymbol{q}_s|}\beta, \quad \text{(D.15)}$$

we derive an explicit expression of

$$\mathrm{E}_{\boldsymbol{h}_k|\boldsymbol{y}_{\mathrm{T}}}[\mathrm{g}_k(\boldsymbol{h}_k)\boldsymbol{h}_k] = \beta\mathrm{E}_{\boldsymbol{h}_k|\boldsymbol{y}_{\mathrm{T}}}\left[\frac{(\boldsymbol{h}_k^{\mathrm{T}}\boldsymbol{q}_s)^*}{|\boldsymbol{h}_k^{\mathrm{T}}\boldsymbol{q}_s|}\boldsymbol{h}_k\right]. \quad \text{(D.16)}$$

Defining the random variable $z_k = \boldsymbol{h}_k^{\mathrm{T}}\boldsymbol{q}_s$ we can rewrite it using the properties of the conditional expectation

$$\mathrm{E}_{\boldsymbol{h}_k|\boldsymbol{y}_{\mathrm{T}}}\left[\frac{(\boldsymbol{h}_k^{\mathrm{T}}\boldsymbol{q}_s)^*}{|\boldsymbol{h}_k^{\mathrm{T}}\boldsymbol{q}_s|}\boldsymbol{h}_k\right] = \mathrm{E}_{\boldsymbol{h}_k,z_k|\boldsymbol{y}_{\mathrm{T}}}\left[\frac{z_k^*}{|z_k|}\boldsymbol{h}_k\right] = \mathrm{E}_{z_k|\boldsymbol{y}_{\mathrm{T}}}\left[\frac{z_k^*}{|z_k|}\mathrm{E}_{\boldsymbol{h}_k|z_k,\boldsymbol{y}_{\mathrm{T}}}[\boldsymbol{h}_k]\right]. \quad \text{(D.17)}$$

Because $\mathrm{p}_{\boldsymbol{h}_k|z_k,\boldsymbol{y}_{\mathrm{T}}}(\boldsymbol{h}_k|z_k,\boldsymbol{y}_{\mathrm{T}})$ is complex Gaussian the conditional mean $\mathrm{E}_{\boldsymbol{h}_k|z_k,\boldsymbol{y}_{\mathrm{T}}}[\boldsymbol{h}_k]$ is equivalent to the (affine) LMMSE estimator

$$\mathrm{E}_{\boldsymbol{h}_k|z_k,\boldsymbol{y}_{\mathrm{T}}}[\boldsymbol{h}_k] = \mathrm{E}_{\boldsymbol{h}_k|\boldsymbol{y}_{\mathrm{T}}}[\boldsymbol{h}_k] + \boldsymbol{c}_{\boldsymbol{h}_k z_k|\boldsymbol{y}_{\mathrm{T}}} c_{z_k|\boldsymbol{y}_{\mathrm{T}}}^{-1}\left(z_k - \mathrm{E}_{z_k|\boldsymbol{y}_{\mathrm{T}}}[z_k]\right) \quad \text{(D.18)}$$

with covariances

$$\begin{aligned}\boldsymbol{c}_{\boldsymbol{h}_k z_k|\boldsymbol{y}_{\mathrm{T}}} &= \mathrm{E}_{\boldsymbol{h}_k,z_k|\boldsymbol{y}_{\mathrm{T}}}\left[\left(\boldsymbol{h}_k - \mathrm{E}_{\boldsymbol{h}_k|\boldsymbol{y}_{\mathrm{T}}}[\boldsymbol{h}_k]\right)\left(z_k - \mathrm{E}_{z_k|\boldsymbol{y}_{\mathrm{T}}}[z_k]\right)^*\right] \\ &= \boldsymbol{C}_{\boldsymbol{h}_k|\boldsymbol{y}_{\mathrm{T}}}\boldsymbol{q}_s^* \quad \text{(D.19)}\\ c_{z_k|\boldsymbol{y}_{\mathrm{T}}} &= \mathrm{E}_{z_k|\boldsymbol{y}_{\mathrm{T}}}\left[|z_k - \mathrm{E}_{z_k|\boldsymbol{y}_{\mathrm{T}}}[z_k]|^2\right] = \boldsymbol{q}_s^{\mathrm{H}}\boldsymbol{C}_{\boldsymbol{h}_k|\boldsymbol{y}_{\mathrm{T}}}^*\boldsymbol{q}_s. \quad \text{(D.20)}\end{aligned}$$

The first order moment in (D.18) is

$$\mu_{z_k|\boldsymbol{y}_\mathrm{T}} = \mathrm{E}_{z_k|\boldsymbol{y}_\mathrm{T}}[z_k] = \hat{\boldsymbol{h}}_k^\mathrm{T}\boldsymbol{q}_s \tag{D.21}$$

with $\hat{\boldsymbol{h}}_k = \mathrm{E}_{\boldsymbol{h}_k|\boldsymbol{y}_\mathrm{T}}[\boldsymbol{h}_k]$. The estimator (D.18) has to be understood as the estimator of $\boldsymbol{h}_k$ given z_k under the "a priori distribution" $\mathrm{p}_{\boldsymbol{h}_k|\boldsymbol{y}_\mathrm{T}}(\boldsymbol{h}_k|\boldsymbol{y}_\mathrm{T})$. Defining $\mathrm{g}_k(z_k) = z_k^*/|z_k|$ and applying (D.18) to (D.17) we obtain

$$\begin{aligned}\mathrm{E}_{\boldsymbol{h}_k,z_k|\boldsymbol{y}_\mathrm{T}}[\mathrm{g}_k(z_k)\boldsymbol{h}_k] &= \mathrm{E}_{z_k|\boldsymbol{y}_\mathrm{T}}[\mathrm{g}_k(z_k)]\,\mathrm{E}_{\boldsymbol{h}_k|\boldsymbol{y}_\mathrm{T}}[\boldsymbol{h}_k] + \\ &+ \boldsymbol{c}_{\boldsymbol{h}_k z_k|\boldsymbol{y}_\mathrm{T}} c_{z_k|\boldsymbol{y}_\mathrm{T}}^{-1}\left(\mathrm{E}_{z_k|\boldsymbol{y}_\mathrm{T}}[|z_k|] - \mathrm{E}_{z_k|\boldsymbol{y}_\mathrm{T}}[\mathrm{g}_k(z_k)]\,\mathrm{E}_{z_k|\boldsymbol{y}_\mathrm{T}}[z_k]\right).\end{aligned} \tag{D.22}$$

With [148] the remaining terms, i.e., the CM of the receivers' model $\mathrm{g}_k(z_k)$ and the magnitude $|z_k|$, are given as

$$\begin{aligned}\hat{g}_k &= \mathrm{E}_{z_k|\boldsymbol{y}_\mathrm{T}}[\mathrm{g}_k(z_k)] \\ &= \frac{\sqrt{\pi}}{2}\frac{|\mu_{z_k|\boldsymbol{y}_\mathrm{T}}|}{c_{z_k|\boldsymbol{y}_\mathrm{T}}^{1/2}}\frac{\mu_{z_k|\boldsymbol{y}_\mathrm{T}}^*}{|\mu_{z_k|\boldsymbol{y}_\mathrm{T}}|}\,{}_1F_1\left(\frac{1}{2}, 2, -\frac{|\mu_{z_k|\boldsymbol{y}_\mathrm{T}}|^2}{c_{z_k|\boldsymbol{y}_\mathrm{T}}}\right),\end{aligned} \tag{D.23}$$

where ${}_1F_1(\alpha,\beta,z)$ is the confluent hypergeometric function [2], and

$$\mathrm{E}_{z_k|\boldsymbol{y}_\mathrm{T}}[|z_k|] = \frac{\sqrt{\pi}}{2}c_{z_k|\boldsymbol{y}_\mathrm{T}}^{1/2}\,{}_1F_1\left(-\frac{1}{2}, 1, -\frac{|\mu_{z_k|\boldsymbol{y}_\mathrm{T}}|^2}{c_{z_k|\boldsymbol{y}_\mathrm{T}}}\right). \tag{D.24}$$

In special cases, the confluent hypergeometric functions can be written in terms of the modified Bessel functions of the first kind and order zero $\mathrm{I}_0(z)$ and first order $\mathrm{I}_1(z)$ [147]:

$${}_1F_1(\frac{1}{2}, 2, -z) = \exp(-z/2)\left(\mathrm{I}_0(z/2) + \mathrm{I}_1(z/2)\right) \tag{D.25}$$

$${}_1F_1(-\frac{1}{2}, 1, -z) = \exp(-z/2)\left((1+z)\mathrm{I}_0(z/2) + z\mathrm{I}_1(z/2)\right). \tag{D.26}$$

Asymptotically the modified Bessel functions are approximated by [2]

$$\mathrm{I}_n(z) = \frac{1}{\sqrt{2\pi z}}\exp(z)(1 + \mathrm{O}(1/z)), \quad z \to \infty. \tag{D.27}$$

D.3 An Explicit Solution for Linear Precoding with Statistical Channel State Information

We assume S-CSI at the transmitter with $\mathrm{E}[\boldsymbol{h}] = \boldsymbol{0}$ and cascaded receivers with a perfect phase compensation (knowing $\boldsymbol{h}_k^\mathrm{T}\boldsymbol{p}_k$) in the first stage and a common scaling β (gain control), which is based on the same CSI as available at the transmitter.

Minimization of the CM estimate of the corresponding average sum MSE (4.24) reads

$$\min_{\boldsymbol{P},\beta} \sum_{k=1}^{K} \mathrm{E}_{\boldsymbol{h}_k}\left[\mathrm{MSE}_k\left(\boldsymbol{P}, \beta\frac{(\boldsymbol{h}_k^{\mathrm{T}}\boldsymbol{p}_k)^*}{|\boldsymbol{h}_k^{\mathrm{T}}\boldsymbol{p}_k|}; \boldsymbol{h}_k\right)\right] \quad \text{s.t.} \quad \|\boldsymbol{P}\|_{\mathrm{F}}^2 \leq P_{\mathrm{tx}}. \tag{D.28}$$

Its explicit solution is derived in the sequel [25]. We write the cost function explicitly

$$\begin{aligned}\mathrm{E}_{\boldsymbol{h}_k}\left[\mathrm{MSE}_k\left(\boldsymbol{P}, \beta\frac{(\boldsymbol{h}_k^{\mathrm{T}}\boldsymbol{p}_k)^*}{|\boldsymbol{h}_k^{\mathrm{T}}\boldsymbol{p}_k|}; \boldsymbol{h}_k\right)\right] &= 1 + |\beta|^2 \sum_{i=1}^{K} \boldsymbol{p}_i^{\mathrm{H}} \boldsymbol{C}_{\boldsymbol{h}_k}^* \boldsymbol{p}_i + |\beta|^2 c_{n_k} \\ &\quad - 2\mathrm{Re}\left(\beta\, \mathrm{E}_{\boldsymbol{h}_k}\left[\frac{(\boldsymbol{h}_k^{\mathrm{T}}\boldsymbol{p}_k)^*}{|\boldsymbol{h}_k^{\mathrm{T}}\boldsymbol{p}_k|}\boldsymbol{h}_k\right]^{\mathrm{T}} \boldsymbol{p}_k\right). \end{aligned} \tag{D.29}$$

With (D.24) the cross-correlation term is (with $\mathrm{I}_0(0) = 1$ and $\mathrm{I}_1(0) = 0$)

$$\mathrm{E}_{\boldsymbol{h}_k}\left[\frac{(\boldsymbol{h}_k^{\mathrm{T}}\boldsymbol{p}_k)^*}{|\boldsymbol{h}_k^{\mathrm{T}}\boldsymbol{p}_k|}\boldsymbol{h}_k\right]^{\mathrm{T}} \boldsymbol{p}_k = \mathrm{E}_{\boldsymbol{h}_k}\left[|\boldsymbol{h}_k^{\mathrm{T}}\boldsymbol{p}_k|\right] = \frac{\sqrt{\pi}}{2}\left(\boldsymbol{p}_k^{\mathrm{H}}\boldsymbol{C}_{\boldsymbol{h}_k}^*\boldsymbol{p}_k\right)^{1/2}. \tag{D.30}$$

The Lagrange function for problem (D.28) is

$$\begin{aligned}\mathrm{L}(\boldsymbol{P},\beta,\mu) &= K + |\beta|^2 \sum_{k=1}^{K} \boldsymbol{p}_k^{\mathrm{H}} \mathrm{E}_{\boldsymbol{h}}\left[\boldsymbol{H}^{\mathrm{H}}\boldsymbol{H}\right]\boldsymbol{p}_k + |\beta|^2 \sum_{k=1}^{K} c_{n_k} \\ &\quad - \sqrt{\pi}\mathrm{Re}(\beta) \sum_{k=1}^{K} \left(\boldsymbol{p}_k^{\mathrm{H}}\boldsymbol{C}_{\boldsymbol{h}_k}^*\boldsymbol{p}_k\right)^{1/2} + \mu\left(\sum_{k=1}^{K} \|\boldsymbol{p}_k\|_2^2 - P_{\mathrm{tx}}\right). \end{aligned} \tag{D.31}$$

The necessary KKT conditions are

$$\begin{aligned}\frac{\partial \mathrm{L}(\boldsymbol{P},\beta,\mu)}{\partial \boldsymbol{p}_k^*} &= |\beta|^2 \mathrm{E}_{\boldsymbol{h}}\left[\boldsymbol{H}^{\mathrm{H}}\boldsymbol{H}\right]\boldsymbol{p}_k - \frac{1}{2}\sqrt{\pi}\mathrm{Re}(\beta)\left(\boldsymbol{p}_k^{\mathrm{H}}\boldsymbol{C}_{\boldsymbol{h}_k}^*\boldsymbol{p}_k\right)^{-1/2}\boldsymbol{C}_{\boldsymbol{h}_k}^*\boldsymbol{p}_k \\ &\quad + \mu\boldsymbol{p}_k = \boldsymbol{0}_M \end{aligned} \tag{D.32}$$

$$\frac{\partial \mathrm{L}(\boldsymbol{P},\beta,\mu)}{\partial \beta^*} = \beta\sum_{i=1}^{K} \boldsymbol{p}_i^{\mathrm{H}} \mathrm{E}_{\boldsymbol{h}}\left[\boldsymbol{H}^{\mathrm{H}}\boldsymbol{H}\right]\boldsymbol{p}_i + \beta\sum_{k=1}^{K} c_{n_k} - \frac{\sqrt{\pi}}{2}\sum_{k=1}^{K}\left(\boldsymbol{p}_k^{\mathrm{H}}\boldsymbol{C}_{\boldsymbol{h}_k}^*\boldsymbol{p}_k\right)^{1/2} = 0 \tag{D.33}$$

$$\mathrm{tr}\left[\boldsymbol{P}^{\mathrm{H}}\boldsymbol{P}\right] \leq 0 \tag{D.34}$$

$$\mu(\mathrm{tr}\left[\boldsymbol{P}^{\mathrm{H}}\boldsymbol{P}\right] - P_{\mathrm{tx}}) = 0. \tag{D.35}$$

Proceeding as in Appendix D.1 we obtain

$$\xi = |\beta|^{-2}\mu = \frac{\sum_{k=1}^{K} c_{n_k}}{P_{\mathrm{tx}}}. \tag{D.36}$$

The first condition (D.32) yields

$$\left(\mathrm{E}_{\boldsymbol{h}}\left[\boldsymbol{H}^{\mathrm{H}}\boldsymbol{H}\right] + \xi \boldsymbol{I}_M\right)\boldsymbol{p}_k = \beta^{-1}\frac{\sqrt{\pi}}{2}\left(\boldsymbol{p}_k^{\mathrm{H}}\boldsymbol{C}_{\boldsymbol{h}_k}^{*}\boldsymbol{p}_k\right)^{-1/2}\boldsymbol{C}_{\boldsymbol{h}_k}^{*}\boldsymbol{p}_k, \tag{D.37}$$

which is a generalized eigenvalue problem. We normalize the precoding vector

$$\boldsymbol{p}_k = \alpha_k \tilde{\boldsymbol{p}}_k \tag{D.38}$$

such that $\|\tilde{\boldsymbol{p}}_k\|_2 = 1$ and $\alpha_k \in \mathbb{R}_{+,0}$. For convenience, we transform the generalized eigenvalue problem to a standard eigenvalue problem[2]

$$\boxed{\left(\mathrm{E}_{\boldsymbol{h}}\left[\boldsymbol{H}^{\mathrm{H}}\boldsymbol{H}\right] + \xi \boldsymbol{I}_M\right)^{-1}\boldsymbol{C}_{\boldsymbol{h}_k}^{*}\tilde{\boldsymbol{p}}_k = \alpha_k\beta\frac{2}{\sqrt{\pi}}\left(\tilde{\boldsymbol{p}}_k^{\mathrm{H}}\boldsymbol{C}_{\boldsymbol{h}_k}^{*}\tilde{\boldsymbol{p}}_k\right)^{1/2}\tilde{\boldsymbol{p}}_k.} \tag{D.39}$$

The eigenvalue corresponding to the eigenvector $\tilde{\boldsymbol{p}}_k$ is defined by

$$\sigma_k = \alpha_k\beta\frac{2}{\sqrt{\pi}}\left(\tilde{\boldsymbol{p}}_k^{\mathrm{H}}\boldsymbol{C}_{\boldsymbol{h}_k}^{*}\tilde{\boldsymbol{p}}_k\right)^{1/2}, \tag{D.40}$$

which allows for the computation of the power allocation

$$\alpha_k = \sigma_k\beta^{-1}\frac{\sqrt{\pi}}{2}\left(\tilde{\boldsymbol{p}}_k^{\mathrm{H}}\boldsymbol{C}_{\boldsymbol{h}_k}^{*}\tilde{\boldsymbol{p}}_k\right)^{-1/2}. \tag{D.41}$$

The scaling in $\boldsymbol{P}$ by

$$\beta = \frac{\sqrt{\pi}}{2}\left(\frac{\sum_{k=1}^{K}\sigma_k^2(\tilde{\boldsymbol{p}}_k^{\mathrm{H}}\boldsymbol{C}_{\boldsymbol{h}_k}^{*}\tilde{\boldsymbol{p}}_k)^{-1}}{P_{\mathrm{tx}}}\right)^{1/2} \tag{D.42}$$

results from the total transmit power $P_{\mathrm{tx}} = \sum_{k=1}^{K}\alpha_k^2$.

To derive the minimum MSE we apply $\boldsymbol{p}_k = \alpha_k\tilde{\boldsymbol{p}}_k$, (D.40), and the solution for β to the cost function (D.28). Additionally we use the definition $\boldsymbol{C}_{\boldsymbol{h}_k}^{*}\tilde{\boldsymbol{p}}_k = \sigma_k(\mathrm{E}_{\boldsymbol{h}}\left[\boldsymbol{H}^{\mathrm{H}}\boldsymbol{H}\right] + \xi\boldsymbol{I}_M)\tilde{\boldsymbol{p}}_k$ of σ_k (D.39), which yields $\sigma_k = \tilde{\boldsymbol{p}}_k^{\mathrm{H}}\boldsymbol{C}_{\boldsymbol{h}_k}^{*}\tilde{\boldsymbol{p}}_k/(\tilde{\boldsymbol{p}}_k^{\mathrm{H}}(\mathrm{E}_{\boldsymbol{h}}\left[\boldsymbol{H}^{\mathrm{H}}\boldsymbol{H}\right] + \xi\boldsymbol{I}_M)\tilde{\boldsymbol{p}}_k)$, and can express the minimum in terms of σ_k

$$\sum_{k=1}^{K}\mathrm{E}_{\boldsymbol{h}_k}\left[\mathrm{MSE}_k\left(\boldsymbol{P}, \beta\frac{(\boldsymbol{h}_k^{\mathrm{T}}\boldsymbol{p}_k)^*}{|\boldsymbol{h}_k^{\mathrm{T}}\boldsymbol{p}_k|}; \boldsymbol{h}_k\right)\right] = K - \frac{\pi}{4}\sum_{k=1}^{K}\sigma_k. \tag{D.43}$$

[2] Although the direct solution of the generalized eigenvalue problem is numerically more efficient.

Therefore, choosing $\tilde{\boldsymbol{p}}_k$ as the eigenvector corresponding the largest eigenvalue σ_k yields the smallest MSE.

D.4 Proof of BC-MAC Duality for AWGN BC Model

We complete the proof in Section 4.5.1 and show the converse: For a given dual MAC model, the same performance can be achieved in the BC with same total transmit power. Using the transformations (4.131) to (4.133) the left hand side of (4.127) reads (cf. (4.49))

$$\begin{aligned}\mathrm{E}_{\boldsymbol{h}_k|\boldsymbol{y}_\mathrm{T}}[\mathrm{COR}_k(\boldsymbol{P}, \mathrm{g}_k(\boldsymbol{h}_k); \boldsymbol{h}_k)] &= 1 + \mathrm{E}_{\boldsymbol{h}_k^\mathrm{mac}|\boldsymbol{y}_\mathrm{T}}\left[|\mathrm{g}_k^\mathrm{mac}(\boldsymbol{h}_k^\mathrm{mac})|^2\right] \mathrm{E}_{\boldsymbol{h}_k|\boldsymbol{y}_\mathrm{T}}\left[|\mathrm{g}_k(\boldsymbol{h}_k)|^2\right] \\ &\times \left(c_{n_k} + \sum_{i=1}^{K} c_{n_k} \xi_i^2 \boldsymbol{u}_i^{\mathrm{mac},\mathrm{T}} \boldsymbol{R}_{\boldsymbol{h}_k^\mathrm{mac}|\boldsymbol{y}_\mathrm{T}} \boldsymbol{u}_i^{\mathrm{mac},*} \right) \\ &- 2\mathrm{E}_{\boldsymbol{h}_k^\mathrm{mac}|\boldsymbol{y}_\mathrm{T}}\left[\mathrm{Re}\left(\mathrm{g}_k^\mathrm{mac}(\boldsymbol{h}_k^\mathrm{mac}) \boldsymbol{h}_k^{\mathrm{mac},\mathrm{T}} \boldsymbol{u}_k^\mathrm{mac} p_k^\mathrm{mac} \right)\right]. \end{aligned} \tag{D.44}$$

With (4.129), (4.127) results in

$$\begin{aligned}&\mathrm{E}_{\boldsymbol{h}_k^\mathrm{mac}|\boldsymbol{y}_\mathrm{T}}\left[|\mathrm{g}_k^\mathrm{mac}(\boldsymbol{h}_k^\mathrm{mac})|^2\right] \xi_k^{-2} p_k^{\mathrm{mac},2} \left(1 + \sum_{i=1}^{K} \xi_i^2 \boldsymbol{u}_i^{\mathrm{mac},\mathrm{T}} \boldsymbol{R}_{\boldsymbol{h}_k^\mathrm{mac}|\boldsymbol{y}_\mathrm{T}} \boldsymbol{u}_i^{\mathrm{mac},*} \right) \\ &= \mathrm{E}_{\boldsymbol{h}_k^\mathrm{mac}|\boldsymbol{y}_\mathrm{T}}\left[|\mathrm{g}_k^\mathrm{mac}(\boldsymbol{h}_k^\mathrm{mac})|^2\right] \left(\|\boldsymbol{u}_k^\mathrm{mac}\|_2^2 + \sum_{i=1}^{K} p_i^{\mathrm{mac},2} \boldsymbol{u}_k^{\mathrm{mac},\mathrm{T}} \boldsymbol{R}_{\boldsymbol{h}_i^\mathrm{mac}|\boldsymbol{y}_\mathrm{T}} \boldsymbol{u}_k^{\mathrm{mac},*} \right),\end{aligned} \tag{D.45}$$

which simplifies to

$$\begin{aligned}p_k^{\mathrm{mac},2} = \xi_k^2 &\left(\|\boldsymbol{u}_k^\mathrm{mac}\|_2^2 + \sum_{i=1}^{K} p_i^{\mathrm{mac},2} \boldsymbol{u}_k^{\mathrm{mac},\mathrm{H}} \boldsymbol{R}_{\boldsymbol{h}_i^\mathrm{mac}|\boldsymbol{y}_\mathrm{T}}^* \boldsymbol{u}_k^\mathrm{mac} \right) \\ &- p_k^{\mathrm{mac},2} \sum_{i=1}^{K} \xi_i^2 \boldsymbol{u}_i^{\mathrm{mac},\mathrm{H}} \boldsymbol{R}_{\boldsymbol{h}_k^\mathrm{mac}|\boldsymbol{y}_\mathrm{T}}^* \boldsymbol{u}_i^\mathrm{mac}.\end{aligned} \tag{D.46}$$

Defining $\boldsymbol{l}^\mathrm{mac} = [p_1^{\mathrm{mac},2}, p_2^{\mathrm{mac},2}, \ldots, p_K^{\mathrm{mac},2}]^\mathrm{T}$ and

$$[\boldsymbol{W}^\mathrm{mac}]_{u,v} = \begin{cases} -p_u^{\mathrm{mac},2} \boldsymbol{u}_v^{\mathrm{mac},\mathrm{H}} \boldsymbol{R}_{\boldsymbol{h}_u^\mathrm{mac}|\boldsymbol{y}_\mathrm{T}}^* \boldsymbol{u}_v^\mathrm{mac}, & v \neq u \\ \|\boldsymbol{u}_v^\mathrm{mac}\|_2^2 + \sum\limits_{\substack{i=1 \\ i\neq v}}^{K} p_i^{\mathrm{mac},2} \boldsymbol{u}_v^{\mathrm{mac},\mathrm{H}} \boldsymbol{R}_{\boldsymbol{h}_i^\mathrm{mac}|\boldsymbol{y}_\mathrm{T}}^* \boldsymbol{u}_v^\mathrm{mac}, & u = v \end{cases} \tag{D.47}$$

we obtain the system of equations

$$\boldsymbol{W}^{\mathrm{mac}}\boldsymbol{\xi}^{\mathrm{mac}} = \boldsymbol{l}^{\mathrm{mac}} \tag{D.48}$$

in $\boldsymbol{\xi}^{\mathrm{mac}} = [\xi_1^2, \xi_2^2, \ldots, \xi_K^2]^{\mathrm{T}}$.

If $\|\boldsymbol{u}_v^{\mathrm{mac}}\|_2^2 > 0$ for all v, $\boldsymbol{W}^{\mathrm{mac}}$ is (column) diagonally dominant because of

$$\left[\boldsymbol{W}^{\mathrm{mac}}\right]_{v,v} > \sum_{u\neq v} \left|\left[\boldsymbol{W}^{\mathrm{mac}}\right]_{u,v}\right| = \sum_{u\neq v} p_u^{\mathrm{mac},2} \boldsymbol{u}_v^{\mathrm{mac},\mathrm{H}} \boldsymbol{R}^*_{\boldsymbol{h}_u^{\mathrm{mac}}|\boldsymbol{y}_{\mathrm{T}}} \boldsymbol{u}_v^{\mathrm{mac}}. \tag{D.49}$$

It has the same properties as $\boldsymbol{W}$ in (4.136).

Identical performance is achieved in the BC with same total transmit power as in the dual MAC:

$$\sum_{k=1}^{K} p_k^{\mathrm{mac},2} = \|\boldsymbol{l}^{\mathrm{mac}}\|_1 = \sum_{k=1}^{K} \xi_k^2 \|\boldsymbol{u}_k^{\mathrm{mac}}\|_2^2 = \sum_{k=1}^{K} \|\boldsymbol{p}_k\|_2^2. \tag{D.50}$$

This concludes the "only-if" part of the proof for Theorem 4.1.

D.5 Proof of BC-MAC Duality for Incomplete CSI at Receivers

To prove Theorem 4.2 we rewrite (4.141) using the proposed transformations (4.145)

$$\begin{aligned}
|\beta_k|^2 \mathrm{E}_{\boldsymbol{h}_k|\boldsymbol{y}_{\mathrm{T}}}\left[|\mathrm{g}_k(\boldsymbol{h}_k)|^2\right] c_{n_k} &+ \sum_{i=1}^{K} |\beta_k|^2 \boldsymbol{p}_i^{\mathrm{H}} \mathrm{E}_{\boldsymbol{h}_k|\boldsymbol{y}_{\mathrm{T}}}\left[\boldsymbol{h}_k^* \boldsymbol{h}_k^{\mathrm{T}} |\mathrm{g}_k(\boldsymbol{h}_k)|^2\right] \boldsymbol{p}_i \\
&= \|\boldsymbol{p}_k\|_2^2 \xi_k^{-2} + \sum_{i=1}^{K} |\beta_i|^2 \xi_i^2 \xi_k^{-2} \boldsymbol{p}_k^{\mathrm{H}} \mathrm{E}_{\boldsymbol{h}_i|\boldsymbol{y}_{\mathrm{T}}}\left[\boldsymbol{h}_i^* \boldsymbol{h}_i^{\mathrm{T}} |\mathrm{g}_i(\boldsymbol{h}_i)|^2\right] \boldsymbol{p}_k.
\end{aligned} \tag{D.51}$$

It can be simplified to

$$\begin{aligned}
\|\boldsymbol{p}_k\|_2^2 = \xi_k^2 |\beta_k|^2 &\left(\mathrm{E}_{\boldsymbol{h}_k|\boldsymbol{y}_{\mathrm{T}}}\left[|\mathrm{g}_k(\boldsymbol{h}_k)|^2\right] c_{n_k} + \sum_{i=1}^{K} \boldsymbol{p}_i^{\mathrm{H}} \mathrm{E}_{\boldsymbol{h}_k|\boldsymbol{y}_{\mathrm{T}}}\left[\boldsymbol{h}_k^* \boldsymbol{h}_k^{\mathrm{T}} |\mathrm{g}_k(\boldsymbol{h}_k)|^2\right] \boldsymbol{p}_i \right) \\
&- \sum_{i=1}^{K} |\beta_i|^2 \xi_i^2 \boldsymbol{p}_k^{\mathrm{H}} \mathrm{E}_{\boldsymbol{h}_i|\boldsymbol{y}_{\mathrm{T}}}\left[\boldsymbol{h}_i^* \boldsymbol{h}_i^{\mathrm{T}} |\mathrm{g}_i(\boldsymbol{h}_i)|^2\right] \boldsymbol{p}_k,
\end{aligned} \tag{D.52}$$

which yields the system of linear equations

$$\boldsymbol{W}\boldsymbol{\xi} = \boldsymbol{l}. \tag{D.53}$$

We made the following definitions

$$\boldsymbol{l} = \left[\|\boldsymbol{p}_1\|_2^2, \|\boldsymbol{p}_2\|_2^2, \ldots, \|\boldsymbol{p}_K\|_2^2\right]^{\mathrm{T}} \tag{D.54}$$

$$[\boldsymbol{W}]_{u,v} = \begin{cases} -|\beta_v|^2 \boldsymbol{p}_u^{\mathrm{H}} \mathrm{E}_{\boldsymbol{h}_v|\boldsymbol{y}_{\mathrm{T}}}\left[\boldsymbol{h}_v^* \boldsymbol{h}_v^{\mathrm{T}} |\mathrm{g}_v(\boldsymbol{h}_v)|^2\right] \boldsymbol{p}_u, & u \neq v \\ |\beta_v|^2 \mathrm{E}_{\boldsymbol{h}_v|\boldsymbol{y}_{\mathrm{T}}}[c_{r_v}] - |\beta_v|^2 \boldsymbol{p}_v^{\mathrm{H}} \mathrm{E}_{\boldsymbol{h}_v|\boldsymbol{y}_{\mathrm{T}}}\left[\boldsymbol{h}_v^* \boldsymbol{h}_v^{\mathrm{T}} |\mathrm{g}_v(\boldsymbol{h}_v)|^2\right] \boldsymbol{p}_v, & u = v \end{cases} \tag{D.55}$$

$$\mathrm{E}_{\boldsymbol{h}_v|\boldsymbol{y}_{\mathrm{T}}}[c_{r_v}] = \mathrm{E}_{\boldsymbol{h}_v|\boldsymbol{y}_{\mathrm{T}}}\left[|\mathrm{g}_v(\boldsymbol{h}_v)|^2\right] c_{n_v} + \sum_{i=1}^{K} \boldsymbol{p}_i^{\mathrm{H}} \mathrm{E}_{\boldsymbol{h}_v|\boldsymbol{y}_{\mathrm{T}}}\left[\boldsymbol{h}_v^* \boldsymbol{h}_v^{\mathrm{T}} |\mathrm{g}_v(\boldsymbol{h}_v)|^2\right] \boldsymbol{p}_i. \tag{D.56}$$

If $c_{n_k} > 0$ and $\mathrm{E}_{\boldsymbol{h}_k|\boldsymbol{y}_{\mathrm{T}}}\left[|\mathrm{g}_k(\boldsymbol{h}_k)|^2\right] > 0$ for all k, $\boldsymbol{W}$ is (column) diagonally dominant:

$$[\boldsymbol{W}]_{v,v} > \sum_{u\neq v} \left|[\boldsymbol{W}]_{u,v}\right| = \sum_{u\neq v} |\beta_v|^2 \boldsymbol{p}_u^{\mathrm{H}} \mathrm{E}_{\boldsymbol{h}_v|\boldsymbol{y}_{\mathrm{T}}}\left[\boldsymbol{h}_v^* \boldsymbol{h}_v^{\mathrm{T}} |\mathrm{g}_v(\boldsymbol{h}_v)|^2\right] \boldsymbol{p}_u. \tag{D.57}$$

Hence, it is always invertible. Because the diagonal of $\boldsymbol{W}$ is positive on its diagonal and negative elsewhere, its inverse has only non-negative entries [21] and $\boldsymbol{\xi}$ is always positive.

The sum of all equations in (D.53) can be simplified and using (4.145) and the total transmit power in the BC turns out to be the same as in the MAC:

$$\sum_{k=1}^{K} \|\boldsymbol{p}_k\|_2^2 = 1 + \sum_{k=1}^{K} |\beta_k|^2 c_{n_k} \xi_k^2 \mathrm{E}_{\boldsymbol{h}_k|\boldsymbol{y}_{\mathrm{T}}}\left[|\mathrm{g}_k(\boldsymbol{h}_k)|^2\right] = \sum_{k=1}^{K} p_k^{\mathrm{mac},2}. \tag{D.58}$$

To prove the "only if" part of Theorem 4.2, we rewrite the left hand side of (4.141) with the transformations (4.145)

$$\begin{aligned} &\mathrm{E}_{\boldsymbol{h}_k|\boldsymbol{y}_{\mathrm{T}}}[\mathrm{MSE}_k(\boldsymbol{P}, \beta_k \mathrm{g}_k(\boldsymbol{h}_k); \boldsymbol{h}_k)] = \\ &\quad p_k^{\mathrm{mac},2} \xi_k^{-2} + \sum_{i=1}^{K} \xi_i^2 \xi_k^{-2} p_k^{\mathrm{mac},2} \boldsymbol{g}_i^{\mathrm{mac},\mathrm{H}} \mathrm{E}_{\boldsymbol{h}_k^{\mathrm{mac}}|\boldsymbol{y}_{\mathrm{T}}}\left[\boldsymbol{h}_k^{\mathrm{mac},*} \boldsymbol{h}_k^{\mathrm{mac},\mathrm{T}}\right] \boldsymbol{g}_i^{\mathrm{mac}}. \end{aligned} \tag{D.59}$$

Together with (4.141) this yields

$$\begin{aligned} p_k^{\mathrm{mac},2} = \xi_k^2 &\left(\|\boldsymbol{g}_k^{\mathrm{mac}}\|_2^2 + \sum_{i=1}^{K} p_i^{\mathrm{mac},2} \boldsymbol{g}_k^{\mathrm{mac},\mathrm{H}} \mathrm{E}_{\boldsymbol{h}_i^{\mathrm{mac}}|\boldsymbol{y}_{\mathrm{T}}}\left[\boldsymbol{h}_i^{\mathrm{mac},*} \boldsymbol{h}_i^{\mathrm{mac},\mathrm{T}}\right] \boldsymbol{g}_k^{\mathrm{mac}} \right) \\ &- \sum_{i=1}^{K} p_k^{\mathrm{mac},2} \xi_i^2 \boldsymbol{g}_i^{\mathrm{mac},\mathrm{H}} \mathrm{E}_{\boldsymbol{h}_k^{\mathrm{mac}}|\boldsymbol{y}_{\mathrm{T}}}\left[\boldsymbol{h}_k^{\mathrm{mac},*} \boldsymbol{h}_k^{\mathrm{mac},\mathrm{T}}\right] \boldsymbol{g}_i^{\mathrm{mac}}. \end{aligned} \tag{D.60}$$

This system of linear equations can be written in matrix notation $\boldsymbol{W}^{\mathrm{mac}} \boldsymbol{\xi}^{\mathrm{mac}} = \boldsymbol{l}^{\mathrm{mac}}$ with

$$[\boldsymbol{W}^{\mathrm{mac}}]_{u,v} = \begin{cases} -p_u^{\mathrm{mac},2} \boldsymbol{g}_v^{\mathrm{mac},\mathrm{H}} \mathrm{E}_{\boldsymbol{h}_u^{\mathrm{mac}}|\boldsymbol{y}_{\mathrm{T}}}\left[\boldsymbol{h}_u^{\mathrm{mac},*} \boldsymbol{h}_u^{\mathrm{mac},\mathrm{T}}\right] \boldsymbol{g}_v^{\mathrm{mac}}, & u \neq v \\ c_v - p_v^{\mathrm{mac},2} \boldsymbol{g}_v^{\mathrm{mac},\mathrm{H}} \mathrm{E}_{\boldsymbol{h}_v^{\mathrm{mac}}|\boldsymbol{y}_{\mathrm{T}}}\left[\boldsymbol{h}_v^{\mathrm{mac},*} \boldsymbol{h}_v^{\mathrm{mac},\mathrm{T}}\right] \boldsymbol{g}_v^{\mathrm{mac}}, & u = v \end{cases}, \tag{D.61}$$

the variance

$$c_v = \|\boldsymbol{g}_v^{\mathrm{mac}}\|_2^2 + \sum_{i=1}^{K} p_i^{\mathrm{mac},2} \boldsymbol{g}_v^{\mathrm{mac},\mathrm{H}} \mathrm{E}_{\boldsymbol{h}_i^{\mathrm{mac}}|\boldsymbol{y}_{\mathrm{T}}}\left[\boldsymbol{h}_i^{\mathrm{mac},*} \boldsymbol{h}_i^{\mathrm{mac},\mathrm{T}}\right] \boldsymbol{g}_v^{\mathrm{mac}}$$

of the output of $\boldsymbol{g}_v^{\mathrm{mac}}$, and $\boldsymbol{l}^{\mathrm{mac}} = [p_1^{\mathrm{mac},2}, p_2^{\mathrm{mac},2}, \ldots, p_K^{\mathrm{mac},2}]^{\mathrm{T}}$. If $\|\boldsymbol{g}_k^{\mathrm{mac}}\|_2^2 > 0$ for all k, $\boldsymbol{W}^{\mathrm{mac}}$ is (column) diagonally dominant due to

$$\begin{aligned} [\boldsymbol{W}^{\mathrm{mac}}]_{v,v} > \sum_{u \neq v} \left| [\boldsymbol{W}^{\mathrm{mac}}]_{u,v} \right| = \\ = \sum_{u \neq v} p_u^{\mathrm{mac},2} \boldsymbol{g}_v^{\mathrm{mac},\mathrm{H}} \mathrm{E}_{\boldsymbol{h}_u^{\mathrm{mac}}|\boldsymbol{y}_{\mathrm{T}}}\left[\boldsymbol{h}_u^{\mathrm{mac},*} \boldsymbol{h}_u^{\mathrm{mac},\mathrm{T}}\right] \boldsymbol{g}_v^{\mathrm{mac}}. \end{aligned} \tag{D.62}$$

With the same reasoning as for the transformation from BC to MAC, we conclude that identical performance can be achieved in the BC as in the MAC with the same total transmit power.

D.6 Tomlinson-Harashima Precoding with Sum Performance Measures

For optimization of THP in Section 5.3 by alternating between transmitter and receiver models, we consider the cost function

$$\begin{aligned} \mathrm{F}^{\mathrm{THP}}(\boldsymbol{P}, \boldsymbol{F}, \beta, \mathcal{O}) = |\beta|^2 \bar{G} + |\beta|^2 \mathrm{tr}\left[\boldsymbol{P}^{\mathrm{H}} \boldsymbol{\mathcal{X}} \boldsymbol{P} \boldsymbol{C}_{\boldsymbol{w}}\right] \\ + \mathrm{E}_{\boldsymbol{w}}\left[\left\| \left(\beta \boldsymbol{\Pi}^{(\mathcal{O})} \boldsymbol{H}_{\mathrm{G}} \boldsymbol{P} - (\boldsymbol{I}_K - \boldsymbol{F})\right) \boldsymbol{w}[n] \right\|_2^2\right] \end{aligned} \tag{D.63}$$

with $\boldsymbol{\mathcal{X}} = \boldsymbol{R} - \boldsymbol{H}_{\mathrm{G}}^{\mathrm{H}} \boldsymbol{H}_{\mathrm{G}}$. The definitions of $\bar{G}$, $\boldsymbol{R}$, and $\boldsymbol{H}_{\mathrm{G}}$ depend on the specific problem addressed in Section 5.3. We would like to minimize (D.63) subject to a total transmit power constraint $\mathrm{tr}[\boldsymbol{P}\boldsymbol{C}_{\boldsymbol{w}}\boldsymbol{P}^{\mathrm{H}}] \leq P_{\mathrm{tx}}$ and the constraint on $\boldsymbol{F}$ to be lower triangular with zero diagonal.

First, we minimize (D.63) w.r.t. $\boldsymbol{F}$. Only the third term of (D.63) depends on $\boldsymbol{F}$ and can be simplified to

$$\sum_{k=1}^{K} \left\| \beta \boldsymbol{\Pi}^{(\mathcal{O})} \boldsymbol{H}_{\mathrm{G}} \boldsymbol{p}_k - \boldsymbol{e}_k + \boldsymbol{f}_k \right\|_2^2 c_{w_k} \tag{D.64}$$

because of (5.26). Taking into account the constraint on its structure, the solution $\boldsymbol{F} = [\boldsymbol{f}_1, \ldots, \boldsymbol{f}_K]$ reads

$$\boxed{\boldsymbol{f}_k = -\beta \begin{bmatrix} \mathbf{0}_{k\times M} \\ \boldsymbol{B}_k^{(\mathcal{O})} \end{bmatrix} \boldsymbol{p}_k,} \tag{D.65}$$

where we require the partitioning of

$$\boldsymbol{\Pi}^{(\mathcal{O})} \boldsymbol{H}_{\mathrm{G}} = \begin{bmatrix} \boldsymbol{A}_k^{(\mathcal{O})} \\ \boldsymbol{B}_k^{(\mathcal{O})} \end{bmatrix} \tag{D.66}$$

with $\boldsymbol{A}_k^{(\mathcal{O})} \in \mathbb{C}^{k\times M}$ and $\boldsymbol{B}_k^{(\mathcal{O})} \in \mathbb{C}^{(K-k)\times M}$.

For minimization of (D.63) w.r.t. $\boldsymbol{P}$, we construct the Lagrange function

$$\mathrm{L}(\boldsymbol{P}, \boldsymbol{F}, \beta, \mathcal{O}, \mu) = \mathrm{F}^{\mathrm{THP}}(\boldsymbol{P}, \boldsymbol{F}, \beta, \mathcal{O}) + \mu \left(\mathrm{tr}\left[\boldsymbol{P}\boldsymbol{C}_{\boldsymbol{w}}\boldsymbol{P}^{\mathrm{H}}\right] - P_{\mathrm{tx}}\right) \tag{D.67}$$

to include the transmit power constraint. Identifying $\boldsymbol{C}_{\boldsymbol{d}} = \boldsymbol{C}_{\boldsymbol{w}}$ and $\mathrm{E}_{\boldsymbol{h}|\boldsymbol{y}_{\mathrm{T}}}[\mathbf{G}(\boldsymbol{h})\boldsymbol{H}] = (\boldsymbol{I}_K - \boldsymbol{F})^{\mathrm{H}} \boldsymbol{\Pi}^{(\mathcal{O})} \boldsymbol{H}_{\mathrm{G}}$ in (D.2), its solution is given by (D.4)

$$\boldsymbol{P} = \beta^{-1} \left(\boldsymbol{R} + \frac{\bar{G}}{P_{\mathrm{tx}}} \boldsymbol{I}_M\right)^{-1} \boldsymbol{H}_{\mathrm{G}}^{\mathrm{H}} \boldsymbol{\Pi}^{(\mathcal{O}),\mathrm{T}} (\boldsymbol{I}_K - \boldsymbol{F}). \tag{D.68}$$

Plugging (D.65) into the kth column of (D.68) we get

$$\boldsymbol{p}_k = \beta^{-1} \left(\boldsymbol{R} + \frac{\bar{G}}{P_{\mathrm{tx}}} \boldsymbol{I}_M\right)^{-1} \boldsymbol{H}_{\mathrm{G}}^{\mathrm{H}} \boldsymbol{\Pi}^{(\mathcal{O}),\mathrm{T}} \left(\boldsymbol{e}_k + \beta \begin{bmatrix} \mathbf{0}_{k\times M} \\ \boldsymbol{B}_k^{(\mathcal{O})} \end{bmatrix} \boldsymbol{p}_k\right), \tag{D.69}$$

which can be solved for the kth column of $\boldsymbol{P}$

$$\boxed{\boldsymbol{p}_k = \beta^{-1} \left(\boldsymbol{\mathcal{X}} + \boldsymbol{A}_k^{(\mathcal{O}),\mathrm{H}} \boldsymbol{A}_k^{(\mathcal{O})} + \frac{\bar{G}}{P_{\mathrm{tx}}} \boldsymbol{I}_M\right)^{-1} \boldsymbol{A}_k^{(\mathcal{O}),\mathrm{H}} \boldsymbol{e}_k.} \tag{D.70}$$

The scaling β is chosen to satisfy the transmit power constraint $\mathrm{tr}[\boldsymbol{P}\boldsymbol{C}_{\boldsymbol{w}}\boldsymbol{P}^{\mathrm{H}}] = P_{\mathrm{tx}}$ with equality

$$\beta^2 = \sum_{k=1}^{K} c_{w_k} \boldsymbol{e}_k^{\mathrm{T}} \boldsymbol{A}_k^{(\mathcal{O})} \left(\boldsymbol{\mathcal{X}} + \boldsymbol{A}_k^{(\mathcal{O}),\mathrm{H}} \boldsymbol{A}_k^{(\mathcal{O})} + \frac{\bar{G}}{P_{\mathrm{tx}}} \boldsymbol{I}_M\right)^{-2} \boldsymbol{A}_k^{(\mathcal{O}),\mathrm{H}} \boldsymbol{e}_k / P_{\mathrm{tx}}. \tag{D.71}$$

For $\boldsymbol{\mathcal{X}} = \mathbf{0}_M$ and high SNR, the matrix inversion lemma A.2.2 has to be applied to (D.70) and (D.71) to ensure existence of the solution.

It remains to find the precoding order $\mathcal{O}$. Applying (D.65), (D.70), and (D.71) to (D.63) we can simplify the expression for its minimum to

$$
\begin{aligned}
&\mathrm{F}^{\mathrm{THP}}(\mathcal{O}) = \\
&= \mathrm{tr}[\boldsymbol{C}_{\boldsymbol{w}}] - \sum_{k=1}^{K} c_{w_k} \boldsymbol{e}_k^{\mathrm{T}} \boldsymbol{A}_k^{(\mathcal{O})} \left(\boldsymbol{\mathcal{X}} + \boldsymbol{A}_k^{(\mathcal{O}),\mathrm{H}} \boldsymbol{A}_k^{(\mathcal{O})} + \frac{\bar{G}}{P_{\mathrm{tx}}} \boldsymbol{I}_M \right)^{-1} \boldsymbol{A}_k^{(\mathcal{O}),\mathrm{H}} \boldsymbol{e}_k .
\end{aligned} \tag{D.72}
$$

In Section 5.3.1 a suboptimum precoding order is derived.

Appendix E
Channel Scenarios for Performance Evaluation

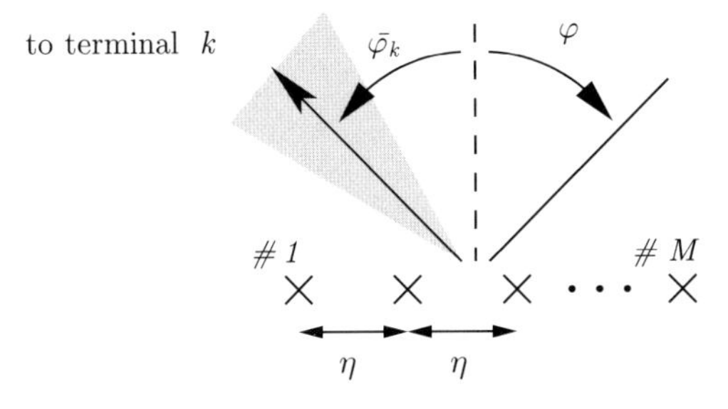

Fig. E.1 Uniform linear array with spacing η and mean azimuth direction $\bar{\varphi}_k$ of propagation channel to terminal k (with angular spread).

For the evaluation of the algorithms presented in this book, the following model for the wireless channel is employed: The Q samples of the $P = MK(L+1)$ channel parameters $\boldsymbol{h}_{\mathrm{T}}[q] \in \mathbb{C}^{PQ}$ (2.8) are assumed to be a zero-mean complex Gaussian distributed $\mathcal{N}_{\mathrm{c}}(\mathbf{0}_{PQ}, \boldsymbol{C}_{\boldsymbol{h}_{\mathrm{T}}})$ and stationary random sequence. They are generated based on the Karhunen-Loève expansion, i.e., $\boldsymbol{h}_{\mathrm{T}}[q] = \boldsymbol{C}_{\boldsymbol{h}_{\mathrm{T}}}^{1/2}\boldsymbol{\zeta}[q]$ with $\boldsymbol{\zeta}[q] \sim \mathcal{N}_{\mathrm{c}}(\mathbf{0}, \boldsymbol{I}_{PQ})$. The structure of $\boldsymbol{C}_{\boldsymbol{h}_{\mathrm{T}}}$ is exploited to reduce the computational complexity.

We assume identical temporal correlations for every element in $\boldsymbol{h}[q]$ (see Section 2.1 for details), which results in $\boldsymbol{C}_{\boldsymbol{h}_{\mathrm{T}}} = \boldsymbol{C}_{\mathrm{T}} \otimes \boldsymbol{C}_{\boldsymbol{h}}$ (2.12) with $\boldsymbol{C}_{\mathrm{T}} \in \mathbb{T}_{+,0}^{Q}$ and $\boldsymbol{C}_{\boldsymbol{h}} \in \mathbb{S}_{+,0}^{P}$.

If not defined otherwise, Clarke's model [206, p. 40] is chosen for the temporal correlations $c_{h}[\ell] = c_{h}[0]\mathrm{J}_0(2\pi f_{\max}\ell)$.[1] Its power spectral density is shown in Figure 2.12. To be consistent with the definition of $\boldsymbol{C}_{\boldsymbol{h}} = \mathrm{E}\left[\boldsymbol{h}[q]\boldsymbol{h}[q]^{\mathrm{H}}\right]$ and to obtain a unique Kronecker model, the first row of the Toeplitz covariance matrix $\boldsymbol{C}_{\mathrm{T}}$ in the Kronecker model is $[c_{h}[0], \ldots, c_{h}[Q-1]]^{\mathrm{T}}/c_{h}[0]$.

[1] $\mathrm{J}_0(\bullet)$ is the Bessel function of the first kind and of order zero.

For frequency flat channels ($L = 0$), we consider spatially well separated terminals with uncorrelated channels, i.e., $\boldsymbol{C}_{\boldsymbol{h}}$ is block diagonal with $\boldsymbol{C}_{\boldsymbol{h}_k} \in \mathbb{S}_{+,0}^{M}, k \in \{1,2,\ldots,K\}$, on the diagonal.[2] To model the spatial correlations $\boldsymbol{C}_{\boldsymbol{h}_k}$ we choose a uniform linear array with M elements which have an omnidirectional characteristic in azimuth and equal spacing η (Figure E.1).[3]

If the narrowband assumption [125] holds, i.e., the propagation time over the array aperture is significantly smaller than the inverse of the signal bandwidth, the spatial characteristic depends only on the carrier wavelength λ_{c}. Mathematically, it is described by the array steering vector

$$\boldsymbol{a}(\varphi) = [1, \exp(-\mathrm{j}2\pi\eta\sin(\varphi)/\lambda_{\mathrm{c}}), \ldots, \exp(-\mathrm{j}2\pi\eta(M-1)\sin(\varphi)/\lambda_{\mathrm{c}})]^{\mathrm{T}} \in \mathbb{C}^M,$$

which contains the phase shifts resulting from the array geometry. A common value of the interelement spacing is $\eta = \lambda_{\mathrm{c}}/2$. The azimuth angle φ is defined w.r.t. the perpendicular on the array's orientation (Figure E.1).

Thus, the spatial covariance matrix is modeled as

$$\boldsymbol{C}_{\boldsymbol{h}_k} = c_k \int_{-\pi/2}^{\pi/2} \boldsymbol{a}(\varphi)\boldsymbol{a}(\varphi)^{\mathrm{H}} \mathrm{p}_{\varphi}(\varphi + \bar{\varphi}_k)\, \mathrm{d}\varphi \tag{E.1}$$

with mean azimuth direction $\bar{\varphi}_k$ and $\mathrm{tr}[\boldsymbol{C}_{\boldsymbol{h}_k}] = M$. A widely used choice for the probability density $\mathrm{p}_{\varphi}(\varphi + \bar{\varphi}_k)$ describing the azimuth spread is a truncated Laplace distribution

$$\mathrm{p}_{\varphi}(\varphi) \propto \exp\left(-\sqrt{2}|\varphi|/\sigma\right), \quad |\varphi| \leq \pi/2, \tag{E.2}$$

where σ is the standard deviation, e.g., $\sigma = 5°$, of the Laplace distribution without truncation [166].[4] The mean directions for all terminals are summarized in $\bar{\boldsymbol{\varphi}} = [\bar{\varphi}_1, \ldots, \bar{\varphi}_K]^{\mathrm{T}}$, whereas σ and the mean channel attenuation $c_k = 1$ (normalized) are equal for all terminals for simplicity. We evaluate the integral (E.1) numerically by Monte-Carlo integration

$$\boldsymbol{C}_{\boldsymbol{h}_k} \approx c_k \frac{1}{W} \sum_{w=1}^{W} \boldsymbol{a}(\varphi_w + \bar{\varphi}_k)\boldsymbol{a}(\varphi_w + \bar{\varphi}_k)^{\mathrm{H}} \tag{E.3}$$

with $W = 100$ pseudo-randomly selected φ_w according to $\mathrm{p}_{\varphi}(\varphi)$ and $\varphi_{k,w} = \varphi_w + \bar{\varphi}_k$.

Only in Section 2.2.4 we consider a frequency selective channel of order $L = 4$ for $K = 1$: $\boldsymbol{C}_{\boldsymbol{h}}$ is block diagonal (2.7) with $\boldsymbol{C}_{\boldsymbol{h}^{(\ell)}} \in \mathbb{S}_{+,0}^{M}, \ell \in \{0,1,\ldots,L\}$, on its diagonal due to uncorrelated scattering [14]. The constant diagonal of $\boldsymbol{C}_{\boldsymbol{h}^{(\ell)}}$ is the variance $c^{(\ell)}$ (power delay spectrum), for which the standard

[2] The model is tailored to the broadcast channel considered in Chapters 4 and 5.

[3] We restrict the model to be planar.

[4] See [66, 240, 73] for an overview over other models and [20] for an interesting discussion regarding the relevance of the Laplace distribution.

model is $c^{(\ell)} \propto \exp(-\ell/\tau)$ and $\sum_{\ell=0}^{L} c^{(\ell)} = 1$. For a symbol period $T_{\mathrm{s}} = 1\,\mu$s, $\tau = 1$ is a typical value [166]. Else $\boldsymbol{C}_{\boldsymbol{h}^{(\ell)}}$ is determined as in (E.1).

The noise variances at the terminals are equal for simplicity: $c_{n_k} = c_n$.

Appendix F
List of Abbreviations

AWGN	Additive White Gaussian Noise
BC	Broadcast Channel
BER	Bit-Error Rate
C-CSI	Complete Channel State Information
CM	Conditional Mean
CSI	Channel State Information
DPSS	Discrete Prolate Spheroidal Sequences
ECME	Expectation-Conditional Maximization Either
EM	Expectation Maximization
EVD	EigenValue Decomposition
EXIP	EXtended Invariance Principle
FDD	Frequency Division Duplex
FIR	Finite Impulse Response
HDS	Hidden Data Space
KKT	Karush-Kuhn-Tucker
LMMSE	Linear Minimum Mean Square Error
LS	Least-Squares
MAC	Multiple Access Channel
MF	Matched Filter
MIMO	Multiple-Input Multiple-Output
ML	Maximum Likelihood
MSE	Mean Square Error
MMSE	Minimum Mean Square Error
O-CSI	Outdated Channel State Information
pdf	probability density function
P-CSI	Partial Channel State Information
psd	power spectral density
RML	Reduced-Rank Maximum Likelihood
Rx	Receiver

SAGE	Space-Alternating Generalized Expectation maximization
S-CSI	Statistical Channel State Information
SDMA	Space Division Multiple Access
SISO	Single-Input Single-Output
SINR	Signal-to-Interference-plus-Noise Ratio
SIR	Signal-to-Interference Ratio
SNR	Signal-to-Noise Ratio
s.t.	subject to
TDD	Time Division Duplex
TDMA	Time Division Multiple Access
TP	ThroughPut
THP	Tomlinson-Harashima Precoding
VP	Vector Precoding
UMTS	Universal Mobile Telecommunications System
w.r.t.	with respect to

References

1. 3GPP: Universal Mobile Telecommunications System (UMTS); Physical Channels and Mapping of Transport Channels onto Physical Channels (FDD). 3GPP TS 25.211, Version 6.6.0, Release 6 (2005). (Online: http://www.etsi.org/)
2. Abramowitz, M., Stegun, I.A.: Handbook of Mathematical Functions with Formulas, Graphs, and Mathematical Tables, 9th edn. Dover, New York (1964)
3. Alkire, B., Vandenberghe, L.: Convex Optimization Problems Involving Finite Autocorrelation Sequences. Mathematical Programming, Series A **93**(3), 331–359 (2002)
4. Anderson, T.W.: Statistical Inference for Covariance Matrices with Linear Structure. In: Proc. of the Second International Symposium on Multivariate Analysis, vol. 2, pp. 55–66 (1968)
5. Anderson, T.W.: Estimation of Covariance Matrices which are Linear Combinations or whose Inverses are Linear Combinations of Given Matrices. In: Essays in in Probability and Statistics, pp. 1–24. The University of North Carolina Press (1970)
6. Arun, K.S., Potter, L.C.: Existence and Uniqueness of Band-Limited, Positive Semidefinite Extrapolations. IEEE Trans. Acoust., Speech, Signal Processing **48**(3), 547–549 (1990)
7. Assir, R.E., Dietrich, F.A., Joham, M.: MSE Precoding for Partial Channel State Information Based on MAC-BC Duality. Tech. Rep. TUM-LNS-TR-05-08, Technische Universität München (2005)
8. Assir, R.E., Dietrich, F.A., Joham, M., Utschick, W.: Min-Max MSE Precoding for Broadcast Channels based on Statistical Channel State Information. In: Proc. of the IEEE 7th Workshop on Signal Processing Advances in Wireless Communications. Cannes, France (2006)
9. Baddour, K.E., Beaulieu, N.C.: Improved Pilot Symbol Aided Estimation of Rayleigh Fading Channels with Unknown Autocorrelation Statistics. In: Proc. IEEE 60th Vehicular Technology Conf., vol. 3, pp. 2101–2107 (2004)
10. Bahng, S., Liu, J., Host-Madsen, A., Wang, X.: The Effects of Channel Estimation on Tomlinson-Harashima Precoding in TDD MIMO Systems. In: Proc. of the IEEE 6th Workshop on Signal Processing Advances in Wireless Communications, pp. 455–459 (2005)
11. Barton, T.A., Fuhrmann, D.R.: Covariance Structures for Multidimensional Data. In: Proc. of the 27th Asilomar Conference on Signals, Systems and Computers, pp. 297–301 (1991)
12. Barton, T.A., Fuhrmann, D.R.: Covariance Estimation for Multidimensional Data Using the EM Algorithm. In: Proc. of the 27th Asilomar Conference on Signals, Systems and Computers, pp. 203–207 (1993)

13. Barton, T.A., Smith, S.T.: Structured Covariance Estimation for Space-Time Adaptive Processing. In: IEEE International Conference on Acoustics, Speech, and Signal Processing, pp. 3493–3496 (1997)
14. Bello, P.A.: Characterization of Randomly Time-Variant Linear Channels. IEEE Trans. on Comm. Systems **11**(12), 360–393 (1963)
15. Ben-Haim, Z., Eldar, Y.C.: Minimax Estimators Dominating the Least-Squares Estimator. In: Proc. of the IEEE Int. Conf. on Acoustics, Speech and Signal Processing, vol. 4, pp. 53–56 (2005)
16. Ben-Tal, A., Nemirovski, A.: Robust convex optimization. Mathematics of Operations Research **23** (1998)
17. Ben-Tal, A., Nemirovski, A.: Robust Optimization - Methodology and Applications. Mathematical Programming Series B **92**, 453–480 (2002)
18. Bengtsson, M., Ottersten, B.: Uplink and Downlink Beamforming for Fading Channels. In: Proc. of the 2nd IEEE Workshop on Signal Processing Advances in Wireless Communications, pp. 350–353 (1999)
19. Bengtsson, M., Ottersten, B.: Optimum and Suboptimum Transmit Beamforming. In: Handbook of Antennas in Wireless Communications, chap. 18. CRC Press (2001)
20. Bengtsson, M., Volcker, B.: On the Estimation of Azimuth Distributions and Azimuth Spectra. In: Proc. Vehicular Technology Conference Fall, vol. 3, pp. 1612–1615 (2001)
21. Berman, A.: Nonnegative Matrices in the Mathematical Sciences. Academic Press (1979)
22. Beutler, F.: Error Free Recovery of Signals from Irregular Sampling. SIAM Rev. **8**, 322–335 (1966)
23. Bienvenu, G., Kopp, L.: Optimality of High Resolution Array Processing Using the Eigensystem Approach. IEEE Trans. Acoust., Speech, Signal Processing **31**(5), 1235–1248 (1983)
24. Boyd, S., Vandenberghe, L.: Convex Optimization. Cambridge University Press (2004)
25. Breun, P., Dietrich, F.A.: Precoding for the Wireless Broadcast Channel based on Partial Channel Knowledge. Tech. Rep. TUM-LNS-TR-05-03, Technische Universität München (2005)
26. Brewer, J.W.: Kronecker Products and Matrix Calculus in System Theory. IEEE Trans. Circuits Syst. **25**, 772–781 (1978)
27. Brown, J.L.: On the Prediction of a Band-Limited Signal from Past Samples. Proc. IEEE **74**(11), 1596–1598 (1986)
28. Brown, J.L., Morean, O.: Robust Prediction of Band-Limited Signals from Past Samples. IEEE Trans. Inform. Theory **32**(3), 410–412 (1986)
29. Burg, J.P., Luenberger, D.G., Wenger, D.L.: Estimation of Structured Covariance Matrices. Proc. IEEE **70**(9), 500–514 (1982)
30. Byers, G.J., Takawira, F.: Spatially and Temporally Correlated MIMO Channels: Modeling and Capacity Analysis. IEEE Trans. Veh. Technol. **53**(3), 634–643 (2004)
31. Caire, G.: MIMO Downlink Joint Processing and Scheduling: A Survey of Classical and Recent Results. In: Proc. of Workshop on Information Theory and its Applications. San Diego, CA (2006)
32. Caire, G., Shamai, S.: On the Achievable Throughput of a Multiantenna Gaussian Broadcast Channel. IEEE Trans. Inform. Theory **49**(7), 1691–1706 (2003)
33. Chaufray, J.M., Loubaton, P., Chevalier, P.: Consistent Estimation of Rayleigh Fading Channel Second-Order Statistics in the Context of the Wideband CDMA Mode of the UMTS. IEEE Trans. Signal Processing **49**(12), 3055–3064 (2001)
34. Choi, H., Munson, Jr., D.C.: Analysis and Design of Minimax-Optimal Interpolators. IEEE Trans. Signal Processing **46**(6), 1571–1579 (1998)

35. Choi, H., Munson, Jr., D.C.: Stochastic Formulation of Bandlimited Signal Interpolation. IEEE Trans. Circuits Syst. II **47**(1), 82–85 (2000)
36. Chun, B.: A Downlink Beamforming in Consideration of Common Pilot and Phase Mismatch. In: Proc. of the European Conference on Circuit Theory and Design, vol. 3, pp. 45–48 (2005)
37. Cools, R.: Advances in Multidimensional Integration. Journal of Computational and Applied Mathematics **149**, 1–12 (2002)
38. Cools, R., Rabinowitz, P.: Monomial Cubature Rules since "Stroud": A Compilation. Journal of Computational and Applied Mathematics **48**, 309–326 (1993)
39. Costa, M.: Writing on Dirty Paper. IEEE Trans. Inform. Theory **29**(3), 439–441 (1983)
40. Cover, T.M., Thomas, J.A.: Elements of Information Theory. John Wiley & Sons (1991)
41. Czink, N., Matz, G., Seethaler, D., Hlawatsch, F.: Improved MMSE Estimation of Correlated MIMO Channels using a Structured Correlation Estimator. In: Proc. of the 6th IEEE Workshop on Signal Processing Advances in Wireless Communications, pp. 595–599. New York City (2005)
42. D. Forney, Jr., G.: On the Role of MMSE Estimation in Approaching the Information-Theoretic Limits of Linear Gaussian Channels: Shannon Meets Wiener. In: Proc. 2003 Allerton Conf., pp. 430–439 (2003)
43. Dahl, J., Vandenberghe, L., Fleury, B.H.: Robust Least-Squares Estimators Based on Semidefinite Programming. In: Conf. Record of the 36th Asilomar Conf. on Signals, Systems, and Computers, vol. 2, pp. 1787–1791. Pacific Grove, CA (2002)
44. Dante, H.M.: Robust Linear Prediction of Band-Limited Signals. In: Proc. of the IEEE Int. Conf. on Acoustics, Speech, and Signal Processing, vol. 4, pp. 2328–2331 (1988)
45. Degerine, S.: On Local Maxima of the Likelihood Function for Toeplitz Matrix Estimation. IEEE Trans. Signal Processing **40**(6), 1563–1565 (1992)
46. Dembo, A.: The Relation Between Maximum Likelihood Estimation of Structured Covariance Matrices and Periodograms. IEEE Trans. Acoust., Speech, Signal Processing **34**(6), 1661–1662 (1986)
47. Dembo, A., Mallows, C.L., Shepp, L.A.: Embedding Nonnegative Definite Toeplitz Matrices in Nonnegative Definite Circulant Matrices, with Application to Covariance Estimation. IEEE Trans. Inform. Theory **35**(6), 1206–1212 (1989)
48. Dharanipragada, S., Arun, K.S.: Bandlimited Extrapolation Using Time-Bandwidth Dimension. IEEE Trans. Signal Processing **45**(12), 2951–2966 (1997)
49. Dietrich, F.A., Breun, P., Utschick, W.: Tomlinson-Harashima Precoding: A Continuous Transition from Complete to Statistical Channel Knowledge. In: Proc. of the IEEE Global Telecommunications Conference, vol. 4, pp. 2379–2384. Saint Louis, U.S.A. (2005)
50. Dietrich, F.A., Breun, P., Utschick, W.: Robust Tomlinson-Harashima Precoding for the Wireless Broadcast Channel. IEEE Trans. Signal Processing **55**(2), 631–644 (2007)
51. Dietrich, F.A., Dietl, G., Joham, M., Utschick, W.: Robust and reduced rank space-time decision feedback equalization. In: T. Kaiser, A. Bourdoux, H. Boche, J.R. Fonollosa, J.B. Andersen, W. Utschick (eds.) Smart Antennas—State-of-the-Art, EURASIP Book Series on Signal Processing and Communications, chap. Part 1: Receiver, pp. 189–205. Hindawi Publishing Corporation (2005)
52. Dietrich, F.A., Hoffmann, F., Utschick, W.: Conditional Mean Estimator for the Gramian Matrix of Complex Gaussian Random Variables. In: Proc. of the IEEE Int. Conf. on Acoustics, Speech, and Signal Processing, vol. 3, pp. 1137–1140. Philadelphia, U.S.A. (2005)

53. Dietrich, F.A., Hunger, R., Joham, M., Utschick, W.: Robust Transmit Wiener Filter for Time Division Duplex Systems. In: Proc. of the 3rd IEEE Int. Symp. on Signal Processing and Information Technology, pp. 415–418. Darmstadt, Germany (2003)
54. Dietrich, F.A., Ivanov, T., Utschick, W.: Estimation of Channel and Noise Correlations for MIMO Channel Estimation. In: Proc. of the Int. ITG/IEEE Workshop on Smart Antennas. Reisensburg, Germany (2006)
55. Dietrich, F.A., Joham, M., Utschick, W.: Joint Optimization of Pilot Assisted Channel Estimation and Equalization applied to Space-Time Decision Feedback Equalization. In: Proc. of the IEEE Int. Conf. on Communications, vol. 4, pp. 2162–2167. Seoul, South Korea (2005)
56. Dietrich, F.A., Utschick, W.: Pilot-Assisted Channel Estimation Based on Second-Order Statistics. IEEE Trans. Signal Processing **53**(3), 1178–1193 (2005)
57. Dietrich, F.A., Utschick, W.: Robust Tomlinson-Harashima Precoding. In: Proc. of the IEEE 16th Int. Symp. on Personal, Indoor and Mobile Radio Communications, vol. 1, pp. 136–140. Berlin, Germany (2005)
58. Dietrich, F.A., Utschick, W., Breun, P.: Linear Precoding Based on a Stochastic MSE Criterion. In: Proc. of the 13th European Signal Processing Conference. Antalya, Turkey (2005)
59. Dimic, G., Sidiropoulos, N.D.: On Downlink Beamforming with Greedy User Selection: Performance Analysis and a Simple New Algorithm. IEEE Trans. Signal Processing **53**(10), 3857–3868 (2005)
60. Dogandzic, A., Jin, J.: Estimating Statistical Properties of MIMO Fading Channels. IEEE Trans. Signal Processing **53**(8), 3065–3080 (2005)
61. Dorato, P.: A Historical Review of Robust Control. IEEE Control Syst. Mag. **7**(2), 44–56 (1987)
62. Doucet, A., Wang, X.: Monte Carlo Methods for Signal Processing: A Review in the Statistical Signal Processing Context. IEEE Signal Processing Mag. **22**(6), 152–170 (2005)
63. Duel-Hallen, A., Hu, S., Hallen, H.: Long-Range Prediction of Fading Signals. IEEE Signal Processing Mag. **17**(3), 62–75 (2000)
64. Ekman, T.: Prediction of Mobile Radio Channels Modeling and Design. Dissertation, Uppsala University (2002). ISBN 91-506-1625-0
65. Erez, U., Brink, S.: A Close-to-Capacity Dirty Paper Coding Scheme. IEEE Trans. Inform. Theory **51**(10), 3417–3432 (2005)
66. Ertel, R.B., Cardieri, P., Sowerby, K.W., Rappaport, T.S., Reed, J.H.: Overview of Spatial Channel Models for Antenna Array Communication Systems. IEEE Personal Commun. Mag. **5**(1), 10–22 (1998)
67. Eyceoz, T., Duel-Hallen, A., Hallen, H.: Deterministic Channel Modeling and Long Range Prediction of Fast Fading Mobile Radio Channels. IEEE Commun. Lett. **2**(9), 254–256 (1998)
68. Fessler, J.A., Hero, A.O.: Space-Alternating Generalized Expectation-Maximization Algorithm. IEEE Trans. Signal Processing **42**(10), 2664–2677 (1994)
69. Fischer, R.: Precoding and Signal Shaping for Digital Transmission. John Wiley & Sons, Inc. (2002)
70. Fischer, R.F.H.: The Modulo-Lattice Channel: The Key Feature in Precoding Schemes. International Journal of Electronics and Communications pp. 244–253 (2005)
71. Fischer, R.F.H., Windpassinger, C., Lampe, A., Huber, J.B.: MIMO Precoding for Decentralized Receivers. In: Proc. IEEE Int. Symp. on Information Theory, p. 496 (2002)
72. Fischer, R.F.H., Windpassinger, C., Lampe, A., Huber, J.B.: Tomlinson-Harashima Precoding in Space-Time Transmission for Low-Rate Backward

Channel. In: Proc. Int. Zurich Seminar on Broadband Communications, pp. 7–1 – 7–6 (2002)
73. Fleury, B.H.: First- and Second-Order Characterization of Direction Dispersion and Space Selectivity in the Radio Channel. IEEE Trans. Inform. Theory **46**(6), 2027–2044 (2000)
74. Forster, P., T. Asté: Maximum Likelihood Multichannel Estimation under Reduced Rank Constraint. In: Proc. IEEE Conf. Acoustics Speech Signal Processing, vol. 6, pp. 3317–3320 (1998)
75. Franke, J.: Minimax-Robust Prediction of Discrete Time Series. Z. Wahrscheinlichkeitstheorie verw. Gebiete **68**, 337–364 (1985)
76. Franke, J., Poor, H.V.: Minimax-Robust Filtering and Finite-Length Robust Predictors. Lecture Notes in Statistics: Robust and Nonlinear Time Series Analysis **26**, 87–126 (1984)
77. Fuhrmann, D.: Progress in Structured Covariance Estimation. In: Proc. of the 4th Annual ASSP Workshop on Spectrum Estimation and Modeling, pp. 158–161 (1998)
78. Fuhrmann, D.R.: Correction to "On the Existence of Positive-Definite Maximum-Likelihood Estimates of Structured Covariance Matrices". IEEE Trans. Inform. Theory **43**(3), 1094–1096 (1997)
79. Fuhrmann, D.R., Barton, T.: New Results in the Existence of Complex Covariance Estimates. In: Conf. Record of the 26th Asilomar Conference on Signals, Systems and Computers, vol. 1, pp. 187–191 (1992)
80. Fuhrmann, D.R., Barton, T.A.: Estimation of Block-Toeplitz Covariance Matrices. In: Conf. Record of 24th Asilomar Conference on Signals, Systems and Computers, vol. 2, pp. 779–783 (1990)
81. Fuhrmann, D.R., Miller, M.I.: On the Existence of Positive-Definite Maximum-Likelihood Estimates of Structured Covariance Matrices. IEEE Trans. Inform. Theory **34**(4), 722–729 (1988)
82. Ginis, G., Cioffi, J.M.: A Multi-User Precoding Scheme Achieving Crosstalk Cancellation with Application to DSL Systems. In: Conf. Record of 34th Asilomar Conf. on Signals, Systems, and Computers, vol. 2, pp. 1627–1631 (2000)
83. Goldsmith, A.: Wireless Communications. Cambridge University Press (2005)
84. Golub, G.H., Loan, C.F.V.: Matrix Computations, 2nd edn. The John Hopkins University Press (1989)
85. Gray, R.M.: Toeplitz and Circulant Matrices: A review. Foundations and Trends in Communications and Information Theory **2**(3), 155–239 (2006)
86. Grenander, U., G. Szegö: Toeplitz Forms and Their Applications. University of California Press (1958)
87. Guillaud, M., Slock, D.T.M., Knopp, R.: A Practical Method for Wireless Channel Reciprocity Exploitation Through Relative Calibration. In: Proc. of the 8th Int. Symp. on Signal Processing and Its Applications, vol. 1, pp. 403–406 (2005)
88. Haardt, M.: Efficient One-, Two, and Multidimensional High-Resolution Array Signal Processing. Dissertation, Technische Universität München (1997). ISBN 3-8265-2220-6, Shaker Verlag
89. Harashima, H., Miyakawa, H.: Matched-Transmission Technique for Channels With Intersymbol Interference. IEEE Trans. Commun. **20**(4), 774–780 (1972)
90. Hassibi, B., Hochwald, B.M.: How Much Training is Needed in Multiple-Antenna Wireless Links? IEEE Trans. Inform. Theory **49**(4), 951–963 (2003)
91. Herdin, M.: Non-Stationary Indoor MIMO Radio Channels. Dissertation, Technische Universität Wien (2004)
92. Higham, N.J.: Matrix Nearness Problems and Applications. In: M.J.C. Gover, S. Barnett (eds.) Applications of Matrix Theory, pp. 1–27. Oxford University Press (1989)

93. Hochwald, B.M., Peel, C.B., Swindlehurst, A.L.: A Vector-Perturbation Technique for Near-Capacity Multiantenna Multiuser Communications—Part II: Perturbation. IEEE Trans. Commun. **53**(3), 537–544 (2005)
94. Hofstetter, H., Viering, I., Lehne, P.H.: Spatial and Temporal Long Term Properties of Typical Urban Base Stations at Different Heights. In: COST273 6th Meeting TD(03)61. Barcelona, Spain (2003)
95. Horn, R.A., Johnson, C.R.: Matrix Analysis, 1st edn. Cambridge University Press (1985)
96. Horn, R.A., Johnson, C.R.: Topics in Matrix Analysis, 1st edn. Cambridge University Press (1991)
97. Huber, P.J.: Robust Statistics. John Wiley & Sons (1981)
98. Hunger, R., Dietrich, F.A., Joham, M., Utschick, W.: Robust Transmit Zero-Forcing Filters. In: Proc. of the ITG Workshop on Smart Antennas, pp. 130–237. Munich, Germany (2004)
99. Hunger, R., Joham, M., Schmidt, D.A., Utschick, W.: On Linear and Nonlinear Precoders for Decentralized Receivers (2005). *Unpublished Technical Report*
100. Hunger, R., Joham, M., Utschick, W.: Extension of Linear and Nonlinear Transmit Filters for Decentralized Receivers. In: Proc. European Wireless 2005, vol. 1, pp. 40–46 (2005)
101. Hunger, R., Joham, M., Utschick, W.: Minimax Mean-Square-Error Transmit Wiener Filter. In: Proc. VTC 2005 Spring, vol. 2, pp. 1153–1157 (2005)
102. Hunger, R., Utschick, W., Schmidt, D., Joham, M.: Alternating Optimization for MMSE Broadcast Precoding. In: Proc. of the IEEE Int. Conf. on Acoustics, Speech, and Signal Processing, vol. 4, pp. 757–760 (2006)
103. Ivanov, T., Dietrich, F.A.: Estimation of Second Order Channel and Noise Statistics in Wireless Communications. Tech. Rep. TUM-LNS-TR-05-09, Technische Universität München (2005)
104. Ivrlac, M.T., Choi, R.L.U., Murch, R., Nossek, J.A.: Effective Use of Long-Term Transmit Channel State Information in Multi-User MIMO Communication Systems. In: Proc. of the 57th IEEE Vehicular Technology Conference, pp. 409–413. Orlando, Florida (2003)
105. Jansson, M., Ottersten, B.: Structured Covariance Matrix Estimation: A Parametric Approach. In: Proc. of the IEEE International Conference on Acoustics, Speech, and Signal Processing, vol. 5, pp. 3172–3175 (2000)
106. Jindal, N.: High SNR Analysis of MIMO Broadcast Channels. In: Proc. of the Int. Symp. on Information Theory, pp. 2310–2314 (2005)
107. Jindal, N.: MIMO Broadcast Channels with Finite Rate Feedback. IEEE Trans. Inform. Theory **52**(11), 5045–5059 (2006)
108. Joham, M.: Optimization of Linear and Nonlinear Transmit Signal Processing. Dissertation, Technische Universität München (2004). ISBN 3-8322-2913-2, Shaker Verlag
109. Joham, M., Brehmer, J., Utschick, W.: MMSE Approaches to Multiuser Spatio-Temporal Tomlinson-Harashima Precoding. In: Proc. of the ITG Conf. on Source and Channel Coding, pp. 387–394 (2004)
110. Joham, M., Kusume, K., Gzara, M.H., Utschick, W., Nossek, J.A.: Transmit Wiener Filter. Tech. Rep. TUM-LNS-TR-02-1, Technische Universität München (2002)
111. Joham, M., Kusume, K., Gzara, M.H., Utschick, W., Nossek, J.A.: Transmit Wiener Filter for the Downlink of TDD DS-CDMA Systems. In: Proc. 7th IEEE Int. Symp. on Spread Spectrum Techniques and Applications, vol. 1, pp. 9–13 (2002)
112. Joham, M., Kusume, K., Utschick, W., Nossek, J.A.: Transmit Matched Filter and Transmit Wiener Filter for the Downlink of FDD DS-CDMA Systems. In: Proc. PIMRC 2002, vol. 5, pp. 2312–2316 (2002)

113. Joham, M., Schmidt, D.A., Brehmer, J., Utschick, W.: Finite-Length MMSE Tomlinson-Harashima Precoding for Frequency Selective Vector Channels. IEEE Trans. Signal Processing **55**(6), 3073–3088 (2007)
114. Joham, M., Schmidt, D.A., Brunner, H., Utschick, W.: Symbol-Wise Order Optimization for THP. In: Proc. ITG/IEEE Workshop on Smart Antennas (2007)
115. Joham, M., Utschick, W.: Ordered Spatial Tomlinson Harashima Precoding. In: T. Kaiser, A. Bourdoux, H. Boche, J.R. Fonollosa, J.B. Andersen, W. Utschick (eds.) Smart Antennas—State-of-the-Art, EURASIP Book Series on Signal Processing and Communications, chap. Part 3: Transmitter, pp. 401–422. EURASIP, Hindawi Publishing Corporation (2006)
116. Joham, M., Utschick, W., Nossek, J.A.: Linear Transmit Processing in MIMO Communications Systems. IEEE Trans. Signal Processing **53**(8), 2700–2712 (2005)
117. Johnson, C.R., Lundquist, M., Nævdal, G.: Positive Definite Toeplitz Completions. Journal of London Mathematical Society **59**(2), 507–520 (1999)
118. Johnson, D.B.W.D.H.: Robust Estimation of Structured Covariance Matrices. IEEE Trans. Signal Processing **41**(9), 2891–2906 (1993)
119. Jongren, G., Skoglund, M., Ottersten, B.: Design of Channel-Estimate-Dependent Space-Time Block Codes. IEEE Trans. Commun. **52**(7), 1191–1203 (2004)
120. Jung, P.: Analyse und Entwurf digitaler Mobilfunksysteme, 1st edn. B.G. Teubner (1997)
121. Kailath, T., Sayed, A.H., Hassibi, B.: Linear Estimation. Prentice Hall (2000)
122. Kassam, A.A., Poor, H.V.: Robust Techniques for Signal Processing: A Survey. Proc. IEEE **73**(3), 433–481 (1985)
123. Kassam, S.A., Poor, H.V.: Robust Techniques for Signal Processing: A Survey. Proc. IEEE **73**(3), 433–481 (1985)
124. Kay, S.M.: Fundamentals of Statistical Signal Processing - Estimation Theory, 1st edn. PTR Prentice Hall (1993)
125. Krim, H., Viberg, M.: Two Decades of Array Signal Processing Research. IEEE Signal Processing Mag. **13**(4), 67–94 (1996)
126. Kusume, K., Joham, M., Utschick, W., Bauch, G.: Efficient Tomlinson-Harashima Precoding for Spatial Multiplexing on Flat MIMO Channel. In: Proc. of the IEEE Int. Conf. on Communications, vol. 3, pp. 2021–2025 (2005)
127. Kusume, K., Joham, M., Utschick, W., Bauch, G.: Cholesky Factorization with Symmetric Permutation Applied to Detecting and Precoding Spatially Multiplexed Data Streams. IEEE Trans. Signal Processing **55**(6), 3089–3103 (2007)
128. Lapidoth, A., Shamai, S., Wigger, M.: On the Capacity of Fading MIMO Broadcast Channels with Imperfect Transmitter Side-Information. In: Proc. of the 43rd Annual Allerton Conference on Communication, Control, and Computing, pp. 28–30 (2005). Extended version online: http://arxiv.org/abs/cs.IT/0605079
129. Larsson, E.G., Stoica, P.: Space-Time Block Coding for Wireless Communications. Cambridge University Press (2003)
130. Laurent, M.: Matrix Completion Problems. In: The Encyclopedia of Optimization, vol. III, pp. 221–229. Kluwer (2001)
131. Lev-Ari, H., Parker, S.R., Kailath, T.: Multidimensional Maximum-Entropy Covariance Extension. IEEE Trans. Inform. Theory **35**(3), 497–508 (1989)
132. Li, C.K., Mathias, R.: Extremal Characterizations of the Schur Complement and Resulting Inequalities. SIAM Review **42**(2), 233–246 (2000)
133. Li, H., Stoica, P., Li, J.: Computationally Efficient Maximum Likelihood Estimation of Structured Covariance Matrices. IEEE Trans. Signal Processing **47**(5), 1314–1323 (1999)
134. Li, Y., Cimini, L.J., Sollenberger, N.R.: Robust Channel Estimation for OFDM Systems with Rapid Dispersive Fading Channels. IEEE Trans. Commun. **46**(7), 902–915 (1998)

135. Liavas, A.P.: Tomlinson-Harashima Precoding with Partial Channel Knowledge. IEEE Trans. Commun. **53**(1), 5–9 (2005)
136. Liu, J., Khaled, N., Petré, F., Bourdoux, A., Barel, A.: Impact and Mitigation of Multiantenna Analog Front-End Mismatch in Transmit Maximum Ratio Combining. EURASIP Journal on Applied Signal Processing **2006**, Article ID 86,931, 14 pages (2006)
137. Lu, J., Darmofal, D.L.: Higher-Dimensional Integration with Gaussian Weight for Applications in Probabilistic Design. SIAM Journal on Scientific Computing **26**(2), 613–624 (2004)
138. Marvasti, F.: Comments on "A Note on the Predictability of Band-Limited Processes". Proc. IEEE **74**(11), 1596 (1986)
139. Mathai, M., Provost, S.B.: Quadratic Forms in Random Variables - Theory and Applications. Marcel Dekker (1992)
140. Matz, G.: On Non-WSSUS Wireless Fading Channels. IEEE Trans. Wireless Commun. **4**(5), 2465– 2478 (2005)
141. Matz, G.: Recursive MMSE Estimation of Wireless Channels based on Training Data and Structured Correlation Learning. In: Proc. of the IEEE/SP 13th Workshop on Statistical Signal Processing, pp. 1342–1347. Bordeaux, France (2005)
142. McLachlan, G., Krishnan, T.: The EM Algorithm and Extensions. Wiley Series in Probability and Statistics. John Wiley & Sons (1997)
143. Medard, M.: The Effect Upon Channel Capacity in Wireless Communications of Perfect and Imperfect Knowledge of the Channel. IEEE Trans. Inform. Theory **46**(3), 933–946 (2000)
144. Meyr, H., Moeneclaey, M., Fechtel, S.A.: Digital Communication Receivers. John Wiley (1998)
145. Mezghani, A., Hunger, R., Joham, M., Utschick, W.: Iterative THP Transceiver Optimization for Multi-User MIMO Systems Based on Weighted Sum-MSE Minimization. In: Proc. of the IEEE 7th Workshop on Signal Processing Advances in Wireless Communications. France (2006)
146. Mezghani, A., Joham, M., Hunger, R., Utschick, W.: Transceiver Design for Multi-User MIMO Systems. In: Proc. of the Int. ITG/IEEE Workshop on Smart Antennas. Reisensburg, Germany (2006)
147. Middleton, D.: Introduction to Statistical Communication Theory. McGraw Hill (1960)
148. Miller, K.S.: Complex Stochastic Processes, 1st edn. Addison-Wesley (1974)
149. Miller, M.I., Fuhrmann, D.R., O'Sullivan, J.A., Snyder, D.L.: Maximum-Likelihood Methods for Toeplitz Covariance Estimation and Radar Imaging. In: Advances in Spectrum Analysis and Array Processing, pp. 145–172. Prentice Hall (1991)
150. Miller, M.I., Snyder, D.L.: The Role of Likelihood and Entropy in Incomplete-Data Problems: Applications to Estimating Point-Process Intensities and Toeplitz Constrained Covariances. Proc. IEEE **75**(7), 892–907 (1987)
151. Miller, M.I., Turmon, M.J.: Maximum-Likelihood Estimation of Complex Sinusoids and Toeplitz Covariances. IEEE Trans. Signal Processing **42**(5), 1074–1086 (1994)
152. Monogioudis, P., Conner, K., Das, D., Gollamudi, S., Lee, J.A.C., Moustakas, A.L., Nagaraj, S., Rao, A.M., Soni, R.A., Yuan, Y.: Intelligent Antenna Solutions for UMTS: Algorithms and Simulation Results. IEEE Commun. Mag. **42**(10), 28–39 (2004)
153. Mugler, D.H.: Computationally Efficient Linear Prediction from Past Samples of a Band-Limited Signal and its Derivative. IEEE Trans. Inform. Theory **36**(3), 589–596 (1990)
154. Neumaier, A.: Solving Ill-Conditioned and Singular Linear Systems: A Tutorial on Regularization. SIAM Rev. **40**(3), 636–666 (1998)

155. Newsam, G., Dietrich, C.: Bounds on the Size of Nonnegative Definite Circulant Embeddings of Positive Definite Toeplitz Matrices. IEEE Trans. Inform. Theory **40**(4), 1218–1220 (1994)
156. Nicoli, M., Simeone, O., Spagnolini, U.: Multislot Estimation of Fast-Varying Space-Time Channels in TD-CDMA Systems. IEEE Commun. Lett. **6**(9), 376–378 (2002)
157. Nicoli, M., Simeone, O., Spagnolini, U.: Multi-Slot Estimation of Fast-Varying Space-Time Communication Channels. IEEE Trans. Signal Processing **51(5)**, 1184–1195 (2003)
158. Nicoli, M., Sternad, M., Spagnolini, U., Ahlén, A.: Reduced-Rank Channel Estimation and Tracking in Time-Slotted CDMA Systems. In: Proc. IEEE Int. Conference on Communications, pp. 533–537. New York City (2002)
159. Otnes, R., Tüchler, M.: On Iterative Equalization, Estimation, and Decoding. In: Proc. of the IEEE Int. Conf. on Communications, vol. 4, pp. 2958–2962 (2003)
160. Palomar, D.P., Bengtsson, M., Ottersten, B.: Minimum BER Linear Transceivers for MIMO Channels via Primal Decomposition. IEEE Trans. Signal Processing **53**(8), 2866–2882 (2005)
161. Paolella, M.S.: Computing Moments of Ratios of Quadratic Forms in Normal Variables. Computational Statistics and Data Analysis **42**(3), 313 – 331 (2003)
162. Papoulis, A.: A Note on the Predictability of Band-Limited Processes. Proc. IEEE **73**(8), 1332 (1985)
163. Papoulis, A.: Levinson's Algorithm, Wold's Decomposition, and Spectral Estimation. SIAM Review **27**(3), 405–441 (1985)
164. Papoulis, A.: Probability, Random Variables, and Stochastic Processes, 3rd edn. WCB/McGraw-Hill (1991)
165. Paulraj, A., Papadias, C.B.: Space-Time Processing for Wireless Communications. IEEE Signal Processing Mag. **14**(6), 49–83 (1997)
166. Pedersen, K.I., Mogensen, P.E., Fleury, B.H.: A Stochastic Model of the Temporal and Azimuthal Dispersion Seen at the Base Station in Outdoor Propagation Environments. IEEE Trans. Veh. Technol. **49**(2), 437–447 (2000)
167. Pedersen, K.I., Mogensen, P.E., Ramiro-Moreno, J.: Application and Performance of Downlink Beamforming Techniques in UMTS. IEEE Commun. Mag. **42**(10), 134–143 (2003)
168. Poor, H.V.: An Intoduction to Signal Detection and Estimation, 2nd edn. Springer-Verlag (1994)
169. Porat, B.: Digital Processing of Random Signals. Prentice-Hall (1994)
170. Porat, B.: A Course in Digital Signal Processing. John Wiley & Sons (1997)
171. Prékopa, A.: Stochastic Programming. Kluwer (1995)
172. Press, W.H., Teukolsky, S.A., Vetterling, W.T., Flannery, B.P.: Numerical Recipes in C, 2nd edn. Cambridge University Press (1999)
173. Proakis, J.G.: Digital Communications, 3rd edn. McGraw-Hill (1995)
174. Psaltopoulos, G., Joham, M., Utschick, W.: Comparison of Lattice Search Techniques for Nonlinear Precoding. In: Proc. of the Int. ITG/IEEE Workshop on Smart Antennas. Reisensburg, Germany (2006)
175. Psaltopoulos, G., Joham, M., Utschick, W.: Generalized MMSE Detection Techniques for Multipoint-to-Point Systems. In: Proc. of the IEEE Global Telecommunications Conference. San Francisco, CA (2006)
176. Quang, A.N.: On the Uniqueness of the Maximum-Likeliwood Estimate of Structured Covariance Matrices. IEEE Trans. Acoust., Speech, Signal Processing **32**(6), 1249–1251 (1984)
177. Remmert, R.: Theory of Complex Functions. Springer-Verlag (1991)
178. Requicha, A.A.G.: The Zeros of Entire Functions: Theory and Engineering Applications. Proc. IEEE **68**(3), 308–328 (1980)

179. Rey, F., Lamarca, M., Vazquez, G.: A Robust Transmitter Design for MIMO Multicarrier Systems with Imperfect Channel Estimates. In: Proc. of the 4th IEEE Workshop on Signal Processing Advances in Wireless Communications, pp. 546 – 550 (2003)
180. Rosen, Y., Porat, B.: Optimal ARMA Parameter Estimation Based on the Sample Covariances for Data with Missing Observations. IEEE Trans. Inform. Theory **35**(2), 342–349 (1989)
181. Schafer, R.W., Mersereau, R.M., Richards, M.A.: Constrained Iterative Restoration Algorithms. Proc. IEEE **69**(4), 432–450 (1981)
182. Scharf, L.L.: Statistical Signal Processing: Detection, Estimation, and Time Series Analysis. Addison-Wesley (1991)
183. Schmidt, D., Joham, M., Hunger, R., Utschick, W.: Near Maximum Sum-Rate Non-Zero-Forcing Linear Precoding with Successive User Selection. In: Proc. of the 40th Asilomar Conference on Signals, Systems and Computers (2006)
184. Schmidt, D., Joham, M., Utschick, W.: Minimum Mean Square Error Vector Precoding. In: Proc. of the EEE 16th Int. Symp. on Personal, Indoor and Mobile Radio Communications, vol. 1, pp. 107–111 (2005)
185. Schubert, M., Boche, H.: Solution of the Multiuser Downlink Beamforming Problem With Individual SINR Constraints. IEEE Trans. Veh. Technol. **53**(1), 18–28 (2004)
186. Schubert, M., Boche, H.: Iterative Multiuser Uplink and Downlink Beamforming under SINR Constraints. IEEE Trans. Signal Processing **53**(7), 2324–2334 (2005)
187. Schubert, M., Shi, S.: MMSE Transmit Optimization with Interference Pre-Compensation. In: Proc. of the IEEE 61st Vehicular Technology Conference, vol. 2, pp. 845–849 (2005)
188. Sharif, M., Hassibi, B.: On the Capacity of MIMO Broadcast Channels with Partial Side Information. IEEE Trans. Inform. Theory **51**(2), 506–522 (2005)
189. Shi, S., Schubert, M.: MMSE Transmit Optimization for Multi-User Multi-Antenna Systems. In: Proc. of the IEEE Int. Conf. on Acoustics, Speech, and Signal Processing, vol. 3, pp. 409–412 (2005)
190. Simeone, O., Bar-Ness, Y., Spagnolini, U.: Linear and Non-Linear Precoding/Decoding for MIMO Systems with Long-Term Channel State Information at the Transmitter. IEEE Trans. Wireless Commun. **3**(2), 373–378 (2004)
191. Simeone, O., Spagnolini, U.: Multi-Slot Estimation of Space-Time Channels. In: Proc. of the IEEE Int. Conf. on Communications, vol. 2, pp. 802–806 (2002)
192. Simeone, O., Spagnolini, U.: Lower Bound on Training-Based Channel Estimation Error for Frequency-Selective Block-Fading Rayleigh MIMO Channels. IEEE Trans. Signal Processing **32**(11), 3265–3277 (2004)
193. Sion, M.: On General Minimax Theorems. IEEE Trans. Inform. Theory **8**, 171–176 (1958)
194. Slepian, D.: On Bandwidth. Proc. IEEE **64**(3), 292–300 (1976)
195. Slepian, D.: Prolate Spheroidal Wave Functions, Fourier Analysis, and Uncertainty—V: The Discrete Case. The Bell System Technical Journal **57**(5), 1371–1430 (1978)
196. Snyders, J.: Error Formulae for Optimal Linear Filtering, Prediction and Interpolation of Stationary Time Series. The Annals of Mathematical Statistics **43**(6), 1935–1943 (1972)
197. Solov'ev, V.N.: A Minimax-Bayes Estimate on Classes of Distributions with Bounded Second Moments. Russian Mathematical Surveys **50**(4), 832–834 (1995)
198. Solov'ev, V.N.: On the Problem of Minimax-Bayesian Estimation. Russian Mathematical Surveys **53**(5), 1104–1105 (1998)
199. Soloviov, V.N.: Towards the Theory of Minimax-Bayesian Estimation. Theory Probab. Appl. **44**(4), 739–754 (2000)

200. Stark, H., Woods, J.W.: Probability, Random Processes, and Estimation Theory for Engineers, 2nd edn. Prentice Hall (1994)
201. Steiner, B., Baier, P.W.: Low Cost Channel Estimation in the Uplink Receiver of CDMA Mobile Radio Systems. Frequenz **47**(11-12), 292–298 (1993)
202. Stenger, F.: Tabulation of Certain Fully Symmetric Numerical Integration Formulas of Degree 7, 9 and 11. Mathematics of Computation **25**(116), 935 (1971). Microfiche
203. Stoica, P., Moses, R.: Spectral Analysis of Signals. Pearson Prentice Hall (2005)
204. Stoica, P., Söderström, T.: On Reparametrization of Loss Functions used in Estimation and the Invariance Principle. Signal Processing **17**(4), 383–387 (1989)
205. Stoica, P., Viberg, M.: Maximum Likelihood Parameter and Rank Estimation in Reduced-Rank Multivariate Linear Regressions. IEEE Trans. Signal Processing **44**(12), 3069–3078 (1996)
206. Stüber, G.L.: Principles of Mobile Communication. Kluwer Academic Publishers (1996)
207. Sturm, J.F.: Using SeDuMi 1.02, a MATLAB toolbox for optimization over symmetric cones. Optimization Methods and Software **11–12**, 625–653 (1999)
208. Sultan, S.A., Tracy, D.S.: Moments of the Complex Multivariate Normal Distribution. Linear Algebra and its Applications **237-238**, 191–204 (1996)
209. Taherzadeh, M., Mobasher, A., Khandani, A.K.: LLL Lattice-Basis Reduction Achieves the Maximum Diversity in MIMO Systems. In: Proc. Int. Symp. on Information Theory, pp. 1300–1304 (2005)
210. Tejera, P., Utschick, W., Bauch, G., Nossek, J.A.: Subchannel Allocation in Multiuser Multiple-Input Multiple-Output Systems. IEEE Trans. Inform. Theory **52**(10), 4721–4733 (2006)
211. Tepedelenlioğlu, C., Abdi, A., Giannakis, G.B., Kaveh, M.: Estimation of Doppler Spread and Signal Strength in Mobile Communications with Applications to Handoff and Adaptive Transmission. Wireless Communications and Mobile Computing **1**(1), 221–242 (2001)
212. Tikhonov, A.N., Arsenin, V.Y.: Solutions of Ill-Posed Problems. V. H. Winston & Sons (1977)
213. Tomlinson, M.: New Automatic Equaliser Employing Modulo Arithmetic. Electronics Letters **7**(5/6), 138–139 (1971)
214. Tong, L., Perreau, S.: Multichannel Blind Identification: From Subspace to Maximum Likelihood Methods. Proc. IEEE **86**(10), 1951–1968 (1998)
215. Tong, L., Sadler, B.M., Dong, M.: Pilot-Assisted Wireless Transmissions. IEEE Signal Processing Mag. **21**(6), 12–25 (2004)
216. Trefethen, L.N., Bau, D.: Numerical Linear Algebra. SIAM - Society for Industrial and Applied Mathematics (1997)
217. Tüchler, M., Mecking, M.: Equalization for Non-ideal Channel Knowledge. In: Proc. of the Conf. on Information Sciences and Systems. Baltimore, U.S.A. (2003)
218. Tugnait, J., Tong, L., Ding, Z.: Single-User Channel Estimation and Equalization. IEEE Signal Processing Mag. **17**(3), 17–28 (2000)
219. Utschick, W.: Tracking of signal subspace projectors. IEEE Trans. Signal Processing **50**(4), 769–778 (2002)
220. Utschick, W., Dietrich, F.A.: On Estimation of Structured Covariance Matrices. In: J. Beyerer, F.P. León, K.D. Sommer (eds.) Informationsfusion in der Mess- und Sensortechnik, pp. 51–62. Universitätsverlag Karlsruhe (2006). ISBN 978-3-86644-053-1
221. Utschick, W., Joham, M.: On the Duality of MIMO Transmission Techniques for Multiuser Communications. In: Proc. of the 14th European Signal Processing Conference. Florence, Italy (2006)
222. Vaidyanathan, P.P.: On Predicting a Band-Limited Signal Based on Past Sample Values. Proc. IEEE **75**(8), 1125 (1987)

223. Vandelinde, V.D.: Robust Properties of Solutions to Linear-Quadratic Estimation and Control Problems. IEEE Trans. Automat. Contr. **22**(1), 138–139 (1977)
224. Vanderveen, M.C., der Veen, A.J.V., Paulraj, A.: Estimation of Multipath Parameters in Wireless Communications. IEEE Trans. Signal Processing **46**(3), 682 – 690 (1998)
225. Vastola, K.S., Poor, H.V.: Robust Wiener-Kolmogorov Theory. IEEE Trans. Inform. Theory **30**(2), 316–327 (1984)
226. Verdú, S., Poor, H.V.: On Minimax Robustness: A General Approach and Applications. IEEE Trans. Inform. Theory **30**(2), 328–340 (1984)
227. Viering, I.: Analysis of Second Order Statistics for Improved Channel Estimation in Wireless Communications. Dissertation, Universität Ulm (2003). ISBN 3-18-373310-2, VDI Verlag
228. Viering, I., Grundler, T., Seeger, A.: Improving Uplink Adaptive Antenna Algorithms for WCDMA by Covariance Matrix Compensation. In: Proc. of the IEEE 56th Vehicular Technology Conference, vol. 4, pp. 2149–2153. Vancouver (2002)
229. Viering, I., Hofstetter, H., Utschick, W.: Validity of Spatial Covariance Matrices over Time and Frequency. In: Proc. of the IEEE Global Telecommunications Conference, vol. 1, pp. 851–855. Taipeh, Taiwan (2002)
230. Viswanath, P., Tse, D.N.C.: Sum Capacity of the Vector Gaussian Broadcast Channel and Uplink-Downlink Duality. IEEE Trans. Inform. Theory **49**(8), 1912–1921 (2003)
231. Weichselberger, W., Herdin, M., Özcelik, H., Bonek, E.: A Stochastic MIMO Channel Model with Joint Correlation of Both Link Ends. IEEE Trans. Wireless Commun. **5**(1), 90–100 (2006)
232. Weingarten, H., Steinberg, Y., Shamai, S.S.: The Capacity Region of the Gaussian Multiple-Input Multiple-Output Broadcast Channel. IEEE Trans. Inform. Theory **52**(9), 3936–3964 (2006)
233. Whittle, P.: Some Recent Contributions to the Theory of Stationary Processes. In: A Study in the Analysis of Stationary Time Series, Appendix 2. Almquist and Wiksell (1954)
234. Wiener, N., Masani, P.: The Prediction Theory of Multivariate Stochastic Processes I. Acta Math. **98**, 111–150 (1957)
235. Wiener, N., Masani, P.: The Prediction Theory of Multivariate Stochastic Processes II. Acta Math. **99**, 93–137 (1958)
236. Wiesel, A., Eldar, Y.C., Shamai, S.: Linear Precoding via Conic Optimization for Fixed MIMO Receivers. IEEE Trans. Signal Processing **54**(1), 161–176 (2006)
237. Windpassinger, C., Fischer, R.F.H., Huber, J.B.: Lattice-Reduction-Aided Broadcast Precoding. IEEE Trans. Commun. **52**(12), 2057–2060 (2004)
238. Xu, Z.: On the Second-Order Statistics of the Weighted Sample Covariance Matrix. IEEE Trans. Signal Processing **51**(2), 527–534 (2003)
239. Yates, R.D.: A Framework for Uplink Power Control in Cellular Radio Systems. IEEE J. Select. Areas Commun. **13**(7), 1341–1347 (1995)
240. Yu, K., Bengtsson, M., Ottersten, B.: MIMO Channel Models. In: T. Kaiser, A. Bourdoux, H. Boche, J.R. Fonollosa, J.B. Andersen, W. Utschick (eds.) Smart Antennas —State-of-the-Art, EURASIP Book Series on Signal Processing and Communications, chap. Part 2: Channel, pp. 271–292. Hindawi Publishing Corporation (2005)
241. Yu, W., Cioffi, J.M.: Trellis Precoding for the Broadcast Channel. In: Proc. of the IEEE Global Telecommunications Conf., vol. 2, pp. 1344–1348 (2001)
242. Zemen, T., Mecklenbräuker, C.F.: Time-Variant Channel Estimation Using Discrete Prolate Spheroidal Sequences. IEEE Trans. Signal Processing **53**(9), 3597–3607 (2005)
243. Zerlin, B., Joham, M., Utschick, W., Nossek, J.A.: Covariance Based Linear Precoding. IEEE J. Select. Areas Commun. **24**(1), 190–199 (2006)

244. Zerlin, B., Joham, M., Utschick, W., Seeger, A., Viering, I.: Linear Precoding in W-CDMA Systems based on S-CPICH Channel Estimation. In: Proc. of the 3rd IEEE Int. Symp. on Signal Processing and Information Technology, pp. 560–563. Darmstadt, Germany (2003)
245. Zhang, X., Palomar, D.P., Ottersten, B.: Robust Design of Linear MIMO Transceivers Under Channel Uncertainty. In: Proc. of the IEEE Int. Conf. on Acoustics, Speech, and Signal Processing, vol. 4, pp. 77–80 (2006)

Index